Animal Environment and Welfare

—Proceedings of International Symposium

October 23 – 26, 2015
Rongchang, Chongqing, China

动物环境和福利
国际研讨会论文集

2015 年 10 月 23 – 25 日

中国 · 重庆 · 荣昌

International Research Center for
Animal Environment and Welfare

Edited by

Ji-Qin Ni, Teng-Teeh Lim, Chaoyuan Wang

China Agriculture Press

Beijing · China

图书在版编目（CIP）数据

动物环境和福利国际研讨论文集：英文 /（美）倪既勤，（美）林挺治，王朝元主编. —北京：中国农业出版社，2015.10

ISBN 978-7-109-21022-6

Ⅰ.①动… Ⅱ.①倪… ②林… ③王… Ⅲ.①动物保护—国际学术会议—文集—英文 Ⅳ.①S863-53

中国版本图书馆 CIP 数据核字（2015）第 238238 号

中国农业出版社出版

（北京市朝阳区麦子店街 18 号楼）

（邮政编码 100125）

责任编辑 刘 玮

中国农业出版社印刷厂印刷 新华书店北京发行所发行

2015 年 10 月第 1 版 2015 年 10 月北京第 1 次印刷

开本：700mm×1000mm 1/16 印张：24

字数：500 千字

定价：110.00 元

Sponsored by

International Research Center for Animal Environment and Welfare
Key Lab of Agricultural Engineering in Structure and Environment, MOA, China
Southwest Collaborative Innovation Center of Swine for Quality and Safety

Hosted by

Chongqing Commission of Science and Technology
Chinese Society of Agricultural Engineering
China Agricultural University (CAU)
Chongqing Academy of Animal Sciences (CAAS)

Symposium Co-Chairs

Dr. Baoming Li, Professor and Associate Dean, CAU, China
Dr. Hongwei Xin, Distinguished and Endowed Professor, Iowa State University, USA

Proceedings Co-Chairs

Dr. Ji-Qin Ni, Associate Professor, Purdue University, USA
Dr. Teng-Teeh Lim, Associate Professor, University of Missouri, USA
Dr. Chaoyuan Wang, Associate Professor, CAU, China

Organizing Committee

Dr. Zuohua Liu, Professor and President, CAAS, China (Chair)
Mr. Yong Huang, Associate Professor, CAAS, China
Dr. Baozhong Lin, Professor, CAAS, China
Dr. Chaoyuan Wang, Associate Professor, CAU, China
Ms. Wen Liu, Associate Professor, CAAS, China
Dr. Liangpeng Ge, Associate Professor, CAAS, China

Acknowledgements

We wish to thank those who reviewed the submitted abstracts and full-length papers for the symposium, and helped to control and improve the quality of the papers. These individuals are: Andre Aarnink, Thomas Banhazi, Robert Burns, Richard S Gates, Liangpeng Ge, Peter Groot Koerkamp, Baoming Li, Wen Liu, Daniella Jorge de Moura, Guoqiang Zhang, Qiang Zhang, Weichao Zheng, Lingying Zhao, Ruiqian Zhao, and Hongwei Xin.

Preface

The 2015 International Symposiums on Animal Environment and Welfare (ISAEW 2015) is the third of a biennial symposium series organized by the International Research Center for Animal Environment and Welfare (IRCAEW), which was initiated by 11 acdemic institutions from five different countries and established in 2011 at Chongqing Academy of Animal Sciences (CAAS), Chongqing, China. The IRCAEW currently has 16 member institutions from eight countries in five continents. The two previous symposia were the International Symposium on Health Environment and Animal Welfare (ISHEAW 2011), October 19–22, 2011 and the ISAEW 2013, October 19–22, 2013, both held in Rongchang, Chongqing, China.

As a tradition of the ISAEW, this year's symposium also brought in diversified speakers, participants, and attendees. The announcement of ISAEW 2015 was responded by submission of 62 abstracts. A total of 44 final full-length papers were accepted in this book. These papers represent the latest research and development in animal environment and welfare, which were conducted and achieved by professionals and graduate students working or studying in various universities, research institutions, livestock farms, companies, and a commodity group from 11 countries in Africa, Asia, Australia, Europe, and North and South Americas.

Since the ending of hunter-gatherer societies and beginning of agricultural societies about 10,000 years ago, humans began domestication of animals. Agriculture and animal production has experienced tremendous development in the past decades. Modern animal agriculture has created a production system, which is highly energy-intensive and driven by advanced technologies and is being adapted on a global scale. Following these revolutionized changes, numerous questions about various aspects of animal environment and welfare have been raised and drawn attentions from the scientific communities, producers, public, and regulatory agencies in many developed and developing nations.

The papers presented in this book responded to part of these questions. They reflect three major areas of research and development: (1) indoor and outdoor environment monitoring, assessment, control, and management; (2) animal behavior, welfare, and health; and (3) animal production systems, management, equipment, techniques, and information resources. Based on these areas and the symposium sessions, the 44 papers are divided into six themes in this book.

This book will be a valuable information source for animal scientists, agricultural engineers, graduate students, and livestock and poultry producers. It can also be a current reference for environmentalists, agricultural and environmental policy makers, and animal production equipment manufacturers.

The publication of the book could not be possible without the 122 authors who shared their research progresses, findings, and experiences in their papers. We are thankful to all the authors who collaboratively conducted revisions of their papers, per reviewers' comments and editors' requests, to meet the quality requirement of the proceedings. We are particularly grateful for the support from the IRCAEW, and the symposium co-chairs and organization committee. We would also like to convey our sincere thanks to all the symposium paper reviewers, who helped to ensure the quality and scientific merit of the papers.

The Editors
September 30, 2015

Table of Contents

Theme I:

Environmental Monitoring and Assessment for Modern Animal Production

Thermal Environmental Index for Productivity of Rabbits in Different Housing Systems

Mohamed H. Hatem *, Badr A. Mohamed

Department of Agricultural Engineering, Faculty of Agriculture, Cairo University, Giza, Egypt

* Corresponding author. Email: hatem@cu.edu.eg

Abstract

This work was carried out at the Rabbit Research Unit, Faculty of Agriculture, Cairo University, Giza, Egypt, during hot months (July and August), 2013 and 2014. The aim of this work was to investigate the effect of two housing systems on the productive and reproductive performance of New Zealand White (NZW) rabbits for predicting the main physiological responses and performance indexes. Indoor and outdoor air temperatures, relative humidity and wind speed were measured over these months. Nonlinear regression approach was used to derive a regression equation to calculate the thermal environmental index for productivity of rabbits (TEIR) and prediction of the main rabbit's physiological responses surface temperature (Tsp) inside the housing systems. In house "A", the roof gained about 78.20% of the total heat. Using 5 cm of compressed straw slabs will decrease the combined UA-value from 2,383.56 to 749.11 W $°C^{-1}$ and this will decrease the heat gain or loss from the building. The data indicate that there is a need to modify the existing structural design, particularly the intense heat load transmission into the house in hot months that will cause heat stress on the rabbits and reduce their productivity.

Keywords: Rabbits, housing, thermal properties, energy balance, production and reproduction.

1. Introduction

At the present time, there are a lot of problems like pig's flu and bird's flu. So, there is a global interest in the alternate animal production resources. The domestic rabbits have been recommended as a good alternative source of dietary protein for the increasing human population in developing countries. Rabbits have significant potential to improve food security. A female rabbit can produce up to 80 kg of meat per year and up to 40 offspring a year, compared to 0.8 and 1.4 for cattle and sheep respectively (FAO, 1999). The thermal comfort zone of rabbits is around 21°C. Animals react to gradual exposition to higher ambient temperatures with internal physiological means to maintain homoeothermic. The development of non-invasive methods for measuring stress-indicating variables has been provided instead of classical descriptive behavioral observations, allowing an evaluation of stress by multiple criteria under different housing conditions and management procedures. The environmental effects on productive and reproductive performances are variable in the extent and direction but the management to keep suitable conditions is still the key for enhancing rabbit productivity (Ahmed et al., 2008). With increasing demand of rabbit's meat, there is an urgent need to continuous rabbit breeding throughout the summer months. These conditions exposed rabbits to high temperature and this caused drastic influence on rabbit's productive and reproductive performance. This situation may expose rabbits to many problems of stillbirth percentage and pre-weaning mortality. Therefore, this study was an attempt to solve the rabbit's problems under Egyptian summer conditions. It was conducted to assess and evaluate the impact of two different housing systems on the thermoregulatory mechanisms as indicators for predicting the productive and reproductive efficiency of the New Zealand White rabbits in Egypt.

2. Materials and Methods

This work was carried out at the Rabbit Research Unit, Faculty of Agriculture, Cairo University, Giza, Egypt. During hot months (July and August) of 2013 and 2014, Environmental measurements, physiological parameters and thermal-reactions were recorded six times daily.

Two housing systems were used to perform the field experiments. House "A" was roofed by 0.5 mm thickness of thin zinc-coated corrugated iron sheets. Its dimensions (length, width and ridge height) were 39.41, 14.66 and 5.20 m, respectively. House "B" was roofed by a 5.1 cm thickness of a polyurethane sandwich panel. Its dimensions were 36.60, 14.90 and 6.00 m, respectively therefore; the main difference between them is the roofing materials. Both houses were naturally ventilated by windows and there were ceiling fans for mixing air. Tri Sense manufacturer was used to measure air temperature (AT), relative humidity (RH) and air velocity (u). Temperature of structures surface for were measured by Infrared thermometer. Physiological parameters and thermal-reactions were measured by using electronic digital thermometer.

2.1. Evaluation of sensible energy balance for rabbits houses

By considering the structure as a control volume (Albright, 1990):

Energy gain – Energy loss = Energy stored

Energy gains: There was no supplemental heating in the houses during the whole experimental work. Hence, two modes of heat sources were considered.

2.1.1. Sensible heat gain from the rabbits, q_s

Nonlinear regression approach was used to derive a regression equation (1), a regression equation was developed by software Data Fit (version 9.0.59) to calculate the rate of sensible heat production (SHP) inside rabbit houses by using reviewed data as reported by Lebas et al. (1997) between ambient air temperature (°C) and release of sensible heat (W kg^{-1}). Coefficient of multiple determination of the regression equation is (R^2) = 0.99 and its P = 0.00015 (Table 1).

$$SHP = a \cdot T_a^3 + b \cdot T_a^2 + c \cdot T_a + d \quad (1)$$

$$q_s = SHP \cdot W \cdot n \quad (2)$$

Table 1. Regression variable results of SHL.

Variable	Value	Standard Error
a	-0.0000088	0
b	0.0031666	0.002
c	-0.2165158	0.041
d	5.7642857	0.205

2.1.2. Solar heat gain, q_{so}

Sensible heat gain from the sun can be gained through windows in a house, and be relatively small and will be neglected.

2.2. Energy losses

The heat losses were considered through surface conductive and/or convective heat exchange, heat exchange through floor and ventilation and/or infiltration heat exchange.

2.2.1. Heat loss through the building surfaces elements

Surface conductive and/or convective heat exchange can be estimated by the using the following equations. Zhang et al. (1989) using a simple model for calculating heat losses through the building surfaces. Ignoring the effect of solar radiation, and using the difference between the inside and outside air temperatures as the driving force, heat losses through the building surfaces can be calculated as the following:

$$q_b = (T_a - T_o)\sum_e U_{se} \cdot A_{se} \quad (3)$$

$$U_{se} = 1/R \quad (4)$$

$$R = x/K \quad (5)$$

$$R_T = R_{si} + R_1 + R_2 + \cdots + R_{so} \tag{6}$$

where q_b is heat loss through the building surfaces elements, W; A_{se} is Surface area of element, m^2; U_{se} is overall unit area thermal conductance of surfaces elements, W m^{-2} °C^{-1}; T_a is air temperature inside the building °C; T_o is outdoor temperature, °C; e is path of heat transfer which may be wall or roof or window or door element; R_1, R_2 are thermal resistance of each layer with thickness x, m^2 W °C^{-1}; R_{si}, R_{So} are thermal resistance of inside and outside air surface of the building element, m^2 W °C^{-1}; K is thermal conductivity, m^{-1} W °C^{-1}; x is thickness, m; UA-value for house "A" equal to 2.4 kW °C^{-1} and that for house "B" equal to 885.4 kW °C^{-1}.

2.2.2. Heat flow through building floor

Heat transfer through floors on grade can be estimated using the following equation, according to Albright (1990).

$$q_f = FP \cdot (T_a - T_o) \tag{7}$$

where qf is heat flow through building floor, W; F is Perimeter heat loss factor, Values of F for an uninsulated and unheated slab floor on grade range between 1.4 and 1.6 W m^{-1} °C^{-1}; P is the building perimeter, m; T_o is outside air temperature, °C.

2.2.3. Ventilation heat flows into and out of the structure (qvent)

Sensible heat contained in the ventilation air entering the house was assumed to be equal to that leaving the house. The heat difference between entering and leaving air streams is the heat lost from the house and was calculated by:

$$q_{vent} = C_{pair} \cdot \rho \cdot V_{total} \left(T_a - T_o\right) \tag{8}$$

where C_{pair} is specific heat of moist air, J kg^{-1} °C^{-1}; ρ is air density, kg m^{-3}; V_{total} is the sum of volumetric air flow rate due to wind and thermal buoyancy, m^3 s^{-1}.

Markus and Morris (1980) reported that the normal value for Cp_{air} ρ is 1.2 kJ m^{-3} °C^{-1}. The overall heat loss was the sum of equations (3), (7) and (8). Natural ventilation due to wind and thermal buoyancy separately and then combine them using the following equations (ASHRAE, 2009).

$$V_{total} = \left(V_{wind}^2 + V_{thermal}^2\right)^{0.5} \tag{9}$$

$$V_{wind} = C_v \cdot A \cdot u \tag{10}$$

$$V_{thermal} = C_d A \sqrt{2g\,\Delta H_{NPL} \left(T_o - T_a\right) / T_o} \tag{11}$$

where V_{wind} is airflow rate (by wind only), m^3 s^{-1}; C_v is effectiveness of openings (C_v is assumed to be 0.5 to 0.6 for perpendicular winds and 0.25 to 0.35 for diagonal winds, (unit less)); A is free area of inlet openings, m^2; u is air velocity, m s^{-1}; $V_{thermal}$ is airflow rate (by thermal forces only), m^3 s^{-1}; C_d is discharge coefficient for opening = 0.65 should then be used (ASHRAE, 2009); ΔH_{NPL} is height from midpoint of lower opening to NPL, m; T_a is indoor temperature, °C; T_o is outdoor temperature, °C; n is number of rabbits.

Available data on the neutral pressure level (NPL) in various kinds of buildings are limited. The NPL in tall buildings varies from 0.3 to 0.7 of total building height (ASHRAE, 2009). Equation (11) applies when $T_a < T_o$, if $T_a > T_o$ replace T_o in the denominator with T_a, and replace $(T_o - T_a)$ in the numerator with $(T_a - T_o)$ (ASHRAE, 2009).

2.3. Heat storage

The heat stored was calculated as: Total heat stored = Total heat gain - Total heat loss

2.4. Thermal environment and productivity of the rabbit

The thermal environment was evaluated in the rabbit housing systems through temperature-humidity index (THI) using the following equation, developed by Marai et al. (2001).

$$THI = t_{db} - \left(0.31 - 0.31 \cdot \frac{RH}{100}\right) \cdot (t_{db} - 14.4) \tag{12}$$

where T_{db} is dry- bulb temperature, °C; RH is Relative humidity, %.

1. Conditions of absence of stress, for THI values lower than 27.8; 2. Moderate stress conditions, for 27.8<THI<28.9; 3. Severe stress for THI between 28.9 and 30; 4. Extremely severe stress conditions for THI values exceeding 30 (Marai et al., 2001).

Medeiros et al. (2005) developed an equation to evaluate the productivity for broiler chickens, using the thermal environmental index for productivity of broilers (TEIbc), in which values between 21 and 24 (comfortable) are associated to the maximum productivity; between 25 and 27 (moderately comfortable), between 28 and 30 (discomfort), between 31 and 34 (extremely discomfort), and for values above of 35 (dangerous). Nonlinear regression approach was used to derive a regression equation to calculate TEIR

$$TEI_R = f(T_a, RH, u) \tag{13}$$

$$TEIR = 1.0903407 \times T_a - 0.0129632 \times RH - 2.2703248 \times u \tag{14}$$

Data Fit software (version 9.0.59) was used to develop a nonlinear regression equation to calculate TEIR. Coefficient of multiple determination of the regression equation is (R^2) = 0.9966425 and its probability (P) = 0.00019 (Table 2).

Table 2. Regression variable results of TEIR.

Variable	Value	Standard Error
a	1.0903407	0.087
b	-0.0129632	0.009
c	-2.2703248	3.20

2.5. Physiological responses

Medeiros (2001) developed equations to predict of the main broilers physiological responses for avian farm broiler chickens surface temperature, T_a ranging from 16 to 36°C, RH of 20% to 90% and mean air velocity of 0.0 to 3.0 m s^{-1}. Determination coefficient (R^2) of 0.96 was found for broiler chickens surface temperature. Prediction of the main rabbits physiological responses surface temperature (average of skin and fur temperatures) (T_{sp}) in rabbit housing systems can be estimated by the nonlinear regression equation (15). T_{sp} can be predicted by a nonlinear regression equation, that developed by Data Fit software (version 9.0.59).

$$T_{sp} = f(T_a, RH, u) \tag{15}$$

Table 3. Regression variable results of surface temperature (T_{sp}).

Variable	Value	Standard Error
a	1.21990371	0.113
b	0.0406879	0.028
c	-8.6782479	1.153
d	-0.1898457	5.786

3. Results and Discussion

3.1. Sensible energy balance for house "A" and house "B":

In house "A", heat gain, loss, and storage were dynamic. During the hottest periods of the day, (08:00-18:00) the heat gain, heat loss, and heat stored ranged from 2.8 at 14:00 to 3.3 at 08:00, ranged from -5, at 08:00 to 54 at 12:00 and ranged from -8.4 at 08:00 to 51.2 kW at 12:00, respectively. The rate of heat flow depends on the amount of temperature difference and the thermal conductivity of the material. The heat is transferred from outside building to inside building from 10:00 to 16:00 because the outside air temperature was higher than inside air temperature in this period, but the outside air temperature at 08:00 and 18:00 was lower than inside air temperature. Thus, larger heat losses were experienced at 10:00 to 16:00 than at 08:00 and 18:00. So, the ventilation inlet openings should be covered with curtains (Foylon with heat transfer coefficient 1.93 W m^{-2} k^{-1}) at 10:00 to 16:00 to decrease the sensible heat gain by the ventilation at these periods and the heat gains resulted in high heat storage of averaged 23 kW (Figure 1). The roof gain about 78.2% of the total heat gained by the building components during hot periods. So, the insulation layer must be added above the roof to minimize the heat gained. A compressed straw slabs with thermal conductivity of 0.03 W m^{-1} °C^{-1} were suggested to be located above the roof. Using 5 cm of compressed straw slabs decreased the combined UA-value from 2.4 to 0.75 kW °C^{-1} and this decreased the heat gain or loss from the building and consequently decreased the total resultant heat storage.

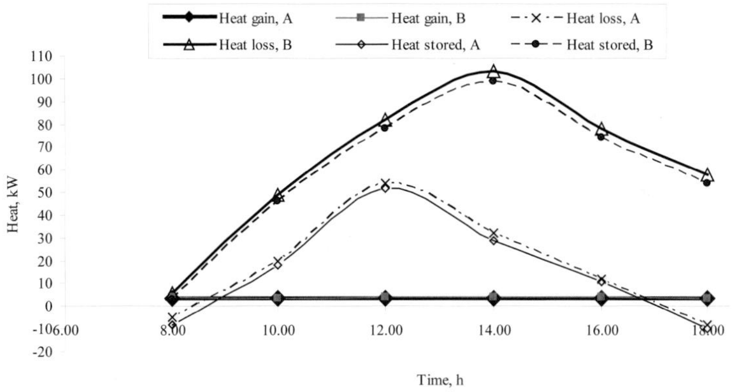

Figure 1. Heat gain, heat loss and heat stored in houses "A" and "B" during July and August.

In house "B", heat gain, loss and storage were also dynamic. During the hottest periods in the day, the heat gain, heat loss, and heat stored ranged from 3.3 at 14:00 to 3.7 at 08:00, ranged from 5.9 at 08:00 to 10.3 at 14:00, and ranged from 2.2 at 08:00 to 0.99 kW at 14:00, respectively. The heat was transferred from outside building to inside building at 08:00 to 18:00 because the outside air temperature was higher than inside air temperature. The heat stored in house "B" was higher than that in house "A" because the difference between inside and outside air temperature in house "B" was higher than in house "A". Accordingly, the driving force was high in house "B". During this hot period, the heat loss was not sufficient to offset the heat gains resulting in high heat storage which averaged 5.9 kW. The consequent dispersion of the stored heat may have resulted in elevated inside air temperature which could potentially stress the rabbits especially during at 08:00 to 18:00. So, the ventilation inlet openings should be covered with curtains from 08:00 to 18:00 to decrease the sensible heat by the ventilation at these periods and the heat gains resulting in high heat storage which averaged 1.4 kW (Figure 2).

3.2. Difference between THI and TEIR in house "A" and house "B":

During July and August 2013 and 2014, TEIR in house "A" was higher than THI at 08:00 to 18:00 (Figure 3). The difference between THI and TEIR varied from 1.3 at 08:00 to 4.9 at 14:00, with an average of 3.7. The maximum value of TEIR was 37.51 at 14:00 this means that inside thermal environment was dangerous and the minimum value of TEIR was 28.3 at 08:00 this means that inside thermal was discomfort. While TEIR in house "B" was lower than THI at 08:00 to 18:00 varied from 0.51 at 08:00 to 3.00 at 14:00, with an average of 2.61 The maximum value of TEIR was 32.74 at 14:00 this means that inside thermal environment was extremely discomfort and the minimum value of TEIR was 26.99 at 08:00 this means that inside thermal was moderately comfortable.

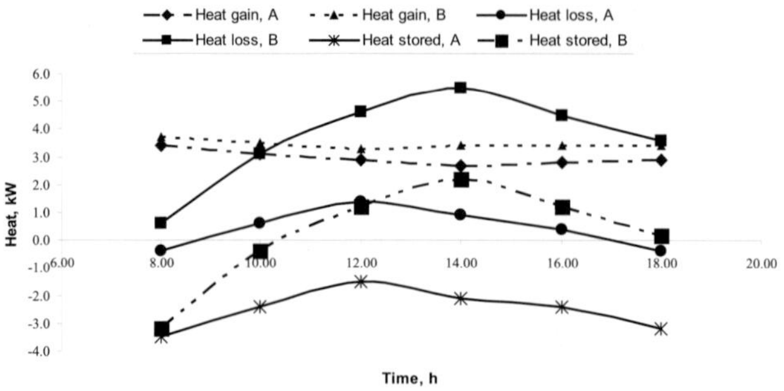

Figure 2. Heat gain, heat loss and heat stored after covering the ventilation inlet openings and using compressed straw slabs in houses "A" and "B" during July and August.

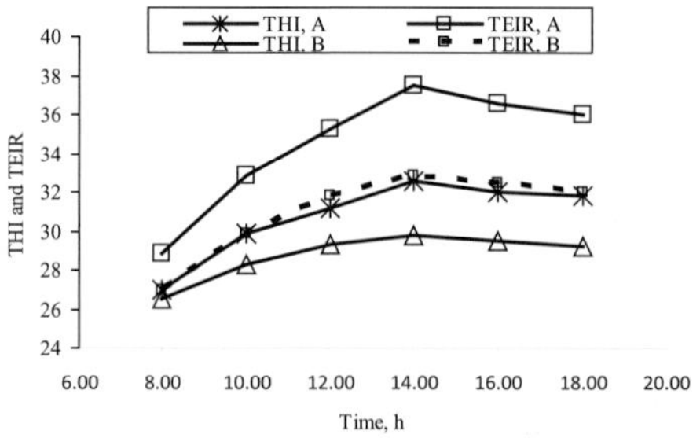

Figure 3. Difference between THI and TEIR in houses "A" and "B" during July and August.

3.3. Difference between *Ts* and *Tsp* in house "A" and house "B":

During hot months, measured surface temperature (T_s) in house "A" was lower than T_{sp} at 10:00 to 18:00, but it was higher than *Tsp* at 08:00 with 0.11°C (Figure 4). The difference between T_s and T_{sp} varied from 0.62°C at 12:00 to 1.2°C at 18:00, with an average of 0.92°C. The maximum value of T_{sp} was 37.5°C at 14:00 and the maximum value of T_s was 36.4°C. While the minimum value of T_{sp} was 29.6°C at 08:00 and the minimum value of T_s was 29.7°C at 08:00. While T_s in house "B" was higher than T_{sp} at 08:00 to 18:00. The difference between *Ts* and *Tsp* varied from 0.25°C at 08:00 to 1.16°C at 18:00, with an average of 0.85°C. The maximum value of T_{sp} was 34°C at 14:00 and the maximum value of T_s was 35°C. While the minimum value of T_{sp} was 28.4°C at 08:00 and the minimum value of T_s was 28.6°C at 08:00.

3.4. Effect of housing system on reproductive performance

Tables 4 to 6 show the effect of housing systems on reproductive performance of NZW rabbit does inside two rabbit housing systems (house "A" and house "B"). It can be observed that the conception rate in house "B" was higher than conception rating in house "A" (Table 4). This is due to severe internal thermal conditions especially high ambient air temperature. Also average of parities in the house "B" was higher than average of parities in house "A". But the numbers of services per conception and gestation length were lower than those in house "A". Similar result with house "B" was found by Morsy (2001) who reported that conception rate of NZW rabbits during hot condition averaged 50%. Housing systems did not show effect on the gestation period. The difference between both systems was 0.6 days. These results agree with the results of Ahmed (2000). It can be noticed that litter size at birth in house "B" was higher than that in house "A" (Table 5). This is due to the severe high ambient air temperature. Litter size at weaning in house "B" was higher than that in house "A". Pre-weaning mortality rate % in house "A" was higher than that in house "B". Similar result with house "B" was found by Morsy (2001) who reported that litter size at birth of NZW rabbits during hot condition averaged

6.5. Similar result with house "A" agree with those of Sallam et al. (1992) who found that litter size at birth was 5.9 in the hot seasons.

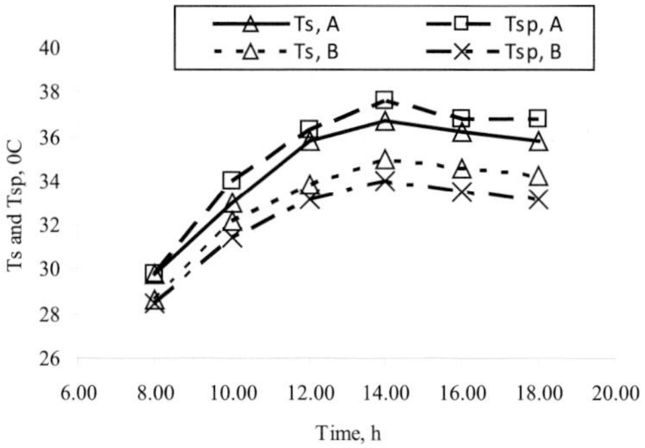

Figure 4. Difference between T_s and T_{sp} in houses "A" and "B" during July and August.

Table 4. Effect of housing system on reproductive performance of NZW rabbit does.

Item	House "A"	House "B"
Conception rate (%)	43.9 ± 1.93	50 ± 1.21
Average of parities/doe/year	5.2 ± 0.62	6.5 ± 0.72
Number of services per conception	2.4 ± 0.29	1.5 ± 0.26
Gestation length (day)	31.8 ± 0.29	31.2 ± 0.26

Table 5. Effect of housing system on Pre-weaning mortality rates in NZW litters.

Item	House "A"	House "B"
Litter size at birth/does/parity	5.9 ± 0.64	6.6 ± 0.52
Litter size at weaning	3.9 ± 0.65	5.2 ± 0.42
Pre-weaning mortality rate (%)	34.18 ± 4.59	21.19 ± 2.5

The higher mortality rate during hot season than the cold season, probably, due to the intrauterine environment of pregnant does that cause hyperthermia throughout the gestation period which inhibits, in some manner, the post-natal viability of the offspring. It also affects negatively the post-partum development of the mammary glands and subsequently milk production (Abd El-Moty et al., 1991). Table 6 shows that bunny weight at birth in house "B" was higher than that in house "A". This may be due to severe internal thermal conditions inside house "A" especially high ambient air temperature. Bunny weight at weaning was higher than that in house "A". Growth rate % in house "B" was higher than growth rate (%) in house "A". The lower bunny weight during the hot season could be attributed to reducing feed intake of does, which results in depressing bunny weight at birth. In addition the less milk supply to the growing bunnies in hot season (Abd El-Moty et al., 1991) affects their weight at weaning. This reduction under hot conditions may be due to the general depression of metabolic activity in hot conditions (Shafie et al., 1984).

4. Conclusions

The data indicates that a need to modify the existing structural design so that the intense heat load transmission into the house in summer will be reduced which otherwise will cause heat stress on the rabbits and reduce their productivity. In house "A", the roof gain is about 78.20% of the total heat gained by the building components during hot periods. Therefore, the insulation layer must be added above the roof to minimize the gained heat by the roof. It can be suggested to cover the roof by several compressed slabs to create insulation. In house "A" and "B", in hot periods it is often and essential to increase the rate of convection loss from the rabbits by increasing air speed at rabbit level with extra fans, horizontal delivery fans are better than ceiling fans in this context, because they are more effective in increasing the air speed at rabbit level in hot weather.

Table 6. Effect of housing system on NZW rabbit offspring's performance.

Item	House "A"	House "B"
Bunny weight at birth (g)	55.3 ± 2.6	60.2 ± 2.77
Bunny weight at weaning (g)	385.9 ± 16.13	442.6 ± 15.5
Bunny daily gain (g)	11.8 ± 0.49	13.7 ± 0.48
Growth rate (%)	149.9 ± 0.76	152.1 ± 1.25

References

Abd El-Moty, A.K., A.A. Abd El-Hakeam, A.M. Abd El-Rahman, 1991. Physiological responses of rabbits to high air temperature. 2. Productive and reproductive efficiency. Egyptian J. Rabbit Sci. 1 (2):146–158.

Ahmed, N. A., 2000. Diurnal and seasonal variations in thermoregulation and physiological reactions of pregnant and non-pregnant rabbits. J. Agric. Sci. Mansoura Univ. 25(6):3221–3233.

Ahmed, N.A., A.M. Hassan, and A.S. Morsy, 2008. Productivity of rabbits under climatic condition in Egypt condition. *1st Egyptian Conf. on Rabbit Sci. Fac. Agric.*, Cairo Univ. Egypt.

Albright, L.D., 1990. Environment Control for Animals and Plants. ASAE. St. Joseph, Mich., USA.

ASHRAE, 2009. Standards, (Natural ventilation Equations). American Society of Heating, Refrigerating and Air-Conditioning Engineers Inc. New York, USA.

Chepete, H. R. Tsheko, 2006. Hot and cold weather heat load dynamics of uninsulated broiler house in Botswana. Agricultural Engineering International: the CIGR Ejournal. Manuscript BC 06 001. Vol. VIII, pp. 1–17.

FAO, 1999. FAO helps Mediterranean countries set up a network for rabbit breeding, FAO's Animal Production Office, Rome, Italy, http://www.fao.org.

Lebas, F., P. Coudert, H. De Rochambeau, R.G. Thébault, 1997. The Rabbit Husbandry, Health and Production, FAO, Animal Production and Health Series No.: 21, Rome, Italy.

Marai, I.F.M., M.S. Ayyat, U.M. Abd El-Monem, 2001. Growth performance and reproductive traits at first parity of New Zealand White female rabbits as affected by heat stress and its alleviation under Egyptian conditions. J. Trop. Anim. Heal. Prod., 33:1–12.

Markus, T.A. E.N. Morris, 1980. Buildings Climate and Energy. Pitman Publishing Ltd., London.

Medeiros, C.M., 2001. Ajuste de modelos e determinação de índice térmico ambiental de produtividade para frangos de corte. DS thesis. Viçosa: Universidade Federal de Viçosa, Departamento de Engenharia Agrícola.

Medeiros, C.M., F.C. Baêta, R.F.M. Oliveira, I.F.F. Tinôco, L.F.T. Albino, P.R. Cecon, 2005. Índice térmico ambiental de produtividade para frangos de corte. Rev. Brasileira de Eng. Agrícolae Ambiental. 9 (4):660–665.

Morsy, A.S., 2001. Reproductive and productive performance of rabbits under different housing systems. M. Sc. Thesis, Fac. Agric., Cairo Univ., Egypt.

Sallam, M.T., F.M.R. El-Feel, H.A. Hasan, M.F. Ahmed, 1992. Evaluation of litter performance and carcass traits of two and three-way crosses of three breeds of rabbits. Egyptian J. Anim. Prod. 29 (2):287–302.

Shafie, M.M., G.A. Kamar, A.M. Borady, H. Hassaneln, 1984. Reproductive performance of Giza rabbit does under different natural and artificial environmental conditions. Egyptian J. Anim. Prod. 24:167–174.

Zhang, J.S., K.A. Janni, L.D. Jacobson, 1989. Modeling natural ventilation induced by combined thermal buoyancy and wind. Trans. ASAE. 32(6):2165–2174.

Preferred Environmental Temperature of Group-Housed Sows Fed with High Heat-Increment Diets

Bernardo Predicala *, **Alvin Alvarado**

Prairie Swine Centre Inc., Saskatoon, Saskatchewan S7H 5N9, Canada

* Corresponding author. Email: bernardo.predicala@usask.ca

Abstract

As the pig industry shifts from stalls towards group housing system for sows, there is a need to re-consider the current lower critical temperature (LCT) values being used in setting temperatures in sow barns considering that sows housed in groups have the potential to exhibit thermoregulatory behavior to maintain comfort even when the temperature in the barn is lowered. In addition, high heat-increment diet increases heat production in sows without increasing digestible energy and can be a means of reducing activity and limiting aggression in group-housed sows under reduced barn temperature. This study aimed to determine the preferred temperature of sows housed in groups and fed with high heat-increment diets. An operant mechanism that allowed the sows to demonstrate their preferred environmental temperature by enabling them to control the operation of the heating and ventilation equipment in the room was designed and developed. Three replicate trials were carried out in two fully controlled-environment chambers. Sows in one chamber were fed with standard gestation diet (Control) while sows in the other chamber were fed with high heat-increment diet (Treatment). Results showed that sows fed with high heat-increment diet could tolerate exposure to lower temperatures better than those fed with a standard gestation diet. Moreover, performance of sows fed with high heat-increment diet seemed to have been less affected by the exposure to colder temperatures compared to those fed with standard diet.

Keywords: Swine, group housing, environmental temperature, operant mechanism, sows

1. Introduction

The recently revised Canadian Code of Practice for the Care and Handling of Pigs mandates that gestation sow housing in all sow farms should be fully converted from stalls to group systems by July 2024 (NFACC, 2014). With this change, a number of considerations have to be taken into account in order to provide a positive and safe environment for sows and barns workers while operating a cost-effective production system. Two such considerations are the temperature requirements of gestating sows and the potential for higher activity levels and aggression among sows housed in groups.

At present, temperatures maintained in gestation barns are based on the reported lower critical temperature (LCT) for individually housed sows such as in stall systems (Geuyen et al, 1984). Allowing the temperature to drop below the LCT (which is currently around 15°C) will require additional feed for the sows to maintain body condition and weight gain over the gestation period (Le Goff et al., 2002). However, it has been suggested that sows in groups, even without bedding, may have LCT values significantly lower than 15°C since sows in groups have the potential to use thermoregulatory behavior, such as huddling, to maintain comfort even if temperatures in the barn are lowered. Thus, if group-housed sows can maintain body condition and weight gain at temperatures lower than currently maintained in sow barns and without the need for additional feed, the potential exists to reduce energy costs for ventilation and heating during winter season.

Aside from temperature requirements, some issues anticipated with group-housed sows include the potential for higher activity levels and aggression among sows. These problems are exacerbated when sows are put on a restricted feeding regime, which is a common practice for gestating sows to maintain optimal body condition. Satiety, the sensation of feeling 'full', is improved with high heat-increment (high-fiber) diets which are known to reduce the urge to feed continuously as well as overall activity and repetitive (stereotypic) behavior in sows. Moreover, high-fiber diets induce an increase in heat production in sows exposed to low temperatures (Close and Cole, 2000). As such, adding fiber to the diet can be a means of reducing activity and limiting aggression in sows housed in groups, as well as contributing to the energy balance of sows under reduced barn temperature.

The general objective of this project is to investigate management practices that will allow pig producers to benefit from potential advantages of housing sows in groups. Specifically, this study aimed to determine the environmental temperature preferred by sows fed diets with high fermentable fiber in terms of metabolic rate, body temperature, and thermoregulatory behavior.

2. Materials and Methods

A preliminary experiment was carried out to develop the instrumentation (operant mechanism) that will allow sows to demonstrate their preferred environmental temperature by enabling them to control the operation of the heating and ventilation equipment installed in the room. The mechanism was then modified and implemented in controlled-environment chambers to determine the preferred temperature of sows housed in groups and to be able to understand the physiological effects of high heat-increment (high-fiber) diet on sow metabolism and thermoregulatory behavior.

2.1. Description of facilities and experimental set-up

Two fully instrumented and controlled-environment chambers at the Prairie Swine Center barn facility in Saskatoon, Saskatchewan, Canada were used in this study. Each chamber has inside dimensions of 4.2 m × 3.6 m × 2.7 m. The pen area has a slatted concrete floor at one end of the pen and a solid floor extending towards the opposite end of the pen. A commercial feeder and nipple drinker were installed on one side of the plastic penning.

The chambers were operated on a negative pressure ventilation system. Fresh air was forced through a filtration unit (Circul-Aire USA-H204-B, Dectron International, Roswell, GA, USA) by a 0.6-m diameter centrifugal fan (Delhi BIDI-20, Delhi Industries Inc., Delhi, ON, Canada) before entering the chamber through an actuated ceiling inlet. The room air was exhausted from the chamber through a sidewall exhaust fan (H18, Del-Air Systems Inc., Humboldt, SK, Canada) into the exhaust duct with a flow measuring device to monitor the airflow rate. A 2-kW in-duct heater was installed in the supply duct to the chamber to add heat when necessary. All these equipment were controlled with a Proportional Environment Control (PEC) system (Phason Electronic Control Systems, Winnipeg, MB, Canada).

To allow the sows to control their own environmental temperature, an operant mechanism was designed and developed, and installed in each chamber. The circuitry of the developed mechanism is shown in Figure 1. The operant mechanism was configured to control the heating system of the chamber as well as a small radiant heater using a switch located on the pen wall which the sows can access. When a sow activates the switch, it operates the existing supplementary heating system for the entire room for 3 minutes as well as the small radiant heater above the location of the switch for 2 minutes as an immediate feedback reward. One of the installed timers was configured to prevent sows from successively activating the heaters by deactivating the switch for a period of 5 minutes after its previous activation, i.e., any switch presses during this 5-minute period will not operate the heaters. In addition to the functional heat control switch, a 'dummy' switch that does not operate the radiant heater (i.e., unrewarded activity) was also installed close to the operant switch to distinguish between deliberate behavior by the sows to control the room temperature and random interaction with the mechanism. Both the operant and dummy switches were fixed onto the penning of each chamber and the radiant heater was installed above the switches as shown in Figure 2. A similar operant mechanism has been used in determining temperature preference of piglets in a previous study (Bench and Gonyou, 2007).

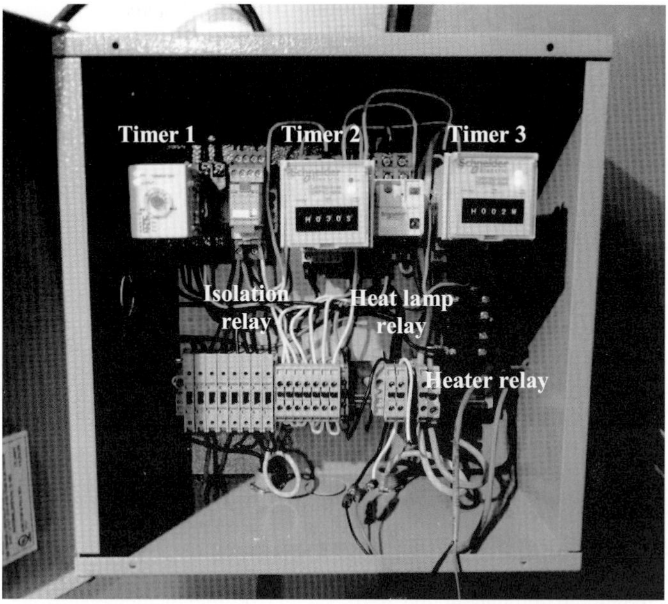

Figure 1. Circuitry of the developed operant mechanism.

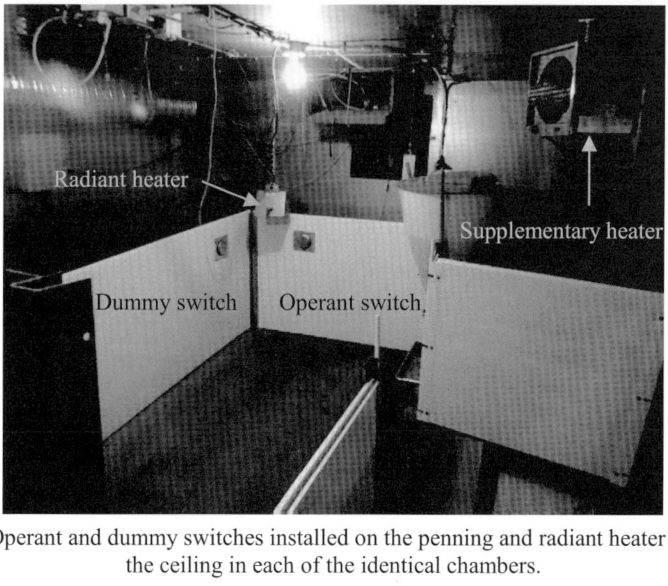

Figure 2. Operant and dummy switches installed on the penning and radiant heater fixed from the ceiling in each of the identical chambers.

In order to encourage the sows to use the operant mechanism, the chambers were operated at a set point temperature of 8°C. To be able to do this, cold ambient air from outside the barn was directly drawn into the chambers. The control system was configured such that the heaters inside the fresh air supply duct were activated when the temperature inside the chamber was below the set point temperature.

2.2. Experimental procedure

A total of 3 replicate trials were carried out in the environmental chambers with each chamber housing 2 sows. The duration of each trial was 3 weeks, with the first week dedicated to allow sows to acclimatize to the chamber and to learn to use the operant mechanism, while the remaining 2 weeks were for data collection. Two experimental diets were used, with sows in one chamber fed with the standard gestation diet (Control) while sows in the other chamber fed with the high heat-increment diet (Treatment). To account for room effect, assignment of diets was interchanged between the two chambers on subsequent trials. During the acclimation period (first week of trial), standard gestation diet was given to all sows.

Throughout the trials, data collection included continuous monitoring of temperature at different locations in the chamber, activation events of the operant and dummy switches, and behavior of sows. Effect on physiology of animals as well as diet digestibility were also assessed through analysis of blood thyroxine, peptide C and leptin levels, and nitrogen and phosphorus in manure, respectively. Additionally, pig performance of sows was assessed by determining their average daily gain (ADG).

3. Results and Discussion

3.1. Chamber temperature

Temperature at the pig level as well as the corresponding operant switch activations by sows and the operation of the radiant and room supplementary heaters in both chambers were continuously monitored throughout the trial. Sample profile of temperature and switch and heater activations during the second trial is presented in Figure 3. Throughout the trial, a pattern was observed where temperature changes occur mainly during the day when sows are mostly active. Correspondingly, most of the switch activations by the sows were made during daytime. Typically, barn operations are carried out between 7 AM to 3 PM; beyond this period, lights in both chambers are turned off. This procedure has been routinely done at the entire facility to provide animals the sense of actual day and night time cycles.

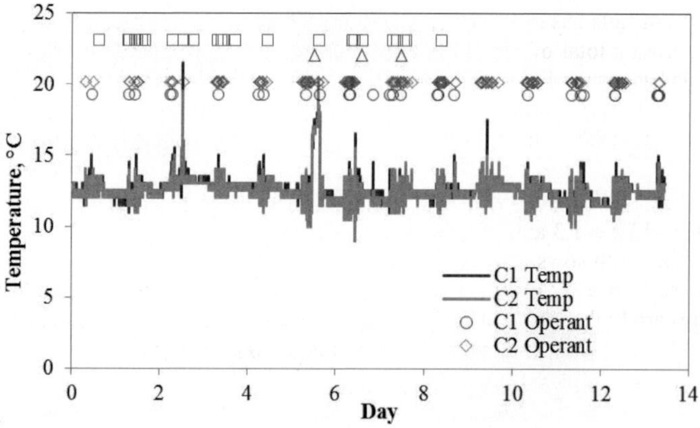

Figure 3. Pig level temperature and corresponding operant and dummy switch activations by sows throughout the second replicate trial in the controlled-environment chambers. Sows in chamber 1 (C1) were fed with standard gestation diet while those in chamber 2 (C2) were fed with high heat-increment diet.

The actual room temperature at the time when sows activated the operant mechanism was also recorded and shown in Figure 4. Most of the time, sows fed with high heat-increment diet activated the operant mechanism at a relatively lower pig level temperature than sows fed with standard gestation diet. Over 3 trials, the average temperature when the operant mechanism was activated by sows fed with high heat-increment diet was 12.5°C while that in the control chamber was higher at 13.4°C. This suggests that sows fed with high heat-increment diet could tolerate lower temperature before calling for supplemental heat than sows fed with standard gestation diet.

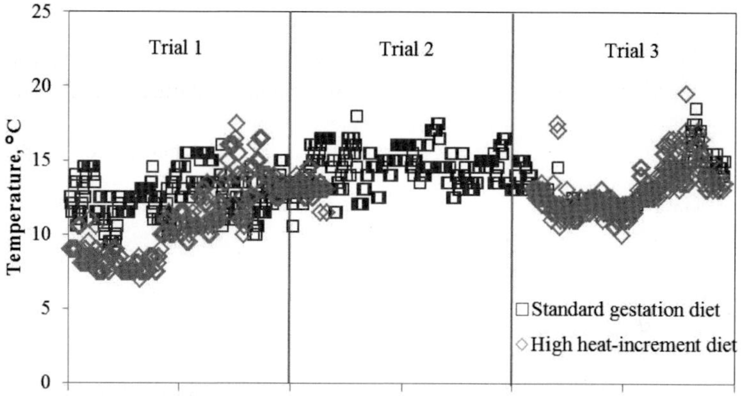

Figure 4. Pig level temperature at the time the operant mechanism was activated by the sows in the two chambers.

As shown in Figure 3, the sows were considerably engaged in the use of the operant mechanism as evidenced in the number of activations. Figure 5 shows one of the instances of deliberate operant activation of one of the sows. It can be observed that dummy switch activations were numerous as well. This could be due to accidental presses by the sows such as during times when they are feeding while standing close to the switch. To minimize cases of accidental activations, guard rails were installed around the switches in subsequent trials. The installation of guard rails has caused significant drop in the number of switch activations of the dummy switch, from a total of 225 presses during the first trial down to 43 presses during the second trial. This has resulted to considerably higher number of switch activations of the operant switch in both chambers (a total of 752 presses over three trials) than those of the dummy switch (407 presses) which indicated the preference of the sows for the operant switch over the dummy switch.

In the chamber with sows fed high-heat increment diet, average day and night time temperatures were 13.4 ± 1.3 and 12.5 ± 1.4°C, respectively; these values were relatively higher than in the chamber with sows fed standard gestation diet where the average day and night time temperatures were 12.6 ± 1.7 and 12.1 ± 1.6°C, respectively, over 3 trials. It was also observed that temperatures are higher during the day when sows are most active (compared to night time) due to greater body heat generated by the sows during periods of higher activity.

3.2. Sow performance and physiological response

Performance of sows was assessed based on average daily gain (ADG) while their physiological responses were measured through blood thyroxine, leptin and peptide C levels. In terms of ADG, a slight difference was observed between the two treatment diets. On average, sows fed with high heat-increment diet had an ADG of 0.2 kg day^{-1} while sows fed with standard gestation diet had 0.16 kg day^{-1}.

Over three trials, leptin level of sows fed with high heat-increment diet was 2.95 ng ml^{-1}; this was about 47% lower than those sows fed with standard gestation diet (4.34 ng ml^{-1}). This indicates that sows fed with high heat-increment diet utilized larger amounts of body lipids (fats) but very little protein stores than those sows fed with standard gestation diet. Cook et al. (undated) has reported that as fat stores are utilized, concentrations of leptin should decrease. Conversely, the level of thyroxine in the blood plasma of sows fed with high heat-increment diet was higher than those sows fed with standard gestation diet. On average, blood thyroxine of sows with high heat-increment diet was 74.2 nmol L^{-1} while that of sows with standard gestation diet was 66.9 nmol L^{-1}. This implies that the rate of metabolism of sows with high heat-increment diet was relatively higher than those sows with standard gestation diet.

Figure 5. Activation of radiant and supplemental heaters as the operant switch was pressed by the sow.

Over three trials, leptin level of sows fed with high heat-increment diet was 2.95 ng ml^{-1}; this was about 47% lower than those sows fed with standard gestation diet (4.34 ng ml^{-1}). This indicates that sows fed with high heat-increment diet utilized larger amounts of body lipids (fats) but very little protein stores than those sows fed with standard gestation diet. Cook et al. (undated) has reported that as fat stores are utilized, concentrations of leptin should decrease. Conversely, the level of thyroxine in the blood plasma of sows fed with high heat-increment diet was higher than those sows fed with standard gestation diet. On average, blood thyroxine of sows with high heat-increment diet was 74.2 nmol L^{-1} while that of sows with standard gestation diet was 66.9 nmol L^{-1}. This implies that the rate of metabolism of sows with high heat-increment diet was relatively higher than those sows with standard gestation diet.

Results from analysis of peptide C as well as test for feed digestibility are still to be analyzed. Wide variability of the data between sows within the same treatment diet was observed, thus, these parameters will be closely monitored in subsequent trials to establish more definitively the impact of sow diet and low temperature housing on sow performance and physiological response.

Table 1. Performance and physiological responses of sows fed with standard gestation diet and high heat-increment diet when subjected to colder temperatures (N = 3).

Parameters	Standard gestation diet	High heat-increment diet
Average daily gain, kg day^{-1}	0.16 ± 0.07	0.20 ± 0.68
Thyroxine, nmol L^{-1}	66.9 ± 7.0	74.2 ± 9.0
Leptin, ng ml^{-1}	4.34 ± 0.80	2.95 ± 0.14

4. Conclusions

Results obtained from three trials conducted in controlled-environment chambers have shown that sows fed with high heat-increment diet tended to maintain relatively lower temperatures (12.5°C on average) in the chamber than those fed with standard gestation diet (13.4°C). Moreover, the exposure of sows fed with high heat-increment diet to relatively colder temperatures had no considerable effect on their performance and physiological response; however, additional trials will be conducted to further assess the overall effect of this treatment approach on sow physiology and overall performance.

Acknowledgements

The research team would like to acknowledge the financial support for this research project from the Saskatchewan Agriculture Development Fund and the Saskatchewan Pork Development Board. The authors also acknowledge the strategic program funding provided by Sask Pork, Alberta Pork, Ontario Pork, the Manitoba Pork Council and the Saskatchewan Agriculture Development Fund to the Prairie Swine Centre. In addition, the authors also wish to express their appreciation for the support of the production and research staff at Prairie Swine Centre in carrying out this study.

References

Bench, C.J. and H.W. Gonyou. 2007. Temperature preference in piglets weaned at 12–14 days of age. Canadian Journal of Animal Science. 87, 299–302.

Close, W.H., and D.J.A. Cole. 2000. *Nutrition of Sows and Boars*. Nottingham University Press, Nottingham. 377p.

Cook, C.M., R.A. Easter, and J.M. Bahr. Undated. Leptin — endocrine signal regulating the onset of post-weaning estrus in pigs. http://livestocktrail.illinois.edu/uploads/porknet/papers/Leptin.pdf. Accessed June 19, 2015.

Geuyen, T.P.A., Verhagen, J.M.F., and M.W.A. Verstegen. 1984. Effect of housing and temperature on metabolic rate of pregnant sows. Animal Production. 38, 477–485.

Le Goff, G., L. Le Groumellec, J. van Milgen, S. Dubois, and J. Noblet. 2002. Digestibility and metabolic utilization of dietary energy in adult sows: influence of addition and origin of dietary fibre. British Journal of Nutrition. 87(4), 325–335.

NFACC. 2014. Code of Practice for the Care and Handling of Pigs. Canada: National Farm Animal Care Council. https://www.nfacc.ca/pdfs/codes/pig_code_of_practice.pdf. Accessed May 15, 2015.

Systematic Evaluation of Heat Stress Impact and Selection of Abatement Systems

Joseph M. Zulovich *, Teng-Teeh Lim

Food Systems and Bioengineering, University of Missouri, Columbia, MO 65211-5200, USA

* Corresponding author. Email: zulovichj@missouri.edu

Abstract

Heat stress caused by hot weather conditions is well known to result in swine production and growth declines. Heat abatement systems commonly applied to provide supplemental cooling include evaporative cooling pad and sprinkler cooling systems. Evaporative cooling pad systems are usually preferred over sprinkler cooling systems because the animal living space stays dryer with pad systems. However, evaporative cooling pad systems do not always provide sufficient cooling in all weather conditions and climatic locations. The challenge has always been to know when a producer should use a sprinkler system instead of an evaporative cooling pad type system. A method has been developed to use location specific weather data to determine the duration and severity of heat stress conditions. This method can also determine the effectiveness of cooling resulting from the use of an evaporative cooling pad system, using a heat stress hours (HSH) procedure. The output from the overall procedure determines one of three operational scenarios for supplemental cooling needs based on weather data for a given location: (1) no supplemental cooling even though heat stress conditions exist in the afternoon; (2) supplemental cooling is needed and can be adequately provided by implementing an evaporative cooling pad system; (3) supplemental cooling is needed and a sprinkler cooling system should be implemented rather that an evaporative cooling pad system. The duration of cooling system use can also be determined by conducting the HSH procedure.

Keywords: Hot weather, supplemental cooling, cooling recommendations, swine production, evaporative cooling

1. Introduction

An ongoing challenge for livestock producers is to determine what kind of heat abatement system should be selected for their operation and how the system should be operated to obtain maximum benefit. Heat stress has been shown to have significant impacts on livestock production and animal welfare as discussed in Chapter 7 "Effects of the Thermal and Gaseous Environment on Livestock" by Hellickson and Walker (1983) and parts of several chapters by DeShazer (2009). However, studies that evaluate the intensity and duration of heat stress and determine the benefits of heat abatement systems in terms of a financial return to swine production are somewhat limited. Bridges et al. (2003) showed the variation in returns as dollars pig^{-1} $year^{-1}$ for a variety of pig placement dates and locations in US. Depending upon the animal capacity of a given swine grow-finish facility, a swine evaporative cooling system can be paid for within a few years to less than a year depending upon the input assumptions. Ulmer and Zulovich (2008) evaluated the impact of cooling during summer by estimating the resulting decrease in financial returns due to pig growth performance decreases when no cooling was provided. When building space availability was limited, adding an evaporative cooling system was very beneficial from a returns perspective. So, the production efficiency improvements and financial rewards for implementing an effective evaporative cooling system for swine can easily be shown.

The challenge is to determine what kind of system and how should it be managed to obtain the maximum benefit for a given operation in a given location. A fairly recent and comprehensive discussion of current heat stress abatement systems for livestock production can be found in Chapter 8 "Environmental Management" in DeShazer (2009). The general operating performances of different cooling systems within general climatic scenarios are presented in this paper. By reviewing the evaporative cooling information, one can strategically determine if indirect cooling (via evaporative cooling pads or some other method) to reduce inlet air temperature should be selected as the preferred method. If indirect cooling does not appear to be adequate, direct cooling (via a sprinkler system) should be selected if the animal type and housing system allows for direct sprinkling of water on animals. The first objective of this paper is to use long-term weather data to assess the intensity and duration of heat stress conditions on swine production. The second objective is to use this assessment and evaluate the heat stress reduction potential by different heat abatement systems.

2. Materials and Methods

The first critical question to answer can be phrased as 'Is indirect cooling adequate for my scenario?' To address this question, one must select a method to determine adequacy from an animal production and welfare perspective. Air temperature alone is typically not sufficient to evaluate the overall thermal environmental impact on an animal. Several key factors, including temperature, humidity, air velocity, and thermal radiation can affect the thermal comfort of the animal living environment. So some type of thermal index must be selected to evaluate the adequacy. A very comprehensive discussion on thermal indices can be found in Chapter 5 "Thermal Indices and Their Applications for Livestock Environments" by DeShazer (2009). However, a simple and straight-forward temperature-humidity index (THI) was selected to be the evaluation parameter for evaporative cooling adequacy from a swine production perspective. From the variety of THI calculation options available, the following THI calculation was selected:

$$THI = 0.72\ (T_{db} + T_{wb}) + 40.6 \quad (1)$$

where: THI is the Temperature-Humidity Index; T_{db} is dry-bulb temperature in °C; T_{wb} is wet-bulb temperature in °C.

The evaporative cooling adequacy can be evaluated by comparing the THI value of conditioned air exiting an evaporative cooling pad with the Livestock Weather Safety Index as presented in Chapter 5 in DeShazer (2009). The "Categories of Livestock Weather Safety Index" based on THI values used to evaluate evaporative cooling adequacy are as follows: Normal Conditions (No heat stress): $THI \leq 74$; Alert Conditions: $74 < THI \leq 78$; Danger Conditions: $78 < THI < 84$; and Emergency Conditions: $84 \leq THI$.

The cooling benefit that can be obtained from an evaporative cooling pad system can be expressed as the Evaporative Cooling Efficiency (ECE) as presented by Albright (1990). Evaporative Cooling Efficiency is defined as the percentage of wet bulb depression, and a given evaporative cooling system can achieve and is typically assumed to be 75%. The wet bulb depression is defined at the difference between the dry-bulb temperature and the wet-bulb temperature. In mathematical terms, the outside dry-bulb temperature is reduced by ECE times the wet bulb depression for a given set of weather conditions. For example, if the outside dry-bulb temperature was 28°C and the wet-bulb depression was 4°C, the dry-bulb temperature leaving an evaporative cooling pad would be 25°C assuming a 75% ECE.

A source of weather data containing both dry-bulb and wet-bulb temperatures for the same time during the day is needed to determine the cooling effectiveness of an evaporative cooling pad system. A new comprehensive weather database ASHRAE (2013b) was released, that includes frequency vectors and joint frequency vectors with a variety of coincident design weather conditions. For example, one joint frequency weather data vectors provide the typical dew point temperatures for the entire range of air temperatures for a given location. This weather database provides basic design weather data and joint frequency matrices for 6,443 weather stations around the world. Weather data from this database was used to provide the long-term weather data for the evaporative cooling evaluation procedure. The weather databases "annual dry-bulb temperature with coincident wet-bulb temperature joint frequency matrix" and the "2% design condition load calculation" were used (ASHRAE, 2013b). The "annual joint frequency matrix" provided the weather data for annual evaluation computations, while the "2% design load calculation" data provided for extreme worse case daily evaluation computations.

The dry-bulb and coincident wet-bulb joint frequency matrix from the ASHRAE weather database provided the dry-bulb temperatures with the coincident wet-bulb temperature for each dry-bulb temperature as well as the typical number of hours per year the dry-bulb temperature is expected for the given location. Several locations in China with different weather/climatic conditions were selected to show results from the presented heat stress evaluation procedure and show how different weather conditions impact the outcome. Table 1 shows the annual summer weather data used in analyses. The dry-bulb temperature (T_{db}°C) is common to all four locations.

Table 1. Summer Weather Data Used in Analyses for Various Locations.

	Chongqing		Shanghai		Harbin		Guangzhou	
T_{db}°C	T_{wb}°C	hours	T_{wb}°C	hours	T_{wb}°C	hours	T_{wb}°C	hours
20.0	18.0	351.50	16.9	300.92	16.2	275.99	16.6	308.64
21.0	18.9	352.27	18.0	342.64	17.0	275.50	17.5	350.48
22.0	19.8	364.23	18.9	363.45	17.7	258.43	18.6	379.99
23.0	20.6	361.31	19.9	366.18	18.2	226.17	19.7	403.09
24.0	21.4	353.69	20.8	348.43	18.7	210.01	21.0	490.99
25.0	22.2	341.04	21.8	345.43	19.3	184.20	22.3	538.16
26.0	22.7	320.16	22.9	339.48	19.6	152.09	23.4	682.21
27.0	23.2	279.40	24.0	328.83	20.0	123.19	24.1	659.76
28.0	23.6	258.94	24.7	326.59	20.3	95.07	24.6	596.52
29.0	24.0	226.50	25.2	268.58	20.5	66.20	24.9	466.20
30.0	24.2	189.32	25.5	189.24	20.8	45.91	25.1	325.90
31.0	24.5	147.86	25.9	137.35	20.5	27.14	25.4	288.04
32.0	24.8	122.08	26.2	104.96	20.0	16.47	25.6	231.32
33.0	25.0	98.82	26.4	79.09	20.6	8.66	25.9	162.20
34.0	25.2	78.97	26.6	54.64	20.6	3.81	26.1	112.78
35.0	25.4	64.77	26.9	30.64	19.9	1.76	26.1	51.10
36.0	25.5	43.79	26.9	17.91	20.6	1.52	26.3	25.74
37.0	25.5	24.95	26.7	5.66	20.4	0.36	26.2	8.33
38.0	25.3	12.59	26.6	1.70	18.0	0.04	26.2	2.28
39.0	25.3	5.19	26.2	0.20	18.0	0.04	25.6	0.58
40.0	25.0	2.80					25.0	0.13
41.0	24.7	0.92						
42.0	25.0	0.13						
43.0	24.0	0.04						

For the evaluation of heat stress on an annual basis, the wet-bulb depression was calculated for dry-bulb temperature from the difference between the dry-bulb temperature and coincident wet-bulb temperature. Next, the THI, using equation (1), was calculated not only using the weather database dry-bulb and wet-bulb temperatures but also using the evaporative cooled/reduced dry-bulb temperature and coincident database wet-bulb temperature for both outside weather conditions and conditions leaving the cool cell system. Finally, the heat stress hours (HSH) is the annual sum of the product of hours per year at a given temperature times only a positive difference between the calculated THI minus the critical THI for a given dry-bulb temperature which is similar to the method as presented by Zulovich et al. (2008).

A daily heat stress evaluation procedure was developed using the "2% design condition load calculation" weather databases which provide the typical hourly extreme dry-bulb and wet-bulb temperature conditions for a 24-hour day for each month on both a dry-bulb and wet-bulb basis.

The dry-bulb basis provides weather conditions to evaluate extreme dry-bulb temperature with coincident wet-bulb temperature. The wet-bulb basis provides weather conditions for extreme wet-bulb temperatures with coincident dry-bulb temperatures. The dry-bulb and wet-bulb basis daily design load databases allow for evaluation of extreme daily temperatures or extreme daily humility levels. The daily heat stress evaluation provides a quantified indication of the intensity of the heat stress for the extreme design day for each month of year. The wet-bulb basis was included because animal cooling systems depend upon the evaporation of water to provide cooling. The wet-bulb basis helps assess the cooling potential on a daily basis during extreme humid weather conditions for the location.

Typically, the weather is not at an extreme for both dry-bulb temperature and humidity at the same time. Depending upon local weather conditions and design challenge, differences in system design can require one uses the dry-bulb or wet-bulb design load data. THI values were calculated for each hour of the day for every month using the same procedure as the annual analysis resulting in four different THI values for each hour: (1) Outside dry-bulb based; (2) Evaporative cooled dry-bulb based; (3) Outside wet-bulb based; and 4) Evaporative cooled wet-bulb based. Then the total heat stress hours was calculated for each hour using a similar method as resented by Zulovich et al. (2008). Finally the HSH were totalled for each day of the month to determine if heat stress conditions existed for the entire day or only during part of the day.

3. Results and Discussion

A summary of the annual total hours the conditions exceeded a critical THI value and HSH hours are provided in Table 2. The "Total Hours Outside Air" column is the hours the outside weather conditions exceed the critical THI value for the location. The "Total Hours with Cooling" column is the total hours the conditioned air from the evaporative cooling system exceed the critical THI value.

Table 2. Total Annual Hours and Total Annual Heat Stress Hours (HSH).

Location	Critical THI Value	Total Hours Outside Air	Total Hours with Cooling	Total HSH Outside Air	Total HSH with Cooling
Chongqing	74	2218.3	1557.1	9734.5	3907.8
Chongqing	78	1018.7	333.0	3197.0	217.5
Harbin	74	390.2	0.0	737.6	0.0
Harbin	78	32.7	0.0	29.0	0.0
Guangzhou	74	4151.3	3613.1	18744.0	10139.9
Guangzhou	78	2271.1	882.5	5598.2	713.8
Lanzhou	74	178.5	0.1	322.2	0.1
Lanzhou	78	15.9	0.0	23.4	0.0
Shanghai	74	2230.3	1884.9	9926.7	5670.1
Shanghai	78	1216.6	621.4	3267.2	718.2

The critical THI of 74 and 78 were selected based on the "Categories of Livestock Weather Safety Index" previously discussed. A THI of 74 or less is considered to be normal conditions which no heat stress by pigs is anticipated. A THI greater than 74 but less than or equal to 78 is considered to be a transition zone with respect to evaporative cooling pad systems used in pig production. Ulmer and Zulovich (2008) reported that the impact of pig performance will be dependent on the length and severity of hot conditions but were not able to quantify the production impact with respect to hot conditions. So, these safety index categories provide a method to quantify heat stress conditions.

Several important values can be obtained from the information in Table 1 to determine if any and what type of heat abatement system should be installed for a given swine facility. The 'Total Hours Outside Air' (THOA) provides an indication of how many hours per year one would need to operate evaporative cooling pads in a swine facility. A high THOA hour number indicates that installation of cooling is beneficial. The 'Total Hours with Cooling' (THwC) provides an indication of how long per year an evaporative cooling pad cannot cool air entering sufficiently to remove heat stress conditions of the inlet air to the swine facility. A low THwC hour number indicates that one should select evaporative cooling pad system to provide cooling. A high THwC hour number indicates one should probably select a sprinkler system to wet animals for cooling instead of an evaporative cooling pad system because the maximum amount of moisture can be absorbed by the outside condition of inlet ventilation air. A high THwC hour number can also be interpreted as a sprinkler cooling system should be used in conjunction with an evaporative cooling pad system. The sprinkler cooling system performance may be reduced somewhat because moisture is added to the inlet air from the evaporative cooling pad system. However, pigs would be provided with additional heat stress relief from the added sprinkler cooling system. The benefit would be a building/facility can be kept dryer when an evaporative cooling pad system is sufficient. The 'Total HSH' columns give an indication of the intensity of heat stress. If HSHs are relatively high even though the THOA seems relatively low, a cooling system should be considered because the weather data indicate the location has short term high intensity heat stress events. These high intensity short term events can be more economically detrimental and production critical because a significant number of animal death losses might occur during the short term, high intensity heat stress events, especially when the production facility is not equipped with any cooling system.

Reviewing the values in Table 2 shows several significant trends that were expected. The expected trends include a reduction in total hours and heat stress hours when evaporative cooling pads are implemented and a reduction in all values when the critical THI was increased from 74 to 78 for all locations. However, the difference between 'Total Hours with Cooling' and 'Total HSH with Cooling' is not consistent for all locations and THI differences. The value for 'Total HSH with Cooling' was expected to be significantly larger than 'Total Hours with Cooling' because the total hours is multiplied by the intensity of heat stress (difference between the critical THI and the calculated THI for a given temperature bin); however, this expected difference does not exist in all scenarios. So, the daily heat stress evaluation procedure was developed and implemented to help provide insight into why the inconsistent differences exist on an annual basis.

A summary of the daily extreme heat stress hours are given in Table 3 for Chongqing, Shanghai and Guangzhou. No daily extreme heat stress hours for any month existed for Harbin and Lanzhou because the cooler nights offset any heat stress conditions in the afternoons. So no cooling system is needed for facilities near Harbin or Lanzhou.

The assessment of heat stress hours on a daily basis helps one determine between an evaporative cooling pad system and a sprinkler cooling system for a given swine facility. In Table 3, one can see when THI = 78 heat stress has almost been eliminated on a daily basis with an evaporative cooling pad system for facilities near Chongqing or with weather similar to Chongqing. However, when THI = 78 near Shanghai or with weather similar to Shanghai, heat stress conditions still remain while using an evaporative cooling pad system, so a sprinkler cooling system should be used instead or in conjunction with an evaporative cooling pad system assuming the facility/operation can tolerate water spraying inside to wet the pigs. Guangzhou is basically in between conditions with respect to Chongqing and Shanghai from a maximum perspective. However, the heat stress 'season' is longest for Guangzhou for the locations evaluated.

The variation in THI for the daily design load conditions in August for Harbin is shown in Figure 1. A visual review of Figure 1 reveals that the area under the THI curves above 74 is less than the area above the THI curves and below 74. When area above the critical THI, 74 in this case, is less than the area under the critical THI, no accumulated heat stress occurs because the pigs can cool off during the cooler night time conditions even though they may get hot during the hot afternoon. However, if the curve areas are reversed, pigs probably do not cool off at night. Greater than 0.0 values in Table 3 also indicate that pigs probably do not cool off sufficiently during the night so production reductions can be expected.

Table 3. Summary of Daily Extreme Heat Stress Hours (HSH).

Location	Month	Dry-bulb Basis Outside Air	Dry-bulb Basis with Cooling	Wet-bulb Basis Outside Air	Wet-bulb Basis with Cooling
Chongqing THI = 74	June	84.9	15.4	79.3	49.1
	July	158.1	75.4	128.9	100.5
	August	159.7	55.8	124.3	89.4
	Sept	98.1	10.1	73.1	34.9
Chongqing THI = 78	June	0.0	0.0	0.0	0.0
	July	62.1	0.0	32.9	4.5
	August	63.7	0.0	28.3	0.0
	Sept	2.1	0.0	0.0	0.0
Guangzhou THI = 74	April	34.0	0.0	31.5	14.0
	May	89.9	28.8	84.9	66.2
	June	142.2	78.5	125.2	106.6
	July	168.6	89.1	133.3	111.2
	August	153.6	74.8	127.5	105.8
	Sept	123.5	50.3	107.0	86.5
	October	34.1	0.0	37.9	17.2
Guangzhou THI = 78	June	46.2	0.0	29.2	0.0
	July	72.6	0.0	37.3	10.6
	August	57.6	0.0	31.5	15.2
	Sept	27.5	0.0	11.0	9.8
Shanghai THI = 74	June	99.4	40.3	98.7	66.1
	July	185.7	110.3	162.7	128.9
	August	179.5	108.9	155.8	121.0
	Sept	122.0	69.9	105.0	82.4
Shanghai THI = 78	June	3.4	0.0	2.7	0.0
	July	89.7	14.3	66.7	32.9
	August	83.5	12.9	59.8	25.0
	Sept	26.0	0.0	9.0	0.0

Reviewing the daily variation in THI can also provide insight on when during the day an evaporative cooling pad system should be operated. An evaporative cooling system should be run at least during the heat stress period of the day and typically should to be operated longer than during heat of day to ensure pigs can fully cool off if they got hot during the day. However, one needs to look closely at the operation during night time hours. For some summer weather conditions like seen in Harbin, operating an evaporative cooling pad at night during humid weather (wet-bulb basis evaluations) provides little or no benefit and probably should be shut off during the night.

4. Conclusions and Recommendations

A method using weather data has been developed to assist with implementation and operation of cooling systems for swine. The recommendations for the selection and operation of cooling systems for swine are as follows:

1. A supplemental cooling system should be considered when a significant number of heat stress hours exist on an annual basis.

2. No supplemental cooling system is required when daily heat stress hours are zero even though a significant number of heat stress hours exist on an annual basis.
3. An evaporative cooling pad system should be used when it is capable of removing all or almost all heat stress hours.
4. A sprinkler cooling system should be used when an evaporative pad cooling system is not capable of providing a sufficient amount of heat stress relief. However, the facility/operation must be capable of wet conditions inside.
5. If an evaporative pad cooling system is not capable of providing a sufficient amount of heat stress relief, a sprinkler cooling system can be added and used only when conditions require more cooling capability.

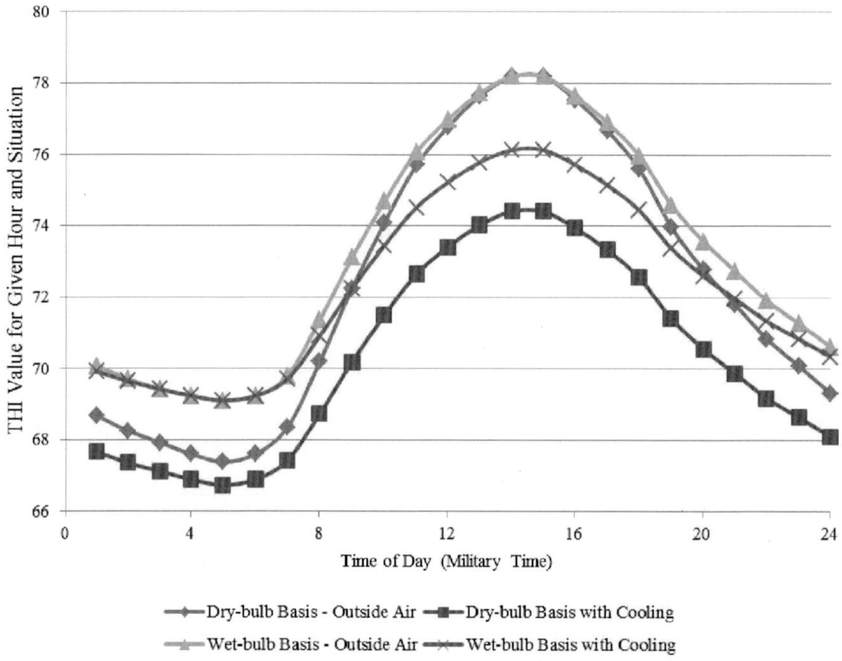

Figure 1. Daily THI variation during August in Harbin.

References

Albright, L.D. 1990. Environment Control for Animals and Plants. American Society of Agricultural Engineers. 453 p.

ASHRAE. 2013a. Chapter 14, Climatic Design Information. In ASHRAE Handbook of Fundamentals. American Society of Heating, Refrigeration and Air Conditioning Engineers, Inc. 47 p.

ASHRAE. 2013b. ASHRAE Weather Data Viewer, Version 5.0. American Society of Heating, Refrigeration and Air Conditioning Engineers, Inc. CD.

Bridges, T.C., L.W. Turner, R.S. Gates and D.G. Overhults. 2003. Assessing the benefits of misting-cooling systems for growing/finishing swine as affected by environment and pig placement date. Applied Engineering in Agriculture 19 (3), 361-366.

DeShazer, J.A. 2009. *Livestock Energetics and Thermal Environmental Management*. American Society of Agricultural and Biological Engineers. 212 p.

Hellickson, M.A. and J.N. Walker. 1983. *Ventilation of Agricultural Structures.* American Society of Agricultural Engineers. 372 p.

Ulmer, A.M. and J.M Zulovich. 2008. Paying for cooling systems in swine production. In *Eighth International Livestock and Environment Symposium.* Iguassu Falls, Brazil, August 31-September 4: St. Joseph, Michigan: ASABE. 709-712.

Zulovich, J.M., J.T. Tyson and M. Brugger. 2008. Improving management of heat abatement systems on US dairy farms. In *Eighth International Livestock and Environment Symposium.* Iguassu Falls, Brazil, August 31-September 4: St. Joseph, Michigan: ASABE. 693-699.

Cooling Effect of Fabric Air Ducting System for Loose Housing Dairy Cows in an Open-Sided Barn

Lina Xie [a,b], Chaoyuan Wang [a,b,*], Zhiyuan Gui [a,b], Lu Zhang [a,b], Zhengxiang Shi [a,b], Baoming Li [a,b], Luyu Ding [a,b], Chuntao Jia [c]

[a] Department of Agricultural Structure and Bioenvironmental Engineering, China Agricultural University, Beijing 100083, China

[b] Key Laboratory of Agricultural Engineering in Structure and Environment, Ministry of Agriculture, Beijing 100083, China

[c] Hai Lin Dairy Farm, Tianjin 301718, China

* Corresponding author. Email: gotowchy@cau.edu.cn

Abstract

Curtain-sided barns with circulation fans above stall are commonly used to house dairy cows in China. Many of the circulation fans currently used are unable to provide appropriate cooling, especially for the naturally ventilated shed, and can result in decreased feed intake and milk production. Designed for alleviating heat stress and improving animal comfort, a system consisting of an air cooler and a 30 m fabric air duct (FAD) was integrated to evenly distribute cooling air to targeted zones above stall beds. The system was evaluated in an open sided dairy barn in Tianjin, China. For the stalls equipped with FAD system, air velocity reached 1.1 m/s and higher at the 0.7 m height plane of the stall space, and was more uniformly distributed. Compared to the stalls equipped with only circulation fans, the FAD system lowered air temperature by 1.5° C, and increased relative humidity by 8.1%. On average, Temperature Humidity Index (*THI*) and Equivalent Temperature Index (*ETI*) were decreased by 0.5 and 0.6, respectively. After a 15-day operation of the system, rectal temperatures of the treated dairy cows were significantly lowered ($P < 0.05$). The result also showed that the treated group had higher milk production. These findings suggest the FAD can be an effective cooling alternative for naturally ventilated dairy barns.

Keywords: Temperature Humidity Index, Equivalent Temperature Index, freestall, cooling, natural ventilation

1. Introduction

Mature dairy cows can adapt to a wide range of climatic condition without significant decline in milk production. However, high environmental temperature with high relative humidity in particular, can lead to heat stress, increase respiration rate, decrease feed intake and alter feeding behaviour, which ultimately results in reduced production efficiency and milk yield (Holter et al., 1997; Nienaber et al., 2003). In order to relieve heat stress, strategies from primary non-evaporative cooling (conduction, convection and radiation) to evaporative cooling are widely investigated and applied to different housing systems in varied climates (Turner et al., 1992; Kadzere et al., 2002; Gebremedhin et al., 2013). The cooling system combining fans and sprinklers or foggers is broadly applied in dairy barns, and has been proven to be one of the most effective methods in alleviating heat stress for dairy cows (Turner et al., 1992; Calegari et al., 2003; Avendaño-Reyes et al., 2006; Khongdee et al., 2006; Arbel et al., 2007; Matarazzo et al., 2007). However, efficiency of the evaporative cooling system strongly depends on the relative humidity of ambient air, and its effectiveness is compromised under humid climate. Brouk et al. (2001) pointed out temperature humidity index (*THI*) could only be lowered by less than 10% with an evaporative cooling system under high relative humidity conditions.

Furthermore, it is difficult to achieve an ideal cooling effect with less energy consumption by evaporation cooling system at house level. Climate in China is typically characterized by hot and humid weather in summer even in the northern regions, which is a really big challenge for dairy production. In China, open or curtain sided barn with outdoor excising yard where the animals have free access is very typical for housing dairy cows. A cooling system combining sprinklers or foggers and circulation fans above stall is currently used, which usually is unable to provide sufficient cooling, and can result in decreasing feed intake and milk production. Dairy cows usually rest for around 10-14 hr a day on the beds in freestall system. In order to improve thermal comfort and alleviate heat stress, a cooling system consisting of an air cooler and a fabric air duct (FAD) was integrated to evenly distribute cooling air to the stall space. In this paper, a field test on the cooling system was conducted, and its performance was evaluated via thermal environment and animal production.

2. Materials and Methods

2.1. Dairy cow barn

The surveyed farm in Tianjin, China, had two naturally ventilated barns housing milking cows. The tests were conducted in the northern barn, which was 174 m in length, 27 m in width, and oriented east-west, having an opening along the ridge. During the experiment, the curtains in the sidewalls were all pulled down, making the barn fully open in two sides. The barn was connected to a fully open exercising yard with a width of 10 m, where the cows were freely accessible. The barn was divided into four sections by a central feeding alley and a pathway to milking parlour, which was perpendicular to the feeding alley (Figure 1). In each section, two-row head to head freestalls, 1.2 m by 2.4 m each, were equipped.

The experiment was conducted in the southwestern section of the barn, where a total of 90 stalls were installed, and 86 Holstein milking cows were housed. There were two groups of freestalls in this section, which were separated by a 12 m cross alley. The west section, where 50 stalls were placed, was installed with the air cooler and FAD system as a treatment group, and the east with 40 stalls was as a control group. For the control group, circulation fans were hung about 2.5 m above the floor surface at an interval of 18 m to accelerate the air speed for the resting cows on the stalls. Sprinklers mounted on the feeding fences combining the circulation fans hanging above were applied to cool the animals during feeding.

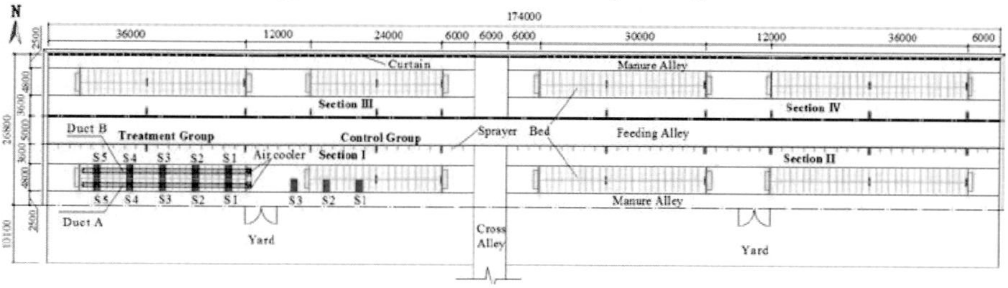

Figure 1. Layout of the open-sided dairy barn surveyed.

2.2. Air cooler and fabric air duct (FAD) cooling system

The cooling system mainly consisted of a FAD and an air cooler. The configuration of the air cooler was 1.17 m (width) x 1.17 m (length) x 0.96 m (height), with a power rating of 1.5 kW, and its maximum airflow rate was 20,000 m^3/h. Underground water from a well in the farm was pumped and circulated inside the air cooler. The fabric air duct was made of permanent flame retardant polyester, a light-weighted material, was 30 m long and 0.8 m in diameter, connected to the outlet of the air cooler for conveying and dispersing cooling air to body surfaces of the dairy cows. Each row of the stalls in the treatment group was installed with a FAD, and the lower edge of the duct was 2.0 m above the bedding. In each FAD (Duct A and Duct B), four-row of orifices with different diameters along the length of the duct were designed. Distributions of the orifices and its specification as well as the airflow rate of the system are shown in Figure 2 and Table 1, respectively. For Duct A, the diameter of the outer orifice was 3.5 cm, and 3.2 cm for inner orifice, while they were 2.6 cm and 2.3 cm for Duct B. Theoretical airflow rates were around 14,680 m^3/h and 14,050 m^3/h for the two FADs. The designed air velocities at the 0.7 m height plane above the bedding, which was documented as the height of the milking cows when lying down (Anonymous, 2002), were 1.0 and 1.5 m/s for FADs A and B, respectively.

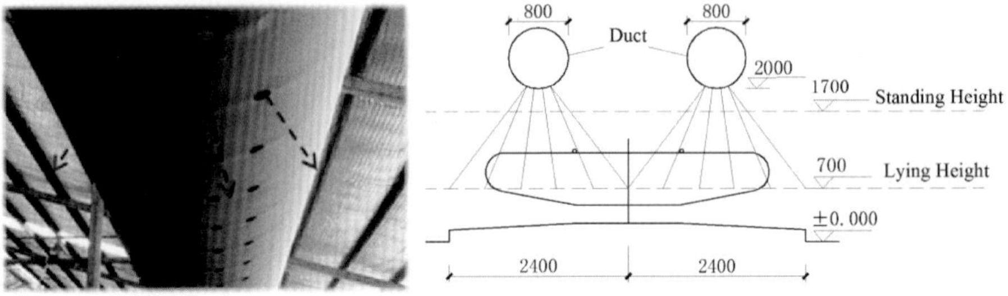

Figure 2. Schematics of the FAD system: (a) distribution of the orifices; and (b) air jetting from the duct to the resting spaces of the head to head stalls (Units of (b) are in mm).

Table 1. Specifications of the orifices, theoretical jetted air velocity and airflow rates of FADs.

FAD	Jetted air velocity (m/s)	Orifice diameter (mm)	Orifice numbers	Orifice spacing (m)	Airflow per orifice (m^3/h)	Airflow (m^3/h)
A	8.25	35	140	0.214	28.6	14,680
		32	140	0.214	23.9	
		32	140	0.214	23.9	
		35	140	0.214	28.6	
B	8.25	26	250	0.120	15.8	14,050
		23	250	0.120	12.3	
		23	250	0.120	12.3	
		26	250	0.120	15.8	

2.3. Data collection

2.3.1. Temperature, relative humidity (RH) and air velocity measurement

The temperature and relative humidity (*RH*) inside the barn were automatically monitored at an interval of 1 min by using HOBO data loggers (Model U14-001, Onset Computer Corporation, MA, USA), which were installed on the neck rails of the stalls, and 1.2 m above the bedding. Measurement range and accuracy of the sensor were -20 to 50°C, ±0.2°C for temperature, and 0% to 100%, ±2.5% for RH, respectively. Air velocities of the stall space and inside FAD as well as the air temperature inside FAD were measured by using a handled anemometer (Model KA41L, Kanomax, Japan). The accuracy for temperature and RH measurement were 0.5°C and ±3% of measured value, respectively.

In order to specify spatial distribution of air velocities, thirteen stalls were selected for measurement including five under each FAD in treatment group and three under the circulation fans in the control group (Figure 1). For each of the stall, air velocity distributions were determined at three planes of 0.7 m (lying height of the cow), 1.2 m, and 1.7 m (standing height) above the bedding. At each plane, air velocities at 28 locations were sampled. The measurements were replicated three times at each sampling location, and the values were averaged as the air velocity of each point. Temperature of the water in air cooler was measured by using a handled thermometer (Model Temp-5, Oakton, USA) with a measuring accuracy of 0.2°C. Air temperature, RH and velocity outside the barn were automatically recorded at an interval of 1 min using a HOBO weather station (Model U21, Onset, MA, USA), and the accuracy were ±0.2°C, ±3.5% and ±0.5 m/s, respectively.

2.3.2. Rectal temperature and respiration rate measurement

For each group, five dairy cows resting on the stall for more than 10 min were selected to measure the rectal temperature (*Tr*) and respiration rate (*RR*), and the measurement started at 4:00 pm. *Tr* was sampled using a handled thermometer (Model Temp-5, Oakton, USA), and *RR* was determined by counting with a stop watch in one minute.

2.3.3. Milk yield recording

Milk yield of the cows was automatically recorded by the milking system (Afimilk Ltd., Kibbutz Afikim, Israel), and the data was used for production comparison.

2.4. Data treatment and analysis

2.4.1. Distribution of air velocity at different height planes

In order to assess the uniformity and distribution of air velocities in stall spaces, coefficient of non-uniformity (*Kv*) was applied in this paper. A smaller *Kv* indicates that air velocity is more evenly distributed in the plane. The *Kv* was determined by using equation (1).

$$Kv = \frac{\sigma_i}{\bar{v}} = \frac{\sqrt{\sum\left(v_i - \bar{v}\right)^2}}{n} \Bigg/ \frac{\sum v_i}{n} \tag{1}$$

where σ_i is root mean square error of air velocity, m/s; $\bar{v}$ is average air velocities of all the locations measured in the plane, m/s; v_i is air velocity at a location, m/s; and n is the number of the locations for air velocity measurement.

2.4.2. Temperature and Humidity Index (THI), and Equivalent Temperature Index (ETI)

THI and *ETI* estimated by meteorological variables are the common indicators to evaluate the degree of thermal stress. Both indexes were also applied to assess the performance of cooling effect of FADs. The equation for *THI* calculation is (National Research Council, 1971):

$$THI=0.81t+(0.99t-14.3)RH+46.3 \tag{2}$$

where *THI* is temperature and humidity index, dimensionless; t is the dry bulb temperature, in °C; and *RH* is the relative humidity, in %.

Compared with *THI, ETI* is a less popular alternative, while it incorporates the effect of air velocity along with air temperature and *RH. ETI* is considered to have better performance in evaluating heat stress for dairy under hot and humid climates (Silva et al., 2007). Equation (3) is used to calculate *ETI* with dry bulb temperature, RH and air velocity (Baeta et al., 1987):

$$ETI=27.88-0.456t+0.010754t^2-0.4905RH+0.00088(RH)^2+1.1507v-0.126447v^2+ \\ 0.019876tRH-0.046313tv \tag{3}$$

Baeta et al. (1987) reported the alert levels of dairy cattle based on *ETI* were safe (16–26.5), caution (26.5–31.5), extreme caution (31.5–37.5), danger (37.5–43.5) and extreme danger (> 43.5). While Silva et al. (2007) suggested four alert categories using *ETI*, including safe (< 30), caution (30–34), extreme caution (34–38) and danger (> 38) for dairy cattle in tropical regions.

2.4.3. Statistical analysis

Statistical analyses were conducted by using Statistical Product and Service Solutions 17.0 (SPSS 17.0, International Business Machines Corporation, Armonk, USA) software. Results were expressed as "mean ± SD", and statistical significance was based on $P < 0.05$.

3. Results and Discussion

3.1. Performance of the FAD system

3.1.1. Temperature of the air jetted from FAD

Before being pumped into the air cooler, underground water from a well was stored in an outside tank, and the temperature of water in air cooler changed within 20–27°C during the period of the experiments. Results showed that the temperature of the air emitted from the duct was significantly correlated with water temperature ($R^2 > 0.87$). During the test, water temperatures were slightly different for the coolers connected to Ducts A and B, resulting in a difference in the air temperature delivered by the two FADs. In summer time, relative lower air temperature of FAD system was able to achieve a better cooling effect and increase thermal comfort of the animals. Generally, the temperature of the air emitted from the duct increased with the distance from the air cooler (Figure 3). Air temperatures from the beginning to the end of the duct, were raised by 1.7 and 2.2°C for FADs A and B on average, respectively.

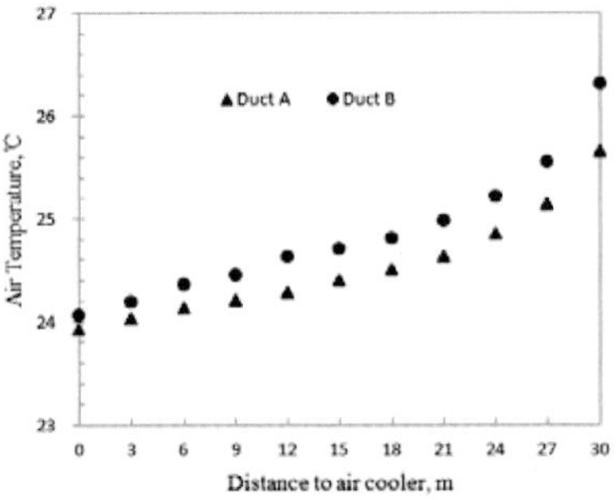

Figure 3. Temperature of the jetted air from the Ducts A and B, increased as the increasing of the distance to air coolers.

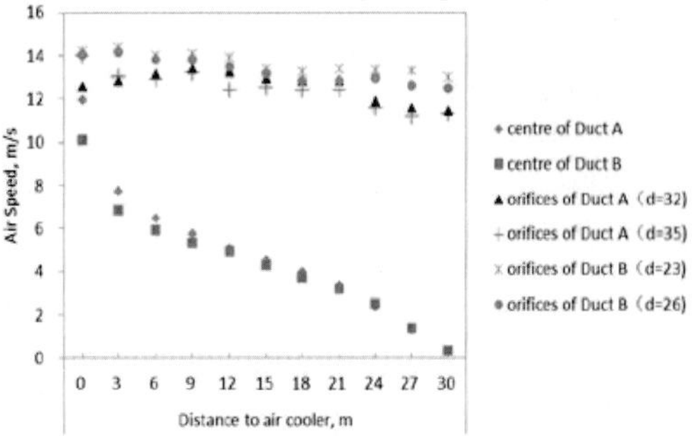

Figure 4. Velocities of the air inside and jetted from Ducts A and B, decreased along the length of the ducts.

Lower temperature of the incoming water to the air cooler is helpful in having a better cooling effectiveness. As the jetted air temperature increases along the length of the duct, the cooling effect was compromised at the end of the FAD, which was constrained by the length of the FAD system. At the end of the system, both the airflow rate and air temperature were adversely affected, making the system could not achieve the expected cooling effect.

3.1.2. Velocity of the air jetted from and inside the Ducts

Air velocities in the center of the ducts and from orifices were plotted in Figure 4. Longitudinal air velocities in the center of the ducts decreased from 10-12 m/s to less than 0.5 m/s from the beginning to the end. Velocities of the air jetted from FADs were kept relatively stable, varying from 11 m/s to 15 m/s. The size of the orifice impacted the jetted air velocity, while no significant effect was found. Jetted air velocity from the ducts was related with the arrays and sizes of the orifices. The smallest orifice had the highest jetted speed, and vice versa.

3.2. Spatial distribution of air velocities above the stalls

3.2.1. Air velocity distribution

As reported, dairy cows rest 12 hr a day or even longer, among which the majority is lying on the stalls. Thus a good spatial thermal environment at the stalls, air velocity at lying height in particular, is critical to animals, and beneficial to milk production as well. Air velocities at three different height levels (0.7 m, 1.2 m, 1.7 m) above the stalls are shown in Table 2. In hot climates, recommended minimum wind speed across the body surface of the dairy cow was 1 m/s. At the lying height of 0.7 m, air velocities in the stall space under the Duct A and Duct B averaged 1.1 m/s and 1.3 m/s, which were close to theoretical values and satisfied the minimum requirements recommended. The velocities were also slightly higher than those in control group with circulation fans ($P > 0.05$). Field test also found that air velocity above the bedding was easily affected by the wind outside the barn because of the open system, and Duct B, closer to the feeding fence, was also impacted by the circulation fans above.

Compared with Duct A, air velocities at both heights of 1.2 m and 1.7 m were higher for the stalls under Duct B, due to the difference in array and size of orifices. At these two heights, air velocities were also significantly different from the control group ($P < 0.05$). Generally, air velocities of Duct A were all approximately 1.1 m/s at the three different heights, while for Duct B they gradually decreased from 1.7 m/s to 1.3 m/s at the heights from 1.7 m to 0.7 m.

Table 2. Air velocities at three different heights of stalls in the treatment and control groups, m/s.

Stall #	1.7 m height			1.2 m height			0.7 m height		
	Duct A	Duct B	Control	Duct A	Duct B	Control	Duct A	Duct B	Control
S1	0.9±0.3 ab	1.8±0.3	0.8±0.5	0.9±0.2	1.4±0.3	0.8±0.2	0.9±0.3	0.9±0.1 b	0.7±0.4
S2	1.4±0.1 a	1.6±0.1	1.0±0.5	1.3±0.0	1.4±0.1	1.2±0.6	1.0±0.1	1.4±0.0 a	1.8±1.1
S3	1.1±0.0 ab	1.7±0.2	1.0±0.1	1.1±0.1	1.5±0.4	0.7±0.2	1.0±0.1	1.5±0.1 a	0.4±0.1
S4	1.1±0.0 b	1.8±0.0		1.1±0.1	1.4±0.1		1.2±0.3	1.3±0.0 a	
S5	0.9±0.1 b	1.5±0.2		1.2±0.1	1.3±0.2		1.2±0.3	1.2±0.1 a	
Mean	1.1±0.2	1.7±0.1	0.9±0.1	1.1±0.1	1.4±0.1	0.9±0.3	1.1±0.2	1.3±0.2	1.0±0.7

3.2.2. Air velocity uniformity

Generally, Duct B had the smallest *Kv*, while it was the highest for the stalls equipped with circulation fans in the control, which indicated Duct B had the best uniformity of air velocity. At the heights of 1.2 m and 1.7 m, no significant difference was found for the treatment and the control groups. While at the lying height of milking cows (0.7 m), *Kv* for Duct B was significantly lower, compared to Duct A and circulation fans, indicating the air velocities for the stall spaces underneath Duct B was much more uniformly distributed.

3.3. Effects of FAD on thermal environment of stall spaces

3.3.1. Air temperature, relative humidity and THI

Results suggested that the ambient temperature for the period of 0:00 to 20:00 inside the barn averaged at 28.1°C during the test, which were significantly higher than air temperature of the treated stall spaces under ducts A and B. Compared to the control group, air temperatures of the stall spaces equipped with Ducts A and B were decreased by 0.6°C and 0.9 ° C, respectively. Average *RH* (75.5%, 75.8%) was increased by 6% because of the evaporative cooling process via the air coolers, while no significant difference was observed ($P > 0.05$). THI for the treated stall spaces was significantly lowered by 1.4 compared to ambient air ($P < 0.05$). Although air temperature was reduced, no difference on THI was found for the treatment and control groups due to the offset of *RH* increasing, and the cooling effect was also somewhat compromised.

Table 3. Average temperature, *RH* and *THI* of stall spaces in Treatment and Control groups.

Index	Outdoor	Circulation Fan	Duct A	Duct B
Temperature, °C	28.1±0.7	27.5±0.6	26.9±0.5	26.6±0.6
RH, %	71.9±6.0	69.7±5.7	75.5±5.5	75.8±4.9
THI	78.0±0.3	77.1±0.3	77.0±0.4	76.6±0.4

Table 4. Average temperature, *RH* and *THI* of the stall spaces under Duct A, Duct B (Treatment) and circulation fans (Control group) for the period of 12:00 to 18:00.

Thermal index	Outdoor	Circulation Fan	Duct A	Duct B
Temperature, °C	35.3±1.2 a	32.9±1.7 b	31.3±1.1 b	31.4±1.1 b
RH, %	41.3±3.6 c	44.3±4.2 b	52.3±2.7 a	52.4±3.0 a
THI	83.3±0.6 a	81.0±1.3 b	80.4±1.1 b	80.5±1.0 b

As temperature increases in the afternoon, dairy cows are usually more stressful. Thermal indexes are illustrated in Table 4 for the mostly challenged period of 12:00 to 18:00. In this period, the average outdoor temperature reached 35.3°C, and air temperature of the stall spaces in FADs was significantly decreased by 4 to 31.4°C, which was also 1.5°C lower than that of the control group. Comparing the stall space under circulation fans, RH was raised by 8.1% to 52.3% for the FAD stall space ($P < 0.05$). The *THI* of the stall spaces in both the treatment and control groups were much lower than outside, which were reduced by 2.9 and 2.3, respectively. Compared to the daily average *THI* reduction (1.2), it suggested that the air cooler and FAD system had a higher cooling efficiency in the hottest period of a day, during which the outdoor RH was relatively lower.

3.3.2. Equivalent temperature index (ETI)

Although *ETI* is less common, it incorporates the effect of air temperature, RH along with air velocity. Based on a survey of 413 dairy cows in tropical regions, Silva et al., (2007) concluded that *ETI* had significant correlations with rectal temperature and respiratory rate of the animals, while the correlation with *THI* was the lowest, and *ETI* was suggested to be a best thermal stress indicator among five indexes for dairy in hot climates. The study also categorized dairy cow in a safe state when *ETI* is less than 30, while in a caution state when it is 30-34. In the period of 12:00-18:00, average *ETI* for the treatment group was mostly within the range of 30-31, showing dairy cow was still under a slight heat stress even with the air cooler and FADs. Average *ETI* reached 31.1 for the control group, was 0.6 higher than the treatment group, but no significant difference was found. *ETI* peak occurred around 13:00-14:00, and dramatically decreased after 17:00, stating the thermal comfort of the cows increasing afterwards.

3.4. Effects of FAD system on respiration rate and rectal temperature

Compared to the control group, respiration rate (*RR*) for the treatment cows averaged 67 min^{-1}, which was 6/min lower and showed no significant difference. Rectal temperatures (*Tr*) of the dairy cattle rested on the stalls under Duct A and Duct B were 38.7°C, and 38.8°C, respectively, which were lowered by 0.3°C and 0.2°C. Significant difference on *Tr* was observed for the control and treatment cows ($P < 0.05$).

Because the system was installed in a naturally ventilated barn with fully open sides, the cooling air velocity was affected by the wind speed of the outdoor ambient air, and drifting may happen. Furthermore, constrained by the underground water temperature and capacity of the air coolers, the cooler and FAD system was unable to significantly cool down the entire spatial environment of the barn. However, the system in this study provided a better resting environment for the cows by retaining the 'cooling properties' and relatively even distribution of the cooler air to the body surfaces and hence improved thermal comfort.

3.5. Effects of FAD system on milk yield

Milk yield of the cows housed in Section I and Section II was recorded and analyzed during the experiment. In Section I, air cooler and ducts were installed and 86 milking cows were housed, while there were 70 cows in section II. Average milk yields were 24.0 kg d^{-1} cow^{-1} and 21.6 kg d^{-1} cow^{-1} for the three days before the FAD system application in Section I and Section II, respectively. After a 15-day operation of the air cooler and FAD system, milk yield for the cows in Section I increased to 24.7 kg d^{-1} cow^{-1}, while it was almost unchanged in Section II.

4. Conclusions

A cooling system consisting of an air cooler and a fabric air duct was developed to evenly distribute cooling air to dairy cattle on the stalls to alleviate heat stress. Field experiment showed that air velocity reached 1.1 and 1.3 m s^{-1} for Duct A and Duct B at the 0.7 m height plane above the bedding, and was more uniformly distributed for the stalls equipped with FAD system. Compared to the stalls equipped with circulation fans, the FAD system lowered air temperature by 1.5°C, and increased relative humidity by 8.1% on average. During the period of 12:00-18:00, *THI* of the treatment group was decreased by 2.9 compared to outdoor ambient air, which was also 0.6 lower than control group. Average *ETI* for the treatment group was within the range of 30-31, about 0.6 lower compared to the control. After a 15-day operation of the system, rectal temperatures of the treated dairy cows were significantly lowered. The result also showed that the treated group had a higher milk production. These findings suggest the FAD can be an effective cooling alternative for naturally ventilated dairy barns.

Acknowledgements

This study was funded by a National 863 project (Grant no. 2013AA10230602) and the China Agricultural Research System (CARS-37) from Ministry of Agriculture.

References

Anonymous, 2002. *Interdisciplinary report: Housing Design for Cattle – Danish Recommendations*. Third edition. Copenhagen, Denmark: The Danish Agricultural Advisory Center.

Arbel, A., M. Barak, and A. Shklyar, 2007. Dairy barns cooling: integrated high pressure fogging system with air ventilation and circulation systems. ASABE Paper No. 074123. St. Joseph, Mich.: ASABE.

Avendaño-Reyes, L., F.D. Alvarez-Valenzuela, A. Correa-Calderón, J.S. Saucedo-Quintero, P.H. Robinson, and J.G. Fadel, 2006. Effect of cooling Holstein cows during the dry period on postpartum performance under heat stress conditions. Livestock Science. 105(1–3), 198–206.

Baeta, F.C., N.F. Meador, M.D. Shanklin, and H.D. Johnson, 1987. Equivalent temperature index at temperatures above the thermoneutral for lactating dairy cows. Paper No. 874015. St. Joseph, Mich.: ASABE.

Brouk, M.J., J.F. Smith, and J.P. Harner, 2001. Efficiency of modified evaporative cooling in Midwest dairy freestall barns. In *Proceedings of the 6th International Livestock Environment Symposium.* Louisville, KY. ASAE, St. Joseph, MI. Eds., R. Stowell, R. Bucklin, and R.W. Bottcher. 412–418.

Calegari, F., L. Calamari, and E. Frazzi, 2003. Effects of ventilation and misting on behaviour of dairy cattle in the hot season in South Italy. In *5th International Dairy Housing Conference*. St. Joseph, Mich.: ASAE. 303–311.

Gebremedhin, K.G., C.N. Lee, J.E. Larson, and J. Davis, 2013. Alternative cooling of dairy cows by udder wetting. Transaction of the ASABE. 56(1), 305–310.

Holter, J.B., J.W. West, and M.L. McGilliard, 1997. Predicting ad libitum dry matter intake and yield of Holstein cows. Journal of Dairy Science. 80(9), 2188–2199.

Kadzere, C., M. Murphy, N. Silanikove, and E. Maltz, 2002. Heat stress in lactating dairy cows: a review. Livestock Production Science. 77(1), 59–91.

Khongdee, S., N. Chaiyabutr, G. Hinch, K. Markvichitr, and C. Vajrabukka, 2006. Effects of evaporative cooling on reproductive performance and milk production of dairy cows in hot wet conditions. International Journal of Biometeorology. 50(5), 253–257.

Matarazzo, S.V., M. Perissinotto, I.J.O. Silva, D.J. Moura, and S.A.A. Fernandes, 2007. Electronic monitoring of behavioral patterns of dairy cows in a cooling freestall. *In 6th International Dairy Housing Conference*. St. Joseph, Mich.: ASABE. 235–238.

National Research Council, 1971. *A Guide to Environmental Research on Animals.* Washington, D.C.: National of Academy Science.

Nienaber, J.A., G.L. Hahn, T.M. Brown-Brandl, and R.A. Eigenberg, 2003. Heat stress climatic conditions and the physiological responses of cattle. ASAE Paper No. 701P0203. St. Joseph, Mich.: ASAE.

Silva, R.G.D., D.A.E.F. Morais, and M.M. Guilhermino, 2007. Evaluation of thermal stress indexes for dairy cows in tropical regions. Revista Brasileira de Zootecnia. 36(4):1192–1198.

Turner, L.W., J.P. Chastain, R.W. Hemken, R.S. Gates, and W.L. Crist, 1992. Reducing heat stress in dairy cows through sprinkler and fan cooling. Applied Engineering in Agriculture. 8(2), 251–256.

Evaluation of Temperature-Humidity Index in a Growing-Finishing Pig Bedding System

Zhongkai Zhou [a,*], Zhu Qin [a], Hui Li [a], Hongru Gu [b], Gang Yu [a], Jie Yang [b]

[a] Institute of Facilities and Equipment in Agriculture, Jiangsu Academy of Agricultural Science, Nanjing, Jiangsu 210014, China

[b] Institute of Animal Science, Jiangsu Academy of Agricultural Science, Nanjing, Jiangsu 210014, China

* Corresponding author. Email: zhouzk@jaas.ac.cn

Abstract

This paper studies the heat stress conditions imposed onto growing-finishing pigs housed in a naturally ventilated bedding system with rice husk during the four seasons from March 2013 to February 2014. The barn is located at the suburbs of Nanjing, a city at central of Jiangsu, east of China. In the region, the main losses in pig production result from the summer period, characterized by high air temperature and relative humidity. Dry-bulb temperature and relative humidity data were obtained in the interior of building, as well as the external conditions. The hourly temperature-humidity index (THI) was calculated for evaluation of thermal comfort. The analysis shows that indoor average daily temperature across all units ranged from 2.3°C to 34.4°C, and average relative humidity from 49.2% to 92.9%. The difference between indoor and outdoor temperature during the four seasons ranged between 0.0°C and 9.0°C, while its mean value equaled to 3.1°C. No significant differences were detected between indoor and outdoor relative humidity levels ($P > 0.05$). The hourly THI values indoor ranged from 34.7 to 99.5, frequently higher than the desirable range in summer. Animals were exposed to heat stress in 10.5%, 31.5% and 4.6% of spring, summer and autumn hours, respectively, whereas no heat stress was identified during winter. Pig are more prone to severe heat stress during summer, as the daily maximum hourly values of THI remained higher than the severe heat stress threshold (THI=79) in 66.3% of the summer days.

Keywords: Pig barn, heat stress, climate, temperature, relative humidity, welfare

1. Introduction

In the east of China meat consumption in general has been growing in recent years. Pig production is not sufficient to cover the internal demand, meaning that there is still scope to increase this activity. This caused public concern over the welfare of animals used for agricultural production has grown over the past years. Pig is considered as a major source of meat in China, studies focused on pig and their interactions with environment can be found in literature.

The region is a characteristic tropical monsoon climate region which the main losses in pig production result from the summer, characterized by high air temperature and relative humidity. Hot weather in summer adversely affects the performance of pig production. The responses of pigs to heat stress is panting and raised body temperature; high level of hormones (such as cortisol) concentration; less locomotion and more lying behaviors; less feed intake and reduced body weight; etc., which may affect the health and welfare of animals (Kuczynski et al., 2011). A widely accepted method to study animals' heat stress is the application of the Temperature Humidity Index (THI). The THI has gained popularity among animal scientists as a means of quantifying discomfort levels caused by heat stress (Nissim, 2000; Lucas et al., 2000). The THI is a function of air temperature and humidity, integrating both of their effects into a single number. Traditionally, it is thought that pig heat stress is occurs when the THI crosses a critical threshold of 75 (70).

This study aims at investigating seasonal heat stress imposed onto growing-finishing pigs housed within a naturally ventilated barn bedding system with rice husk and having no thermal insulation.

2. Materials and Methods

2.1. Description of the pig building

This study is based on measurements recorded at an experimental pig building located at the suburbs of Nanjing, a city at central of Jiangsu, east of China (32° 29′ N, 118° 37′ E, 15m amsl; coordinates refer to the building's location). The pig barn has an overall width, and length, of 9 m, and 30 m, respectively, and the total surface area of the floor of the barn was 225 m^2 and a volume of 664 m^3, the height of building was 3.5 m as shown in Figure 1a. The naturally ventilated pig barn was oriented in an east-west axis facing south direction. The positions of the ventilation openings (doors and air inlet) are shown in Figure 1a. The dimensions (width × height) of D1 – D2 are 1.0 × 2.0 m. The width of O1 – O3 is 30 m, while their height ranges between 0 and 0.6m. The control of ventilation rate was performed by the simultaneously opening and closing of openings, through which the air exhausts. During the colder days, D1 – D2 and half of the air inlet were opened for 2 – 3 h in the morning and for 1 h in the afternoon. During the rest of the days, all doors and air inlet were opened during daytime.

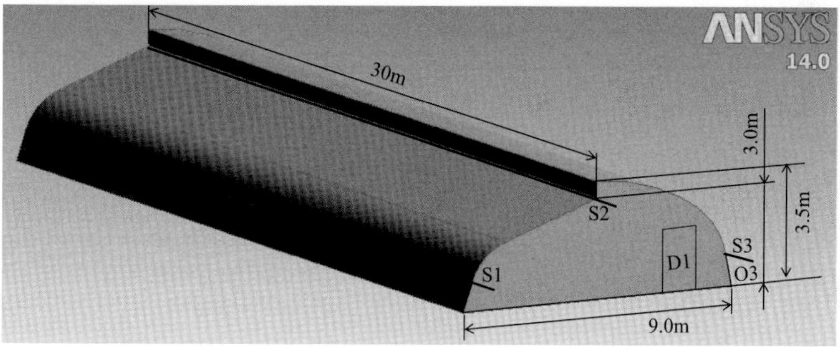

a. Pig barn's dimensions and positions of ventilation openings. D1 and O1 – O3 stand for doors and air inlet, respectively. S1 – S3 represents the trellis shutter.

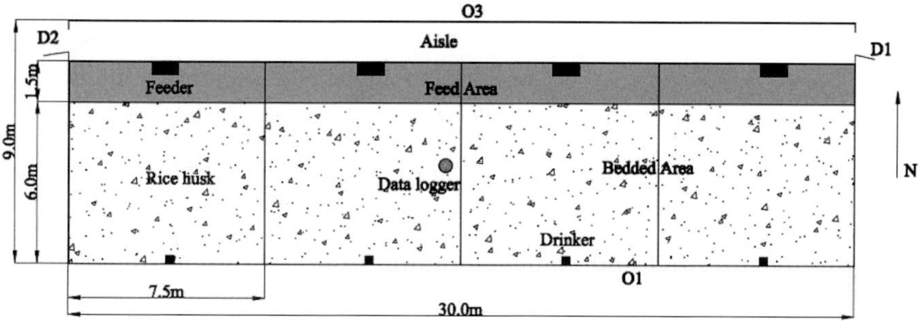

b. Pig barn's plan view and measurement position.

Figure 1. Schematic depiction of the pig barn.

The side walls and ceiling were made sunshade nets and plastic film. Approximately 80 animals were hosted; the vast majority of them were growing-finishing Pig. The pig barn were divided to 4 sections (7.5 m × 4 m) arranged on the east-west axis. Each pen contained one feeder and one drinker and occupied by a total of 20 animals. Rice husk was used as the bedding material, which was replaced every 12 months (Figure 1b).

2.2. Measurements

The air temperatures and relative humidity inside and outside the pig building were measured using an AZ® 8829 Data Logger (AZ Instrument Corp., China) with accuracy of ±0.6°C and ±3%, respectively. Data logger and sensor is placed in a plastic shield with lateral slots in order to be protected from water and indoor dust. Indoor temperature and relative humidity were observed every 30 min, from the 1th of March 2013 until the 28th of February 2014. Outdoor climatic variables were also observed every 30 min. Data values were analyzed statistically using the statistical package SAS 9.2 (SAS Institute Inc., Cary, NC, USA). Seasons were defined in the following way: spring (March to May), summer (June to August), autumn (September to November) and winter (December to February) according to the meteorological seasons.

2.3. Temperature-humidity indices

Heat stress is assessed by means of the hourly temperature-humidity index (THI) (Wegner et al., 2014; Haeussermann et al., 2007a; 2007b). Dry-bulb temperature and relative humidity were measured continuously with a combined sensor at one measuring point in the incoming air as well as inside the compartment at a height of 1.5m. Average and maximum daily THI were calculated according to the following formula of the National Weather Service Centre Region (NWSCR, 1976).

$$THI = [(1.8T) + 32] - [0.55(RH/100)] \cdot [((1.8T) + 32) - 58] \quad (1)$$

where *THI* is temperature-humidity index; *T* is temperature, °C; *RH* is relative humidity, %.

Four heat stress categories, presented in Table 1, are defined (NWSCR, 1976).

Table 1. Heat stress classes as defined by THI values.

THI value	Heat stress condition
THI<75	Absence of heat stress
75≤THI≤78	Moderate heat stress
79≤THI≤83	Severe heat stress
THI≥84	Extreme severe heat stress

3. Results and Discussion

3.1. Climate conditions

Dry-bulb temperature and relative humidity levels inside and outside the pig building are presented in Figure 2, while descriptive statistics are given in Tables 2. Figure 2 refers to the period from the 1th of March 2013 until the 28th of February 2014, when data from both measuring points were available. Figure 2a shows that indoor temperature levels exhibited a significant seasonal variability, a fact that reflects their strong relation with the climate conditions that prevailed outdoors (Spring: r =0.9566, P < 0.0001; Summer: r=0.9544, P < 0.0001; Autumn: r = 0.9693, P < 0.0001; Winter: r = 0.9566, P < 0.0001). The analysis shows that indoor average daily temperature across all units ranged from 2.3°C to 34.4°C (Table 2). The mean indoor temperature during winter was 8.1°C became 9.5 and 11.0°C higher during spring and autumn respectively and reached 28.8°C in summer. The difference between indoor and outdoor temperature during the four seasons ranged between 0.0°C and 9.0°C, while its mean value equaled to 3.1°C. It is clear that temperature levels during summer were higher than the other three seasons (P < 0.05). No significant differences were detected between spring and autumn pig barn's temperature value (P > 0.05). According to Figure 2a, the mean indoor temperature during winter was 8.1°C, being 4.9°C higher than the corresponding outdoor values. On the contrary, doors and air inlet were opened all day during summer, thus, indoor and outdoor temperatures were very similar, their difference being about 1.4°C.

It is clear that relative humidity was higher levels in winter than the other three seasons (P < 0.05). During winter relative humidity inside the pig house varied between 63.4% and 92.9% with an average of 81.0±8.1%. No significant differences were detected between indoor and outdoor relative humidity levels (P > 0.05), and the average relative humidity was 75.1±4.6%, 74.4±5.4% respectively. During experimental period relative humidity outdoor the pig house varied between 43.1% and 97.1%. The average value of indoor relative humidity varied between 49.2% (spring) and 92.9% (winter).

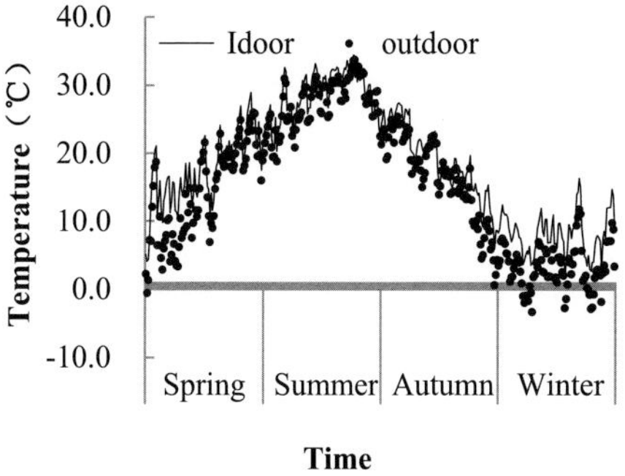

a. Temperature

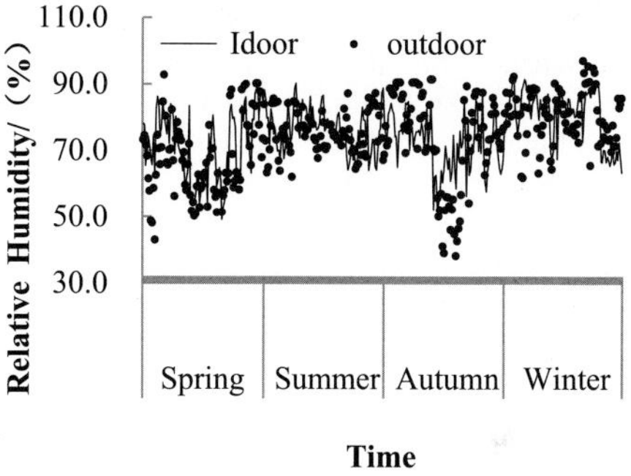

b. Relative humidity

Figure 2. Variation of air temperature and relative humidity.

Table 2. Average air temperature and relative humidity values recorded for the experimental seasons from March 2013 to February 2014 (Hourly values).

Season	location	Temp (°C)				RH (%)			
		Mean*	SD	Max	Min	Mean*	SD	Max	Min
Spring	Indoor	17.6bc	5.2	28.9	4.1	71.1 ef	10.5	89.1	49.2
	Outdoor	14.2d	6.8	26.0	-0.5	67.6 f	11.5	92.8	43.1
Summer	Indoor	28.8a	3.5	34.4	20.3	76.5 bc	6.4	90.3	63.7
	Outdoor	27.4a	3.6	33.7	18.8	75.8 cd	6.2	87.3	62.1
Autumn	Indoor	19.1b	5.2	27.4	6.0	72.0 de	8.3	87.3	52.3
	Outdoor	16.5c	6.0	25.4	0.7	73.6 cde	14.7	91.5	38.3
Winter	Indoor	8.1e	3.1	16.4	2.3	81.0 a	8.1	92.9	63.4
	Outdoor	3.2f	3.2	11.8	-3.3	80.6 ab	8.0	97.1	61.9

*: Means in the same column followed by different letters are significantly different ($P < 0.05$).

3.2. Temperature-humidity index

Distribution of hourly mean THI values for different months indoor and outdoor is given in Table 3. Hourly mean THI values inside the pig building are presented in Figure 3. Table 3 shows the average monthly THI for each month during the experimental period March, 2013 to February, 2014. The mean THI values for different seasons indoor and outdoor were calculated as 36.5 – 99.5 and 33.3 – 88.1 for spring; 57.5 – 98.3 and 56.4 – 100.4 for summer; 42.4 – 88.3 and 34.9 – 87.1 for autumn; 34.7 – 73.7 and 31.6 – 68.5 for winter, respectively, whereas the corresponding mean values were 62.2 and 58.4, 73.2 and 72.1, 63.4 and 60.5, 51.9 and 47. The hourly THI values indoor ranged from 34.7 to 99.5 during the experimental period, frequently higher than the desirable range in summer. In the months of December, January and February the maximum THI values indoor is below the critical THI, i.e., 75, no heat stress during winter. Data presented in the Figure 3a shows the heat stress as depicted by the daily maximum THI values in the months from March to November. The average diurnal variation of THI during the four seasons is presented in Figure 3b. It is obvious that during summer, THI dropped below the heat stress threshold (i.e., 75.0) from 00:00 h to 07:00 h, 17:00 h to 23:00 h, while it remained higher than the heat stress threshold (i.e., 75.0) from 08:00 h to 16:00 h.

Table 3. Average thermal-humidity index values recorded for the experimental months from March 2013 to February 2014 (Hourly values).

Months	Indoor THI				Outdoor THI			
	Mean	SD	Max	Min	Mean	SD	Max	Min
March	57.2	4.6	78.7	36.5	50.9	5.2	80.6	33.3
April	62.1	4.4	88.1	47.3	58.9	5.5	85.9	41.5
May	67.2	4.1	99.5	52.6	65.3	3.4	88.1	50.2
June	69.8	3.6	93.7	57.5	68.4	3.5	87.8	56.4
July	74.4	2.5	92.5	65.4	73.4	2.7	95.8	58.4
August	75.3	4.0	98.3	62.3	74.6	4.1	100.4	60.8
September	68.4	2.7	80.8	57.5	65.7	2.6	87.1	57.2
October	63.8	3.2	88.3	50.4	62.4	3.2	79.7	48.9
November	58.0	3.5	82.4	42.4	53.4	4.5	76.0	34.9
December	51.9	2.5	70.3	38.9	46.3	3.0	61.0	33.1
January	52.6	2.4	73.7	38.4	47.7	2.9	68.5	32.8
February	51.4	3.9	70.2	34.7	47.0	4.3	65.8	31.6

*: Means in the same column followed by different letters are significantly different (P < 0.05).

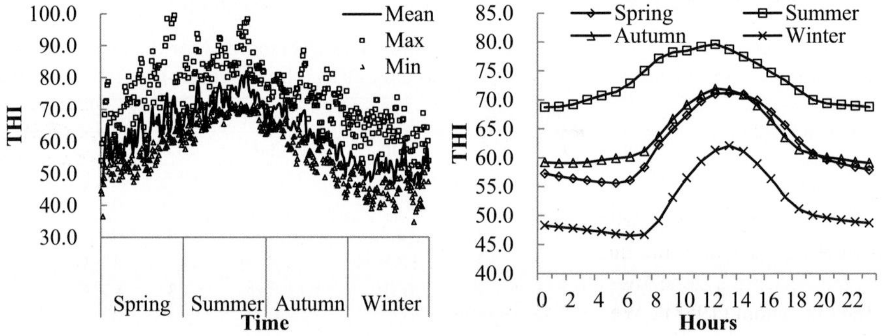

a. Diurnal variation of daily THI b. Diurnal variation of average hourly THI

Figure 3. Average diurnal variation of THI.

Figure 4 shows that frequency of occurrence (%) of THI classes. Pig was exposed to heat stress (THI ≥75) in 10.5%, 31.5% and 4.6% of spring, summer and autumn hours, respectively, whereas no heat stress was identified during winter. Furthermore, severe heat stress (THI ≥ 79) was identified in 6.0%, 15.1% and 0.9% of spring, summer and autumn hours, respectively. Pig are more prone to severe heat stress during summer, as the daily maximum hourly values of THI remained higher than the severe heat stress threshold (THI=79) in 66.3% of the summer days, whereas severe heat stress was 26.0%, 6.6% during Spring and Autumn, respectively.

A series of indexes has been suggested to describe the influence of thermal environment on animal response (Hahn et al., 2009; 2003). These include utilization of the temperature humidity index (THI), which was originally developed by Thom (1958) as a "discomfort index" to estimate the levels of discomfort for human beings during summer months. The temperature humidity index (THI) combines temperature and humidity into a single value and may be calculated by utilizing different formulas and thresholds (Hahn et al., 2009). For growing-finishing pigs, the THI has been used so far mainly to assess the duration and extent of climatic heat stress periods in different regions (Lucas et al., 2000). In particular, even if breed, nutrition, housing type and other factors can modify the susceptibility of animals to hot conditions, numerous studies have been undertaken to establish thresholds for heat stress in animal on the basis of THI values (Gaughan et al., 2009).

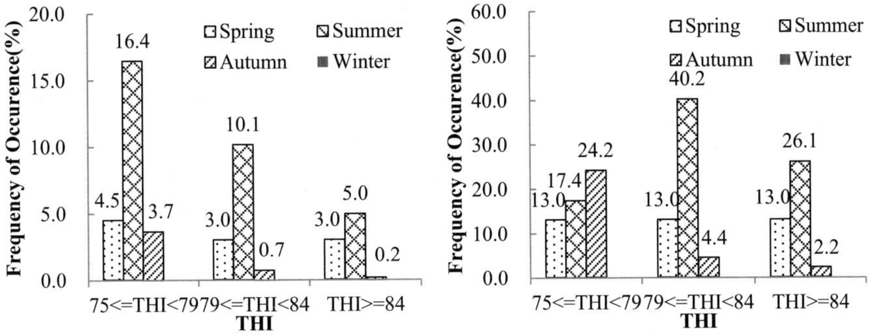

a. Hourly THI b. Daily maximum hourly THI values

Figure 4. Frequency of occurrence (%) of THI classes.

According to these thresholds, a THI less than or equal to 74 can be considered normal. Alert situations occur for THI values equal to or above 75, whereas THI values from 79 to 83 are dangerous. THI values equal to or above 84 are defined as emergency situations (NWSCR, 1976). Even if with some limitations, such as that it does not permit the evaluation of other atmospheric variables impacting on animals (i.e., wind, solar radiation, etc.), the THI is still widely considered a useful tool to predict the effects of environmental warmth on farm animals (Segnalini et al., 2013; 2011). The relevant results showed that: during July – September period, the hottest season in most parts of China, the average Temperature-Humidity Index (THI, as defined by Nissim et al., 2000) the value of pig breeder houses was usually over 80 (Kuczynski et al., 2011). Sales et al. (2008) was concluded through statistical analysis that better comfort for sows and piglets occurred within the thermal comfort zone ($61<\text{THI}\leqslant 65$); better comfort for piglets occurred within the intermediate zone ($65<\text{THI}\leqslant 69$), which is still bearable for sows; and, thermal stress zone ($69<\text{THI}\leqslant 73$) negatively affected both sows and piglets [THI, as defined by Thom (1958)]. Wegner et al. (2014) reported that temperature and THI affected the reproductive performance of the sows and the survival of the piglets in different ways. While increased climatic values at the time of breeding positively affected the total number of piglets born, increased values at the time of farrowing had negative impacts on the reproductive performance of the sows. Lucas et al. (2000) reported that using the temperature-humidity index equations revealed several periods with THI above 84 associated with conditions, which may be related with higher losses in pig production.

This study is evident that under summer conditions animals are apparently exposed to severe heat stress potentially burdening their welfare. Pig was exposed to severe heat stress ($\text{THI} \geq 84$) was identified 8.2% (Total 718 h) in years from 2013 to 2014. These results can assist in the improvement of barn covering material for pig housing, providing appropriate pig comfort. Moreover, during severe heat stress, high-pressure fogging systems combined with fans in naturally ventilated pig barn (Gates et al., 1991; Bridges et al., 1992; Haeussermann et al., 2007a; 2007b) could be adopted as a practical and financially viable solution for the maintenance of indoor conditions.

4. Conclusions

Dry-bulb temperature and relative humidity data were obtained in the interior of building, as well as the external conditions. Hourly values of THI, average diurnal and monthly variations were calculated. Results show that the variation of outdoor climatic conditions affects the indoor environment parameters. Indoor temperature and relative humidity levels were strongly dependent to the outdoor climate conditions. No significant differences were detected between indoor and outdoor relative humidity levels.

Pigs were frequently exposed to potential heat stress conditions during the summer. The daily maximum hourly values of THI remained higher than the heat stress threshold (THI=75) in 83.7% of the summer days, whereas heat stress was 39.0%, 30.8% during spring and autumn, respectively. Heat stress occurs between 8:00 h to 16:00 h of the day, while the night decreases. No heat-stress was detected during the winter.

These results can assist in the improvement of barn covering material for pig housing, providing appropriate pig comfort. Future work will need to evaluate animal physiological responses and animal behavior in Bedding System.

Acknowledgements

Funding was provided by the Agriculture Science and Technology Innovation Foundation of Jiangsu Province [CX (12) 1001-04] and the Science and Technology Agency of Jiangsu Province (BE2014342-1).The authors would like to thank all farmers who participated in this study.

References

Bridges, T.C., R.S. Gates, L.W. Turner, 1992. Stochastic assessment of evaporative misting for growing-finishing swine in Kentucky. Applied Engineering in Agriculture. 8(5), 685–693.

Gaughan, J., N. Lacetera, S.E. Valtorta, H.H. Khalifa, L. Hahn, T. Mader, 2009. Response of domestic animals to climate challenges. In: *Biometeorology for Adaptation to Climate Variability and Change*. Eds. Ebi, K.L., I. Burton, G.R. McGregor. Springer, Heidelberg, pp 131–170.

Gates, R.S., J.L. Usry, J.A. Nienaber, L.W. Turner, T.C. Bridges, 1991. An optimal misting method for cooling livestock housing. Transactions of the ASAE. 34(5), 2199–2206.

Haeussermann, A., E. Vranken, J.M. Aerts, E. Hartung, T. Jungbluth, D. Berckmans, 2007 a. Evaluation of control strategies for fogging systems in pig facilities. Transactions of the ASABE. 50(1), 265–274

Haeussermann, A., E. Hartung, T. Jungbluth, E. Vranken, J.M. Aerts, D. Berckmans, 2007 b. Cooling effects and evaporation characteristics of fogging systems in an experimental piggery. Biosystems Engineering. 97, 395–405.

Hahn, G.L., J.B. Gaughan, T.L. Mader, R.A. Eigenberg, 2009. Thermal indices and their applications for livestock environments. In Livestock energetics and thermal environmental management (ed. JA DeShazer), ASABE, St. Joseph, MI, USA. 113–130

Hahn,G.L., T.L. Mader, R.A. Eigenberg, 2003. Perspective on development of thermal indices for animal studies and management. In Interactions Between Climate and Animal Production, N. Lacetera, U. Bernabucci, H.H. Khalifa, B. Ronshi, and A. Nardone. Eds., EAAP Technical Series No. 7. Wageningen, The Netherlands: Wageningen Academic Publishers. 31–44.

Huynh, T.T.T., A.J.A. Aarnink, M.W.A. Verstegen, W.J.J. Gerrits, M.J.W. Heetkamp, B. Kemp, T.T. Canh, 2005. Effects of increasing temperatures on physiological changes in pigs at different relative humidities. Journal of Animal Science. 83,1385–1396.

Kuczynski, T., V. Blanes-Vidal, B. Li, R.S. Gates, I.d.A. Nääs, D.J. Moura, D. Berckmans, T.M. Banhazi, 2011.Impact of global climate change on the health, welfare and productivity of intensively housed livestock. International Journal of Agricultural and Biological Engineering. 4(1), 1–22.

Lucas, E.M., J.M. Randall, J.F. Meneses, 2000. Potential for Evaporative Cooling during Heat Stress Periods in Pig Production in Portugal (Alentejo). J. agric. Engng Res. 76, 363–371

Nissim, S., 2000. Effects of heat stress on the welfare of extensively managed domestic ruminants. Livestock Production Science. 67, 1–18.

NWSCR, 1976. Livestock hot weather stress. Regional operations manual letter C-31–76. National Weather Service Central Region, USA

Sales, G.T., E.T. Fialho, T. Yanagi Jr, R.T.F.D. Freitas, V.H. Teixeira, R.S. Gates, G.B. Day, 2008. Thermal Environment Influence on Swine Reproductive. Livestock Environment VIII. Iguassu Falls, Brazil, 31 August – 4 September. St. Joseph, Mich: ASAE. 767–772.

Segnalini, M., U. Bernabucci, A. Vitali, A. Nardone, N. Lacetera, 2013. Temperature humidity index scenarios in the Mediterranean basin. International Journal of Biometeorology. 57, 451–458.

Segnalini, M., A. Nardone, U. Bernabucci, A. Vitali, B. Ronchi, N. Lacetera, 2011. Dynamics of the temperature-humidity index in the Mediterranean basin. International Journal of Biometeorology. 55, 253–263.

Thom, E.C., 1958. Cooling degrees-days air conditioning, heating, and ventilating. Transactions of the ASAE. 55(7), 65–72.

Wegner, K., C. Lambertz, G. Daş, G. Reiner, M. Gauly, 2014. Climatic effects on sow fertility and piglet survival under influence of a moderate climate. Animal. 8(9), 1526–1533.

Odour Concentrations and Emissions at Two Manure-Belt Egg Layer Houses in the U.S.

Lingying Zhao [a,*], Lara Jane S. Hadlocon [a], Roderick B. Manuzon [a], Matthew J. Darr [b], Xinjie Tong [a], Albert J. Heber [c], Ji-Qin Ni [c]

[a] Department of Food, Agricultural and Biological Engineering, The Ohio State University, Columbus, OH 43210, USA

[b] Department of Agricultural and Biosystems Engineering, Iowa State University, Ames, IA 50011, USA

[c] Department of Agricultural and Biological Engineering, Purdue University, West Lafayette, IN 47907, USA

* Corresponding author. Email: zhao.119@osu.edu

Abstract

Odour emissions from poultry facilities have caused significant public concerns, but very limited odour data are available to facilitate scientific understanding and effective mitigation of the odour concerns. The objective of this study is to characterize odour emissions from representative layer houses and quantify the odour concentrations and emission rates. Odour samples were taken at the air inlets and exhausts of two identical manure-belt layer houses once a month for a year. The samples were collected using a vacuum chamber and sent within 24 hours of collection to the Purdue University olfactometry laboratory for evaluation of odour characteristics, hedonic tone, intensity, and concentration. Results showed that annual mean odour concentration and emission rates were 335 ± 73 OU_E m^{-3} and 0.14 ± 0.11 OU_E s^{-1} hen^{-1}, respectively. There were significant seasonal variations in odour concentrations and emission rates with high concentrations in summer and winter and high emission rates in warmer months. The findings can be used to assess odour problems associated with layer houses and develop effective mitigation technology and management plans.

Keywords: Indoor air quality, odour emissions, pollution control, poultry, animal feeding operations

1. Introduction

Odour remains a significant nuisance issue associated with animal feeding operations (AFOs). Some nearby surrounding communities of AFOs have reported that odour degraded their quality of life and well-being (Parker, 2008; Donham et al., 2007). Odour nuisance complaints have escalated with the growth of AFOs (Schiffman, 2000), and could significantly affect the expansion and sustainability of the AFOs.

It is challenging to measure odour compared with other air contaminants from animal operations, because livestock odour is not caused by a single compound but a mixture of odourous gases and volatile organic compounds. Another reason is subjective nature of odour, in which annoyance can vary among individuals being exposed to the odour. Currently, a standard method used in the U.S. scientific community to quantitatively assess odour from animal buildings is the triangular forced-choice olfactometry (ASTM 2001; CEN, 2003) in which trained human panellists are used for odour measurement. Odour concentration is a common parameter to quantify odour, which can be reported as odour units per cubic meter air (OU m^{-3}). When odour concentration is expressed using the European standard (CEN, 2003), the unit OU_E m^{-3} is applied. In addition, hedonic tone and intensity are used to measure the offensiveness and strength of the odour (Parker et al., 2008; Nicell, 2009).

Odour from livestock facilities have been reported with large variations in both concentrations and emission rates (Guo et al., 2007; Yu et al., 2010). Factors that may affect odour emissions include animal species, housing systems, feed formula, manure management practices, and weather conditions (Mielcarek and Rzeznik, 2015; Carey et al., 2004). Odour emissions from poultry broiler facilities in European countries were found to have a range of 0.1-0.97 OU_E s^{-1} hen^{-1} (Ogink and Groot Koerkamp, 2001; Robertson et al., 2002). Large egg layer operations in the U.S. also emit significant odourous volatile organic compounds and ammonia gas and have caused significant public concerns (Wang-Li et al., 2013; Hadlocon et al., 2015); but very limited odour data have been reported for representative U.S. poultry facilities. There is a need to quantify odour to help facilitate the adaptation of effective odour mitigation technologies and manure management practices for poultry facilities.

Therefore, this study aims to quantify odour concentrations and emission rates at two typical manure-belt egg layer houses. The specific objectives of this study are to, through monthly odour sampling and analysis, (1) determine odour hedonic tone and intensity, (2) quantify odour concentrations and emission rates, and (3) characterize seasonal variations of odour emission rates.

2. Materials and Methods

2.1. The manure-belt egg layer houses studied

A commercial poultry farm (Figure 1) consisting of four manure-belt layer houses was selected for this study. The four houses were identical in building design, ventilation system, and manure management, but housed laying hens at different ages. Two of the four houses: houses 3 and 4 were selected for a large-scale aerial emission monitoring study. Each layer house was 121.9 m long and 19.5 m wide. The height was 7.7 m at the ridge. The houses had mechanical ventilation systems consisting of 24 ventilation fans on two sidewalls and an additional 24 exhaust fans on two end walls. In each house, there are eight rows and eight tiers of layer cages with manure belts underneath each tier of cages. The cages were about 100 m long. Each house had approximately 160,000 Lohmann white hens of 20 to 109 weeks old with an initial average weight of about 1.5 kg at the beginning of each production period. Automatic feeding, watering, and egg and manure collecting systems were used in the houses. Manure was removed from the houses twice a week by a manure belt conveyor system.

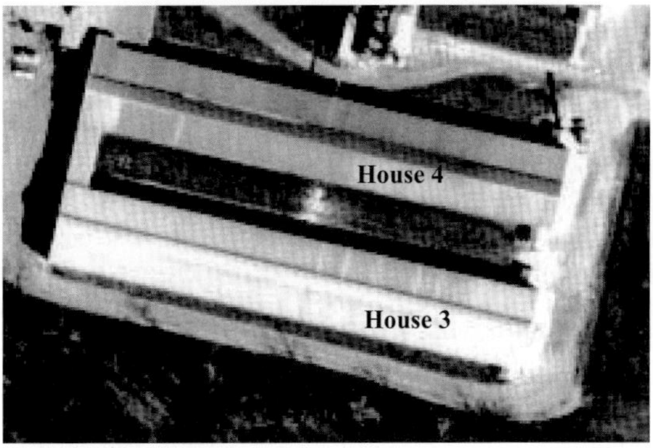

Figure 1. Satellite image of the layer houses (from Google Earth).

2.2. Odour sampling and measurement

Odour samples were collected around noon on the sampling day once a month from March 2007 to April 2008 using 10-L Tedlar bags, which were flushed with either compressed air or nitrogen gas at least three times prior to sampling. Replicate samples were obtained from the inlet and the southeast exhaust of house 3 and northeast exhaust of house 4. These samples were analysed for odour concentrations (OU m^{-3}) within 30 h of collection with a dynamic olfactometer (AC'SCENT® International Olfactometer, St. Croix Sensory, Inc., Stillwater, MN, USA) at Purdue University odour laboratory. The intensity and hedonic tone of odour samples were evaluated according to published methods (Lim et al., 2001).

2.2.1. Measurement of odour concentration, intensity, and hedonic tone

Odour concentration, or odour detection threshold, was measured following U.S. olfactometry standards (ASTM, 1991). The olfactometer delivers mixtures of sample and dilution air at various dilution ratios. The odour panel (ASTM, 1986) consisted of eight human subjects who, one at a time, sniffed three sequential sample coded gas streams in which only one gas stream had the odour with a known dilution ratio, while the other two gas streams were odour-free. The dilution ratio gradually decreased until the panellist was able to correctly detect the odourous sample. The reported odour concentrations are the geometric means of the last non-detectable and the first detectable dilution ratios of the individuals (Lim et al., 2001).

The odour intensity was evaluated by the odour panel via objectively matching the intensity of a sample to one of a series of n-butanol solutions prepared according to a reference scale method (ASTM, 1992). The sample bag was compressed which allow the sample air to flow into a glass funnel. Panellists were asked to sniff the unknown air sample from the funnel and then to sniff the reference solutions starting from the lowest reference scale. They could match the unknown air sample to reference scale solutions as many times as needed. The reported odour intensities were the geometric means of panellists' ratings (Lim et al., 2001).

Hedonic tone was evaluated by the odour panel at the same time as intensity. The hedonic tone scale ranged from –10 (extremely unpleasant) to 0 (neither pleasant nor unpleasant) to +10 (extremely pleasant) (McGinley et al., 2000). The reported hedonic tone is the arithmetic mean of panellists' ratings (Lim et al., 2001).

2.2.2. Measurement of environmental conditions

Ambient temperatures and relative humidity were measured every minute during the odour sampling event using an electronic RH/temp transmitter (Model HMW61, Vaisala Inc., Columbus, OH, USA). The indoor air temperatures of houses 3 and 4 are the average exhaust air temperatures. The relative humidity was measured at the east end of each house. All on-line sensor signals were acquired and pre-processed using an on-site computer system (Ni et al., 2009).

2.3. Air flow monitoring and estimation

Airflow was calculated using data of fan activity, the differential static pressures of the layer houses, and fan airflow curves (Equation 1). The fan operations (on or off) were monitored by vibration sensors (Darr et al., 2007). The differential static pressure was measured using a digital manometer (Model 267MR, Setra Systems Inc., Boxborough, MA). The on-site airflow curves of the ventilation fans were determined using a portable fan test unit (FANS) developed by University of Kentucky (Gates et al., 2004). A degradation factor K, measuring fan performance differences between the manufacturer's published data and the on-site measurement data, was generated using regression analysis.

$$Q = S \times V \times K \tag{1}$$

where: Q is fan airflow, $m^3\ h^{-1}$; S is operation status of a fan monitored by a vibration sensor, % of operation time; V is airflow rate according to the manufacturer's published fan curve, $m^3\ h^{-1}$; and K is degradation factor of a fan, %.

2.4. Odour emission calculations

The hen-specific odour emission rates were calculated using Equation 2:

$$ER = \frac{C_{exhaust} - C_{inlet}}{N} \times Q \tag{2}$$

where $C_{exhaust}$ is the exhaust odour concentration at the house ($OU_E\ m^{-3}$); C_{inlet} is the inlet odour concentration ($OU_E\ m^{-3}$); and N is the total number of hens.

2.5. Statistical data analysis

The data were analysed using JMP® 9.0 (SAS Institute Inc., Cary, NC, USA). The seasonal means were compared using Tukey-Kramer Honest Significant Difference (HSD) test for the odour hedonic tone, intensity, concentrations, and emission rates. An analysis of variance (ANOVA) was used to evaluate the effects of environmental conditions on all these odour parameters. The coefficient of correlation (r) was used to examine the relationship between odour emission rates and the associated environmental conditions. All analyses were conducted at a significance level of 0.05.

3. Results and Discussion

3.1. Odour characteristics

The ambient air near the layer houses was perceived by the panellists to have an earthy and sweaty smell during the fall. The smell of barn, sulphur, and faecal matter was also detected during spring, summer, and winter. Inside the houses, the exhaust air was perceived to have a stinky, animal faeces, rotten meat, and barn smell throughout all seasons. More pronounced odour descriptions inside the houses were selected by the panellists, such as the smell of urine, ammonia, farm, rotten meat, faeces, hay, sulphur, and moulds. All of these descriptions indicate clear odour related to a typical animal feeding operation.

3.2. Odour hedonic tone

As shown in Table 1, the hedonic tones were all negative for the exhaust air of layer houses. This suggested that odour emitted from a layer house was unpleasant. The mean hedonic tone of ambient air ranged from -0.4 ± 0.6 to -2.1 ± 2.4 (average ± standard deviation), which was in the slightly unpleasant range. The overall average hedonic tones of air exhausted from houses 3 and 4 were -4.2 ± 0.2 and -4.1 ± 0.5, respectively, which indicated more unpleasant. The mean hedonic tones observed for houses 3 and 4 were not significantly different ($p > 0.05$). There were no significant differences in the mean hedonic tone between different seasons ($p > 0.05$).

Table 1. Average hedonic tones of inlet and exhaust air at houses 3 and 4.

Season	Inlet air	Exhaust air		
		House 3	House 4	Average [a]
Spring	-2.1 ± 2.4	-4.2 ± 1.8	-4.1 ± 2.0	-4.2 ± 1.2 A
Summer	-1.8 ± 1.4	-4.4 ± 1.7	-3.4 ± 1.3	-3.9 ± 1.3 A
Fall	-1.5 ± 1.7	-4.4 ± 2.6	-4.3 ± 2.8	-4.4 ± 1.6 A
Winter	-0.4 ± 0.6	-3.9 ± 2.0	-4.7 ± 2.9	-4.3 ± 2.3 A
Average[a]	-1.5 ± 0.7 B	-4.2 ± 0.2 C	-4.1 ± 0.5 C	-4.2 ± 0.2

[a] Averages with the same superscript letter are not significantly different.

3.3. Odour Intensity

Table 2 shows the seasonal and overall average odour intensities observed at the inlet and exhaust locations. The odour intensity of inlet air was about 1.4 ± 0.4, which indicated barely perceivable to faint but identifiable odour. In exhaust air, the overall averages were 3.1 ± 0.0 and 3.2 ± 0.3, which indicated easily perceivable odour. There was no significant difference in odour intensity between season or houses ($p > 0.05$).

3.4. Odour concentration

Figure 2 shows the monthly mean odour concentrations at the layer house inlet and exhaust air. The standard deviations indicate measurement variations among the replicates. The exhaust air odour concentrations for house 3 ranged from 176 to 555 OU_E m^{-3}, with an average of 359 ± 123 OU_E m^{-3}, while the odour concentrations at house 4 ranged from 141 to 481 OU_E m^{-3}, with an average of 312 ± 93 OU_E m^{-3}. The time series plot of the odour concentrations shows strong seasonal variations with higher concentrations observed in summer and winter while lower concentrations were observed in spring and fall. An ANOVA revealed that ventilation rate had no significant effect on observed odour concentrations, F(1,22) = 0.54, p = 0.47, which indicated that the increase in house ventilation rate during the peak of summer in July did not significantly decrease odour concentrations. This may be due to an increased generation of odourous compounds from manure in summer as higher temperature triggers higher microbial rates. However, odour concentration increased with the decrease in ventilation rate.

Table 2. Average odour intensities of the inlet and exhaust air at houses and 4.

Season	Inlet air	Exhaust air		
		House 4	House 4	Average [a]
Spring	1.7 ± 1.1	3.1 ± 1.0	2.9 ± 1.1	3.0 ± 0.1 A
Summer	1.0 ± 0.9	3.0 ± 1.1	3.5 ± 1.2	3.3 ± 0.4 A
Fall	1.8 ± 0.9	3.1 ± 0.9	3.3 ± 0.8	3.2 ± 0.1 A
Winter	1.2 ± 0.9	3.0 ± 0.9	2.8 ± 0.9	2.9 ± 0.1 A
Average [a]	1.4 ± 0.4 B	3.1 ± 0.0 C	3.2 ± 0.3 C	3.1 ± 0.2

[a] Averages with the same superscript letter are not significantly different.

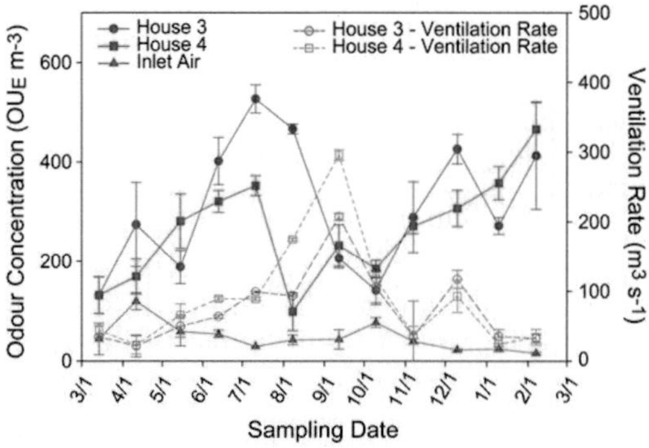

Figure 2. Odour concentrations in the inlet and exhaust air of houses 3 and 4.

Table 3 summarizes the seasonal and overall averages of odour concentrations observed at the inlet and exhaust air. The seasonal averages were calculated assuming that Winter, Spring, Summer and Fall are represented by sampling during December to February, March to May, June to August, and September to November, respectively. An ANOVA revealed a significant effect of season on the mean odour concentrations, F(3,44) = 7.01, p = 0.0006. A Post-hoc Tukey test showed that the mean odour concentrations in spring and fall were not significantly different, neither were the mean odour concentrations in summer and winter. However, the mean odour concentrations in summer and winter were significantly higher than the means in spring and fall. An ANOVA showed that the overall mean odour concentrations at house 3 (359 ± 114 OU_E m^{-3}) and house 4 (312 ± 58 OU_E m^{-3}) were not significantly different, t(46) = -1.46, p = 0.15. Odour concentrations in house 3 showed significant seasonal differences, F(3,20) = 12.17, p < 0.0001. The summer mean odour concentration of 506 ± 50 OU_E m^{-3} in house 3 was the highest seasonal average, followed by the winter mean odour concentration of 391 ± 83 OU_E m^{-3} in house 4. No significant odour concentration differences were observed between spring and fall in house 3. There were no significant seasonal variations in odour concentrations observed at house 4, F(3,20) = 2.86, p = 0.06. The annual average odour concentration of the houses was 335 ± 73 OU_E m^{-3}.

3.5. Odour emission rates

The hen-specific odour emission rates (OU_E s^{-1} hen^{-1}) of houses 3 and 4 are plotted in Figure 3. The odour emission rates from house 3 ranged from 0.03 to 0.32 OU_E s^{-1} hen^{-1} with an average of 0.16 ± 0.12 OU_E s^{-1} hen^{-1}. The odour emission rates of house 4 ranged from 0.02 to

0.45 OU_E s^{-1} hen^{-1} with an average of 0.14 ± 0.12 OU_E s^{-1} hen^{-1}. The time series plot of the odour emission rates does show a general trend of higher emission rates in warmer months in comparison with those of cold months.

Table 3. Average odour concentrations (OU_E m^{-3}) of inlet and exhaust air at houses 3 and 4.

Season	Inlet air	Exhaust air		
		House 3	House 4	Average [a]
Spring	74 ± 40	272 ± 111	268 ± 84	270 ± 3 A
Summer	41 ± 13	506 ± 50	299 ± 137	403 ± 146 A
Fall	53 ± 21	265 ± 56	283 ± 25	274 ± 13 A
Winter	20 ± 4	391 ± 83	397 ± 77	394 ± 4 A
Average [a]	47 ± 23 B	359 ± 114 C	312 ± 58 C	355 ± 73

[a] Averages with the same superscript letter are not significantly different.

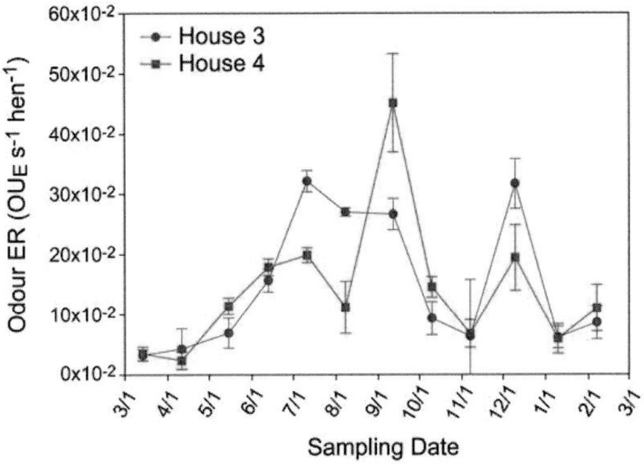

Figure 3. Hen-specific odour emission rates at houses 3 and 4.

Table 4 summarizes the seasonal and overall average odour emission rates of houses 3 and 4. An ANOVA did not reveal a strong significant effect of season on the mean odour emission rates, $F(3,20) = 2.70$, $p = 0.07$. The highest seasonal average odour emission rates was 0.25 ± 0.08 OU_E s^{-1} hen^{-1} from house 3 in summer and 0.22 ± 0.20 OU_E s^{-1} hen^{-1} from house 4 in fall. The comparison of the mean odour emission rates of house 3 (0.15 ± 0.09 OU_E s^{-1} hen^{-1}) and house 4 (0.13 ± 0.08 OU_E s^{-1} hen^{-1}) revealed no significant differences between the layer houses, $t(22) = -0.16$, $p = 0.87$.

Table 4. Average odour emission rates (OU_E s^{-1} hen^{-1}) at houses 3 and 4.

Season	House 3	House 4	Average[a]
Spring	0.05 ± 0.02	0.06 ± 0.05	0.05 ± 0.03 A
Summer	0.25 ± 0.08	0.16 ± 0.05	0.21 ± 0.08 A
Fall	0.14 ± 0.11	0.22 ± 0.20	0.18 ± 0.15 A
Winter	0.16 ± 0.14	0.12 ± 0.07	0.14 ± 0.10 A
Average [a]	0.15 ± 0.11 B	0.14 ± 0.11 B	0.14 ± 0.11

[a] Averages with the same superscript letter are not significantly different.

An increase in odour emission rates from house 3 might be because new hens were introduced in June 2007. Young hens receive feeds that are higher in nitrogen content, which might result in higher emissions of odour-causing gases (Wang et al., 2014). The annual odour emission rate calculated using the entire data from house 3 and 4 was 0.14 ± 0.11 OU_E s^{-1} hen^{-1}. The hen-specific odour emission rates reported in this study were within the range of those reported from broiler production facilities in European countries ranged between 0.06 and 0.97 OU_E s^{-1} hen^{-1} (Ogink and Groot Koerkamp, 2001; Robertson et al., 2002).

4. Conclusions

The odour characteristics of layer house exhaust air were strongly associated with barns and faecal matter. The offensiveness of odour was rated to be in the moderately unpleasant range (hedonic tone = -4.2 on a scale of -10 to +10). The odour intensity was in the easily perceivable range (intensity = 3.2 on a scale of 0 to 5). There were no significant differences in either hedonic tone or odour intensity with respect to seasons or layer houses. The annual mean odour concentration was estimated to be 335 ± 73 OU_E m^{-3}, and the annual mean odour emission rate 0.14 ± 0.11 OU_E s^{-1} hen^{-1}. This emission rate was within the range of odour emission rates reported from poultry facilities in other countries. This study contributed to scientific assessment of odour concerns and will be useful to develop effective odour mitigation technology and management plans associated with egg layer operations.

Acknowledgements

This study was supported by the National Institute of Food and Agriculture, U.S. Department of Agriculture, under award number 2005-35112-15422. Appreciation is also expressed to the participating farm and the staff at the egg production facility for their support of the data collection for this project.

References

ASTM. 1991. E 679–91. Standard Practice for Determination of Odor and Taste Thresholds by a Forced–Choice Ascending Concentration Series Methods of Limits. Vol. 15.07:35–39. Philadelphia, PA.: American Society for Testing and Materials.

ASTM. 1992. E 544–75. Standard Practices for Referencing Suprathreshold Odor Intensity. Vol. 15.07:21–30. Philadelphia, PA.: American Society for Testing and Materials.

ASTM. 2001. EP379.5: Standard Practice for Determining Odor and Taste Thresholds by Force-Choice Concentration Series Methods of Limits. Philadelphia, PA: ASTM.

ASTM. 1986. Special Publication ASTM 758. Guidelines for Selection and Training of Sensory Panel Members. Philadelphia, PA.: American Society for Testing and Materials.

Carey, J. B., Lacey, R. E., Mukhtar, S., 2004. A review of literature concerning odors, ammonia, and dust from broiler production facilities: 2. Flock and house management factors. Journal of Applied Poultry Research, 13(3), 509–513.

CEN. 2003. Air quality: Determination of odour concentration by dynamic olfactometry (EN13725). Brussels, Belgium: European Committee for Standardization.

Darr, M. J., Zhao, L. Y., Ni, J. Q., Gecik, C., 2007. A robust sensor for monitoring the operational status of agricultural ventilation fans. Transactions of the ASABE 50(3): 1019-1027.

Donham, K. J., Wing, S., Osterberg, D., Flora, J. L., Hodne, C., Thu, K. M., Thorne, P. S., 2007. Community health and socioeconomic issues surrounding concentrated animal feeding operations. Environmental Health Perspectives, 115(2), 317–320.

Gates, R. S., Casey, K. D., Xin, H., Wheeler, E. F., Simmons, J. D., 2004. Fan assessment numeration system (FANS) design and calibration specifications. Transactions of the ASAE 47(5): 1709-1715.

Guo, H., Dehod, W., Agnew, J., Feddes, J. R., Pang, S., 2007. Daytime odor emission variations from various swine barns. Transactions of the ASABE, 50(4), 1365–1372.

Hadlocon, L. J. S., Manuzon, R. B., Zhao, L., 2015. Development and evaluation of a full-scale spray scrubber for ammonia recovery and production of nitrogen fertilizer at poultry facilities. Environmental Technology, 36(4), 1–12.

Jongebreur, A. A., Monteny, G. J., and Ogink, N. W. M., 2003. Livestock production and emission of volatile gases. In International Symposium on Gaseous and Odour Emissions from Animal Production Facilities. 1-4 June, Horsens, Denmark, 11-30. CIGR, EurAgEng, NJF.

Lim, T. T., Heber, A. J., Ni, J. Q., Sutton, A. L., Kelly, D. T., 2001. Characteristics and emission rates of odor, Transactions of the ASAE, 44(5), 1275–1282.

McGinley, C. M., and M. A. McGinley. 2000. Odor intensity scales for enforcement, monitoring, and testing. In Proc. Annual Conference of the Air and Waste Management Association. Salt Lake City, Utah. 18–22 June. Pittsburgh, PA.: Air and Waste Management Association.

Mielcarek, P., Rzeźnik, W., 2015. Odor emission factors from livestock production. Polish Journal of Environmental Studies, 24(1), 27–35.

Ni, J.-Q., Heber, A. J., Darr, M. J., Lim, T. T., Diehl, C. A., Bogan, B. W., 2009. Air quality monitoring and on-site computer system for livestock and poultry environment studies. Transactions of the ASABE, 52(3), 937–947.

Nicell, J. A. 2009. Assessment and regulation of odour impacts. Atmospheric Environment, 43(1), 196–206.

Ogink, N. W. M., and Groot Koerkamp, P. W. G., 2001. Comparison of odour emissions from animal housing systems with low ammonia emissions. In: Proc. of the 1st IWA International Conference on Odour and VOCs: Measurement, Regulation and Control Techniques. University of New South Wales, Sydney, Australia, March

Parker, D. B. 2008. Reduction of odor and VOC emissions from a dairy lagoon. Applied Engineering in Agriculture, 24(5), 647–655.

Robertson, A. P., Hoxey, R. P., Demmers, T. G. M., Welch, S. K., Sneath, R. W., Stacey, K. F., Fothergill, A., Filmer, D., Fisher, C., 2002. Commercial-scale studies of the effect of broiler-protein intake on aerial pollutant emissions. Biosystems Engineering, 82(2), 217–225.

Schiffman, S. S., Walker, J. M., Dalton, P., Lorig, T. S., Raymer, J. H., Shusterman, D., Williams, C. M., 2000. Potential health effects of odor from animal operations, wastewater treatment, and recycling of byproducts. Journal of Agromedicine, 7(1), 7–81.

Stevens, S. S. 1957. On the psychophysical law. The Psychological Review, 64(3), 153–181.

Wang-Li, L., Li, Q. F., Chai, L., Cortus, E. L., Wang, K., Kilic, I., Bogan, B.W., Ni, J.-Q., Heber, A. J., 2013. The National Air Emissions Monitoring Study's southeast layer site: Part 3. Ammonia concentrations and emissions. Transactions of the ASABE, 56(3), 1185–1197.

Wang, S., L. Y. Zhao, X. Wang, R. Manuzon, M. Darr, H. Li, H.M. Keener. 2014 Estimation of ammonia emission from manure belt poultry layer houses using an alternative mass-balance method. Transactions of the ASABE. 57 (3): 937-947.

Yu, Z., Guo, H., 2010. Livestock odor dispersion modeling: a review. Transactions of ASABE, 53(4), 1231–1244.

Effects of Air Temperature, Velocity and Floor Type on Gaseous Emissions from a Scale Model of Dairy Open Lots

Luyu Ding [a,b], Guoqiang Zhang [c], Simon Kristensen [c], Wei Cao [a,b], Zhengxiang Shi [a,b], Baoming Li [a,b], Chaoyuan Wang [a,b,*]

[a] Department of Agricultural Structure and Bioenvironmental Engineering, China Agricultural University, Beijing 100083, China

[b] Key Laboratory of Agricultural Engineering in Structure and Environment, Ministry of Agriculture, Beijing100083, China

[c] Department of Engineering, Aarhus University, Blichers Allé 20, P.O. Box 50, 8830 Tjele, Denmark

* Corresponding author. Email: gotowchy@cau.edu.cn

Abstract

In typical Chinese dairy housing and manure management system, outdoor open lot is an important emission source of contaminant gases. In this study, dairy open lot scale models were built to evaluate effects of air temperature (15 to 35°C), surface velocity (0.4 to 1.2 m s^{-1}) and floor type (unpaved soil floor or brick-paved floor) on CO_2, CH_4, N_2O and NH_3 emissions. Generally, CO_2, N_2O and NH_3 emissions increased with the increasing of air temperature and velocity on both unpaved soil and brick-paved floors. Velocity showed different impacts on CH_4 emissions from different floor types. There was a drop on CH_4 emissions from brick floor when velocity kept increasing and the drop would be delayed by air temperature. No similar drop was found from soil floor under the examined velocities. Except methane, other gaseous emissions were more sensible to changes at lower velocity (0.4 to 0.7 m s^{-1}) or lower air temperature (15 to 25°C). Gaseous emissions from soil floor were more sensible to velocity changes while the emissions from brick floor were more sensible to air temperature changes. Higher NH_3 emissions were measured on brick-paved floor. In comparison with NH_3 emission, CO_2 and CH_4 emissions were less affected by floor type.

Keywords: Wind tunnel, unpaved soil floor, brick paved floor, sensitivity analysis

1. Introduction

Dairy open lots are essential pollutant emission sources in typical housing and manure management system in China and some regions in the USA (Ding et al., 2015; Bjorneberg et al., 2009). Except methane from enteric fermentation, gaseous emissions include ammonia (NH_3), volatile organic compounds (VOCs), carbon dioxide (CO_2), methane (CH_4) and nitrous oxide (N_2O) from open lots mainly come from the manure and floor surface after nutrients penetrate into the ground.

Air temperature, wind turbulence, floor type, manure characters such as pH, moisture, degradable volatile solid (VS) and manure management were found to affect the gaseous emissions from manure or floor surfaces (Gonzalez-Avalos et al., 2001; Saha et al., 2010; Pereira et al., 2011, 2012; Owen and Silver, 2014). In the outdoor open lots, these factors are quite different from those indoor. For instance, the most commonly studied air emissions under indoor conditions were emissions from well mixed slurry or liquid manure, which has higher water content than that in open lots. The indoor air velocity normally range 0.1–0.5 m s^{-1} , which are much lower than that of outdoor open lots (Arogo et al., 1999; Ye et al., 2008; Saha et al., 2010).

It is costly, labor intensive, time consuming and technically complicated to conduct a direct field measurement. Data collected from field measurements are affected by multi factors. It's also difficult to differentiate the effect of each single factor. For these reasons, a lab scale experiment is needed to simplify the impact factors for estimating and characterizing gaseous emission from the open lots.

The objective of this research was to experimentally evaluate the impacts of air temperature, velocity and floor types on CO_2, CH_4, N_2O and NH_3 emissions from ground level dairy open lots using a scale model of open lot. Data obtained by this experiment will be used for modeling ground level ammonia and methane emission from dairy open lots.

2. Materials and Methods

2.1. Experiment setup

Experiments were conducted in the Air Physics Lab, Research Center Foulum of Aarhus University (AU, Foulum) in Denmark. Experimental setup consists of an air pre-heating system, a wind tunnel, a suction fan, pipe connections and a scale model of dairy open lot (Figure 1). Air pre-heating system, which mainly composed of electrical heating rods and a mixing fan, was used to warm up the incoming air to the expected air temperatures. The power of electrical heating rods was adjustable from 0 to 10 kW.

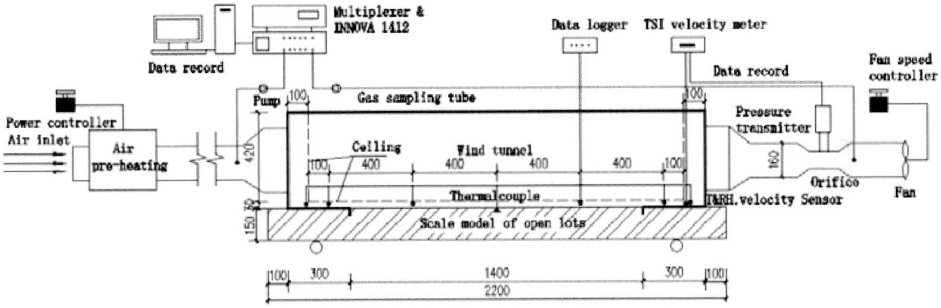

Figure 1. Schematic diagram of experiment setup (all dimensions are in mm).

The wind tunnel was 2 m long, 0.5 m wide and 0.45 m high, with a working area of 0.7 m^2 (1.4×0.5 m, Figure 1). To get the expect surface velocities at working area and a better concentration difference between inlet and outlet, a ceiling was built inside the wind tunnel with a height of 0.03 m (Figure 1). Airflow through the wind tunnel was supplied by a suction fan (RS 160, Ruck Ventilatoren, Denmark) whose airflow rate could be adjusted by a frequency controller.

A scale model of dairy open lot with unpaved soil floor and brick-paved floor was built for this experiment. The unpaved soil open lot scale model was a wooden box filled with soil. It was 2.2 m long, 0.72 m wide and 0.15 m deep. Soil in wooden box was compacted and its bulk density was 2.1 g cm^{-3}. The depth of compact soil in wooden box was about 0.12 m. For the brick-paved open lot scale model, commercial red bricks for building were put on the top of soil filled box.

2.2. Urine and feces

Feces and urine were collected separately from 28 subgroup lactating cows with normal ration (68.39% silage, 18.86% rolled barley, 2.57% rapeseed, 8.57% soybean meal, 0.93% minerals, 0.47% urea and 0.21% supplement) in Danish Livestock Research Center (AU Foulum) every week. Newly collected urine and feces were kept at 4°C and one small dose of feces for each experiment was conditioned at room temperature (15°C) for 8 hours before used for the experiment. Characteristics of urine and feces used for experiment are described in Table 1.

Table 1. Characteristics of feces and urine collected from lactating cow (N=6).

	pH	DM,%	VS,%DM	Total N, g kg^{-1}	TAN, g kg^{-1}
Urine	6.23±0.13	13.7±0.9	87.3±1.2	5.02±0.45	1.11±0.17
Feces	8.38±0.11	6.5±0.3	55.5±2.8	11.16±0.71	1.00±0.21

2.3. Treatment

Before each experiment, floor surfaces of the open lot scale models were fouled with slurry for two months to promote the development of urease or microbes activities. Old Slurry on the floor surface was replaced once a week with new slurry on the surface. For each experiment, 15 kg feces and 1.5 L urine (10:1) were put on the floor surface to form an emitting layer about 3 cm in thickness for 24-h measurement.

The examined air temperature, velocities, floor types and experimental plan were summarized in Table 2. Each run was 24-h measurement which conducted at one temperature and two different surface velocities. The first 12 hours was to get a relatively stable gaseous emission and last 12 hours was used for gas emission tests at two varied surface velocities (e.g., 6 hours ran at 15°C, 0.4 m s^{-1} and 6 hours ran at 15°C, 0.7 m s^{-1}). Each run had three replicates.

Table 2. Examined floor types, air temperature (T_{air}) and velocities in this study.

Floor Type	T_{air}, °C	Velocity			
		0.4 m s^{-1}	0.7 m s^{-1}	1.0 m s^{-1}	1.2 m s^{-1}
Soil Floor	15	Run 1	Run 1	Run 2	Run2
	25	Run 3	Run 3	Run 4	Run 4
	35	Run 5	Run 5	Run 6	Run 6
Brick Floor	15	Run 7	Run 7	Run 8	Run 8
	25	Run 9	Run 9	Run 10	Run 10
	35	Run 11	Run 11	Run 12	Run 12

2.4. Measurement

2.4.1. Airflow rate and surface velocity

Airflow rate through the wind tunnel was set according to the expected surface velocities of 0.4, 0.7, 1.0 and 1.2 m s^{-1}, which were 6.5, 11, 16, 20 l s^{-1}, respectively. Airflow rate was measured by an orifice plates (DIRU 160, Lindab, Denmark). The pressures between upstream and downstream side of orifice were measured by a TSI velocity meter which integrated with a differential pressure transmitter, a hot wire anemometer and an air temperature and relative humidity sensor (Model 9565, TSI, USA). Measurement range and accuracy of the differential pressure was 0–300 Pa and ±0.7%, respectively. Airflow rate through the wind tunnel was calculated by:

$$q = 3.5 \cdot \sqrt{\Delta p} \tag{1}$$

where q is the airflow rate, l s^{-1}; Δp is the pressure difference between upstream and downstream side of orifice, Pa.

Surface air velocities during experiments were measured with the same TSI velocity meter (Model 9565, TSI, USA) right after the ceiling close to wind tunnel outlet (Figure 1) at 0.5 cm high. Surface velocity and differential pressure were logged every 1 minute.

2.4.2. Gas concentration and emission

Ammonia, water vapor, CO_2, CH_4 and N_2O concentrations at air inlet and outlet (Figure 1) was measured continuously by a photoacoustic multi-gas monitor (INNOVA 1412i, LumaSense Technologies A/S, Denmark) and multiplexer (Type 1309, LumaSense Technologies A/S, Denmark) every 5 minutes with 5 measurements. Sampling duration for each gas measurement was 40 s, followed by 20 s flushing. The last 3 measurements were used to calculate gas concentration difference between inlet and outlet. Gaseous emission rates were estimated by the airflow rate and the concentration difference between inlet and outlet.

2.4.3. Temperature and humidity

Air temperature and humidity in wind tunnel was measured at the same position for measuring surface velocity with the same integrated instrument every 1 minute (Model 9565,

TSI, USA). Air temperature at working area was measured by thermocouples at five different positions (Figure 1) at 3 cm high and logged every 5 minutes. Manure temperature was measured by embedding a thermocouple inside the manure and logged every 5 minutes.

2.5. Statistical analyses

The SPSS (version 20.0) was used to do statistical analyses. Repeated measures ANOVA and pairwise comparison had been done to analyze the impacts of air temperature, velocity and floor type on gaseous emission. Statistical significance was based on $P < 0.05$. Results were expressed as "mean±SD".

3. Results and Discussion

3.1. System temperatures

Expected air temperatures (15, 25, and 35°C) were set at the inlet of wind tunnel. Temperature differences between inlet and outlet of wind tunnel due to water evaporative of manure in working area were measured. Therefore, average air temperatures in wind tunnel were slightly lower than the expected. Table 3 shows the manure temperature and average air temperature in the wind tunnel and outlet air humidity. Manure temperature was much lower than air temperature. The air temperature rise was about 8°C in each temperature change level during experiment and led to about 6°C rise of manure temperature.

Table 3. Manure temperature (T_{man}), average air temperature (T_{ave}) in wind tunnel and outlet air humidity (RH) at expect air temperature (T_{exp}) in experiment on soil floor and brick floor.

T_{exp}, °C	$T_{ave\ soil}$, °C	$T_{ave\ brick}$, °C	$T_{man\ soil}$, °C	$T_{man\ brick}$, °C	RH_{soil}, %	RH_{brick}, %
15	14.5±0.8	16.3±0.5	11.1±1.0	13.2±0.8	51.0±5.8	52.8±7.2
25	23.2±1.7	24.5±1.5	17.2±1.2	18.4±1.1	33.1±4.5	33.4±7.4
35	31.2±2.5	31.6±3.0	24.3±1.2	25.1±1.7	27.3±3.5	28.0±4.8

3.2. Carbon emissions

3.2.1. CO_2 emissions

Average CO_2 emissions from the dairy open lot scale model ranged from 34.4 to 160.1 mg kg^{-1} h^{-1} for unpaved soil floor and from 44.9 to 202.6 mg kg^{-1} h^{-1} for brick paved floor at different air temperatures and surface velocities (Figure 2a, 2b). The highest emission was measured at 35°C, 1.2 m s^{-1} and the lowest emission was measured 15°C, 0.4 m s^{-1} at both soil and brick floor.

The increases of air temperature and surface velocity led to an increase of CO_2 emissions ($P < 0.05$), and there were similar increasing patterns on both soil and brick floor.

Floor types showed different impacts on CO_2 emissions at different temperatures. Emissions from soil and brick floor were close to each other at 15°C and 25°C ($P > 0.05$), while higher CO_2 was emitted from brick floor at 35°C ($P < 0.05$).

3.2.2. CH_4 emissions

Average methane emissions from the dairy open lot scale model ranged from 0.3 to 3.2 mg kg^{-1} h^{-1} for unpaved soil floor and from 0.4 to 2.1 mg kg^{-1} h^{-1} for brick paved floor at different air temperatures and surface velocities (Figure 2c, 2d). The lowest emissions were measured at 15°C, 0.4 m s^{-1} for both soil and brick floor while the highest emissions were measured 35°C, 1.2 m s^{-1} for soil floor and 35°C, 1.0 m s^{-1} for brick floor.

Air temperature and surface velocity have significant impacts on CH_4 emissions from the dairy open lot scale model ($P < 0.05$), and there are interactions between air temperature and surface velocity ($P < 0.05$). Air temperature and velocity showed different effects on methane emission from different floor types.

Methane emissions increased with air temperature increase from brick floor and this was in agreement with other researches from concrete floor (Pereira et al., 2011, 2012). With the increased velocity, CH_4 emissions from brick increased at the beginning and there was a drop if the velocity kept increasing. The drop would be delayed at higher air temperature (Figure 3d). Surface crust observed from 1.0 m s^{-1} at 25°C and from 0.4 m s^{-1} at 35°C. The crust would suppress methane emission (Husted, 1994; Petersen et al., 2005) and might be responsible for the drop of methane emissions at 25°C.

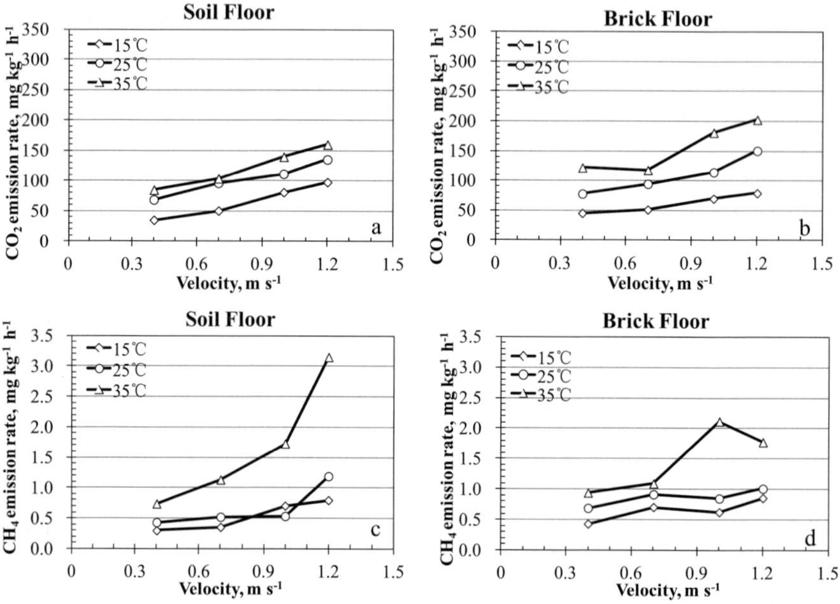

Figure 2. Carbon emissions at different temperatures (15–35°C), velocities (0.3–1.2 m s^{-1}) and floor types (N=3).

For soil floor, methane emissions increased with air temperature increases from 15°C to 35°C and air velocity increases from 0.4 to 1.2 m^{-1}. No similar drop was found as emissions from brick floor. CH_4 emissions only increased by about 29% from 15°C to 25°C but double or tripled from 25°C to 35°C at the same surface air velocity. Emissions were close to each other when velocity lower than 1.0 m s^{-1} at 25°C and this caused the lower emission at 25°C, 1.0 m s^{-1} than that at 15°C, 1.0 m s^{-1}.

It's difficult to explain the different impact of air temperature and velocity on methane emissions from soil and brick floor because soil could be both a source and a sink of methane emissions (Mer and Roger, 2001; Topp and Pattey, 1997). Generally, methane uptake negatively associated with soil moisture and positively with soil temperature (Luo et al. 2013). If the soil is flooded, for instance rice field or swamp, soil would be a major methane emission source (Mer and Roger, 2001). CH_4 emission from soil floor is more complex and more information was needed, e.g., soil temperature and moisture etc.

3.3. Nitrogen emissions

3.3.1. NH_3 emissions

Average NH_3 emissions from dairy the open lot scale model ranged from 46.9 to 166.0 mg h^{-1} for unpaved soil floor and from 55.2 to 261.7 mg h^{-1} for brick paved floor at different air temperatures and surface velocities (Figure 3a, 3b). Similar to CO_2 emissions, the highest emission was measured at 35°C, 1.2 m s^{-1} and the lowest emission was measured 15°C, 0.4 m s^{-1}

at both soil and brick floor. Air temperature and velocity had significant impacts on NH_3 emission ($P < 0.05$), and emissions from brick floor were significantly higher than that from soil floor at the same air temperature and surface velocity ($P < 0.05$).

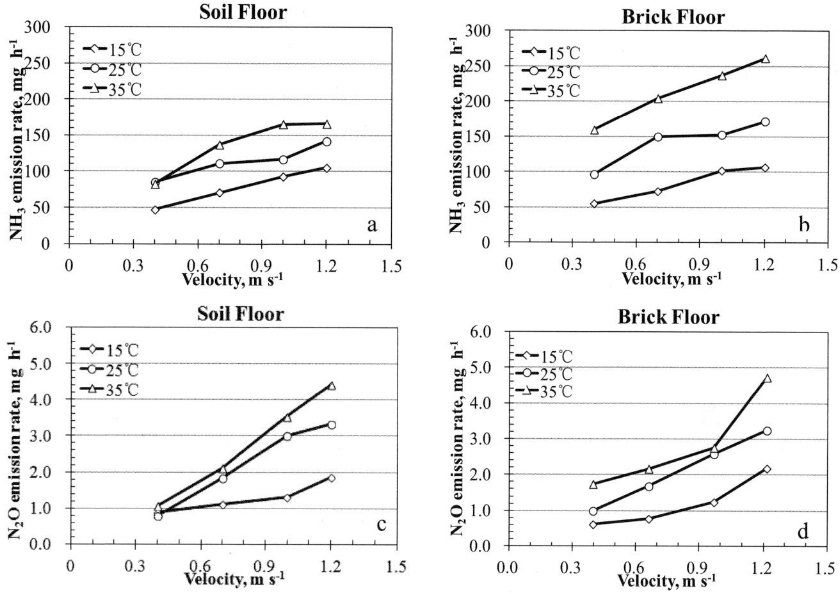

Figure 3. Nitrogen emissions at different temperatures (15–35°C), velocities (0.3 – 1.2 m s^{-1}) and floor types (N=3).

Arogo et al. (1999) examined the overall ammonia mass transfer coefficient from liquid swine manure and aqueous solution at different air temperature, velocity and liquid temperature. They found that ammonia mass transfer coefficient from liquid manure was positively correlated to air velocity but negatively correlated to air temperature. Similar results reported in an earlier research conducted by Halsam et al. (1924). Different with their results, measured NH_3 emissions from dairy open lot scale model increased both with the air temperature and velocity. But this agrees with normal transport phenomena which mass transfer is positively related to air temperature (Ni, 1999).

The NH_3 emissions increased by 38% and 49%, 20% and 19%, 9% and 12% when surface velocity increased from 0.4 to 0.7 m s^{-1}, 0.7 to 1.0 m s^{-1}, 1.0 to 1.2 m s^{-1}, respectively. Impacts of air velocity on NH_3 emission weakened with the increasing of surface air velocity. This agrees with Arogo's (1999) results.

3.3.2. N_2O emissions

Average N_2O emissions from dairy open lot scale model ranged from 0.8 to 4.4 mg h^{-1} for unpaved soil floor and from 0.6 to 4.7 mg h^{-1} for brick paved floor at different air temperatures and surface velocities (Figure 3c, 3d). The lowest emission from soil floor was measured at 0.4 m s^{-1} due to close emissions at different air temperatures while it was measured at 15°C and 0.4 m s^{-1} from brick floor. The highest N_2O emission was measured at 35°C, 1.2 m s^{-1} both for soil and brick floor.

For both soil and brick floor, air temperature and velocity had significant impacts on N_2O emission ($P < 0.05$), and there was an interaction between temperature and velocity ($P < 0.05$). N_2O emissions increased with air temperature and velocity. No significant difference was found on N_2O emissions between soil and brick floor ($P > 0.05$) at 25°C and 35°C.

3.4. Sensitivity

Total carbon emissions were calculated based on the emissions of CO_2 and CH_4 as well as their carbon molecular weight (CO_2-C + CH_4-C) at different air temperature, velocity and floor types. Similarly, total *N* emissions (NH_3-N + N_2O-N) were also calculated.

When air temperature and surface velocity changed from 15°C, 0.4 m s^{-1} to 35°C, 1.2 m s^{-1}, about 3.8 times more carbon and 2.6 times more nitrogen emitted from soil floor. Similarly, 3.5 times more carbon and 3.8 times more nitrogen emitted from brick floor at 35°C, 1.2 m s^{-1}. Generally, soil floor was more sensible to velocity changes while brick floor was more sensible to air temperature changes both on carbon and nitrogen emissions. Figure 4 shows the sensitivities of nitrogen emissions from soil and brick floor to air velocity and temperature changes.

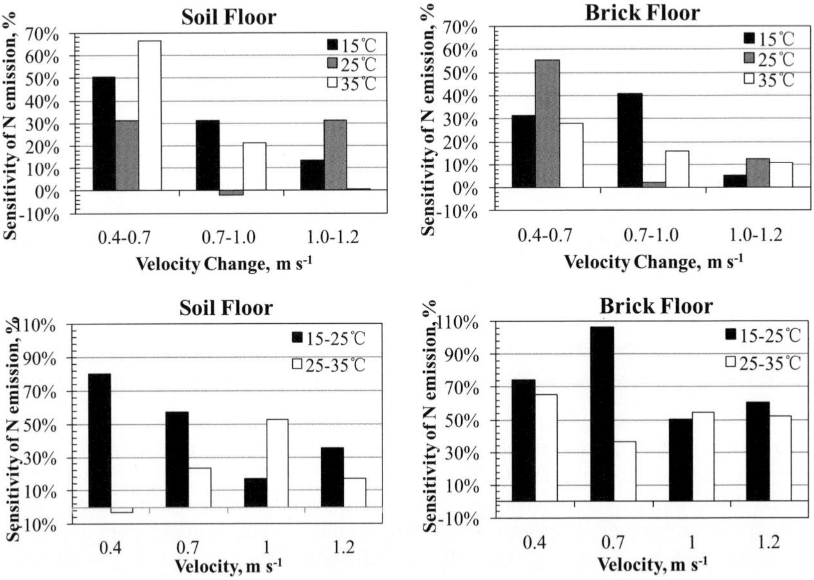

Figure 4. Sensitivities of nitrogen emissions from soil and brick floor to air velocity and temperature changes

Total nitrogen emissions from both soil floor and brick floor were more sensible to changes at lower velocity (0.4 to 0.7 m s^{-1}) or lower air temperature (15 to 25°C). This agrees with the sensitivity analysis of ammonia mass transfer from liquid manure to air velocity and temperature by Arogo et al. (1999).

Different from nitrogen emissions, total carbon emissions were more sensible to velocity changes from 0.7 to 1.0 m s^{-1}at 25°C and 35°C. Methane emissions were more sensible to air temperature changes from 25°C to 35°C. But total carbon emissions were still more sensible to air temperature changes in lower temperature (15°C to 25°C) due to higher CO_2 emissions than methane.

4. Conclusions

Ammonia, CO_2, CH_4 and N_2O emissions from a dairy open lot scale model were measured at varied air temperatures (15°C to 35°C), air velocities (0.4 to 1.2 m s^{-1}) and floor types (unpaved soil floor and brick paved floor). Higher NH_3, CO_2 and N_2O emissions from both soil floor and brick floor were observed with the increased air temperature or velocity. Methane emissions also increased with air temperature. However, air velocity showed different impacts on methane emissions from different floor types. There was a drop on CH_4 emissions from brick floor when air velocity increased while no similar drop was found from soil floor at the examined air velocities. Except methane, effects of air temperature and velocity changes weakened with the increasing temperature and velocity. Soil floor was more sensible to velocity changes while brick floor was more sensible to air temperature changes in terms of both nitrogen and carbon emissions. Brick floor had higher NH_3 emission than soil floor.

Acknowledgements

This study was funded by the National Nature Science Foundation of China (Grant No. 31172244), China Agricultural Research System (CARS-37), and CSC Scholarship (201406350160). Thanks for all the helps from air physics lab and biogas lab in Aarhus University, Foulum, Denmark.

References

Arogo, J., R.H. Zhang, G.L. Riskowski, L.L. Chiristianson, D.L. Day, 1999. Mass transfer coefficient of ammonia in liquid swine manure and aqueous solutions. Journal of Agricultural Engineering Research. 73(1), 77–86.

Bjorneberg, D.L., A.B. Leytem, D.T. Westermann, P.R. Griffiths, L. Shao, M.J. Pollard, 2009. Measurement of atmospheric ammonia, methane, and nitrous oxide at a concentrated dairy production facility in southern Idaho using open-path FTIR spectrometry. Transactions of the ASABE. 52(5), 1749–1756.

Ding, L., Q. Lu, C. Wang, Z. Shi, W. Cao, B. Li, 2015. Effect of configuration and headspace mixing on the accuracy of closed chambers for dairy farm gas emission measurement. Applied Engineering in Agriculture. 31(1), 153–162.

Gonzalez-Avalos, E., L.G. Ruiz-Suarez, 2001. Methane emission factors from cattle manure in Mexico. Bioresource Technology. 80(1), 63–71.

Husted, S., 1994. Seasonal variation in methane emission from stored slurry and solid manures. Journal of Environmental Quality. 23(3), 585–592.

Luo, G.J., R. Kiese, B. Wolf, K. Butterbach-Bahl, 2013. Effects of soil temperature and moisture on methane uptake and nitrous oxide emissions across three different ecosystem types. Biogeosciences. 10(5), 3205–3219.

Mer, J.L., P. Roger, 2001. Production, oxidation, emission and consumption of methane by soils: a review. European Journal of Soil Biology. 37(1), 25–50.

Ni, J., 1999. Mechanistic models of ammonia release from liquid manure: a review. Journal of Agricultural Engineering Research. 72(1), 1–17.

Owen, J.J., W.L. Silver, 2012. Greenhouse gas emissions from dairy manure management: a review of field - based studies. Global change biology. 21(2), 550–565.

Pereira, J., T.H. Misselbrook, D.R. Chadwick, J. Coutinho, H. Trindade, 2012. Effects of temperature and dairy cattle excreta characteristics on potential ammonia and greenhouse gas emissions from housing: A laboratory study. Biosystems Engineering. 112(2), 138–150.

Pereira, J., D. Fangueiro, T.H. Misselbrook, J. Coutinho, H. Trindade, 2011. Ammonia and greenhouse gas emissions from slatted and solid floors in dairy cattle houses: A scale model study. Biosystems Engineering. 109(2), 148–157.

Petersen, S.O., S.G. Sommer, 2011. Ammonia and nitrous oxide interactions: Roles of manure organic matter management. Animal Feed Science and Technology. 166, 503–513.

Saha, C.K., G. Zhang, J. Ni, 2010. Airflow and concentration characterisation and ammonia mass transfer modelling in wind tunnel studies. Biosystems engineering. 107(4), 328–340.

Topp, E., E. Pattey, 1997. Soils as sources and sinks for atmospheric methane. Canadian Journal of Soil Science. 77(2), 167–177.

Ye, Z., G. Zhang, B. Li, J.S. Srøm, G. Tong, P.J. Dahl, 2008. Influence of airflow and liquid properties on the mass transfer coefficient of ammonia in aqueous solutions. Biosystems Engineering. 100(3), 422–434.

An Innovative Ventilation Monitoring System at a Pig Experimental Building

Ji-Qin Ni [a,*], Claude A. Diehl [a], Shule Liu [a], Igor M. Lopes [a], John S. Radcliffe [b], Brian Richert [b]

[a] Dept. of Agricultural and Biological Engineering, Purdue University, West Lafayette, IN 47907, USA

[b] Dept. of Animal Sciences, Purdue University, West Lafayette, IN 47907, USA

* Corresponding author. Email: jiqin@purdue.edu

Abstract

Ventilation rate is one of the key variables to quantify aerial emissions from animal feeding operations but is technically challenging to measure. This research aimed at developing a new system to reliably and continuously monitor ventilation rate at a state-of-the-science pig environmental building, which consists of 12 rooms of 11.0 × 6.1 × 2.7 m (L × W × H) each and with a 1.8-m deep manure pit under slatted floor. Each room has a capacity of 60 finishing pigs and is equipped with four ventilation fans. The two pit fans are variable-speed 25-cm diameter tube-mounting fans. The two wall-fans, one of 36-cm diameter and another of 51-cm diameter, are single-speed fans at the room end wall. A Pit Exhaust Airflow Measurement Assembly (PEAMA) using an impeller anemometer to measure ventilation rate was developed and installed in each pit fan. The PEAMA was calibrated in a laboratory. Twenty-four PEAMAs provide continuous and real-time ventilation outputs in m^3 min^{-1} to an on-site computer system (OSCS) in the building. Wall fan operations are monitored using unregulated AC to DC power supply adapters, which provide fan on-off signals as well as fan power voltages to the OSCS. Room static pressures are measured with pressure transducers. A Wall Fan Tester was developed to test the wall fans and obtain fan airflow-pressure models. Wall fan ventilation rates are indirectly obtained during post-measurement data processing using the monitoring data and fan models. The performance results demonstrated that this innovative system improves reliability and accuracies of aerial emission monitoring.

Keywords: Airflow measurement, air quality, animal building, indoor climate, room ventilation

1. Introduction

Many studies on animal environment and welfare require appropriate facilities to conduct experiments and reliable technologies to perform measurements. Ventilation, or air exchange, is essential to control the indoor animal building environment. Therefore, monitoring building ventilation rate is essential for comprehensive animal environment and welfare studies.

Ventilation also controls air pollutant emissions. Ventilation rate is one of the two key variables to quantify aerial emissions from animal feeding operations because emission rate is a product of the ventilation rate and the aerial pollutant concentration difference between inlet and outlet where air exchange takes place [Eqn. (1)]. However, ventilation rate measurement typically introduces the greatest uncertainties in air pollutant emission estimations (Wathes et al., 1998).

$$E = Q \cdot (C_2 - C_1) \tag{1}$$

where E is aerial pollutant emission rate, mass time^{-1}; Q is air exchange rate, volume time^{-1}; C_1 and C_2 are aerial pollutant concentrations at the inlet and outlet of the air exchange space, respectively, mass volumn^{-1}.

Animal buildings can be designed to have mechanical ventilation, natural ventilation, or mixed ventilation systems. Technical difficulties in ventilation rate measurement at animal buildings for environmental studies have been noticed since the 1960s. Generally, monitoring buildings with natural ventilation is more difficult than those with mechanical ventilation. However, even for mechanically-ventilated buildings, it is still technically challenging.

Ventilation rate measurement can be direct, in which the output of measurement device is in volume time^{-1}, or indirect, in which ventilation rate requires post-measurement data processing and calculation based on other measurement variables, e.g., building static pressures, ventilation fan characteristic models, tracer gas concentrations. Direct and continuous measurement using ventilation sensors (Berckmans et al., 1991) has been realized for many research projects in Europe. However, these sensors are limited to fan diameters < 90 cm and only suitable for chimney/duct ventilation buildings, which are not popular in the U.S. animal industry.

For larger fans with diameters from 122 to 137 cm in the U.S., a portable device was developed in the late 1990s and refined to perform real-time traverse of the airflow entering fans. The device used a horizontal array of anemometers, which could move up-and-down to cover the entire fan opening, It has been tested and applied in many large field research projects (e.g., Burns et al., 2007; Lim et al., 2010). For small size pit fans of 61 cm diameter, Lim et al. (2010) designed and evaluated a reduced device using the same principle. These two devices were used for sporadic fan tests. Post-test data processing was needed to obtain fan airflow rates.

The most frequently used indirect ventilation determination techniques employed fan on-off monitoring for mechanically-ventilated animal buildings. Examples included sail switches and fan rotational methods (Hoff et al., 2009), fan vibration sensors (Chen et al., 2010), and low voltage DC electrical circuit (Calvet et al., 2010). This method usually also depends on the known fan characteristics, which include fan airflow rates under different pressures, power supplies, etc. Mass or heat balance techniques for both mechanically- and naturally-ventilated buildings were also applied (e.g., Xin et al., 2009). Indirect methods are used in research projects when there are technical or budgetary reasons that prevent from the use of direct measurement methods.

A state-of-the-science Swine Environmental Research Building (SERB) for environmental research was built at Purdue University, Indiana, USA in 2004. Multiple research projects have been conducted in the building with comprehensive environmental monitoring on pig production and the related building environment and aerial emissions. The ventilation monitoring methodology in the building has been under development since 2004. An innovative ventilation measurement system was designed and installed in 2011 to improve the research quality. The objectives of this paper were to (1) present the design of the ventilation monitoring system and the methodology of its operation, and (2) evaluate the system performance.

2. Materials and Methods

2.1. Experimental building

2.1.1. Building design

The SERB is located at the Animal Research and Education Center, Purdue University, West Lafayette, Indiana, USA. It has 12 pig rooms on north and south sides of the building (Figure 1). Each room measures 11.0 × 6.1 × 2.7 m (L × W × H) with a capacity of housing 60 finishing pigs. There are two rows of 6 pens each row and an isolated 1.8-m deep manure pit under slatted concrete floor in each room.

The indoor temperatures and ventilations of each room are independently controlled with a Fancom controller (Model FCTC, Fancom BV, Panningen, The Netherlands). Fresh air enters into the building via two air inlets located above the east and west doors. The air inlets are covered with curtains to adjust the openings depending on the seasons and ventilation needs. The east side curtain is controlled automatically with a static pressure controller (Model SP-2, Airstream Ventilation Systems, Assumption, IL, USA) and the west side curtain is controlled manually. Fresh air is supplied to each room from the baffled ceiling and the alley way. There are three baffled V-shaped ceiling air inlets installed along the length of each room. Air from alley way is evenly distributed to each pen via a perforated 38-cm diameter PVC inlet air pipe along the room about 20 cm below the ceiling (Figure 1).

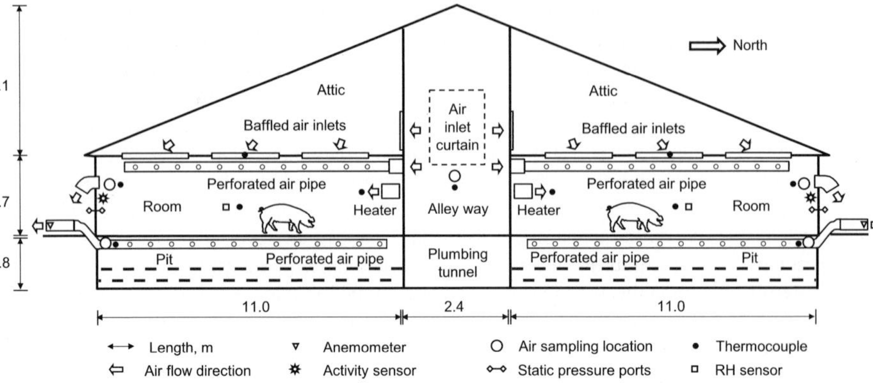

Figure 1. Cross-sectional east side view of the swine environmental research building.

Two propane heaters close to the building east and west air inlets pre-heat the air during cold weather. Each room is also equipped with a propane heater (Model Guardian 60, L.B. White Co. Onalaska, WI, USA) to heat the room air for piglets when the temperature is too low.

2.1.2. Building ventilation system

There are four ventilation fans in each room: two pit fans and two wall fans. The two pit fans are variable-speed 25-cm diameter tube-mounting fans (Model P4E30, Multifan, Bloomington, IL, USA) installed outdoors. Each pit fan inlet is connected to a 25.4-cm diameter perforated PVC exhaust air pipe in the pit. The two wall fans, one of 36 cm diameter (Model V4E35, Multifan) and another of 51 cm diameter (Model V4E50, Multifan), are single-speed fans installed at the north or south end wall, depending on the location of the room. The capacities of the 25-cm, 36-cm, and 51-cm diameter fans at 50 Pa pressure are 28.3, 51.0, and 120.3 m^3 min^{-1}, respectively, according to the manufacturer's specifications.

2.1.3. Building environmental monitoring and data acquisition

The building is equipped with analyzers and sensors for measurement of concentrations of gases (including ammonia, hydrogen sulfide, and greenhouse gases), temperatures, relative humidity, animal activities, room static pressures, and fan operations. An on-site computer system (OSCS), which consists of a personal computer, custom software AirDAC, and data acquisition and control hardware, is installed in the laboratory in the building. This system acquires data from all on-line gas analyzers and other sensors at 1 Hz, converts the signals to engineering units, averages them over 1 min intervals, and records the means into computer files. It also automatically processes daily measurement data on the following day and delivers the results to researchers via email (Ni and Heber, 2010).

2.2. Direct pit fan ventilation measurement

2.2.1. Pit Exhaust Airflow Measurement Assembly

Ventilation measurements for pit fans are realized by using an innovative Pit Exhaust Airflow Measurement Assembly (PEAMA) combined with laboratory PEAMA and anemometer calibrations. The PEAMA was designed to continuously measure the airflow rates of pit fans. One PEAMA was installed at each pit fan.

The PEAMA is composed of a 121.9 cm long and 30.5 cm inside diameter (ID) PVC pipe, a 61-cm long flow straightener made of aluminum sheet, and an 18-cm diameter impeller anemometer (Model 27106RS, RM Young, Traverse City, MI, USA). One end of the PVC pipe is connected airtight to the pit fan exhaust so that all air from the fan can flow through the pipe. Inside the other end of the pipe the flow straightener and the anemometer are installed.

2.2.2. Laboratory calibration of the measurement assembly

A laboratory calibration test rig with a duplicated pit fan system was setup to investigate the PEAMA characteristics. The test rig is composed of a pit fan (Model P4E30, Multifan), a 69.9-cm long and 25.4-cm ID tilted fan inlet pipe, a 88.3-cm long and 25.4-cm ID partial horizontal inlet pipe, an adjustable power supply to the fan, and multiple sensors. The end of the inlet pipe is reduced to 19.5-cm ID to house the impeller of an anemometer. During the laboratory study, the PEAMA was installed at the fan exhaust the same way as it is at the pig building. The adjustable fan power supply voltages, fan rotational speeds, air temperatures and relative humidity, fan inlet static pressures, cross-sectional air speeds at the inlet, and PEAMA anemometer voltage outputs in the test rig were continuously recorded. Airflow speed across the19.5-cm ID inlet was measured with a Model 27106RS anemometer and converted to airflow rate. Travers method with a hotwire anemometer was also used to verify the airflow. The inlet airflow rates were compared with the PEAMA voltage outputs under different test conditions. A linear equation was obtained from the calibration to describe the relationship between the pit fan airflow rate and the anemometer rotational speed [Eq. (2)].

$$Q_{PF} = 0.018035 \cdot A \cdot V \tag{2}$$

where Q_{PF} is airflow rate of the pit fan, m^3 min^{-1}; A is coefficient obtained from device calibration, m^3 min^{-1} VDC^{-1}; V is output of anemometer, VDC.

2.2.3. Real-time pit fan ventilation rate monitoring

To realize the direct pit fan ventilation measurement, the combined coefficients $0.018035 \cdot A$ in Eq. (2) for each PEAMA are entered into the table of All-data Display and Dynamic Run-time Configuration in AirDAC (Ni and Heber, 2010). A build-in special function in AirDAC performs calculation using the variable V and Eq. (2) to display real-time pit fan ventilation rates at 1 Hz in the PC. The 1 Hz data are averaged each min and saved in a data file.

2.3. Indirect wall fan ventilation measurement

2.3.1. Wall fan operation monitoring

Continuous ventilation measurement for the 24wall fans is realized by monitoring the fan operational on/off status and room static pressures. Each fan is individually monitored by recording the on/off and voltages of the fan power supply using a 240-VAC to 6-VDC unregulated power supply adapter that is connected in parallel to the fan. All adapter VDC outputs are connected to a digital input data acquisition module (USB DIO 96 H, Measurement Computing Co., Norton, MA, USA) to record the on/off status of the fans. The 1-Hz digital signals are converted into percentage of time that each fan operated, e.g., if the fan operated for 30 sec during 1 min, a datum of 50% was recorded into 1 min data file as the fan operational time. The 24 adapter voltage outputs are also connected to two analog input modules (FP-AI-112, National Instrument Co., Austin, TX, USA) to monitor the fan power voltage fluctuation.

2.3.2. Wall fan ventilation rate determination

An improved portable Wall Fan Tester (WFT) was designed based on the tester developed by Lim et al. (2010). The fan tester has an aluminum frame with an opening area of 66 cm x 66 cm (4356 cm^2). Nine, instead of three as used by Lim et al. (2010), anemometers (Model 27106RS, RM Young) using impellers of larger diameter (20-cm) were installed in the frame. The anemometers were spaced in equal distance; each represents a 484-cm^2 opening sub-area. The mean wind speed in each sub-area is represented by that of the 314 cm^2 area covered by the anemometer impeller with a representative ratio of 65%.

According to the manufacturer, the wind speed and its induced anemometer rotational speed has the relationship shown in Eq (3), in which the AR is also proportional to the anemometer output shown in Eq. (4).

$$S = 0.300 \cdot AR \tag{3}$$

$$AR = A \cdot V \tag{4}$$

where S is wind speed, m min^{-1}; AR is anemometer rotational speed, RPM.

The airflow rate in each sub-area is obtained by multiplying the mean wind speed that the anemometer measures with the 484-cm^2 sub-area. The airflow rate through the entire tester is the sum of the airflow rates in the nine sub-areas [Eq. (5)].

$$Q_{FT} = \sum_{i=1}^{9} 0.3 \cdot A_i \cdot V_i \cdot 0.0484 \tag{5}$$

where Q_{FT} is airflow rate through the fan tester, m^3 min^{-1}; A_i and V_i are coefficient and voltage output of the *i-th* anemometer, respectively.

The WFT was periodically set up at the inlet of each wall fan for ventilation rate measurement at five different differential pressures, ranging from -5 to -110 Pa. A linear model of ventilation rates vs static pressures was developed for each fan. The models were used to calculate fan ventilation rates in post-measurement data processing. They were also used to analyze characteristics of the fans.

2.3.3. Wall fan ventilation rate calculation

Ventilation rates through the two wall fans in the room are calculated using the static pressures across the end wall and the linear models obtained during the wall fan tests with Eq. (6). The static pressures and fan operation times are recorded in the AirDAC data files with 1-min time resolution.

$$Q_{W,R} = \left(a_{W51} \cdot P_W + b_{W51}\right) \cdot t_{51} + \left(a_{W36} \cdot P_W + b_{W36}\right) \cdot t_{36} \tag{6}$$

where $Q_{W,R}$ is wall fan ventilation rate in the room, m^3 min^{-1}; a_{W51} and a_{W36} are slopes of the linear models for the 51-cm and 36-cm diameter wall fans, respectively; b_{W51} and b_{W36} are intercepts of the linear models for the 51-cm and 36-cm diameter wall fans, respectively; P_W is the room static pressure, Pa; t_{51} and t_{36} are operation duration for the 51-cm and 36-cm diameter wall fans recorded in AirDAC, respectively, % min^{-1}.

2.4. Calibration of anemometers and pressure transducers

The coefficient A in Eqs. (2) and (4) for all anemometers in the PEAMAs and WFT were calibrated using a Model 18802 Selectable Speed Anemometer Drive. Calibrations were all conducted with 9 different RPM from 0 to 2400 with 300 RPM increments. Additionally, the anemometer sensitivities were tested using an Anemometer Torque Disc from RM Young.

The Pressure Transducers (Model 260, Setra, Boxborough, MA, USA) for room static pressure measurement were periodically calibrated using a Precision Multifunction Calibrator (Model 726, Fluke Co., Everett, WA, USA) and a Pressure Module with a pressure range of 0 to 2.54 cm water (Model 750-P00, Fluke Co.). Calibrations were conducted at seven different pressures within a range from 0 to -150 Pa.

3. Results and Discussion

3.1. Pit fan ventilation rate

3.1.1. Characteristics of PEAMA under laboratory study

Laboratory study of the duplicated pit fan system with the PEAMA installed at the fan exhaust showed that the rotational speeds of the fan were linearly correlated to the system ventilation rates (R^2 = 0.999, Figure 2, left). The pressures created by the fan were perfectly correlated to the fan speeds with a power function (R^2 = 1.000). The PEAMA voltage outputs were also highly correlated to the system ventilation rates, demonstrating an excellent performance for pit fan ventilation measurement (R^2 = 0.997, Figure 2, right). However, at higher ventilation rate of >25.6 m^3 min^{-1}, non-linear variations in the PEAMA outputs were observed. This was probably due to the higher turbulences in the exhaust air at high fan speeds.

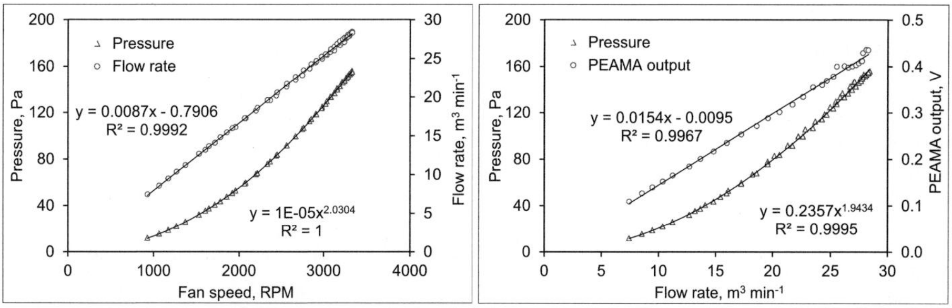

Figure 2. Results of laboratory study. Left: Pressures and airflow rates under different fan rotational speeds. Right: Pressures and PEAMA signal outputs at different airflow rates.

3.1.2. Field performance of PEAMA

The PEAMA have been under field use for about 3 years (Figure 3, left). Maintenances have been performed only on the anemometers, in which the two bearings and the generator were subject to wear and tear. However, it was found that when new bearings and generators were used, the anemometers could reliably and continuously operate for more than a couple of years although the pit exhausts air contained corrosive gases, e.g., hydrogen sulfide.

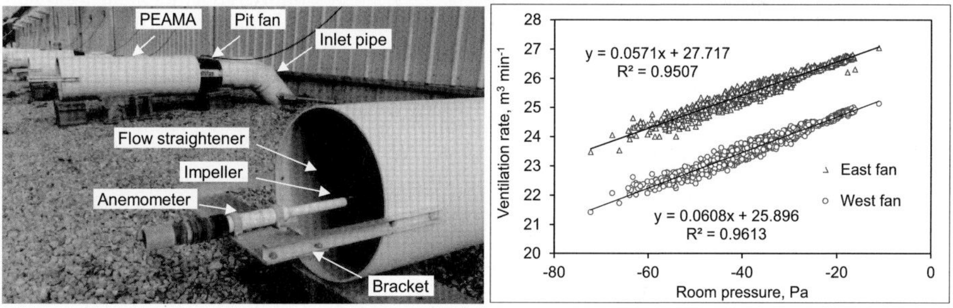

Figure 3. Left: Multiple PEAMA at pit fan exhausts. Right: A typical day of pit ventilations measured in a room under relatively constant fan speeds and different pressures.

The PEAMA was found very sensitive to room static pressures. At relatively constant fan power supplies, the ventilation rates measured with PEAMA demonstrated a good linear relationship with room pressures (Figure 3, right). Figure 3 (right) also shows, as an example that although identical and in the same room, the two pit fans were controlled separately and could provide considerably different ventilations for the room. Therefore, they needed to be monitored individually.

For ventilation fans in many animal buildings, there are often technical difficulties related to the accessibility to perform airflow measurements. Installing a measurement device at outdoor fan exhaust allows sufficient space for the device and its maintenance. The PEAMA provides a solution for convenient and reliable fan airflow monitoring. The development of this method has significantly improved the accuracy of the pit ventilation measurement in the building.

3.2. Wall fan ventilation

3.2.1. Fan status monitoring

The wall fan status monitoring method using unregulated AC-DC power supply adapters has been in use since 2004 in the building. Compared with our previously reported fan status monitoring methods, e.g., fan vibration sensors (Chen et al., 2010) and magnetic proximity sensors (Ni et al., 2012), this method has proved its advantages of low cost, low maintenance, and high reliability. No hardware repair and replacement have been required for all the 24 fans during more than 10 years of operation. Moreover, no errors of fan on/off signals directly from the adapters have been detected.

The recently added fan power voltage monitoring provided additional information about the wall fan ventilation dynamics. Diurnal fluctuations and variations in fan power voltages for different fans (Figure 4, left), which can significantly affect the ventilation rates, were demonstrated and continuously recorded in AirDAC. These data are important and can be used to improve ventilation rate calculations.

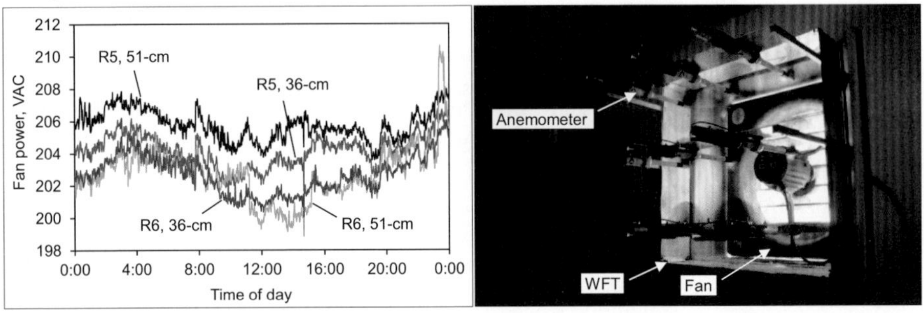

Figure 4. Left: A typical day of wall fan power voltages in Room 5 (R5) and Room 6 (R6). Right: Testing at the inlet of a 51-cm wall fan with the Wall Fan Tester (WFT).

3.2.2. Fan ventilation

Testing with the WFT at fan inlets (Figure 4, right) under different room static pressures demonstrated good linear correlations between ventilation rates and static pressures for all the 24 fans (Table 1). The mean R^2 of the linear models for the 51-cm and 36-cm fans were 0.997 and 0.978, respectively. However, the slopes and intercepts of different fans also revealed that variations among the fans existed. Additionally, due to limitations of fan tests conducted so far, Eq. (5) and the coefficients in Table 1 have not yet included the effect of fan power voltage supplies. This should be improved in the future studies.

Table 1. Coefficients obtained from wall fan tests for fan ventilation calculations.

Room	51-cm fan			36-cm fan		
	Slope (a_{W51})	Intercept (b_{W51})	R^2	Slope (a_{W36})	Intercept (b_{W36F})	R^2
1	0.5106	106.8	0.994	0.4640	59.4	0.960
2	0.5487	128.3	0.998	0.4348	58.9	0.966
3	0.5844	131.4	0.998	0.4086	60.1	0.978
4	0.5195	116.2	0.995	0.4189	61.0	0.983
5	0.6145	117.4	0.996	0.4687	68.9	0.974
6	0.5508	132.3	0.994	0.4086	60.8	0.984
7	0.6147	127.4	0.999	0.4847	65.2	0.992
8	0.5643	133.3	0.996	0.3801	59.3	0.992
9	0.6380	126.7	0.997	0.4459	61.2	0.974
10	0.5517	129.5	0.999	0.4156	59.5	0.952
11	0.6549	131.4	0.998	0.3838	62.4	0.996
12	0.5240	131.3	0.997	0.4275	60.2	0.986
Mean	0.5730	126.0	0.997	0.4284	61.4	0.978
SD	0.0459	7.9	0.002	0.0313	2.8	0.013

The wall fan ventilation rates measured at field conditions were lower than what the fan manufacturer specified. This was normal because as fan age increased, variability in fan speeds of older fans increased due to factors such as variation in fan maintenance and fan and motor condition. Additionally, the elbow-shaped wall fan exhaust cones installed at the building (Figure 1) could also have lowered the efficiencies of fan ventilation rates.

4. Conclusions

Reliability and accuracy of fan ventilation rate measurement can be improved in both direct and indirect methods. The PEAMA for variable-speed pit fan monitoring allowed real-time and direct ventilation monitoring, and convenient installation at fan exhausts. Single-speed wall fan status monitoring using unregulated AC-DC adapters to generate digital input signals was a low-cost and reliable method. Fan power supply voltage measurement generated important information that can improve ventilation monitoring accuracies. The WFT provided an effective method to study wall fan characteristics and assist in post-measurement ventilation calculation.

Acknowledgements

This research was conducted with partial supports from Purdue University, USA; Danisco Animal Nutrition, U.K.; and USDA-University of Arkansas (project UA AES 91088-02). Daniel Kaelin conducted laboratory PEAMA calibration. Caitlin Vonderohe, Ping Zheng, and Qiuju Xie assisted in wall fan testing. Brian Ford managed the swine building daily operations.

References

Berckmans, D., P. Vandenbroeck, V. Goedseels, 1991. Sensor for continuous measurement of the ventilation rate in livestock buildings. Indoor Air. 3 (4), 323–336.

Burns, R., H. Xin, R. Gates, H. Li, S. Hoff, L. Moody, D. Overhults, J. Earnest, 2007. Tyson Broiler Ammonia Emission Monitoring Project: Final Report. Iowa State University and University of Kentucky. 37 p.

Calvet, S., M. Cambra-Lopez, V. Blanes-Vidal, F. Estelles, A.G. Torres, 2010. Ventilation rates in mechanically-ventilated commercial poultry buildings in Southern Europe: Measurement system development and uncertainty analysis. Biosystems Engineering. 106 (4), 423–432.

Chen, Y., J.-Q. Ni, C.A. Diehl, A.J. Heber, B.W. Bogan, L. Chai, 2010. Large scale application of vibration sensors for fan monitoring at commercial layer hen houses. Sensors. 10 (12), 11590–11604. http://dx.doi.org/10.3390/s101211590.

Hoff, S.J., D.S. Bundy, M.A. Nelson, B.C. Zelle, L.D. Jacobson, A.J. Heber, J.Q. Ni, Y.H. Zhang, J.A. Koziel, D.B. Beasley, 2009. Real-time airflow rate measurements from mechanically ventilated animal buildings. Journal of the Air & Waste Management Association. 59 (6), 683–694. http://dx.doi.org/10.3155/1047-3289.59.6.683

Lim, T.-T., J.-Q. Ni, A.J. Heber, Y. Jin, 2010. Applications and calibrations of the fans and traverse methods for barn airflow rate measurement. In: *Int. Sym. on Air Quality and Manure Manag. for Agri.* Dallas, Texas, September 13-16. St. Joseph, Mich.: ASABE.

Ni, J.-Q., A.J. Heber, 2010. An on-site computer system for comprehensive agricultural air quality research. Computers and Electronics in Agriculture. 71 (1), 38–49. http://dx.doi.org/10.1016/j.compag.2009.12.001.

Ni, J.-Q., L. Chai, L. Chen, B.W. Bogan, K. Wang, E.L. Cortus, A.J. Heber, T.-T. Lim, C.A. Diehl, 2012. Characteristics of ammonia, hydrogen sulfide, carbon dioxide, and particulate matter concentrations in high-rise and manure-belt layer hen houses. Atmospheric Environment. 57, 165–174. http://dx.doi.org/10.1016/j.atmosenv.2012.04.023.

Wathes, C.M., V.R. Phillips, M.R. Holden, R.W. Sneath, J.L. Short, R.P. White, J. Hartung, J. Seedorf, M. Schroder, K.H. Linkert, S. Pedersen, H. Takai, J.O. Johnsen, P.W.G.G. Koerkamp, G.H. Uenk, J.H.M. Metz, T. Hinz, V. Caspary, S. Linke, 1998. Emissions of aerial pollutants in livestock buildings in Northern Europe: Overview of a multinational project. Journal of Agricultural Engineering Research. 70 (1), 3–9.

Xin, H., H. Li, R.T. Burns, R.S. Gates, D.G. Overhults, J.W. Earnest, 2009. Use of CO_2 concentration difference or CO_2 balance to assess ventilation rate of broiler houses. Transactions of the ASABE. 52 (4), 1353–1361.

Design and Validation of an Artificial Reference Cow for Breath Measurements of Cows in the Cubicles

Liansun Wu [a,*], Peter W. G. Groot Koerkamp [a,b], Nico Ogink [b]

[a] Farm Technology Group, Wageningen University and Research Center, Wageningen, Gelderland, The Netherlands

[b] Wageningen UR Livestock Research, Wageningen University and Research Center, Wageningen, Gelderland, The Netherlands

* Corresponding author. Email: liansun.wu@wur.nl

Abstract

To mitigate methane production from dairy cows, the primary issue is to evaluate the efficacy of measures to reduce methane production. To evaluate e.g., genetic differences and feeding measures in practice, we need a technique to assess the individual methane production from a large number of cows in the barn. First, we designed and constructed an artificial reference cow (ARC) that simulated exhalation and eructations of cows with known methane production rates. The methane mass balance of the ARC was extensively tested and results showed a strong linear relation between controlled and measured methane mass. Methane concentration release patterns with a sinusoidal curve produced by five simulated cows were compared to patterns measured from real cows. Results showed small differences between simulated and real concentrations with respect to time interval of eructations and lowest and peak concentrations. We concluded that the ARC can be used as a known reference source to develop methane measurement methods. Second, a cubicle hood sampler (CHS) was designed to extract all breath air from a dairy cow that is lying in a cubicle. The CHS's performance was tested by the ARC under laboratory conditions. With the flow rate of 200 $m^3 h^{-1}$, the CHS with fume sample hood, a top curtain, and a cow panel could capture 99.6% emitted methane by the ARC, regardless of nose positions. We concluded that the newly designed CHS was capable of measuring methane production from individual cows in cubicles.

Keywords: Methane, dairy cows, artificial reference cow, cubicle hood sampler

1. Introduction

Efforts to mitigate methane emission from dairy cows are critical to reduce the dairy industry's contribution to the production of greenhouse gases and subsequently to global warming. Dairy cows have been identified as the major producer of methane emission as they account for 15% of the global methane emission budget (Lassey, 2007). On average, a dairy cow produces 250 to 400 gram methane per day (Bannink et al., 2011). Mitigation of methane emission from dairy cows by nutritional and microbial manipulation has already been extensively studied (Boadi et al., 2004). Recently, a wide interest has developed in reducing methane emission by breeding that is inexpensive and provides a long-term effect. Approximately 10-15% variations of methane production were reported between and within cows (Johnson and Johnson, 1995), which provide the potential to select dairy cows with low methane yield. To carry out the genetic selection directly in practice, methane production from individual cows must be accurately and precisely quantified first. Therefore, we need measurement technology capable of evaluating methane production from a large number of cows in the barns.

To develop a methane measurement method, respiration calorimetric chamber is always considered as the reference method because of its accurate measurement results (Blaxter et al., 1972). However, a small and restricted chamber can modify a cow's behavior, reduce their feed intake and consequently influence methane production. Moreover, a respiration chamber is limited by the time it takes to train an animal for measurements, the number of animals available, and the large expense of building and maintaining a chamber.

Given this, an artificial mechanical cow would be a good alternative reference method for the following three reasons. First, an artificial cow has known and controlled methane production rates so that different methane emission levels of cows can be created to calibrate and validate measurement methods. Second, it can automatically operate accurately and stable for a long period without adjustment or maintenance because it is a machine. Third, it can produce the same methane concentration patterns under different circumstances.

In free stall barns, cows typically spend up to 14 hours lying, resting, and ruminating in their cubicles, while generally remaining in a stable position. Considering this, this period could be used to measure the methane flux emitted from the lying cows by placing a sample hood near their heads. Like other methods, the effectiveness of this approach is related to the sample hood's design. Thus, research is needed to explore how a sample hood design can capture all the emitted methane in the cubicle.

The objectives of the study were (1) to design and construct an artificial reference cow (ARC) with known methane production rates and cow specific release patterns, and (2) to design and assess a cubicle hood sampler (CHS) with the ARC, representing a measurement method to measure methane fluxes emitted by cows in a cubicle;

2. Materials and Methods

2.1. ARC's working principle

The ARC (Figure 1) was designed and constructed to simulate a cow's methane production and exhalation procedures during respiration and eructation. The ARC consisted of an aluminum cylinder (40 × 20 cm, 12 l) to provide a cow's tidal and residual volume during respiration. A rubber piston connected to an actuator was placed inside the cylinder. The actuator was operated by compressed air that drove the piston horizontally up and down to simulate a cow's inhaling and exhaling processes.

The ARC also contained two mass flow controllers (MFC). These instruments controlled the ARC's CH_4 and CO_2 production rates from two attached pure CH_4 and CO_2 cylinders. A 20 × 62 cm silicon heating mat covered the cylinder's left side so that the gas inside was uniformly warmed to the temperature of the cow's breath. A 110 cm long plastic tube was connected to the right side of the cylinder's middle to simulate the cow's respiration tract. At the end of the tube, two round openings with 4 cm diameter mimicked the cow's nose. In addition, four methane sensors were located at different positions in the ARC: one inside the cylinder, one in the nose, and two externally on the outside of the cylinder (Figure 1). Each methane sensor was set to sound an emergency alarm if the methane concentration was too high because the controls had failed or because the ARC was leaking methane.

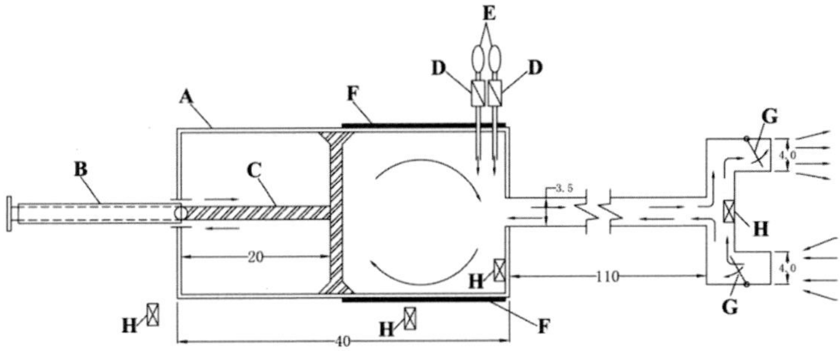

Figure 1. Schematic overview of the ARC: (A) a cylinder, (B) an actuator, (C) a stroke, (D) two mass flow controllers, (E) two pure CH_4 and CO_2 cylinders, (F) a heating mat, (H) four methane sensors, (G) two one-way valves in the nose only used for mass balance experiment.

2.2. Methane mass balance and concentration pattern of the ARC

The methane mass balance experiment was conducted in the lab with the ARC set at different methane injection rates. Two one-way valves (Figure 1) were mounted to the nose of the ARC. One checking valve was open only during inhalation and the other one was open only during exhalation. Thus, the ARC's inhalations and exhalations could be analyzed separately. A 3 m tube with a 50 mm diameter was fixed to the exhaling nose. The exhaled gas was continuously sampled and analyzed for the methane concentration by a portable multicomponent Fourier transform infrared spectroscopy (FTIR) gas analyzer. The methane's injection flow rate was tested from 0.05 to 0.40 l min^{-1} at 8 levels. Each level of methane flow rate was conducted for about 8 minutes and repeated 4 times. During the experiment, the tidal volume and breath frequency of the ARC were controlled at 6 L and 30 times per minute. The methane injected by MFC and inhaled from the ambient environment was the ARC's methane input, and the exhaled air was the ARC's methane output. Therefore, the overall methane mass in the input and output of the ARC during a sampling period can be analyzed and compared for each measurement.

The methane concentration pattern of the ARC was examined with five simulated cows (A-E). The cows' daily methane production rates were 200, 250, 300, 350, and 400 g day^{-1}, respectively. The ARC's nose was placed into a mimic feed bin. A sampling tube was positioned 5 cm away from the nose. Thus, the ARC's breath was continuously sampled at a rate of 4 l min^{-1} and analyzed by FTIR analyzer one value per 2 or 3 seconds.

2.3. CHS composition and layout

The CHS was designed and constructed to capture the methane that cows emit when they are lying in their cubicles. The CHS consists of a sample hood, a rectifier, a flow meter, an air sample point connected to a FTIR gas analyzer, a valve, and a fan. At the front of the cubicle, the sample hood creates an isolated area around the lying cow's head where an airflow generated by a fan captures the cow's released methane. A tube-integrated flow meter measures the flow rate through the system. The measured flow rate is then multiplied by the methane concentration in the sample hood to calculate the captured methane flux after the methane concentration has been corrected for the barn's background concentration.

We constructed two different sample hoods (Figure 2). The basic sample hood (1.2 × 0.6 × 2.0 m) consisted of a top sample hood (A_1) with an open inlet to the sampling tube, a front panel (A_2, 1.2 × 1.2 m), and two side panels (A_3, 1.2 × 0.6 m). The second sample hood was a modification of the basic sample hood. Here two inside panels were added to the front and top segments of the basic sample hood. Each panel had three separate gas inlet areas: at the front (B_1, 1.2 × 0.28 m), at the middle top (B_2, 1.2 × 0.06 m), and at the upper top (B_3, 1.2 × 0.035 m). Each inlet consisted of a perforated panel with round holes (ø 0.54 cm) that were evenly distributed over each gas inlet area. The total inlet surface in each of these areas was 120.9 cm^2 at the front, 31.1 cm^2 at the middle top, and 9.6 cm^2 at the upper top. The fume sample hood could be extended and tested with different auxiliaries to improve the system's performance. A top curtain (C, 1.2 × 0.8 m) could be mounted over the whole width of the front open area of the fume sample hood to decrease the sample hood's front open area from 1.44 m^2 to 0.96 m^2.

2.4. Evaluation of the methane recovery rate in the laboratory

Several factors may affect the CHS's methane recovery rate, including flow rate through the hood, sample hood type, the hood's open area, the cow's body size and its nose position. We performed an experiment in the air laboratory of Wageningen UR Livestock Research to assess the influence of these factors on the CHS's methane recovery rate.

(1) Five flow rates (40, 80, 120, 160, and 200 m^3 h^{-1}) were each tested four times at nose position B at 10, 50, and 100 cm in the basic sample hood (Figure 3). The direction of the exhaled air from the nose was always perpendicular to the back plate as indicated by the arrows

in Figure 3. The selected horizontal positions were restricted to the sample hood's left side as it was assumed that the airflow pattern on both sides of the nose was equal.

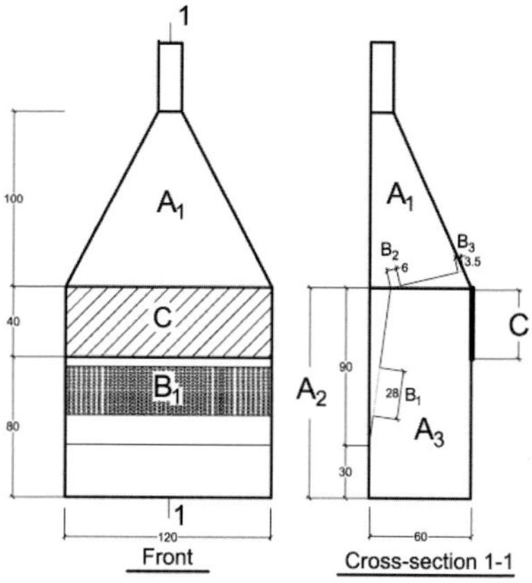

Figure 2. The front (left) and cross-section 1-1 (right) view of the sample hood. The basic sample hood consists of a top sample hood (A_1), a front panel (A_2), and two side panels (A_3). The fume sample hood has two additional panels at the front and top inside the basic sample hood. Each panel contains three gas inlets: the front (B_1), the middle top (B_2), and the upper top (B_3). Each gas inlet was made of a perforated panel with round holes. The fume sample hood was also extended with a top curtain (C).

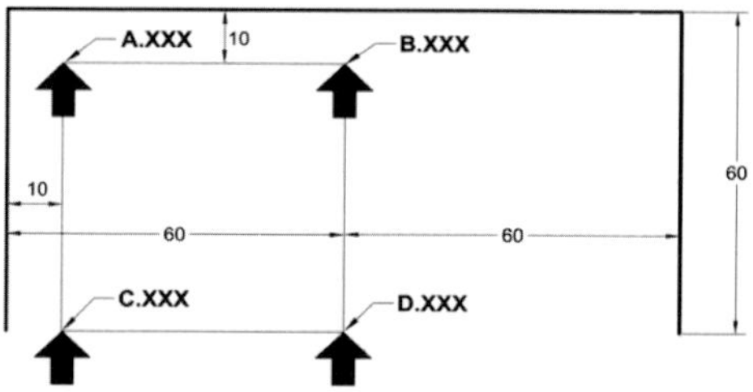

Figure 3. Top view of the sample hood with four nose positions (A-D). This layout was used for tests at a height of 10, 50, and 100 cm, indicated by 'XXX'. For instance, the position of B.050 means the position B at the height of 50 cm.

(2) After analyzing how the flow rate affected the basic sample hood, we assessed the performance of the fume sample hood. The fume sample hood captures the released methane mainly through the front panel, while the basic sample hood sucks the air from the top (Figure 2). The fume sample hood was tested four times at one nose position (B.050) at flow rates of 120 and 200 $m^3\ h^{-1}$.

(3) The hood's front open area was decreased by a top curtain. To analyze effects of the top curtain, the fume sample hood with the top curtain was tested four times at one nose position (B.050) at flow rates of 120 and 200 $m^3\ h^{-1}$.

(4) When a cow was lying under the hood, the cow's body size would reduce the front open area. During the laboratory experiments, a cardboard obstacle ($0.6 \times 0.3 \times 0.85$ m) called the 'cow panel' was placed in front of the sample hood to simulate the effects of a cow's body. The fume sample hood with the cow panel and the top curtain was tested four times at one nose position (B.050) at flow rates of 120 and 200 $m^3\ h^{-1}$.

(5) The nose position was expected to vary when a cow was lying under the hood. We selected four nose positions to assess the effects on methane recovery rates. The four nose positions of A-D are shown in Figure 3 at a height of 50 cm, where a cow's head is normally situated. Each nose position was tested four times in the fume sample hood with the top curtain and the cow panel at flow rates of 120 and 200 m^3 h^{-1}.

3. Results and Discussion

3.1. Performance assessment of the ARC

3.1.1. Overall methane mass balance between input and output

The overall methane mass balance between input and output was the crucial factor to evaluate the system's performance. The output methane mass was strongly positively related to the input methane mass (Figure 4). In regression analysis, the fitted term with methane input and constant had a smaller standard error (0.0135 versus 0.0160) than with only methane input. Therefore, the best fitted model was expressed as: $M_{out} = M_{in} \times 1.01 - 0.01$ ($R^2 = 0.999$, $P < 0.001$). The model predicted that methane output would rise by 1.01 units with each additional methane input unit. Standardized residuals of the model were evenly distributed around the zero line as depicted in Figure 4. The differences of methane input and out may be caused by the MFC on controlling the methane's injection flow rate and the FTIR on analyzing the methane concentration of the exhaled gas. Yet, with about 1% difference between methane input and output, it is demonstrated that the ARC could accurately control the methane mass production.

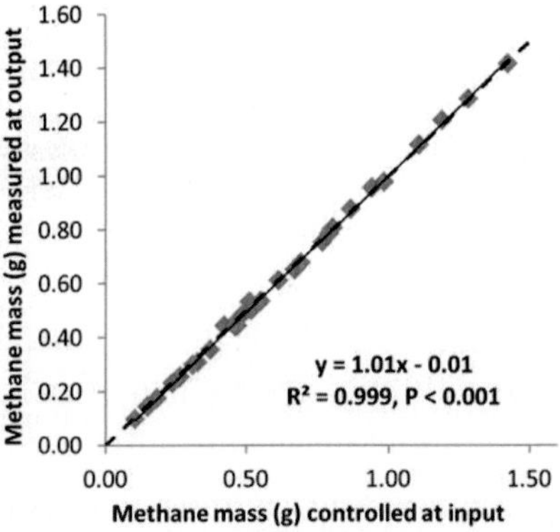

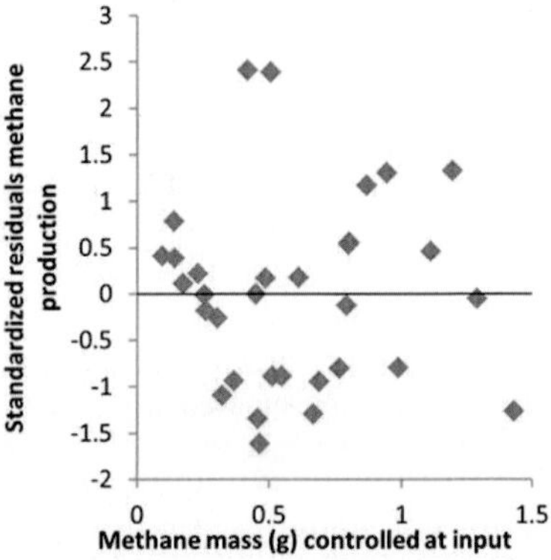

Figure 4. Methane mass (g) controlled at input versus measured at output (left) and standardized residuals (right) versus methane production (g) during each 8 minutes experiment; the dotted line represents the line of equivalence.

3.1.2. Methane concentration pattern measured from the ARC

The five simulated cows produced similar methane concentration patterns under the laboratory conditions as real cows (Figure 5): methane concentration fluctuated as sinusoidal cycles during the measurement period. Each fluctuation cycle has a quick concentration rise and then a longer decay, which corresponded with the patterns measured from real cows. One fluctuation cycle time was controlled precisely at 36.3 s (s.d. 0.2) between five simulated cows, which was 0.6 s longer than the mean level measured in the real cow data set. The average methane concentrations from five simulated cows at start, peak, and end were 593.6, 1098.2, and 603.0 ppm, which were 443.7, 262.6, and 402.3 ppm higher than measured in the dataset with real cow cycles. Higher methane concentrations were to be expected because less dilution factors (e.g., airflow pattern and cows' head movement) were present in the laboratory. However, maximum methane concentrations measured from real cows at start, peak, and end

(537.3, 1365.4, and 611.9 ppm) were in the range of the simulated laboratory values. These maximum values were probably derived from low dilution conditions with low ambient airflow and close sampling to the cow's nose, as is the case for the simulated cows in the laboratory.

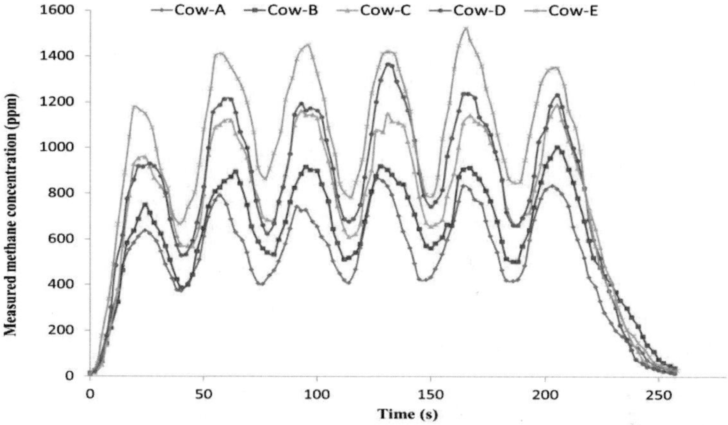

Figure 5. Typical example of the measured methane concentration (ppm) during one measurement period (216 s) for five simulated cows. Methane production rates of cow A to E: 200, 250, 300, 350, and 400 g day^{-1}.

3.2. Performance assessment of the CHS in the laboratory

3.2.1. Flow rate and basic sample hood

Figure 6 shows methane recovery rates of the basic sample hood at three heights of nose position B at five flow rates. Methane recovery rates of the basic sample hood were significantly affected by flow rates, and location and height of the nose positions ($P < 0.001$). Mean methane recovery rates increased from 39.9% at the flow rate of 40 $m^3\ h^{-1}$ to 84.1% at the flow rate 200 $m^3\ h^{-1}$. However, mean methane recovery rates did not further increase between flow rates of 160 and 200 $m^3\ h^{-1}$ ($P > 0.05$), indicating that the effect of flow rate is nearing its maximum at these levels. As nose position B.100 was nearest to the air inlet, it had higher methane recovery rates than the other two positions, especially at a high flow rate.

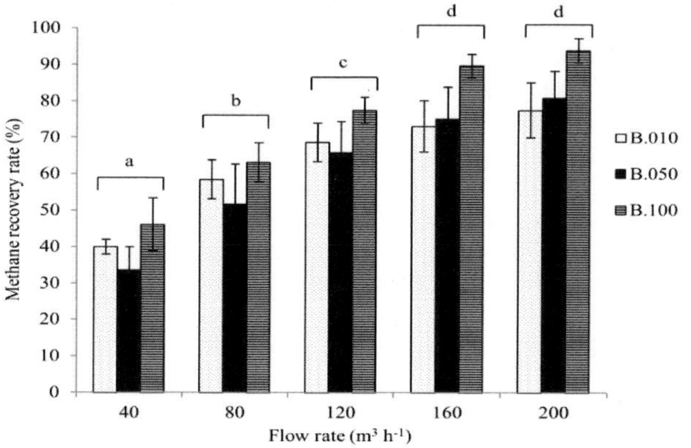

Figure 6. Methane recovery rates of the basic sample hood at three nose positions at five flow rates. Error bars show the standard deviation of each test ($n = 4$). Different letters show significant difference of mean methane recovery rates at different flow rates ($P < 0.05$).

3.2.2. Fume sample hood with top curtain and cow panel

Methane recovery rates of the fume sample hood were compared to those of the basic sample hood at nose position B at two flow rates (Figure 7). At the height of 50 cm, the fume sample hood performed better at the imposed flow rates of 120 ($P = 0.015$) and 200 m^3 h^{-1} ($P = 0.004$). This improvement can be attributed to the implementation of the front gas inlet in the fume sample hood at this height (Figure 2). As a result, the nose position B.050 was much closer to the air inlet of the fume sample hood than it was in the basic sample hood. With a shorter distance to the air inlet, the negative air pressure was also being expected to be higher.

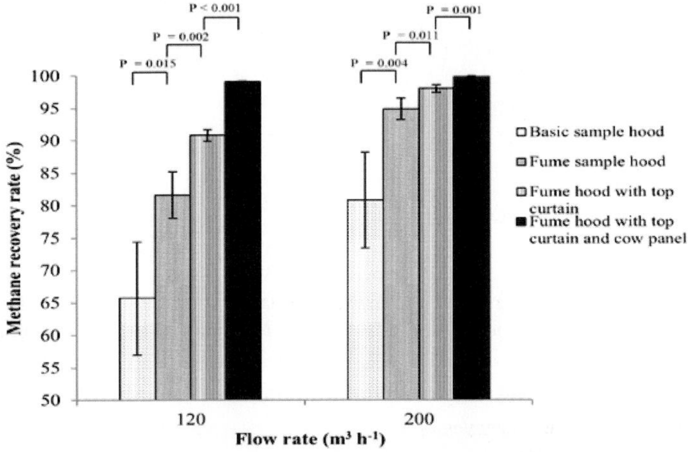

Figure 7. Methane recovery rates of the fume hood with a top curtain and a cow panel compared to the basic sample hood at the flow rates of 120 and 200 m^3 h^{-1}.

A top curtain increased the methane recovery rates at the flow rates of 120 and 200 m^3 h^{-1} (Figure 7). At the flow rate of 120 m^3 h^{-1}, the fume sample hood with a top curtain had a methane recovery rate that was 8.3% higher ($P = 0.002$) than the fume hood without the top curtain. At the flow rate of 200 m^3 h^{-1}, the additional top curtain increased methane recovery rates by 2.0% ($P = 0.011$), although the methane recovery rates were already close to 100%. The improvement caused by the top curtain can be explained by the following two reasons: (1) the top curtain decreases the front open area from 1.44 m^2 to 0.96 m^2, resulting in higher air velocities near the exhalation position, and thus improving the CHS's capture efficiency; (2) the top curtain may also prevent the emitted methane from leaving the sample hood. With the curtain, the breath air that escapes from the top can flow back and be recaptured.

The cow panel also increased the methane recovery rates at flow rates of 120 and 200 m^3 h^{-1} (Figure 7). At the flow rate of 120 m^3 h^{-1}, the cow panel increased methane recovery rate by 7.4% ($P < 0.001$). At the flow rate of 200 m^3 h^{-1}, the methane recovery rate increased by 1.8% ($P = 0.001$), reaching a 99.8% methane recovery rate. This performance improvement can be explained by the decreased size of the front open area; the cow panel reduces the inlet surface from 0.92 m^2 to 0.48 m^2, creating the same effect as the top curtain. Consequently, a cow's body naturally increases the CHS's methane recovery rate when the cow is lying under the sample hood.

3.2.3. Nose position effects on the fume hood with top curtain and cow panel

Methane recovery rates did not differ between four nose positions at the flow rate of 200 m^3 h^{-1} ($P = 0.454$), but they were different at the flow rate of 120 m^3 h^{-1} ($P < 0.001$; Figure 8). At the flow rate of 200 m^3 h^{-1}, the CHS had enough capacity to uniformly capture methane from different nose positions under the sample hood area. At the flow rate of 120 m^3 h^{-1}, however, the capture capacity decreased and could not evenly cover the whole area under the sample hood. Thus methane recovery rates varied in different locations under the sample hood at a lower flow rate.

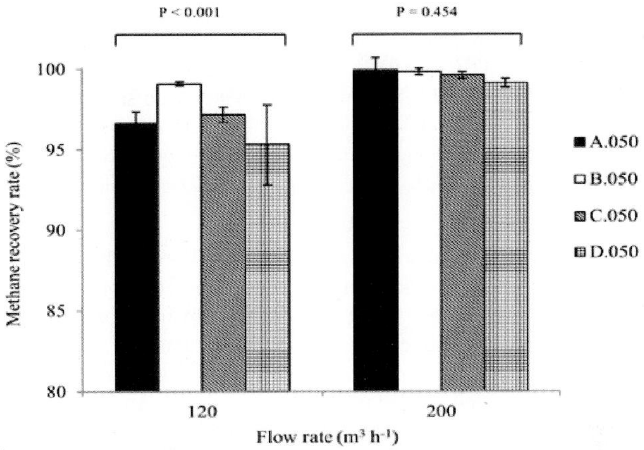

Figure 8. Methane recovery rates for the fume hood with a top curtain and a cow panel at four nose positions at the flow rates of 120 and 200 m^3 h^{-1}.

4. Conclusions

The ARC designed and constructed in this study properly represented the methane production of a cow. The system is capable of precisely controlling methane concentration and production at pre-set reference values. By controlling the methane injecting flow rates, tidal volume, and respiration frequency, the ARC can simulate different levels of methane production by real cows. The ARC is capable of producing methane concentration patterns that are representative of real cows.

The CHS designed and constructed in this study is capable of measuring methane production from individual cows in cubicles. The performance of the CHS was affected by the flow rates, hood types, adding top curtain and cow panel. With the flow rate of 200 m^3 h^{-1}, the CHS with fume sample hood, a top curtain, and a cow panel could capture 99.6% emitted methane by the ARC, regardless of nose positions in steady conditions.

Acknowledgements

This study was funded by Wageningen UR Livestock Research, the Ministry of Economic Affairs (through KB Sustainable Agriculture), and the China Scholarship Council.

References

Bannink, a., M.W. van Schijndel, and J. Dijkstra. 2011. A model of enteric fermentation in dairy cows to estimate methane emission for the Dutch National Inventory Report using the IPCC Tier 3 approach. Anim. Feed Sci. Technol. 166-167:603–618.

Blaxter, A.B.K.L., J.M. Brockway, and A.W. Boyne. 1972. A new method for estimating the heat production of animals. Q. J. Exp. Physiol. 57:60–72.

Boadi, D., C. Benchaar, J. Chiquette, and D. Massé. 2004. Mitigation strategies to reduce enteric methane emissions from dairy cows: Update review. Can. J. Anim. Sci. 84:319–335.

Johnson, K. a, and D.E. Johnson. 1995. Methane emissions from cattle Methane Emissions from Cattle. J. Anim. Sci. 2483–2492.

Lassey, K.R. 2007. Livestock methane emission: From the individual grazing animal through national inventories to the global methane cycle. Agric. For. Meteorol. 142:120–132.

Computational Fluid Dynamics Analyzing the Air Velocity in a Mechanically-Ventilated Broiler Housing

Daniella Jorge de Moura [a,*], Thayla Morandi Ridolfi de Carvalho-Curi [a], Juliana Maria Massari [a], Angelica Signor Mendes [b], Marcio Mesquita [c]

[a] College of Agricultural Engineering/UNICAMP, 501 Candido Rondon Avenue, Campinas, Brazil

[b] College of Agricultural Engineering/UTFPR, 3165 Sete de Setembro Avenue, Curitiba, Brazil

[c] College of Agricultural Engineering/UFPEL, 177 Marechal Floriano Street, Pelotas, Brazil

* Corresponding author. Email: daniella.moura@feagri.unicamp.br

Abstract

With the improvements that the environmental control of broiler houses in Brazil there is a need for an alternative and more precise ventilation system. The objective of this study is to evaluate the current ventilation system to a commercial farm for broilers and propose a more efficient ventilation system to determine how many exhaust fans and in which positions should be located for a homogenous and efficient ventilation inside the barn, using the software Ansys® CFX 14. The study was conducted in a broiler house located in Amparo/Sao Paulo. The study was done in a tunnel ventilated broiler house totally enclosed with brick side walls. The climatic and environmental conditions were monitored at 2 PM when the birds were 42 days in summer conditions of 2012. Data were collected at 27 equidistant points such as: dry bulb temperature and air velocity, through a hot wire anemometer, relative humidity with a hygrometer and static pressure with a manometer. The control conditions are: inlet air velocity, inlet air temperature, outlet air velocity, outlet air temperature and the static pressure next to the exhaust fans. The geometry was made using the software Ansys® Workbench 14.0, the meshes were evaluated by the elements, numbers of node, aspect ratio, element quality and skewness. The validation was done, and a normalised mean square error value (NMSE) of 0.18 was obtained, meaning a good indicator. The Computational Fluid Dynamics (CFD) technique was an effective and feasible method to display the airflow pattern under different operating conditions for exhaust fans in 2D and 3D, as well as to determine the best arrangement of exhaust fans in operation to avoid the areas with low speed and turbulence zone. The best results were observed in simulation 6 where almost all the fans were turned on in the bottom of the building.

Keywords: Air velocity, broiler, environmental control, ventilation rate

1. Introduction

Genetic selection in broilers reduces birds' ability to maintain energy, heat and water balance, leading to less their capability to overcome heat stress event. The thermal stressed broilers showed lower results for feed intake, weight gain and viability, mainly under high temperature (Boiago et al., 2013; Oba et al., 2012). The temperature, relative humidity and wind speed are the main environmental factors affecting the thermal comfort of the birds, providing limits for adequate production (Nascimento et al., 2011; Amaral et al., 2011). The high temperature might cause mortality in birds (Julian, 2005; Toyomizu et al., 2005). This explains the need to use ventilation system which has thermal function, sanitary and physiological. With the recent increase in environmental control for broiler houses in Brazil, the design of ventilation systems in broiler houses has been a major challenge. Different simulation methods have been used as a help for designers and researchers to achieve an exact control of the indoor conditions in broiler houses. The geostatistics have been used to study the spatial variability of the environment variables inside the buildings (Chowdhury et al., 2013). Computational fluid dynamics is a simulation technique that can efficiently develop both spatial and temporal field solutions of fluid pressure, temperature and velocity, and has proven its effectiveness in system design and optimization within the chemical, aerospace, and hydrodynamic industries (Norton,

2014). The aim of this study was to evaluate the ventilation system in a tunnel ventilated broiler house in relation to air velocity patterns in order to improve the environmental control using computational fluid dynamics (CFD).

2. Materials and Methods

- Broiler house description: This trial was carried out in a tunnel ventilated broiler house with the following size: 3 × 20 × 120 m (high × width × length). The walls were built with concrete bricks, the roof have fiber cement tiles and polyethylene film as a ceiling. The birds density was 13 birds m^{-2} (Figure 1).

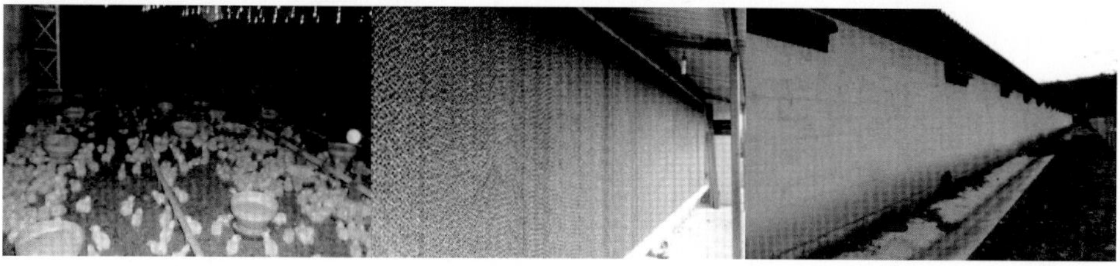

(a) (b) (c)

Figure 1. The broiler house with negative ventilation system. The view inside the broiler house (a), cooling panel (b) and outside view (c).

- Data acquisition: The dry bulb temperature (Tbs, °C) and air velocity (Var, m s^{-1}) were taken at 27 equidistant points (3 columns and 9 lines) inside the building and at the birds height when the birds reached 42 days old. These data were collected with a hot wire anemometer, VelociCalc® (TSI_{TM} manufacturer, Shoreview, USA). The relative humidity (UR, %) was registered by a hygrometer THDL 400 (Instrutherm manufacturer, Sao Paulo, Brazil). The static pressure of the exhaust fans was measured with a manometer, MN 2150 (ICEL® manufacturer, Shropshire, United Kingdom).

- Data analysis: Data were simulated by the CFD modeling to analyze the airflow pattern in the building and propose new exhaust fans positions and operation, if it's necessary.

The software Ansys® CFX 14.0 was used for the simulation that was done in 4 steps: First step: Problem definition; Second step: Pre Processor including mesh creation; Fluid properties; Boundary conditions; Third step: Solver (solution) - Solution control. Fourth step: Post-processor (showing the results)

The geometry was built in Ansys® Workbench 14.0, scale 1:1. The mesh hexahedral was chosen, since the tetrahedral mesh requires a greater amount of elements to provide the same number of nodes (Joaquim Junior *et al.*, 2007). The mesh was evaluated by the element, numbers of node, aspect ratio, element quality and skewness Table 1.

Table 1. Mesh characteristics.

Mesh dimensions	0.23 m
Number of elements	706,390
Number of nodes	729,949
Element quality	0.914
Aspect ratio	2.020
Skewness	0.100

The following boundary conditions were taking in account: inlet air velocity, inlet air temperature, outlet air velocity, outlet air temperature, air density and viscosity and static pressure at the outlet (exhaust fans).

It was assumed a steady state condition, incompressible and turbulent flow, convergence criterion with root mean square of less than 10^{-4} to converger. In the simulation, different positions of the exhaust fans were used in order to obtain the best combination to make the airflow more homogeneous and became soften the turbulence area near the inlets (Figure 2).

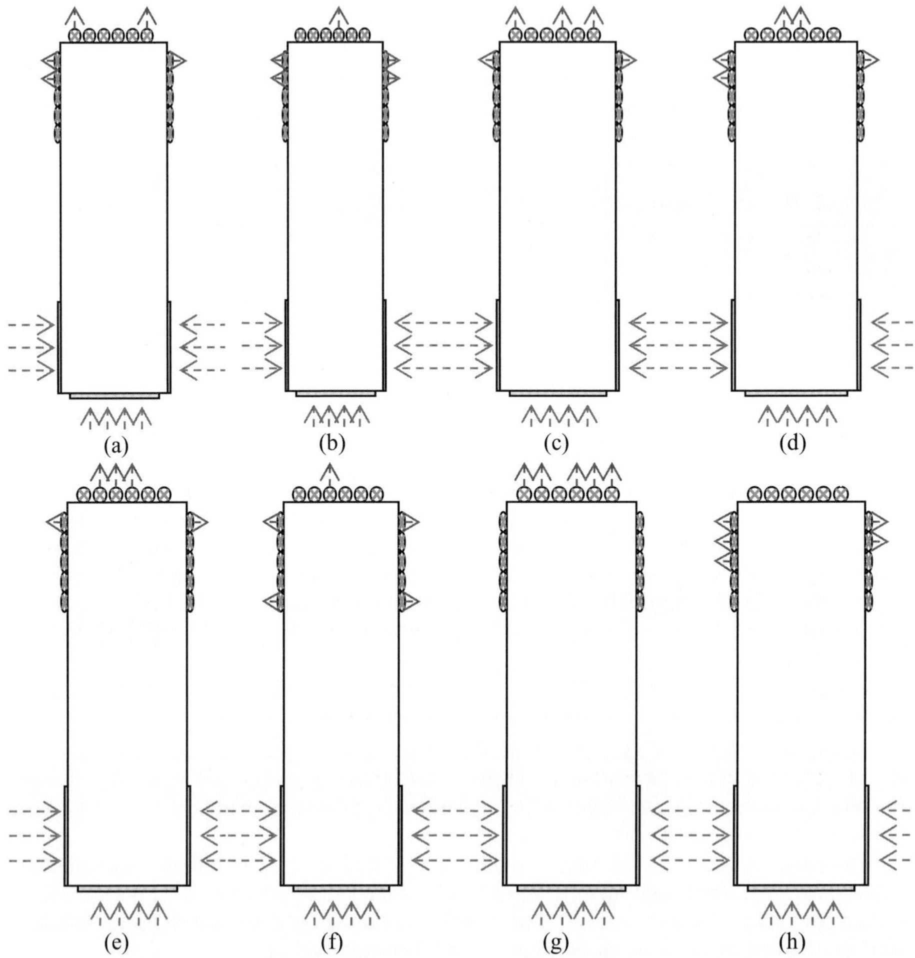

Figure 2. The original design and the simulations with 5 exhaust fans working together. Original 1 (a); Simulation 1 (b); Simulation 2 (c); Simulation 3 (d); Simulation 4 (e); Simulation 5 (f) ; Simulation 6 (g) and Simulation 7 (h).

It was used the closure model of first order standard k-Ɛ (CFX, 2013). The NMSE values smaller than 0.25 were accepted as good indicators of assent (Anderson et al., 1992). The validation was based on the methodology used by Blanes-Vidal et al. (2008). This methodology use 27 points inside the broiler house to measurement the air velocity at birds height, and spend 6 minutes for each point to measure wind speed.

3. Results and Discussion

The advantage of CFD simulations is that they provide easily understandable graphic representation, providing airflow patterns, represented by different colors or vectors in different scenarios.

In relation to the Original Simulation one (Figure 3), the best simulation was 6 (Figures 4 and 5) in relation to the homogeneity of the air velocity and also reaching values within the range considered optimal in the literature, or between 1.5 and 2.5 m s^{-1} (Yahav et al., 2001).

It is observed that a zone of turbulence occurs near the air inlets by the convergence of the airflow entering through the evaporative panels, which increases the value of the air velocity above the recommended levels in the literature, between 1.5 and 2.5 m s^{-1} (Yahav et al., 2001). The Table 2 presents de air velocity values of the different simulations. This area of turbulence may interfere with the normal behavior of the birds, since the speed of the air impact performance, feed and water consumption of the birds (May et al., 2000).

Table 2. The Wind speed values (m s^{-1}) inside the broiler house under different simulations.

Air velocity (m s^{-1})	Maximum	Minimum	Average	Difference
Original simulation	2.80	0.04	1.42	2.76
Simulation 1	2.78	0.03	1.40	2.75
Simulation 2	2.78	0.04	1.41	2.74
Simulation 3	2.78	0.03	1.40	2.75
Simulation 4	2.89	0.04	1.46	2.85
Simulation 5	2.78	0.03	1.40	2.75
Simulation 6	2.86	0.00	1.43	2.86
Simulation 7	3.10	0.03	1.56	3.07

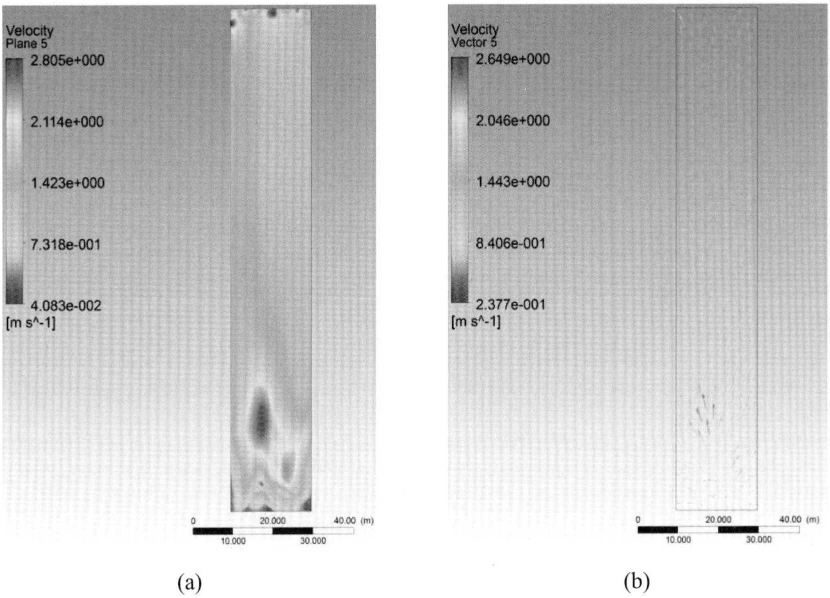

Figure 3. Air velocity distribution at the birds high for Original (a) and the vectorial representation of the airflow inside the building at birds height (b).

Figure 4 shows the transversal view of the airflow, at 90.0m far from the air inlet, where the air velocity in the simulated condition (b) showed a higher ventilation in almost all the building when compared to the original condition (a).

Areas with low air exchange (< 0.24 m s^{-1}) were observed near the bottom of the building and also near to the exhaust fans that were not turned on in the simulation. These points are also known as "dead" spots which the air speed approaches 0 m s^{-1}. The difference between regions with high and low values of wind speed may cause heterogeneity of the flock, influencing in the final weight gain since the air speed affects the thermal sensation and heat exchange of the birds and the thermal comfort condition (Tao and Xin, 2003; Bucklin et al., 2009). Thus, it is noted that the regions of the air inlet and outlet are critical, while the intermediate zone has a softer and flat airflow.

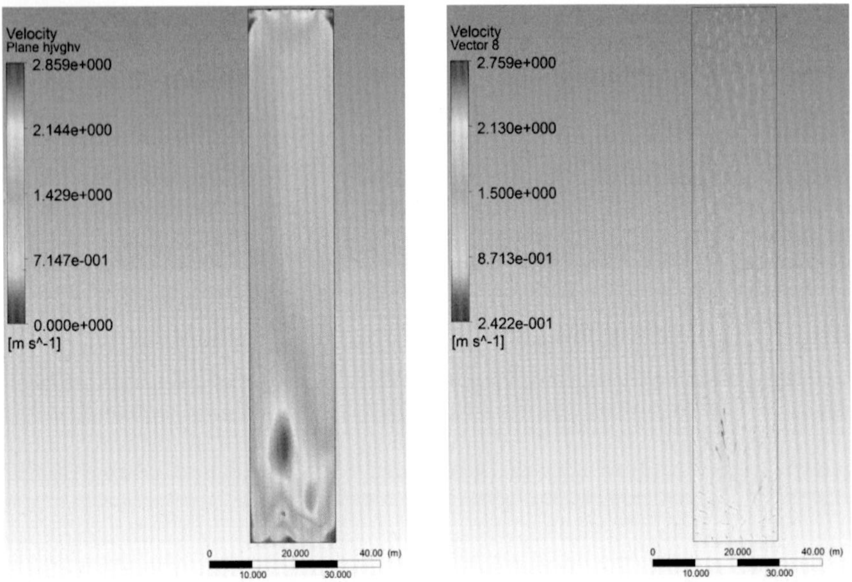

(a) (b)

Figure 4. Air velocity distribution at the birds high for Simulation 6 (a) and the vectorial representation of the airflow inside the building at birds height (b).

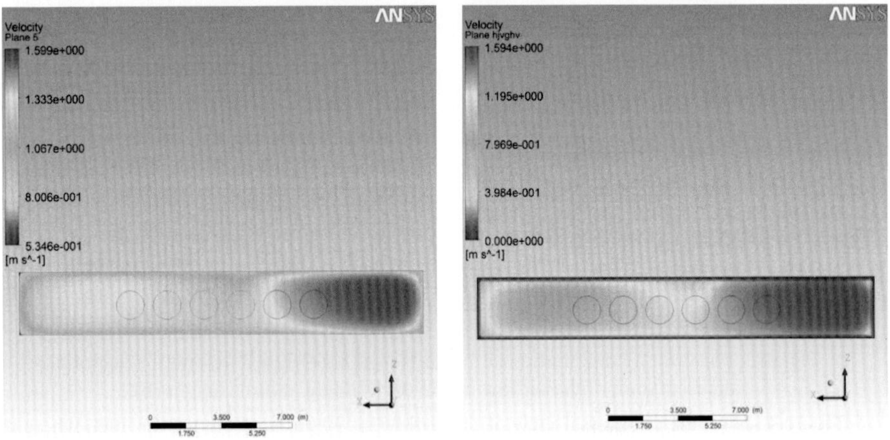

(a) (b)

Figure 5. Air velocity distribution at transversal view for the Original (a) and for Simulation 6 (b) at 90m from the air inlet .

4. Conclusions

The CFD technique is an effective and feasible method to display the airflow pattern under different operating conditions exhaust fans in 2D and 3D, as well as to determine the best arrangement of exhaust fans in operation to prevent the areas with low speed air and turbulence. The best results were observed in simulation 6 where almost all the fans were turned on in the bottom of the building.

Acknowledgements

To Fapesp for the funding – Proc. 2011/07545-7 and CNPq for the PhD Scholarship.

References

Amaral, A.G.; Yanagi Jr, T.; Lima, R.R.; Teixeira, V.H.; Schiassi, L. 2011. Effect of the production environment on sexed broilers reared in a commercial house. Arquivo Brasileiro de Mediciona Veterinária e Zootecnia, v.63, n.3, p.649–658.

Anderson, M.P.; Woessner, W.W. 1992. Applied groundwater modeling simulation of flow and advective transport. 2.ed. New York: Academic Press.

Blanes-Vidal, V.; Guijarro, E.; Balasch, S.; Torres, A.G. 2008. Application of computational fluid dynamics to the prediction of airflow in a mechanically ventilated commercial poultry building. Biosystems Engineering, v.100, n.1, p.105–116.

Boiago, M.M.; Borba, H.; Souza, P.A.; Scatolini, A.M.; Ferrari, F.B.; Giampietro-Ganeco, A. 2013. Desempenho de frangos de corte alimentados com dietas contendo diferentes fontes de selênio, zinco e manganês, criados sob condições de estresse térmico. Arquivo Brasileiro de Medicina Veterinária e Zootecnia, v.65, n.1, p.241–247.

Bucklin, R.A.; Jacob, J.P.; Mather, F.B.; Leary, J.D.; Nääs, I.A. , 2009. Tunnel Ventilation of Broiler Houses. IFAS Extension, v.1, n.1, p.1–3.

CFX, 2013. User manual, ANSYS-CFX.

Chowdhury, S.; Themudo, G.E.; Sandberg, M.; Ersbøll, A.K. 2013. Spatio-temporal patterns of Campylobacter colonization in Danish broilers. Epidemiology and Infection, v.141, n.5, p.997–1008.

Joaquim Junior, C.F.; Nunhez, R.; Celinski U. 2007. Agitação e Mistura na Indústria. Rio de Janeiro, LTC.

Julian, R.J. 2005. Production and growth related disorders and other metabolic diseases of poultry – a review. The Veterinary Journal, v.169, n.3, p. 350–369.

May, J.D.; Lott, B.D.; Simmons, J.D. 2000. The effect of air velocity on broiler performance and feed and water consumption. Poultry Science, v.79, p.1396–1400.

Nascimento, G.R.; Pereira, D.F.; Nääs, I.A.; Rodrigues, L.H.A.Thermal comfort fuzzy index for broiler chickens. Engenharia Agrícola, Jaboticabal, v.31, n.2, p.219–229, 2011.

Oba, A.; Lopes, P.C.F.; Boiago, M.M.; Silva, A.M.S.; Montassier, H.J.; Souza,P.A. 212. Características produtivas e imunológicas de frangos de corte submetidos a dietas suplementadas com cromo, criados sob diferentes condições de ambiente. Revista Brasileira de Zootecnia, v.41, n.5, p.1186–1192.

Tao, X.; Xin, H. 2003. Acute synergistic effects of air temperature, humidity, and velocity on homeostasis of market-size broilers. Transactions of the ASAE, v.46, n.2, p.491–497.

Toyomizu, M.; Tokuda, M.; Mujahid, A.; Akiba, Y. 2005. Progressive alteration to core temperature, respiration and blood acid-base in broiler chickens exposed to acute heat stress. The Journal of Poultry Science, v.42, p.10–118, 200.

Yahav, S.; Straschnow, A.; Vax, E.; Razpakovski, V.; Shinder, D. 2001. Air velocity alters broiler performance under harsh environmental conditions. Poultry Science, v.80, p.724–726.

Measurement of Thermal Environment Uniformity within High Density Stacked-Cage Layer House

Qiongyi Cheng, Baoming Li *

Key Laboratory of Structure and Environment in Agricultural Engineering, Ministry of Agriculture, China Agricultural University, Beijing 100083, China

* Corresponding author. Email: libm@cau.edu.cn

Abstract

High density laying hen production system using eight stacked cages and manure belt is becoming popular in China. Although the system is widespread in USA and some other western countries, it is relatively new in China. The distributions of ventilation and thermal environment within the production facilities are still of big concerns due to the extensive barn size and high bird density. In this study, a field test was carried out to measure the temperature, wind speed, carbon dioxide concentration and their distributions within a newly built barn with cross ventilation system in Shan'xi Province in January 2015. The thermal environment was evaluated based on these measurements. The results showed that for horizontal direction, the average temperature variations were 5.67±0.46°C, 2.76±0.29°C, and 3.11±0.49°C at the first, fifth, and seventh cage heights, respectively. For vertical direction, the temperature variations between first and 8th cage heights were 3.63±0.72°C, 1.30±0.46°C, and 4.09±1.44°C at the cooling pad, middle of the house, and exhaust fan sections, respectively. The highest diurnal temperature difference was as high as 9°C. The average wind speed at the exhaust fan and first cage height areas were too high: 0.36±0.07 m/s and 0.32±0.07 m/s respectively, which were both higher than the recommended threshold of 0.25 m/s. This paper provides an evaluation of environmental control technology and management in a high density stacked-cage-raising layer house.

Keywords: Temperature variation, air velocity, carbon dioxide, comfort

1. Introduction

High density stacked-cage-raising system is relatively new in China, furthermore, research about uniformity of environment variables within high density stacked-cage-raising layer house is rare, especially for the domestic Chinese layer houses. Distributions of airflow and temperature in house have great influence on production performance of laying hens. Webster and Czarick (2000) found that inside closed hen house, because of fluctuation of temperature, egg size at the coolest site (house center) in house was larger than other sites, and excessive variation of temperature resulted in marked variation in feed consumption, which led to difference in egg size. Abbas et al. (2011) studied effects of ambient temperature and humidity on performance of laying hens inside closed system, and the results showed that birds in central section of the house where temperature and humidity were 14 to 28°C and 22% to 90%, respectively, exhibited better performance in daily weight gain at 19 weeks, feed conversion ratio and egg production than locations near cooling pad and exhaust fan, where temperature were 18 to 30°C and 20 to 30°C, and humidity were 30% to 80% and from 34% to 84%, respectively. Emmans (1977) also reported that temperature was an important factor because of its effect on feed consumption and egg size, while during cold temperature, great fluctuation resulted in poor feed conversion ratio and health problem. Kuczynski et al. (2002) reported that compared with humidity, temperature was more important in affecting production performance and hen health. In this study, temperature, wind speed, carbon dioxide concentration and their distribution in a high density laying hen house with eight stacked cages and manure belt was measured. The main purpose of this study was to provide a systematic evaluation of environmental control technology and management, find out major problems existing in the environmental control in high density stacked-cage-raising layer house.

2. Materials and Methods

2.1. Description of laying hen house and management

A newly built manure-belt laying hen house, which had an east-west orientation and dimensions of 93 m in length, 14.3 m in width, and 7.2 m eave height, was used for this study. The house was located at Yangquan City, Shan'xi Province, northern China, and had eight stacked-cage-raising system, a tunnel ventilation system with 28, 1.3-m diameter fans in the northern end-wall, 13.1 m × 3 m cooling pad at 0.6 m height from the floor in southern end-wall, 1.9 m × 2.5 m cooling pad at 2.75 m height from floor in both longitudinal wall, and two rows of continuous slot ceiling inlets (6.8 m from each sidewall), Figure 1. Exhaust fans in the lateral wall created negative pressure. Fans were divided into 7 groups. When indoor temperature was less than 18°C, 1 fan was always kept on to maintain minimum ventilation rate and 1 fan was cycled by operating for 250s, 50s. When ambient temperature in the house was between 18°C and 20°C, 1 more group of fan was switched on per 0.5°C increase.

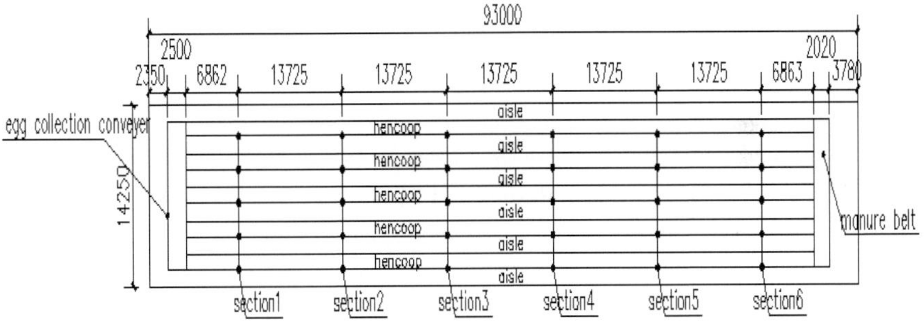

a. Plan view of environment parameters measuring point in measurement in horizontal direction.

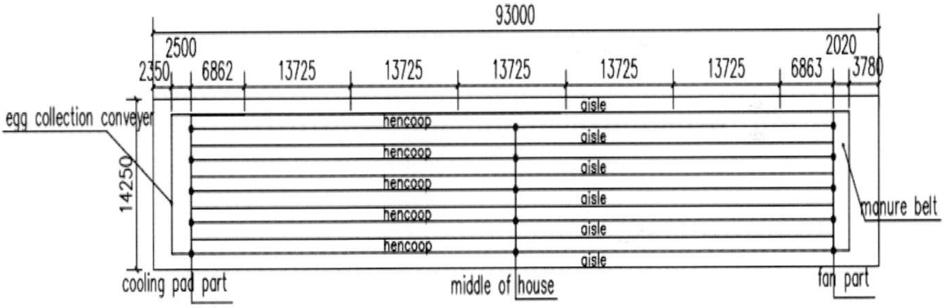

b. Plan view of environment parameters measuring point in vertical direction.

Note: The dots indicate measuring locations of temperatures, wind speeds, and carbon dioxide concentrations.

Figure 1. Plan view of environmental parameters measurement within the layer house.

Outdoor air entered ceiling through the eave, preheated in roof ceiling, then entered laying hen house through stripe inlet in ceiling of roof. The eight stacked cages were arranged in five rows with manure belt under each row. Hen feces fell directly on the belt underneath the cages and were removed 8:00 a.m. every day. Chain feeding and nipple drinkers supplied hens with fodder and water. The house was stocked with 80,000 Hy-Line variety brown laying hens. At the onset of experiment in January 2015, hens were at 17 weeks of age, photoperiod remained 11L:13D, during the experiment period from January 5 to January 24, photoperiod increased by half an hour per week.

2.2. Measurement and methods

This experiment was divided into two parts. Environment variables were measured horizontally and vertically. Measurement items included temperature, wind speed and carbon dioxide concentration.

In the first part, environment measurement was conducted in horizontal direction. Test instruments included 90 (30 per height) calibrated thermometers (U23-001 Hobo, Onset Inc., Pocasset, MA, USA), sampling at 5 minutes intervals. They were installed inside the laying hen house as shown in Figure 1a, which shows the plan view at one height. The installation heights were first, 5th and 7th layers at top of cages. Wind speed was measured with an anemograph (KA41L, Kanomax, Shenyang, China). Carbon dioxide concentrations were measured with a carbon dioxide detector). Measurements of this part of experiment were conducted from January 5 to January 17.

The second part environment measurement was conducted in vertical direction. The test instruments were 45 (15 per height) calibrated hygrometers showed in Figure 1b, which shows plain installation site at one height. The installation heights were the first, 4th and 8th layer cages. During this experiment part, the hen house was divided into three areas: the cooling pad area, middle of house, and fan areas. The time of this experiment was from January18 to January 24. The measurement of wind speed and carbon dioxide concentration in was from 8:00 to 20:00 at 2-hour intervals. The measurement point was the same as the temperature point.

3. Results and Discussion

3.1. Uniformity of environment variables along horizontal direction

In this part, temperature, carbon dioxide concentration and wind speed were measured in the first, fifth and seventh cage heights. It is shown in Figure 2, which illustrates average temperature along horizontal direction in different cage heights. The temperature differences between section 1 and section 2 at three heights were higher than another areas, namely temperature in section close to the cooling pad had significant difference ($P < 0.01$) compared with other sections. At the same height, the largest difference between different sections were 5.67±0.46°C, 2.76±0.29°C, and 3.11±0.49°C at the first, fifth, and seventh cage heights, respectively.

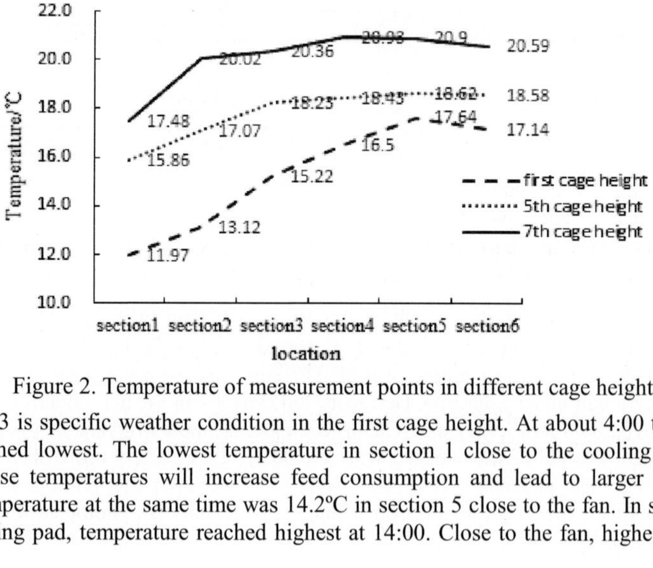

Figure 2. Temperature of measurement points in different cage heights.

Figure 3 is specific weather condition in the first cage height. At about 4:00 temperature in house reached lowest. The lowest temperature in section 1 close to the cooling pad was 9°C. Lower house temperatures will increase feed consumption and lead to larger egg size. The highest temperature at the same time was 14.2°C in section 5 close to the fan. In section 1 close to the cooling pad, temperature reached highest at 14:00. Close to the fan, highest temperature appeared at 18:00. The highest diurnal temperature difference was as high as 9°C, which could affect the hen egg production and mortality (Warren, 1950).

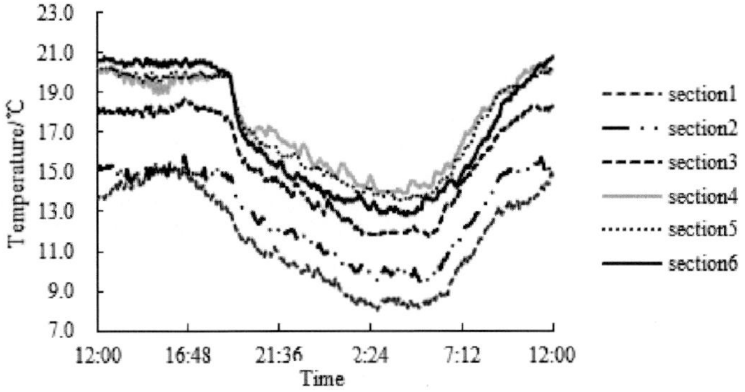

Figure 3. Temperature in first cage height.

Figure 4 is specific weather condition in fifth cage height. At 4:00 temperature in house reached lowest in section 1 near the cooling pad. Lowest temperature was 11.1°C. At same time in section 5 near fan, it had highest temperature 14°C in house. From 12:00 to 19:00, temperature in house was steady and more than 16°C. The highest diurnal temperature difference was 8 °C.

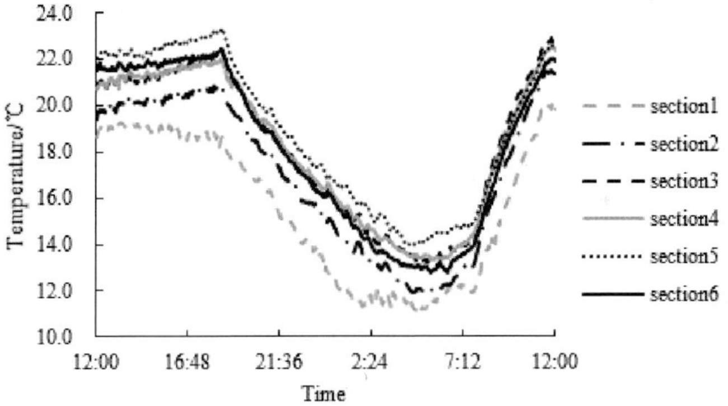

Figure 4. Temperature in fifth cage height.

Figure 5 is specific weather condition in seventh cage height. At about 4:00 temperature in house reached lowest. The lowest temperature was 12.3°C near the cooling pad and the highest temperature was 16.1°C at the same time. Temperature in house reached highest at about 14:00. Specific variation trend was showed in Figure 5. The highest diurnal temperature difference was 9°C.

Table 1 summarizes carbon dioxide concentrations and wind speed in first cage height. The carbon dioxide concentration was within appropriate range. The maximal allowed concentration is 3000 ppm according to Talukder et al. (2010). The wind speed was higher than standard, which is less than 0.25 m/s in winter (MAPRC, 1999). Its trend was increasing firstly and then decreasing. In section 4 it reached highest, which was 0.43±0.18 m/s.

Table 2 is carbon dioxide concentration and wind speed in fifth cage height. Wind speed and carbon dioxide concentration were within appropriate range. Their variation trend increased from section 1 (close to the cooling fan) to section 6 (close to the fan). Wind speed was less than

0.25 m/s except for section 6. Carbon dioxide concentrations in each location were less than 3000 ppm.

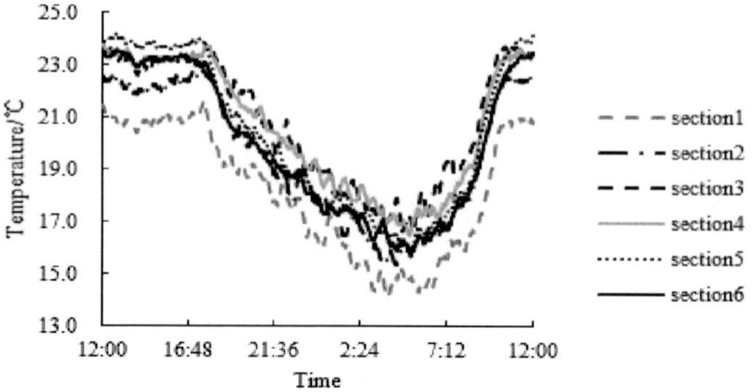

Figure 5. Temperature in seventh cage height.

Table 1. Wind speed and carbon dioxide concentration in first cage height.

	Wind speed (m/s)	Carbon dioxide (ppm)
Section1	0.23±0.07	1295±171
Section2	0.29±0.04	1384±122
Section3	0.37±0.07	1464±65
Section4	0.43±0.18	1821±120
Section5	0.35±0.04	1730±94
Section6	0.27±0.13	1691±66

Table 2. Wind speed and carbon dioxide concentration in fifth cage height.

	Wind speed (m/s)	Carbon dioxide (ppm)
Section1	0.1±0.01	1611±111
Section2	0.14±0.01	1704±125
Section3	0.15±0.02	1842±96
Section4	0.19±0.04	1902±192
Section5	0.23±0.02	1947±91
Section6	0.31±0.05	1975±93

Table 3 is carbon dioxide concentration and wind speed in seventh cage height. Variation trend of wind speed and carbon dioxide concentration was the same with fifth cage height and were all within appropriate range.

Table 3. Wind speed and carbon dioxide concentration in seventh layer cage.

	Wind speed (m/s)	Carbon dioxide (ppm)
Section1	0.08±0.01	2296±194
Section2	0.11±0.01	2523±88
Section3	0.13±0.05	2555±106
Section4	0.13±0.02	2681±210
Section5	0.15±0.02	2539±125
Section6	0.19±0.03	2632±183

3.2. Uniformity of thermal environment along vertical direction

Table 4 is temperature of measurement point in different areas. Temperature increased with height. In the cooling pad area, average temperature in first cage height was 10.82±3.28°C, far below tolerable temperature range of 15 to 27°C (Talukder et al., 2010). Temperature variations between first and 8th cage heights were 3.63±0.72°C, 1.30±0.46°C, and 4.09±1.44°C at the cooling pad, middle of the house, and exhaust fan sections, respectively.

Table 4. Temperature (°C) of measurement point in different areas of house.

	1st cage height	4th cage height	8th cage height
Cooling pad area	10.82±3.28	12.59±3.28	14.45±2.94
Middle of house	15.35±2.88	16.33±2.90	16.69±3.13
Fan area	13.91±4.19	17.38±3.18	18.00±3.24

Figure 6 is temperature in cooling pad area at different heights. At 4:00 temperature in house reached lowest; at 14:00 it reached highest but was generally below tolerable temperature range.

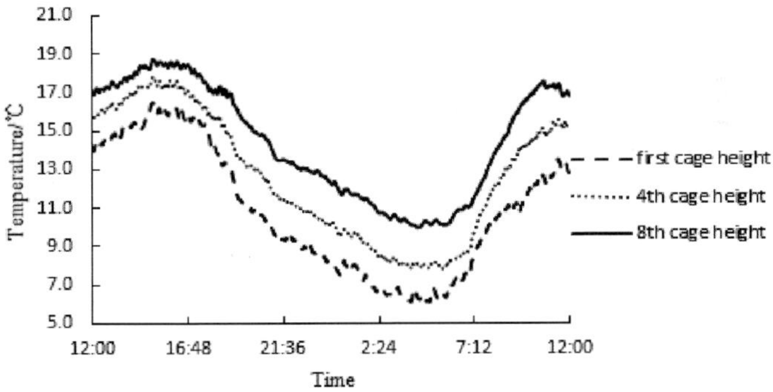

Figure 6. Temperature in cooling pad area.

Figure 7 is temperature in middle of house at different heights. Average temperature difference between 4th and 8th cage heights was -0.4°C. The lowest temperature at first cage height was 11.0°C. Compared with first cage height at cooling pad area, it was higher by 4.9°C.

Figure 8 is temperature in fan area at different heights. Temperature in first cage height had highly significant difference ($P < 0.01$) with 4th and 8th cage heights. The lowest temperature in first cage height was 8.7°C, which was far below suitable temperature for laying hen production.

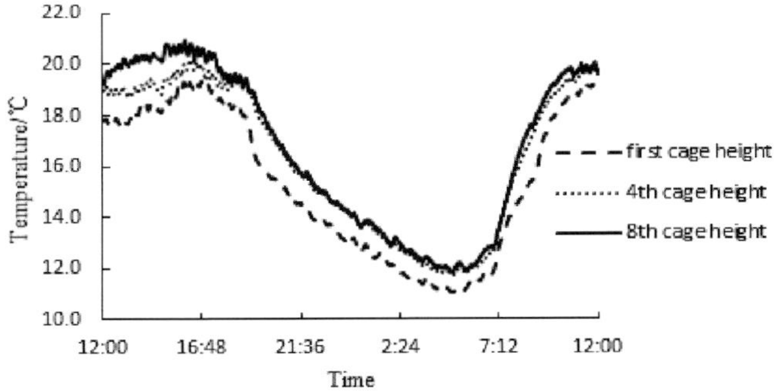

Figure 7. Temperature in middle of house.

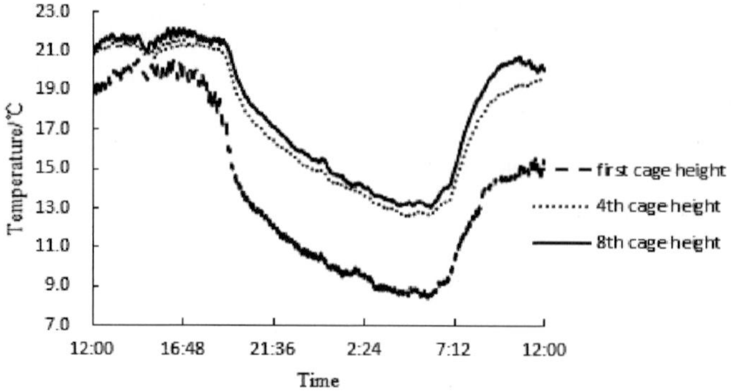

Figure 8. Temperature in fan area.

Table 5 is carbon dioxide concentration in different areas and different cage heights. In the middle of house, carbon dioxide concentrations were higher than in the cooling pad area and fan area, but were all less than lower boundary for laying hen production (3000 ppm).

Table 5. Carbon dioxide concentrations (in ppm) in different areas and different cage heights.

Cage height	Cooling pad area	Middle of house	Fan area
1st	1931±227	2547±239	1544±227
2nd	2217±151	2477±237	1736±231
3rd	2394±295	2694±174	1945±365
4th	2238±268	2582±225	1968±202
5th	2439±119	2641±148	2153±158
6th	2382±153	2683±334	2090±121
7th	2503±224	2851±180	2144±259
8th	2432±263	2445±313	1758±96

Table 6 is wind speed in different areas and different cage heights. Wind speed in fan area was significantly higher than the cooling pad area and middle of house, and was higher than upper boundary of 0.25 m/s.

Table 6. Air velocity (in m/s) in different areas and different cage heights.

Cage height	Cooling pad area	Middle of house	Fan area
1st	0.21±0.1	0.29±0.06	0.31±0.04
2nd	0.19±0.1	0.29±0.09	0.30±0.05
3rd	0.19±0.07	0.29±0.05	0.34±0.07
4th	0.13±0.04	0.27±0.06	0.27±0.08
5th	0.06±0.03	0.14±0.06	0.37±0.08
6th	0.09±0.02	0.15±0.04	0.37±0.13
7th	0.10±0.04	0.13±0.06	0.36±0.12
8th	0.22±0.08	0.17±0.07	0.52±0.15

4. Conclusions

Along horizontal direction, temperature differences between the cooling pad area and the fan area were 2.7 to 5.6°C. Moreover, the temperature difference in the bottom was higher, in the upper was slightly small, and in the middle of house was minimum. Along vertical direction, temperature differences between the first and the 8th cage height were 1.3 to 4.1°C. Temperature difference in sections close to cooling pad and fan were higher. In the middle of house, the difference was minimum. The highest diurnal temperature difference was 9°C.

Wind speed was higher than 0.25 m/s at sections which were close to the fans at the first cage height. One of the most important reasons was that the hen house was not air tight. Air leakages were considered to be the main reason for poor environmental conditions during the cold season (Seo et al., 2012). Air infiltration resulted in unexpected air flow patterns, especially in areas near the cooling pad. Outdoor cold air entered the house through gaps between the shifting board behind cooling pad, which led to decreasing inlet wind speed and static pressure in house. In future study, how air leakage influence environment condition in hen house and hen production could be further researched. Carbon dioxide concentrations in house were lower than 3000 ppm, which were in the suitable range of laying hen production.

References

Abbas, T.E., M.M. Yousuf, M.E. Ahmed, A.A. Hassabo, 2011. Effect of fluctuating ambient temperature on the performance of laying hens in the closed poultry house. Research Opinions in Animal and Veterinary Sciences. 254–257.

Emmans, G.C., and D.R. Charles, 1977. Climatic environment and poultry feeding in practice. In *Nutrition and the Climatic Environment.* Butterworth, London, Eds., W. Haresign, H. Swan and D. Lewis. 31–49.

Kuczynski, T., 2002. The application of poultry behavior responses on heat stress to improve heating and ventilation systems efficiency. Agricultural Engineering. 5(1), 01.

MAPRC, 1999. *Environmental quality standard for the livestock and poultry farm. NY/T388—1999.* The Ministry of Agriculture of the People's Republic of China, Beijing, China.

Seo, I.H., I. B. Lee, O.K. Moon, S.W. Hong, H.S. Hwang, J.P. Bitog, J. W. Lee, 2012. Modelling of internal environmental conditions in a full-scale commercial pig house containing animals. Biosystems Engineering. 111(1), 91–106.

Talukder, S., T. Islam, S. Sarker, M.M. Islam, 2010. Effects of environment on layer performance. Journal of the Bangladesh Agricultural University. 8(2), 253–258.

Warren, D.C., R. Conrad, A.E. Schumacher, and T.B. Avery, 1950. Effects of fluctuating environment on laying hens. Technical Bulletin. Kansas Agricultural Experiment Station, 43 p.

Webster, A.B., and M. Czarick, 2000. Temperatures and performance in a tunnel-ventilated, high-rise layer house. The Journal of Applied Poultry Research. 9(1), 118–129.

Spatial Distribution of Ammonia Emission Density As Impacted by Poultry and Swine Production in North Carolina of the U.S.

Yijia Zhao, Lingjuan Wang-Li *

Department of Biological and Agricultural Engineering, North Carolina State University, Raleigh, North Carolina, USA

* Corresponding author. Email: lwang5@ncsu.edu

Abstract

Emission, fate, transport and transformation of atmospheric ammonia (NH_3) have become an increasing concern due to the associated adverse effects on air/water quality, soil health, and ecosystem. This paper reports an investigation of the spatial distribution of top-ranking NH_3 emission sources, poultry and swine operations, and their associated NH_3 emission spatial distribution in North Carolina (NC). Google Earth, Google Map, and ArcMap were used to identify poultry and swine farms. Regional NH_3 emission distribution map was developed by (1) incorporating emission factors reported in the literature and identified animal feeding operation (AFO) farm spatial distribution data; (2) using Carnegie Mellon University (CMU) NH_3 Emission Inventory model. Geographic information system (GIS) mapping of swine and poultry facilities and spatial variation of NH_3 emission in NC were developed. It was found that the total numbers of poultry and swine farms were 6605, of which 4390 were poultry farms and 2215 were swine farms. Approximately 56,583 tons of NH_3 per year were emitted from poultry production alone and, 78,084 tons per year were emitted from swine production. In total, 134,667 tons per year of NH_3 was emitted from the combination of the poultry and swine farms. Poultry and swine production and corresponding NH_3 emissions were concentrated in the Wilmington and Fayetteville regions of NC. These findings will help advance our understanding of AFO NH_3 emission impact on regional scale that may lead to the improved regulation and control strategies of NH_3 emission.

Keywords: Animal feeding operation, ammonia annual emission, GIS, ArchMap

1. Introduction

As a primary gaseous pollutant emitted from animal feeding operations (AFOs), ammonia (NH_3) has become an increasing concern due to its adverse effects on air and water quality, soil health, and ecosystems (Paerl, 1997; Asman et al., 1998; Aneja et al., 2000). Once emitted, NH_3 may be removed from atmosphere through dry or wet depositions, and or through chemical transformation (Asman et al., 1998). Fate and transport of NH_3 emissions may cause aquatic eutrophication, soil acidification, defoliation and discoloration of the foliage, leading to ecosystem degradation. In addition, NH_3 may react with acidic gas species to form secondary inorganic fine particulate matter (iPM_{fine}, a.k.a. $iPM_{2.5}$).

In the United States (U.S.), AFOs represent the largest source of NH_3 emissions (EPA, 2004), contributing to approximately 71% of total NH_3 emissions in the nation. In animal production systems, NH_3 is a common by-product of animal waste (Gay and Knowlton, 2009). Most NH_3 volatilization comes from urea or uric acid of animal urine (McCubbin et. al, 2002). Ammonia emission from animal production systems depends on factors such as the amount of nitrogen in feed, size and species of the animals, housing conditions (e.g., humidity, temperature), and animal waste handling practices (Anderson et. al, 2003).

In an effort to quantify the nation's NH_3 emission inventory, *Carnegie Mellon University (CMU) Emission Inventory Model* was developed and has been widely used (Strader et al., 2002). The CMU Model is intended to estimate NH_3 emission by county in each state. This model accounts for emissions from 82 individual sources including livestock (Kirchstetter et al., 2003). It incorporates activity data, emission factors as well as temporal aspects.

In North Carolina (NC), the major types of AFOs are poultry and swine. Swine farms are concentrated in the middle area (Wilmington Region and Fayetteville Region) of the state. More than 90% of the swine production of NC resides in the Coastal Plain Region, and this phenomenon tends to have great impacts on coastal estuaries (Walker et al., 2000). The six most highly populated counties in this region (Duplin, Sampson, Greene, Wayne, Bladen, and Lenoir) have an average swine density of about 528 hogs km^{-2}, and the remainder of the region has a density of 65 hogs km^{-2} (Walker et al., 2000). Ammonia emission from the six Coastal Plain counties accounts for 36% of the statewide emission. Ammonia emission from swine production accounts from 77% of the total NH_3 emission in this region (Aneja et al., 2000).

Although it has not been holistically investigated, NH_3 emissions from AFOs have been perceived to be an important contributor to the atmospheric $PM_{2.5}$ (Paulot and Jacobson, 2014). Due to lack of temporal and spatial monitoring data of $iPM_{2.5}$ precursor gases (i.e., sulfur dioxide (SO_2), mono-nitrogen oxides (NO_x) and NH_3) in animal production areas where most NH_3 emissions are generated, contributions of AFO NH_3 emissions to the atmospheric $PM_{2.5}$ remains unknown and needs to be systematically and experimentally investigated. Figure 1 illustrates a holistic understanding of the pathway from NH_3 emissions to atmospheric $iPM_{2.5}$. On the one hand, the spatial and temporal variations of AFO NH_3 emissions may have spatial and temporal effects on ambient concentrations of precursor gases, thus, on atmospheric $iPM_{2.5}$ concentrations; on the other hand, different fate and transport (deposition and dispersion) patterns of different precursor gases, and the dynamics of particle-gas phase partitioning under different atmospheric conditions may alter the effects of AFO NH_3 emissions on ambient $iPM_{2.5}$ spatially and temporally. It is likely that fast removal of NH_3 through dry and wet deposition may limit AFO NH_3 emission contributions to the formation of secondary $iPM_{2.5}$ in the area far away from emission sources where significant SO_2 or NO_x may present in the air.

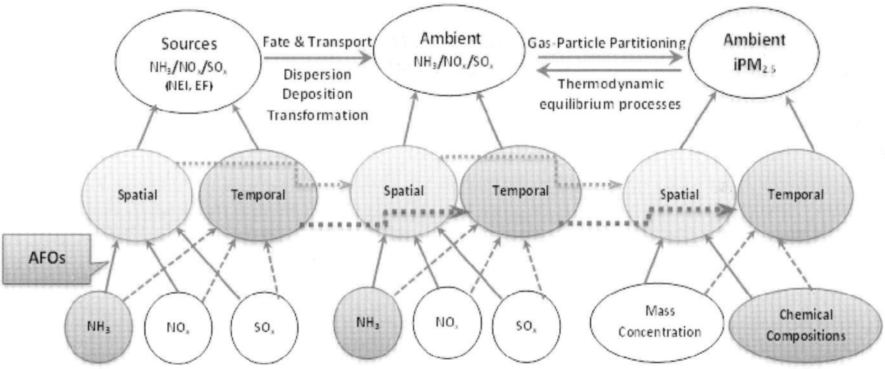

NEI: national emission inventory; EF: emission factor

Figure 1. Interactive pathways from precursor gas emissions to ambient $iPM_{2.5}$ (Wang-Li, 2015)

To systematically assess the contributions of AFO NH_3 emissions to the ambient $iPM_{2.5}$, it is essential to characterize (1) spatial and temporal variations of precursor gas emissions (i.e., NH_3, NO_x, SO_x); (2) depositions and transport of NH_3, NO_x, SO_x as impacted by emission source characteristics, land uses and management practices; (3) spatial and temporal variations of ambient precursor gas concentrations in response to the fate and transport of the emissions; (4) dynamics of gas-particle phases (NH_3/NH_4^+) partitioning in response to the spatial and temporal variations of atmospheric NH_3, NO_x, SO_x concentrations as well as meteorological conditions. This study was to investigate the inter-relationship between the distribution of AFOs, particularly poultry and hog, and ammonia emission within NC. This is the first step to obtain holistic understanding of AFO ammonia emissions impact on ambient $iPM_{2.5}$ on a regional scale.

2. Materials and Methods

This project started with identification of the spatial distributions of poultry and swine operation farms in NC using Google Earth, Google Maps, and ArcMap. Spatial distribution of the NH_3 emissions density in the whole state was determined by converting the farm numbers into emission rates through two approaches (1) using published emission factors and (2) applying CMU Emission Inventory model based upon identified spatial distribution of the poultry and swine farms in the state

2.1. Farm scanning and mapping in Google Earth

The top-ranking emission sources of NH_3 are confined as poultry and swine farms in NC. The spatial distribution of poultry and swine farms was determined by using Google Earth and Google Maps. To be specific, the locations of poultry and swine farms were identified by manually scanning the entire satellite map of NC. Each identified farm was recorded in an Excel spreadsheet by its assigned ID, county name, site name, physical address, latitude/longitude, animal type, house type/house number/dimensions, and ridge vent numbers per house, if applicable, land application type, lagoon number, total house number, size factor, and extra notes.

In order to ensure the accurate mapping results, the spatial distribution of searched swine facilities was verified by referring to the list of permitted swine facilities created by NCDENR (NCDENR, 2015). The information retrieved from the list of permitted facilities included facility name, animal category, permit type, animal type and stage, and animal activity. Over 90% of the permitted facilities were already found through manual scanning of the Google Earth map. Through comparing the farm list of the original Google Earth searching results to the list of permitted facilities, missing animal production facilities and associated farm information were added into the farm list in this study. Since the poultry operations in NC are not required to obtain any regulatory permit and also due to the confidentiality of the poultry facility, this study was not able to verify the accuracy of source mapping of the poultry facilities. However, this problem was minimized by comparing the USDA's statistic reports and the survey results (USDA, 2012).

2.2. Converting farm spatial distribution to NH_3 emission spatial distribution

In this study, swine and poultry farm spatial distributions were converted to the NH_3 emission density spatial distribution through two approaches (1) using most up-to-date reported emission factors and the farm inventory data (e.g., the farm spatial locations, and their activity data, numbers of houses per farm, numbers of animals per house) identified through the Google Earth scanning and (2) using the CMU model simulation.

2.2.1. Emission factor based assessment

In the emission factor approach, NH_3 emission of each production farm was calculated using reported emission factors and the farm activity (number of houses, number of animals per house, etc.) data. In order to calculate the annual NH_3 emission, several important assumptions were made beforehand.

For poultry operations, since it is impossible to distinguish different types of poultry (e.g., broilers, layers, turkeys) through Google Earth searching, also because broiler operations are lot more than layers and turkeys in NC, all the identified poultry farms were treated as broiler operations. Furthermore, it was assumed that a single broiler flock lasted for 49 days and there were totally five complete flocks in a year with one flock on new litter and four flocks on built-up litter. After a thorough literature search for data completeness and representativeness, NH_3 emission rates reported by Burns et al. (2007) were used to estimate NH_3 emissions from the identified broiler farms in NC. Specifically, the emission rates for summer and winter flocks were 0.021 and 0.055 kg bird^{-1} flock^{-1}. The NH_3 emission rate of broiler house on new litter was 0.49 g d^{-1} bird^{-1}. The emission rate of broiler house on built-up litter was 0.62 g d^{-1} bird^{-1}. In addition, in calculating annual NH_3 emission, a bird density of 12.3 bird m^{-2} was applied, as suggested by the study of Burns et al. (2007). Using these reference numbers, the annual total NH_3 emission from poultry production and emissions were calculated and plotted based on the spatial distribution of emission sources in this study.

For swine operations, NH_3 emission rates for finishing barns, sow barns, nursery barns and lagoons from two NC swine sites of the National Air Emission Monitoring Study (NAEMS) (Bogan et al., 2010) were used to calculate the farm level emissions. Specifically, the NH_3 emission rates were 2.7 kg yr^{-1} pig^{-1} for sow barns, 3.0 kg yr^{-1} pig^{-1} for finishing barns, 1.5 kg yr^{-1} pig^{-1} for nursery barns. Based on the reported data of animal characteristics in the NAEMS's NC3B/NC4B summary reports, the stocking densities were assumes to be 1.1 pigs m^{-2} for finish /nursery barns and 0.4 pig m^{-2} for sows. Based on NC4B (sow), the lagoon requirement was assumed to be 12.25 m^2 pig^{-1}. Based on NC3A (finisher), the lagoon requirement was assumed to be 2.37 m^2 pig^{-1}. The annual NH_3 emission rate from swine lagoon was 0.7 kg m^{-2}.

2.2.2. CMU Ammonia emission model simulation

The two animal categories in the CMU model used by this study were poultry and swine. Under the category of poultry, several types of poultry production were listed as broilers, layers with dry manure, layers with wet manure, turkeys, and poultry composite. Under each type of poultry, the emission activities were divided into confinement, storage, and land application. Under the category of swine production, there were housing operation, lagoon operation, outdoor operation, and swine composite. For each type of swine operation, NH_3 emission could come from confinement, storage, and land application.

The results of annual NH_3 emission from poultry production from two methods (emission factor *versus* CMU modeling) were compared. Since the CMU was developed in 2004 and the emission factor approach reflecting recent farm activity levels and emissions, comparison of these two approaches may provide some indications about the changes of spatial distribution of the NH_3 emissions from swine and poultry operations.

2.3. Emission mapping in ArcMap

ArcMap (ESRI 2011, Redlands, CA) technology was applied to generate the AFO farms and associated ammonia emission distribution maps. The data from Google Earth was simply transferred to ArcMap, which automatically generated the collective maps utilized in the Results section, which displays NC maps of farm distributions and NH_3 emissions.

3. Results and Discussion

3.1. Spatial distribution of the poultry and swine farms

In NC, the total numbers of poultry and swine farms were found to be 6605 of which 4390 were poultry farms and 2215 were swine farms. Figure 2 illustrates the spatial distribution map of the poultry and swine farms in NC. The majority of farms concentrates in the middle-west part of NC, specifically Fayetteville Region and Wilmington Region. Among all the counties within NC, the top counties having the largest numbers of farms were found to be Duplin (924 farms, 13.9% of the total), Sampson (746, 11.2%), Union (345, 5.2%), Wayne (296, 4.5%), Randolph (272, 4.1%), Wilkes (252, 3.8%), Alexander (226, 3.4%), and Bladen (202, 3%). As shown in this figure, the spatial distribution of AFOs was spreading out the entire NC. However, approximately 25% of the total number of AFOs was located in Duplin and Sampson counties. This agrees with the livestock inventory in the 2012 Census of Agriculture created by US Department of Agriculture (USDA) (USDA, 2012).

Figure 3 provides information about poultry farms spatial distribution in NC. Out of the 6634 farms, there were 4390 poultry farms found in NC. The counties that contained the largest numbers of poultry farms were found to be Duplin (362 farms, 8.2% of the total number of poultry farms), Union (335, 7.6%), Sampson (296, 6.7%), Randolph (263, 6%), and Wilkes (251, 5.7%).

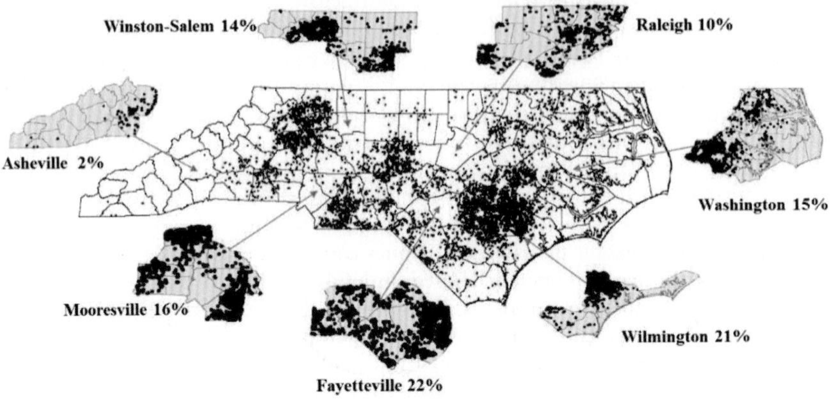

Figure 2. Spatial distribution of poultry and swine farms separated by region in NC.

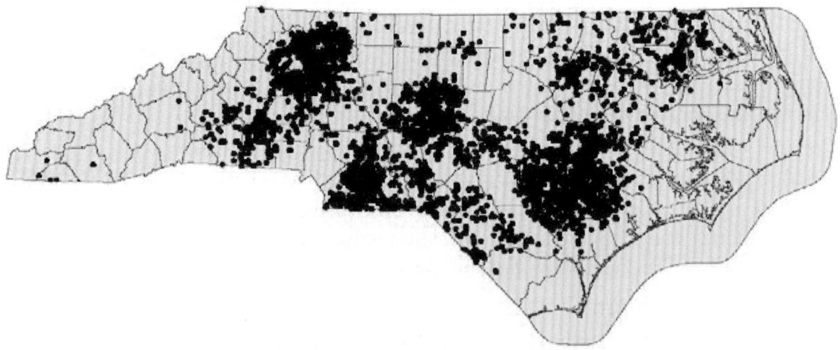

Figure 3. Spatial distribution of the poultry farms in NC.

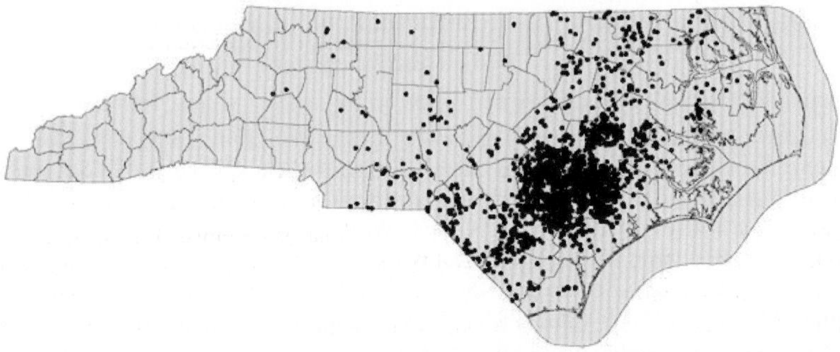

Figure 4. Spatial distribution of the swine farms in NC.

Figure 4 provides information of swine farm spatial distribution in NC. In brief, 2215 swine farms were found in NC. Unlike poultry farms, which were spreading several regions across the entire NC, the distribution of swine farms was found to be quite concentrated in two counties in central regions: Duplin (Wilmington Region) and Sampson (Fayetteville Region). The top counties discovered to have the largest numbers of swine farms are Duplin (562 farms, 25.4% of the total number of swine farms), Sampson (447, 20.2%), Bladen (150, 6.8%), Wayne (126, 5.7%), and Greene (111, 5%).

3.2. Annual NH_3 emission mapping

For the poultry operation, the emission factor method predicted the annual emission from poultry farms of the entire state to be 56,583 tons, whereas the total annual emission predicted by the CMU model was 50,645 tons. Figure 5 shows the comparison of the annual NH_3 emission maps by the two methods. It is noticed that the two methods for NH_3 emission spatial distributions are comparable. In the first method, the top counties having the highest annual NH_3 emissions were Duplin (5457.3 tons, 9.6% of the total state emission), Sampson (4530.3 tons, 8%), Union (3858.6 tons, 6.8%), and Wilkes (3083.8 tons, 5.5%). In the CMU model, the top counties were Union (4815.6 tons, 9.5%), Duplin (4464.4 tons, 8.8%), Wilkes (3893.9 tons, 7.2%), and Sampson (3622.9 tons, 7.2%). The primary factor determining the county level emissions is the farm density. Therefore, the area with high farm density had a high level of NH_3 emissions. Comparing the emissions of Duplin and Sampson counties, the prediction by the emission factor method was 20 to 30% higher than that from the CMU model. Since the CMU model was created based on the livestock inventory prior to 2004, the difference could be due to the increase of poultry production in recent years.

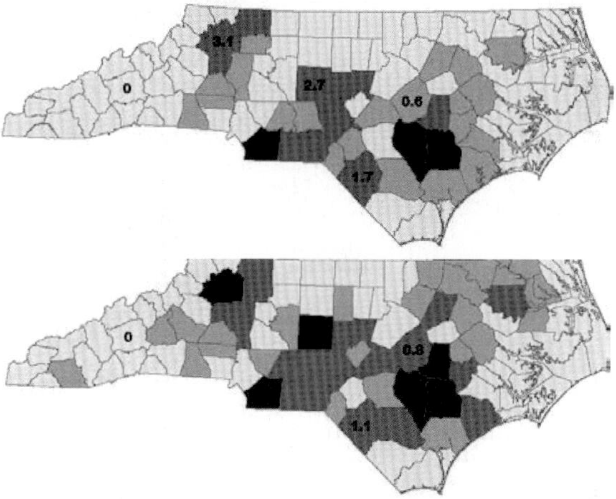

Figure 5. Comparison of annual NH_3 emissions (1×10^3 tons yr^{-1}) from poultry operations by the two methods (top: emission factor method; bottom: CMU model method).

For the swine operations, Figure 6 shows the comparison of the annual NH_3 emission maps by the two methods. Once again, the NH_3 emission spatial distributions assessed by the two methods are comparable. According to the CMU model, total annual NH_3 emission from swine production for the entire state is 78,084 tons. The top counties having the highest annual emissions were Duplin (16,742.6 tons, 21.4% of the total state emission), Sampson (15,475.3 tons, 19.8%), Bladen (7072.4 tons, 9.1%), and Wayne (4372.7 tons, 5.6%). The total emission of these four counties accounts for 43% of the entire state. This finding is in accordance with the spatial distribution of swine farms shown in Figure 4, in which Duplin, Sampson, Bladen, and Wayne have the highest numbers of swine farms.

Annual NH_3 emission maps from the combination of poultry and swine farms in NC are illustrated in Figure 7. Total emission of combination of poultry and swine production was 158,019.6 tons yr^{-1}. As shown in Table 1, the top five counties having the largest amounts of NH_3 emission were found to be Duplin, Sampson, Bladen, Wayne, and Union. Compared to the poultry production, swine production contributes much more to the combined emission. With poultry production added on into the combine emission, the spatial distribution of NH_3 emission was spreading from Wilmington Region towards Fayetteville Region and Winton-Salem Region.

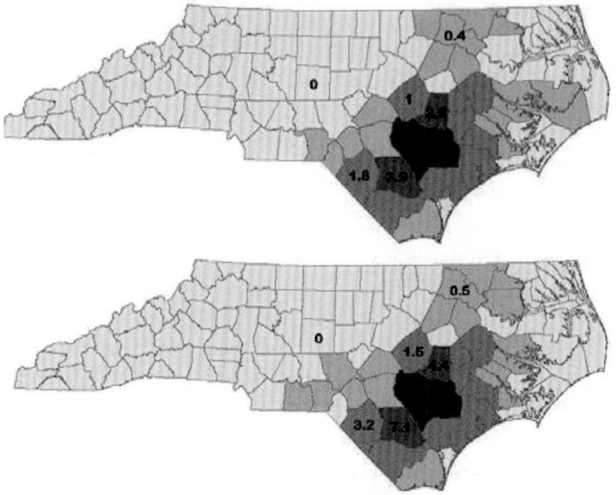

Figure 6. Comparison of annual NH_3 emissions (1×10^3 tons yr^{-1}) from swine operations by the two methods (top: emission factor method; bottom: CMU model method).

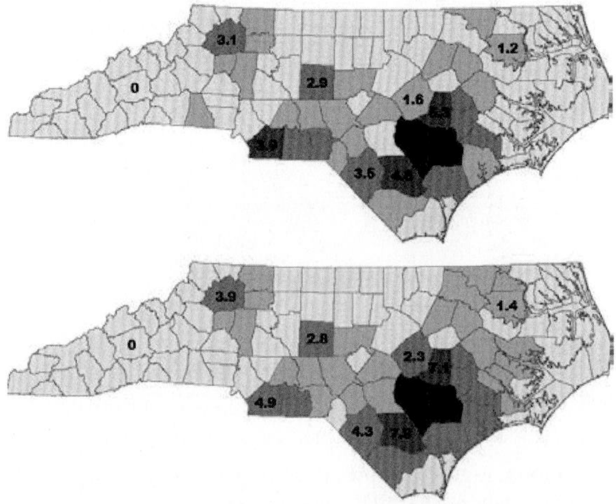

Figure 7. Comparison of annual NH_3 emissions (1×10^4 tons yr^{-1}) from poultry and swine combined by the two methods (top: emission factor method; bottom: CMU model method).

Table 1. Counties with the largest amounts of NH_3 emission.

Rank #	County name	NH_3 emission (ton yr^{-1})	% of the total
1	Duplin	21,207	13.4%
2	Sampson	19,098	12.1%
3	Bladen	7838	5.0%
4	Wayne	7101	4.5%
5	Union	4923	3.1%

4. Conclusions

The spatial distributions of the swine and poultry operations and their associated NH_3 emissions distributions were investigated in this study. In NC, the total numbers of poultry and swine farms were found to be 6605 and out of this total, there were 4390 poultry farms and 2215 swine farms. Approximately 56,583 tons of NH_3 per year were emitted from poultry production alone and, 78,084 tons per year were emitted from swine. In total, 134,667 tons per year of NH_3 was emitted from combination of the poultry and swine farms. The majority of poultry and swine farms are located in Fayetteville Region and Wilmington Region. For each spatial location, the higher the farm numbers, the higher the NH_3 emission density, therefore, the regions with high AFO farms numbers also have the high NH_3 emissions.

References

Anderson, N., R, Strader, C. Davidson, 2003. Airborne reduced nitrogen: ammonia emissions from agriculture and other sources. Environment International. 29 (2), 277–286.

Aneja, V. P., J.P. Chauhan, J.T. Walke,. 2000. Characterization of atmospheric ammonia emissions from swine waste storage and treatment lagoons. Journal of Geophysical Research: Atmospheres (1984–2012). 105 (D9), 11535–11545.

Asman, W. A., M.A. Sutton, J.K. Schjørring, 1998. Ammonia: emission, atmospheric transport and deposition. New Phycologist. 139 (1), 27–48.

Bogan, B.W., K. Wang, W.P. Robarge, J. Kang, and A.J. Heber, 2010. National Air Emissions Monitoring Study: Emissions Data from Three Swine Finishing Barns in North Carolina - Site NC3B. Final Report. Purdue University, West Lafayette, IN, July 2.

Burns, R., H. Xin, R. Gates, H. Li, S. Hoff, L. Moody, J. Earnest, 2007. Tyson broiler NH_3 emission monitoring project. Final report to Tyson Foods. Iowa State University, Ames, IA.

EPA. 2004. Estimating Ammonia Emissions from Anthropogenic Nonagricultural Sources - Draft Final Report.

Gay, S. W., K.F. Knowlton, 2009. Ammonia emissions and animal agriculture. Virginia Cooperative Extension. 442 (110).

Kirchstetter, T. W., C.R. Maser, N.J. Brown, 2003. Ammonia emission inventory for the state of Wyoming. Lawrence Berkeley National Laboratory.

McCubbin, D. R., B.J. Apelberg, S. Roe, F. Divita, 2002. Livestock ammonia management and particulate-related health benefits. Environmental Science & Technology. 36 (6), 1141–1146.

NCDENR, 2015. Animal Feeding Operations Active Permits. Available at: http://portal.ncdenr.org/web/wq/aps/afo/perm. Accessed on August 31, 2015.

Paerl, H. W., 1997. Coastal eutrophication and harmful algal blooms: Importance of atmospheric deposition and groundwater as "new" nitrogen and other nutrient sources. Limnology and oceanography. 42 (5), 1154–1165.

Paulot, F., and D. J. Jacob, 2014. Hidden cost of U.S. agricultural exports: particulate matter from ammonia emissions. Environmental Science & Technology. 48 (2014), 903–908

Strader, R., Anderson, N., Davidson, C., 2001. User Guide—CMU NH_3 Inventory Version 2.0. Carnegie Mellon University, Pittsburgh, PA. http://www.cmu.edu/ammonia/.

USDA NASS, 2012. Census of Agriculture. http://www.agcensus.usda.gov/Publications/2012/. Accessed on September 18, 2015.

Walker, J. T., V.P. Aneja, D.A. Dickey. 2000. Atmospheric transport and wet deposition of ammonium in North Carolina. Atmospheric Environment. 34 (20), 3407–3418.

Wang-Li, Li, 2015. Insights to the Formation of Secondary Inorganic $PM_{2.5}$: Current Knowledge and Future Needs. International Journal of Agricultural and Biological Engineering. 8 (2), 1–13.

Theme II:

Environmental Control for Modern Animal Production

On-farm Evaluation of Wood Bark-based Biofilters for Reduction of Odor, Ammonia, and Hydrogen Sulfide Emissions

Lide Chen *, **Gopi Krishna Kafle, Howard Neibling, Brian He**

Department of Biological and Agricultural Engineering, University of Idaho, Twin Falls, Idaho 83301, USA

* Corresponding author. Email: lchen@uidaho.edu

Abstract

Odor and gas emissions from confined swine facilities can be of serious concern for swine farmers, and the general public. Biofiltration has been recognized as one of the most promising and cost-efficient technologies for mitigating the odor and gas emissions. Two down-flow wood bark-based biofilters were built at a commercial swine nursery facility to reduce odor and gas emissions from the ventilation exhaust. The biofilters were evaluated for their effectiveness in mitigating odor, ammonia (NH_3), and hydrogen sulfide (H_2S) emissions under actual swine farm conditions. The test results showed that the highest reductions in odor, NH_3, and H_2S at empty bed residence times (EBRT) of 1.6–3.1 s, were 73–76%, 95–98%, and 96–100%, respectively. A minimum media depth of 254 mm, moisture content between 35–55%, and EBRT greater than 2–3 s are recommended for effective operation of wood bark-based biofilters at swine farms.

Keywords: Down-flow biofilter, mitigation, gas concentrations, swine, abatement

1. Introduction

The trend toward high-density, confinement swine operations has increased dramatically in recent years to satisfy higher demand of meat products. As the number and size of confinement facilities have increased, air pollution concerns due to odors and gas emissions have also increased. Sources of odor and gas emissions include land application events, manure storage facilities, and barn exhaust. Ventilation air is typically exhausted into the atmosphere without treatment. This exhaust air contains odorous gases and particulate matters that can represent a concentrated odor source (Hoff and Harmon, 2006). Besides the nuisance associated with odors, they have also been shown to have a negative impact on the public health of those living in close proximity to the swine farms. Therefore, effective methods of odor control will have to be put into operation.

Biofiltration has been recognized as the most promising and cost-efficient technology for treating air streams containing odorous compounds (Estrada et al., 2012). Biofilter media selection is critical in biofilter design and performance. The media must have a high porosity for minimizing pressure drop across the biofilter, good moisture holding capacity, and a sufficiently long useful life (Chen and Hoff, 2009). The mixture of wood chips and compost (70:30 to 50:50 on a mass basis) has been recommended for biofilters at animal production sites (Nicolai and Janni, 2001). However, special care is needed to screen fine particles from compost/wood chip mixtures to reduce operating pressure drop. Using only wood chips as biofilter media can reduce the pressure drop without specialty fans (Dumont et al., 2014), which results in fewer construction and operating costs.

Although a number of laboratory scale studies have shown that biofilters could be a promising technology for mitigating aerosol emissions from confined swine production facilities, the use of on-farm biofilters continues to be low in the United States. This is due in part to the lack of information pertaining to on-farm biofilter performance. This study aimed to investigate performance of two on-farm down-flow biofilters: single state biofilter 1 (BF1) and two stages biofilter 2 (BF2). The objectives of this research were: (1) to determine the optimum water supply rate at different airflow rate conditions; (2) to evaluate the performance of the biofilters in odor, ammonia (NH_3), and hydrogen sulfide (H_2S) reduction; and (3) to evaluate the effects of media moisture content (MC) and depth on NH_3 and H_2S reductions.

2. Materials and Methods

2.1. Description of experiment site

This study was conducted at a commercial swine nursery facility, which consisted of four 4.3 m × 12.8 m nursing rooms, near Kimberly, Idaho, USA. Each room had an independent tunnel ventilation system. Air entered the rooms through an inlet located on the south wall of each room. There were two variable speed exhaust fans (primary and secondary) in the north wall of each room. A shallow pit with a depth of 0.6 m was constructed below the slatted floor to collect manure and washing water. Around 60–70% of total volume in the shallow pit was drained to a lagoon every 5 days. Small pigs were moved in at about 5–7 kg and were raised to approximate 64–68 kg at the nursery facility. Two vertical down-flow biofilters (Figure 1) were installed in the front of rooms 7 and 8 (N7 and N8) to treat the exhaust air from the 406-mm first stage fans (Multifan, 4E40Q, Schuyler, NE, USA).

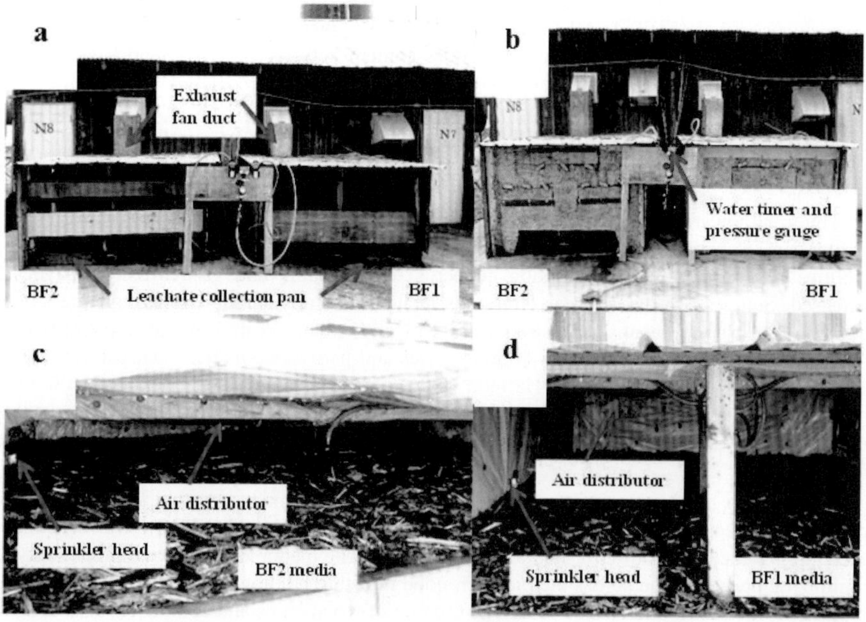

Figure 1. Two vertical down-flow biofilters: (a) biofilter 1 (BF1) and biofilter 2 (BF2) with front door open; (b) BF1 and BF2 with front door closed; (c) media and water supply system in BF2; and (d) media and water supply system in BF1.

2.2. Biofilter details

The biofilters were 2.15 m × 2.26 m in cross-section and 1.22 m in height. Both of the biofilters had the media depth of 381 mm. The biofilters were connected to the primary exhaust fan of their respective rooms using a rectangular duct made of plywood (Figure 1). Metal mesh (10 mm diameter holes) with support on the ground was built to support the biofilter media. In BF2, two horizontal metal mesh surfaces were constructed with a vertical spacing of 381 mm.

Tap water was supplied to the biofilter media via four customized sprinkler heads in each biofilter. The water supply time was controlled by a timer (Model# 6020P/61512, Drip Inc. Concho, AZ, USA). The water supply pressure was controlled within a range of 262–283 kPa by a pressure gauge (Pro Plumber, PP100G, Mansfield, Ohio, USA). Water was supplied for

20–30 s every 15–240 min, depending upon the duration of the exhaust fan operations and environmental conditions.

2.3. Biofilter media selection and arrangement

Based on our lab tests for three different media pressure drop and water holding capacity (WHC), the combination of shredded wood bark (SWB) (species-Thuja plicata) and medium wood bark (MWB) were chosen for the field scale biofilters in this study. The biofilter media composition was SWB: MWB = 1:2 on a volume basis. SWB was used for the top layer (127 mm) and MWB for the bottom layer (254 mm).

Wood bark porosity was measured using the Bucket Method (Nicolai and Janni, 2001). Wood bark WHC was determined by soaking the barks in water for 24 h followed by a gravimetric method which involved oven-drying the chip samples for 48 h at 105°C. The wood bark WHC was calculated as (initial weight (before drying) - final weight (after drying))/final weight. The wood bark MC was determined by weight loss after drying overnight at 105°C.

2.4. Gas and odor measurements

Odor samples were collected using 10 L Tedlar® bags after the biofilters had been in operation for about two months. Vacuum pumps and vacuum boxes were used to simultaneously fill the bags from the inlets and outlets of the biofilters. All samples were analyzed within 24 h of collection. A dynamic forced-choice olfactometer (AC'SCENT International Olfactometer; St. Croix Sensory, Inc. Stillwater, MN) was used to evaluate odor concentration based on ASTM E679-04 (ASTM, 2004). The NH_3 and H_2S concentrations were measured with detection tubes and a gas sampler (Gastec Co. Ltd.). The detection ranges of the NH_3 and H_2S gas tubes were 0.25–75.00 ppm and 0.1–4 ppm with a detecting limit of 0.05 ppm, respectively.

2.5. Media sampling and moisture content (MC) analysis

In BF1, media samples were collected from depths of 25, 127 and 254 mm. In BF2, media samples were collected from 25 and 125 mm in both the first and the second stages. Three replications were taken from different positions at each collection depth. The media MC was calculated by taking the average of sample MC from different depths.

2.6. Test conditions and limitations of the study

Tests were conducted under actual field conditions. BF1 and BF2 were connected to different rooms. The ventilation rate of exhaust fans and the number of swine in each room were different, resulting in different exhaust fan operation duration. These differences led to differences in the test conditions (EBRT, inlet odor, NH_3, H_2S concentrations) for BF1 and BF2. The experimental conditions during odor sample collections are shown in Table 1.

Table 1. Test conditions during odor sample collections.

Location	Date (in 2013)	Media MC[1] (%)	EBRT[2] (s)	Pressure drop (Pa)	Ambient	
					T[3] (°C)	R.H.[4] (%)
BF1[5]	23-Oct	64.9(1.4)	1.6(0.1)	68.8(0.5)	20.0–20.1	52.7–53.7
	13-Nov	50.8(9.3)	3.9(0.7)	16.4(8.5)	13.3–13.4	64.0–65.1
BF2[6]	23-Oct	63.9(3.1)	3.1(0.2)	20.8(0.5)	20.1–20.7	53.0–53.5
	13-Nov	22.4(10.1)	2.7(0.0)	25.5(4.2)	13.3–13.4	65.1–65.8

Data presented in parenthesis is the standard deviation; [1]Moisture contents; [2]Empty bed residence time; [3]Temperature; [4]Relative humidity; [5]Biofilter 1; [6]Biofilter 2.

2.7. Statistical analysis and statistical indicators

The significance of differences in odor, NH_3 and H_2S concentrations, odor, NH_3 and H_2S reductions, and media MC were determined by single factor ANOVA (Analysis of Variance) with P-values of 0.05 in Excel software 2007. Fisher's least significant difference (Fisher's LSD) was calculated to judge whether two or more averages were significantly different.

3. Results and Discussion

3.1. Media characteristics

The characteristics of media (MWB and SWB) used in this study are shown in Table 2.

Table 2. Characteristics of wood barks used in this study.

	Units	Medium wood bark	Shredded wood bark
Moisture content (MC)	% (w.b.[1])	11.3	13.5
Porosity	%	59.9	68.4
Density	kg/m^3	244.3	200.8
Water holding capacity	% (w.b.)	50.8	64.1

[1]Wet basis

3.2. Effects of ventilation duration on water supply rate

Previous studies (Yang et al., 2012; Hartung et al., 2001) suggested that the 40.0–50.0% media MC is suitable for successful biofilters in terms of odor and gas reductions. The ventilation duration in this study was dependent on the size (age or weight) of the swine and their population in the barn. With continuous ventilation (24 h d^{-1}) and EBRT of 2.8–3.3s, the media MC in BF2 was maintained around 50% when the water supply rate was maintained at 227 L m^{-3} d^{-1} (watering duration of 20 s every 30 min) at ambient temperatures of 24.8–25.0°C. At this water supply rate, the leachate from BF2 was around 54% of the supply. At ventilation duration of 12 h d^{-1} and 2.4 h d^{-1}, water supply rates of 113 and 28 L m^{-3} d^{-1} were found sufficient to maintain the media MC above 50%.

With continuous ventilation and EBRT of 1.5–1.8 s, the media MC in BF1 was maintained above 50% (i.e., 56.1%) when the water supply rate was maintained at a 556 L m^{-3} d^{-1} (water supply timing of 20 s every 10 min) at 24.8–25.0°C. These results showed that under continuous ventilation (24 h d^{-1}) the water supply rate needs to be doubled to maintain the same media MC when EBRT was reduced by 50%. When ventilation duration was less than 1 h d^{-1}, the water supply rate of 3.8–4.6 L m^{-3} d^{-1} was found sufficient to maintain media MC > 50–60%. Statistical analysis showed no significant difference in the average media MC when water supply rate ranged from 227 to 556 L m^{-3} d^{-1} for continuous ventilation at the EBRT range of 1.5–3.3 s; however, the stepwise reduction in water supply rate below 227 L m^{-3} d^{-1} showed significant decrease in media MC for each drop in water supply rate. When the exhaust fan was operated less than 1 h d^{-1} no significant difference was found in media MC for water supply rate in the range of 23–453 L m^{-3} d^{-1} at EBRT of 1.5–3.3 s. The results indicate that media moisture removal rate increases with an increase in ventilation duration and air flow rate. Thus, based on the ventilation duration and EBRT, water supply rate needs to be adjusted to meet the required media MC. Lim et al. (2012) maintained MC of their wood chip media (127–254 mm depth) in the range of 40.3–64.3% when suppling water for 3–5 min every 3–4 h (i.e, 24–30 min/d) at an EBRT of 0.3–0.6 s.

3.3. Odor and gas reductions

The untreated (inlet) and treated (outlet) odor, NH_3 and H_2S concentrations and their removal percent in BF1 and BF2 under the maximum possible average MC (63.9–64.9%), average medium MC (50.8%) and average low MC (22.4%) are shown in Figure 2 a and b, respectively. The average MC at different media depths of BF1 and BF2 are also shown in Figures 2a and 2b, respectively. The average inlet odor concentrations from BF1 and BF2 were in the range of 2958–4432 OUE m^{-3} and 2853–3770 OUE m^{-3}, respectively. The odor concentrations of the treated gas were significantly lower ($P < 0.01$) than untreated gas for both BF1 and BF2. Statistical analysis did not show significant differences in the outlet odor concentrations from BF1 and BF2. Among the three levels of MC, the highest odor, NH_3 and H_2S reductions were found with the highest MC (63.9–64.95%). The highest odor, NH_3 and H_2S reductions obtained in this study at an EBRT of 1.6–3.1s were 73.5–76.9%, 95.2–97.9% and

95.8–100.0%, respectively (Figure 2). Statistically, no significant differences were found in reductions of odor, NH_3 and H_2S from both BF1 and BF2 at the high range of media MC.

The test results for BF1 showed no significant increase in odor, NH_3 and H_2S reductions with an increase in media MC from the medium range (50.8%) to the high range (64.9%). The results from BF1 also showed that the reduced EBRT (from 3.9 s to 1.6 s) showed no significant differences ($P < 0.01$) in odor, NH_3 and H_2S reductions if MC is maintained in the high range. The test results for BF2 showed that at similar EBRTs (2.7–3.1 s), the biofilter performance on reductions in odor, NH_3 and H_2S were significantly reduced ($P < 0.01$) when average MC changed from high (63.9%) to low (22.4%). Although the BF2 performance was significantly decreased ($P < 0.01$) at the low range of media MC, the odor, NH_3 and H_2S reductions in BF2 were >69.0%, which was unexpected. Figure 2 shows that the MC at different media depths in BF2 was in the range of 14.5–23.2% for depths of 25–152 mm, but the MC was 37.4% at a depth of 254 mm and >40% at a depth of 381mm (bottom of BF2). These results show that a media depth of 127–254 mm could be adequate to maintain higher reduction efficiencies (>69.0%) if the media MC is maintained in the high range (55–65%) for wood bark based biofilters.

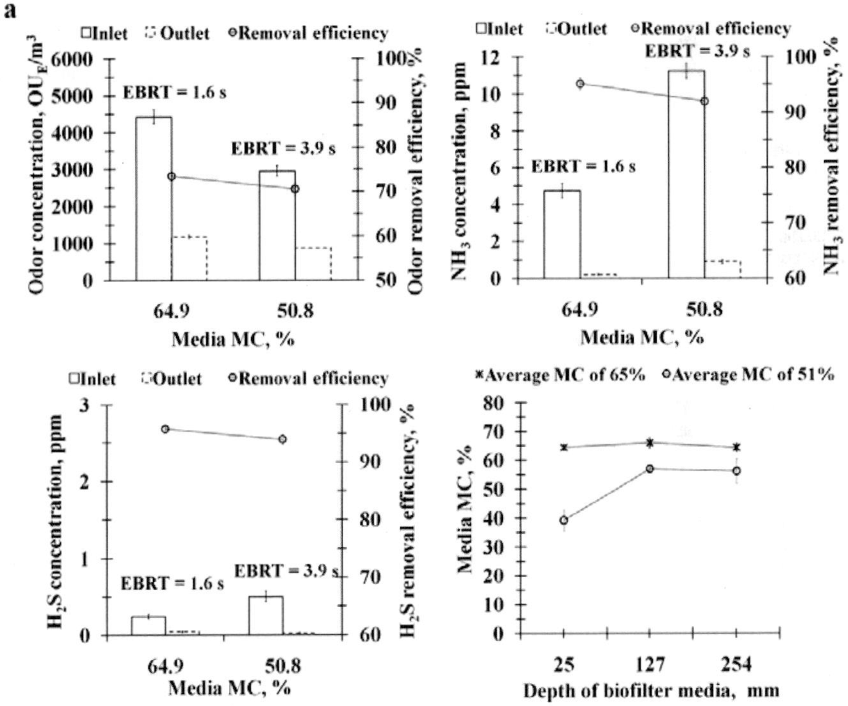

Figure 2a. Odor, NH_3 and H_2S reduction efficiency and moisture distribution at different media depths of biofilter 1 (BF1).

3.4. Effects of media depth and MC on NH_3 and H_2S reductions

Figure 3a and 3b show the NH_3 and H_2S reductions from the first stage of BF2 (SWB-127 mm) with EBRT of 0.9–1.0 s and the second stage of BF2 (MWB-254 mm) with EBRT of 1.8–2.0s at different media MC, respectively. The first stage of BF2 with 127 mm SWB showed NH_3 and H_2S reductions in the range of 56.1–84.2% and 65.2–99.0%, respectively, when MC was maintained in the high range. But decreasing the media MC to the low range resulted in lower reductions or negative reductions for both NH_3 (-22.4–6.7%) and H_2S (-13.0–14.4%) at EBRT of 0.9–1.0 s (Figure 3). The second stage of BF2 with 254 mm MWB showed NH_3 reductions in the range of 81.5–89.2%, 89.2–96% and 53.7–92.8% for the high, medium and low MC ranges, respectively. Similarly, H_2S reductions were in the range of 77.5–100%, 79.1–96.7% and 14.2–69.5% for the high, medium and low MC ranges, respectively. The second stage of BF2 with 254 mm media depth had higher NH_3 and H_2S reduction efficiencies in each range of the media MC than those in the first stage of BF2 (127mm media depth). The reason for this higher removal efficiency in the second stage of BF2 was due to the higher media depth (more media sustain more microorganisms) and longer EBRT which allow a better biodegradation. Theoretically, pollutants in the gas phase first need to be transferred to liquid phase, where they can be degraded by the microorganisms living in the biofilter. Therefore, a sufficient EBRT and microorganisms are necessary to allow the transfer and degradation of pollutants to occur, which makes EBRT a critical design and operating parameter (Hartung et al., 2001).

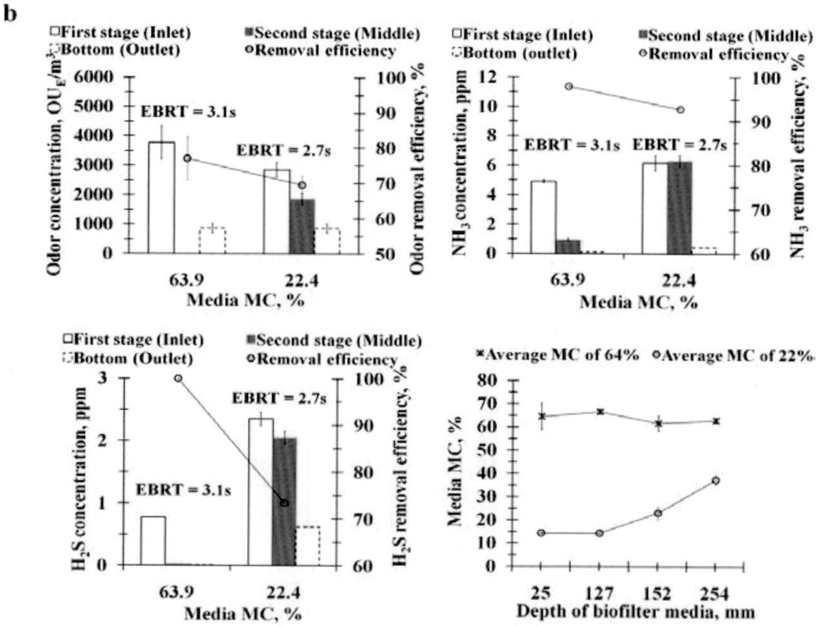

Figure 2b. Odor, NH_3 and H_2S reduction efficiency and moisture distribution at different media depths of biofilter 2 (BF2).

Figure 4 shows the NH_3 and H_2S reductions for the first and the second stages of BF2 with EBRT of 2.7–3.1 s, at different media MC, respectively. The NH_3 reductions were found in the ranges of 93.8–100%, 93.4–98.1% and 53.7–95.1% for the high, medium and low MC ranges, respectively. Similarly, H_2S reductions were found in the ranges of 98.3–100%, 92.9–97.0% and 3.1–73.4% at the high, medium and low MC ranges, respectively. No significant difference in NH_3 and H_2S reductions were detected between the medium and the high MC ranges for the media depth of 381 mm. On the other hand, NH_3 and H_2S reductions were significantly reduced ($P < 0.01$) at the low MC range.

The results of BF2 showed that MC had a greater effect on H2S removal efficiency, while media depth (EBRT) had more effect on NH3 removal efficiency. Hartung et al. (2001) also reported that efficiency of NH_3 reduction was mainly influenced by the air flow rate (EBRT). Our study showed that a minimum media depth of 254 mm and EBRT of 2–3 s are essential for a successful biofilter in terms of NH_3 and H_2S reductions. To keep proper function of the biofilter with a media depth of 127 mm, a high MC and EBRT >3.0 s should be maintained. And for biofilters with media depths of 254 mm and 381 mm, the medium MC range could be sufficient if the EBRT is maintained >2–3 s. A reasonable EBRT is closely related to the media MC and pollutant loading. Higher MC and lower pollutant loadings result in shorter EBRTs (Chen and Hoff, 2009). Lim et al. (2012) maintained a very low EBRT (0.3–0.6 s) when using a wood chip media based biofilter at a swine farm. Sheridan et al. (2002) explained that an EBRT of 1.5–2.7 s would be sufficient for the transfer of ammonia from the gas phase to the liquid.

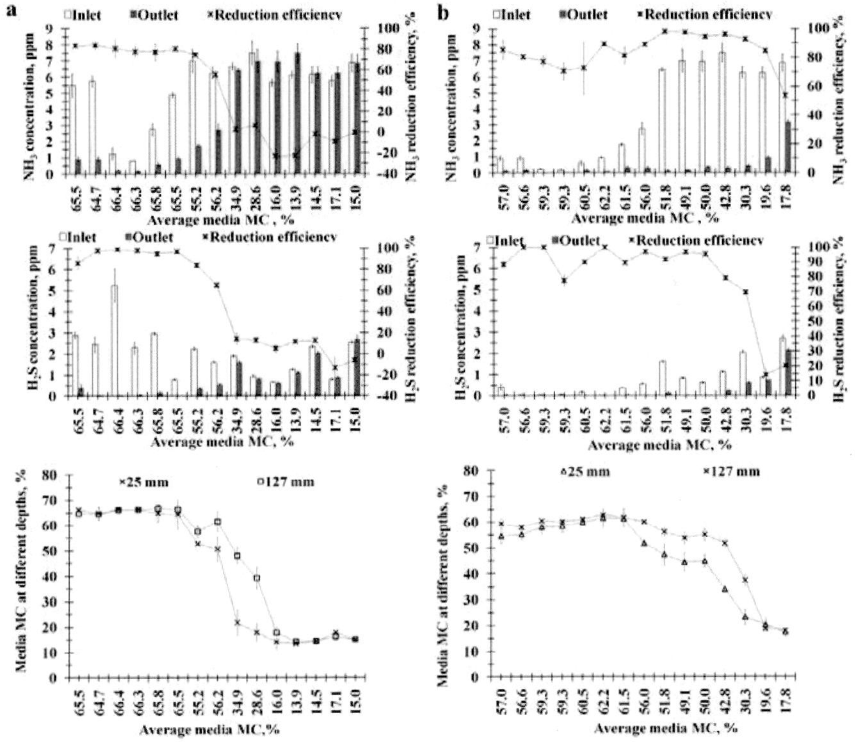

Figure 3. NH_3 and H_2S reduction efficiency for (a) first stage of biofilter 2 (BF2) at different media moisture contents (MC); and (b) second stage of BF2 at different MC.

4. Conclusions

In this study, two field scale down-flow biofilters using wood bark as media were evaluated for odor, NH_3 and H_2S mitigation under swine farm conditions. The results of this study suggested a water supply rate in the range of 227–556 L m^{-3} d^{-1} to maintain media MC of 50% or greater when the ventilation system is continuously operated at an EBRT of 1.5–3.3 s and ambient temperatures are approximately 25°C. The water supply rate needs to be doubled to maintain the same media MC when EBRT is reduced by 50%. The greatest NH_3 and H_2S (95.0% or greater), and odor (73.5–76.9%) emission reductions were obtained when media MC was maintained in the high range (63.9–64.9%). A minimum media depth $\geq$ 254 mm, MC $\geq$ 35–55% and EBRT $\geq$ 2–3 s are recommended for successful operation of wood bark based biofilters on swine farms. The reduction efficiency and pressure drop (not shown in this paper) obtained with the wood bark based biofilters in this study indicate the feasibility of their applications on swine farms.

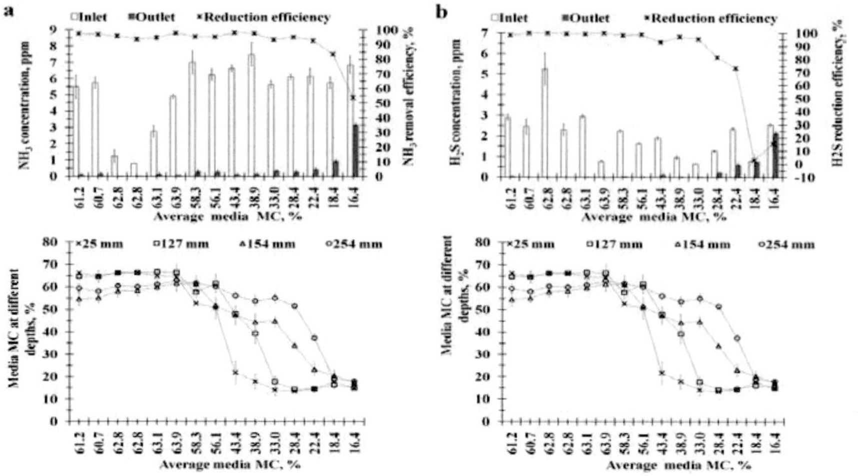

Figure 4. (a) NH_3; (b) H_2S reduction efficiency of biofilter 2 (BF2) at different media MC.

Acknowledgements

This project was partially funded by the USDA Natural Resources Conservation Service through a Conservation Innovation Grant.

References

ASTM. 2004. E679-04 Standard practice for determination of odor and taste thresholds by a forced-choice ascending concentration series methods of limits. West Conshohocken, Pa.: ASTM International.

Chen, L., S. J. Hoff, 2009. Mitigation odors from agricultural facilities: a review of literature concerning biofilters. Appl. Eng. Agric. 25, 751–766

Dumont, E., L. Hamona, S. Lagadec, P. Landrain, B. Landrain, Y. Andrès. 2014. NH_3 biofiltration of piggery air. J. Environ. Manage. 140, 26–32.

Estrada, J. M., N. J. R. Kraakman, R. Lebrero, R. Muñoz. 2012. A sensitivity analysis of process design parameters, commodity prices and robustness on the economics of odour abatement technologies. Biotech. Adv. 30, 1354–1363.

Hartung, E., T. Jungbluth, W. Büscher. 2001. Reduction of ammonia and odor emissions from a piggery with biofilters. Trans. ASABE 44(1), 113–118.

Hoff, S. J., J. D. Harmon. 2006. Biofiltration of the critical minimum ventilation exhaust air. Workshop on Agricultural Air Quality, June 5–8, Potomac, Maryland, USA.

Lau, A., K. Cheng. 2007. Removal of odor using biofilter from duck confinement buildings. J. Environ. Sci. Health A Toxic/Hazardous Substtances and Environ. Eng. 42(7), 955–959.

Lim, T.T., Y. Jin, J. Q. Ni, A. J. Herber. 2012. Field evaluation of biofilters in reducing aerial pollutant emissions from a commercial pig finishing building. Biosys. Eng. 112, 192–201.

Mayehofer, M., J. Govaerts, N. Parmetier, H. Jeanmart, L. Helsen. 2011. Experimental investigation of pressure drop in packed beds of irregular shaped wood particles. 2011. Powder Technol. 205, 30–35.

Nicolai, R.E., K. A. Janni. 2001. Biofilter media mixture ratio of wood chips and compost treating swine odors. Water Sci. Technol. 44(9) 261–267.

Sheridan, B., T. Curran, V. Dodd, J. Colligan. 2002. Biofiltration of odor and ammonia from a pig unit-a pilot -scale study. Biosys. Eng. 82(4), 441–453.

Yang, L., X. Wang, T. L. Funk, S. Shi, R. S. Gates, Y. Zhang. 2012. Moisture monitoring and control in gas phase biofilters to achieve high ammonia removal efficiency and low nitrous oxide generation. Trans. ASABE 56(5), 1895–1903.

Some Approaches for Reducing Methane Emissions from Ruminants

Safwat Mohammed Abdelrahman, Yujie Lou *

College of Animal Science and Technology, Jilin Agricultural University, Changchun, China

* Corresponding author. Email: lyjjlau@163.com

Abstract

Fermentation of the animal feed liberates greenhouse gases including carbon dioxide, methane, and nitrous oxide (CO_2, CH_4, and N_2O) which are not environmental benign. Methanogenesis is a process resulted from the digestion of the food to discard the metabolic hydrogen in the rumen system. Subsequently, myriad efforts have been dedicated to reduce the methanogenesis process. Inophores, plant bioactive compounds, essential oils, lipids, and feed management can have substantial effects not only minimizing the methanogenesis but also improving the nutrition uptake. Inophores and essential oils needed long term of in vivo studies, also some kinds of tannin showed good mitigation effects. Certain feed management strategies like "Total Mixed Ration" still needs more researches in this point, but for concentrates inclusion it was shown its ability for mitigation, but it cannot be generalized under all breeding systems. This paper reviews and discusses some of common approaches with advantages and disadvantages.

Keywords: Mitigation, abatement, effective feed conversion, environmental effects, essential oil, ration

1. Introduction

The livestock sector is facing huge challenge of meeting demand of animal products while improving its environmental sustainability. Its expected in the future to be more Increasing demand for the animal products can reach 70% more than current, as a result of the expected higher population around the world, urbanization and higher incomes (FAO, 2011). Every year ruminants produce almost 80 million tons of methane (CH_4) around the world, this emissions represent mostly 28% of the anthropogenic emissions (Beauchemin et al., 2008). Enteric methane is produced by Methanogenic Achaea and symbiotic associations of protozoa to dispose metabolic hydrogen during feed fermentation (Leng, 2008).

There are three main greenhouse gases (GHG), CH_4 is one of them, in addition to carbon dioxide (CO_2), and nitrous oxide (N_2O). The impact of greenhouse gases which are emitted from livestock on the climate changes has more concerning around the world nowadays (Steinfeld et al., 2006). Rising incomes has important effect on the dietary patterns among the less incomes population (Delgado, 2003). There are trends now for reducing greenhouse gases (GHG) from all sectors of the economy. There are legislative appeals for reducing the GHG emissions (20% less than 1990 levels by 2020) in the European union (Oberthür and Roche Kelly, 2008).

The environmental affect for both of animal production and agriculture in general is an important issue for sustaining the debate on the local and international levels (Steinfeld et al., 2006; Capper et al., 2009). The emissions of GHG excluding CO_2 must be related to Animal production according to some global analyses (Gerber et al., 2011). The ruminal microorganisms degrade the nutrients in the rumen then syntheses the volatile fatty acids (VFA) as energy source and microbial protein as protein source. However, this fermentation processing considered as energy losses in the form of CH_4, also it represents wasting source for protein in the form of ammonia as end product, these both products are carrying environmental problems and lead to limitation in productivity of the animals (Calsamiglia et al., 2007).

Methane is contributing as a greenhouse gas 21 times more than CO_2 in global warming (Gerber et al., 2011). Ruminant nutritionists have studied essential oils (Calsamiglia et al.,

2007), plant secondary metabolites (Woodward et al., 2001; Patra et al., 2006) and dietary lipids (Beauchemin et al., 2007) as rumen manipulators to improve ruminal fermentation characteristics as well as potential anti-methanogenic feed additives, which are feed management to reduce emissions of CH_4. Mc Geough et al. (2010) investigated the effect of Total Mixed Ration (TMR) on CH_4, in addition to the inclusion of concentrates (Mc Geough et al., 2010; Aguerre et al., 2011) for the same purpose. Through the light of these previous facts this paper has been prepared to present some of the approaches which are submitted as a method for CH_4 mitigation through last researches that have submitted as method for solving this problem with clarifying to the common conflicts which are taken against all these approaches.

2. The Methods for Reducing Methane

Mainly there are two common lines to reduce CH_4 release in the rumen it can be summed up in these points.

2.1. Feed additives

Inophores: Inophores were approved by the U.S. FDA in the mid-1970s as feed additives, and since then, their use has become routine in the feeding of growing ruminants. Inophores are chemical feed additives which increase productivity (weight gain per unit of feed intake) by adjusting several fermentation pathways. Reductions in CH_4 production (of up to 25%) have been observed, but the persistence of this reduction is unproven. The use of chemicals and antibiotics to increase animal productivity is increasingly becoming unpopular to the consumers of animal products. It is therefore considered that the use of ionophores to reduce CH_4 production is not a viable option (Van Nevel and Demeyer, 1992).

Monensin: Monensin has been the most studied Inophores and is routinely used in beef production and more recently in dairy cattle nutrition in North America. Inophores are banned in the European Union even though there is no evidence of genes coding for their resistance as there are with other feed-administered antibiotics (Russell and Houlihan, 2003). There have been a number of experiments with Monensin as a rumen modifier in various production systems, where enteric CH_4 production was studied as a main objective either from mitigation or from energy loses perspective. Although some studies reported a long-term mitigating effect of Monensin on CH_4 production (Odongo et al., 2007), overall, the effect of the Inophores appears to be inconsistent. In a meta-analysis of 22 controlled studies, Monensin (given at 32 mg/kg DM) reduced CH_4 emissions and CH_4 conversion rate in beef steers fed total mixed rations by 19 ± 4 g/animal per d and $0.33 \pm 16\%$, respectively (Appuhamy et al., 2013). The corresponding reductions in dairy cows were 6 ± 3 g/animal per day, and 14% for Monensin that was given at a dose of 21 mg/kg DM. Overall, the conclusion of this analysis was that Monensin had stronger anti-methanogenic effect in beef steers than dairy cows, but the effects in dairy cows can potentially be improved by dietary modifications and increasing Monensin dose. On the other hand, Inophores have some resistance worry as a feed additives, some researchers have demonstrated that the beneficial effects of Monensin can be decreased through long time usage (Poos et al., 1979; Rogers and Bestbier, 1997), but there is another opinion that in spite of the widespread of ionospheres and indiscriminate use as growth promotants in ruminants and monogastric animals in the united states and Europe, highly resistant isolates have been only rarely isolated from animals (Aaerstrup; Aarestrup et al., 2000). But according to this conflict ionospheres still need more research to in this point especially.

Meta-analyses have shown Monensin to produce improvement in feed efficiency in feedlot cattle by 7.5% (Goodrich et al., 1984); growing cattle on pasture by 15% (Potter et al., 1986); and dairy cows by 2.5% (Duffield et al., 2008), which may lead to reduce enteric CH_4 emissions. Moreover, another meta-analysis has also shown a consistent decrease in acetate: propionate (Ac:Pr) ratio with Monensin addition in high grain diets fed to beef cattle (Ellis et al., 2012), which may lead to a reduction in CH_4 emission per unit of feed. Our conclusion is that Inophores, through their effect on feed efficiency and reduction in CH_4 per unit of feed, would likely have a moderate CH_4 mitigating effect in ruminants fed high grain or mixed grain-forage diets. The effect is dose-, feed intake-, and diet composition dependent. The effect is less consistent in ruminants that are mainly fed pasture.

Saponins and Tannins: Saponins are secondary compounds found in many plants. They form stable foam in aqueous solutions, as soap does, hence the name "saponins". Saponins are structurally diverse molecules that are divided into two groups: triterpene and steroid glycosides (Vincken et al., 2007). Saponins or saponin-like substances have been reported to suppress CH_4 production, reduce rumen protozoa counts, and modulate fermentation pattern (Blümmel et al., 1997; Makkar et al., 1998; McAllister et al., 1999). China is one of the largest producers of tea globally and after extraction for oil, tea seed meal contains large amount of underutilized tea saponins. In some studies, saponin extracted from tea seeds had the potential to manipulate rumen fermentation, depress protozoa and reduce CH_4 release in-vitro (Liu et al., 2003; Wei-lian et al., 2005). Tea saponin significantly reduced CH_4 production by 8%, and also increased the molar proportion of propionate from 21.5% to 24.1% in addition to reducing the ratio of acetate to propionate from 3% to 2.6% and The relative abundance of protozoa and anaerobic fungal populations were significantly decreased by 51% and 79% respectively (Guo et al., 2008). The number of Cellulolytic bacteria decreased by 30%when 0.5 mg/ml Yucca extract was added to 5 g chopped Lucerne hay and 5 g concentrate in an in vitro fermentation (Wang et al., 1998). Cellulolytic bacteria were more susceptible to Yucca extract than amylolytic bacteria (Wang et al., 2000). Addition of tea saponins (TS) in animal diets may be an effective way to inhibit methanogenesis and hence has implications not only for global environmental protection but also for efficient animal production (Zhou et al., 2011). The effects of saponin was investigated through adding (0.6 g/L), nitrate (5 mm) and sulfate (5 mm), alone and in combinations, on methanogenesis, rumen fermentation, microbial community, and abundances of select microbial populations using in vitro rumen culture (Li and Powers, 2012). Combinations of nitrate with saponin and/or sulfate additively suppressed CH_4 production, with the lowest reduction (nearly 46%) observed for the combination of all the three inhibitors (Patra and Yu, 2013).

Tannins will inevitably be anti-nutritional when dietary CP concentrations are limiting production because they reduce absorption of AA (Waghorn, 2008). In addition, digestibility of diets containing condensed tannins at high levels is common (Waghorn, 2008; Patra and Saxena, 2010). Certain kinds of saponins may exert toxic effects in the body, which must be tested in vivo in long-term experiments (Patra and Saxena, 2009).

Essential Oils: Essential oils are a mixture of volatile lipophilic (fat loving, i.e., soluble in fat) constituents, most commonly sourced from leaf, twig, wood pulp or bark tissue of higher plants, but also widely found in bryophytes, such as the liverworts (Asakawa et al., 2012). Essential oils have antimicrobial activities against both gram-negative and gram-positive bacteria, a property that has been attributed to the presence of terpenoid and phenolic compounds (Conner, 1993). A number of in vitro studies have demonstrated that essential oils (EO) or their components have the potential to favorably alter rumen metabolism (McIntosh et al., 2003; Busquet et al., 2006; Sallam et al., 2009). When an EO is used to inhibit CH_4 production, the hydrogen produced by fiber digestion will accumulate in rumen fluids. An alternative pathway allowing for the capture of reducing equivalents spared from methanogenesis is, therefore, needed. Fumarate is the mid-product of rumen acetogenesis, and reduction of fumarate to succinate can draw electrons away from ruminal methanogenesis; this is seen as a potential method of decreasing CH_4 production in ruminants (Ungerfeld et al., 2007; Lin et al., 2012).

The commercial blend of EO could inhibit the rate of deamination of amino acids and the number of hyper-ammonia-producing bacteria in *in-vitro* batch cultures (McIntosh et al., 2003). Essential oils from a variety of sources have been shown to alter the bacterial growth and metabolism of several types of bacteria, including rumen bacteria (Wallace, 2004). The garlic oil was reported to alter fermentation by reducing the proportion of acetate and increasing that of propionate in a manner similar to Monensin (MO) in a continuous culture (Busquet et al., 2005). The antimicrobial activity of EO has prompted interest in whether these compounds could be used to inhibit methanogenesis in the rumen. The challenge is to identify EO that reduces CH_4 production without a concomitant reduction in feed digestion. Chiquette and Benchaar (2005) showed inhibitory effects of garlic EO on the production of CH_4 in vitro. Kamra (2005) and Kamra et al. (2008) investigated methanol and ethanol extracts of various spices, including fennel, clove, garlic, onion, and ginger for effects on CH_4 production in vitro.

It appears that some essential oils (e.g., garlic and its derivatives and cinnamon) could reduce CH_4 production in vitro. However, there is a need for in vivo investigation to determine whether these compounds can be used successfully to inhibit rumen methanogenesis. The challenge remains to identify essential oils that selectively inhibit rumen methanogenesis at practical feeding rates, with lasting effects and without depressing feed digestion and animal productivity (Benchaar and Greathead, 2011). These kind of products have good characteristics as CH_4 inhibitors, but it may cause some decline in DMI, Essential Oilshave reported by (Sallam et al., 2010). Because of some volatile components inside these oils are responsible for some aromatic characteristics may have negative effects on DM intake, it was reported by Manh et al. (2012) that EUCAP affected on feed intake not digestibility.

Lipids: There is a large body of evidence that lipids (vegetable oil or animal fat) suppress CH4 production. The effects of lipids on rumen Archaea are not isolated from their overall suppressive effect on bacteria and protozoa, but the usage of lipids as an additive in Ruminants rations may have some warning. Woodward (2006) examined the effect of vegetable and fish oils on milk production and CH_4 emission after 14 d and again after 12 wk. Lipids significantly decreased CH_4 production in the short term, but this effect was not observed after 11 wk of feeding lipids, also Grainger and Beauchemin (2011) examined 6 long-term studies (6 to 36 wk, mostly with dairy cows) and concluded that the effect of dietary fat on CH_4 production persists but the effect is not consistent among studies, the matter intake also was reported by Eugène et al. (2008) that 6% reduction in DMI in dairy cows due to lipid supplementation of the diet, Some fats such as coconut oil, for example, can severely depress feed intake, fiber digestibility, and, consequently, milk production and cause milk fat depression in dairy cows (Hristov et al., 2004; Hristov et al., 2011; Lee et al., 2011; Hollmann and Beede, 2012) although they may be still beneficial as CH_4 mitigating agents (Machmüller and Kreuzer, 1999; Machmüller, 2006).

In particular, it is well established that supplementation of diets with lipid sources reduces enteric CH_4 emissions (Boadi et al., 2004). Sources of long-chain fatty acids (FA) such as sunflower oil and canola oil reduced CH_4 emissions from cattle fed high-forage diets by up to 22% of GE intake when added at 45 g kg^{-1} of dry matter (DM) (McGinn et al., 2004; Beauchemin and McGinn, 2006). Adding lipid sources to the diet increased dietary crude fat content from 24.6 to 58.5 g kg^{-1} of DM whereas sunflower oil and seeds decreased (21 g 100 g^{-1} of FA) (Beauchemin et al., 2007). Eugène et al. (2008) reported a 9% reduction in enteric CH_4 production in dairy cows due to lipid supplementation of the diet, but this was accompanied by a 6.4% reduction in DMI, which resulted in no difference in CH_4 per unit of DMI. However, these authors also reported no effect on 4% fat-corrected milk (FCM) which, combined with the reduced DMI, resulted in a trend for increased feed efficiency with oil supplementation. Further, when calculated per unit of FCM, CH_4 production decreased with lipid supplementation (0.82 vs 0.75 MJ CH_4 energy/kg 4-% FCM; $P = 0.04$, $n = 25$).

In some in vitro studies it was reported that the addition of 53 g kg^{-1} DM of palm kernel oil, coconut oil, or canola oil, CH_4 production was reduced by 34, 21 or 20%, respectively (Dohme et al., 2000). The use of various plant oils to reduce enteric CH_4 emissions via ruminal defaunation, toxic effects on methanogens, or decreased diet digestibility has been widely reported (Dohme et al., 2000; Abo-Donia et al., 2009).

2.2. Feed management

TMR and Feeding Frequency: Very Little research is available on the effect of feeding system on CH_4 production (Hristov et al., 2013). Total Mixed Ration as a feeding system for the sheep has shown to decrease CH_4 production (per animal and per unit of DMI) with the free choice feeding system (Yurtseven and Ozturk, 2009). A study by Bargo et al. (2002) compared three feeding systems for lactating dairy cows: pasture plus concentrate, pasture plus TMR (supplementation of pasture with a TMR, i.e., partial TMR), and TMR (non-pasture). Cows fed TMR consumed more feed and produced more milk than cows fed pasture or partial TMR. Feed efficiency estimated from the published data was higher for TMR compared with pasture and partial TMR (1.37 vs 1.25 and 1.23 kg FCM/kg DMI).

Some non-traditional approaches, such as fermenting the TMR before feeding, appeared to be effective in mitigating enteric CH_4 (Cao et al., 2010), hypothetically through increasing lactate intake and conversion of lactate to propionate in the rumen. Apparently, this is a common practice in Japan, aimed at preserving TMR with high inclusion rate of wet feeds.

A series of trials in the 1980s from the laboratory of M. Kirchgessner in Germany found that frequent feeding did not improve dietary energy use but did increase CH_4emissionwhen concentrate was fed more often and separately from forage or with higher CP diets (Muller et al., 1980; Roehrmoser et al., 1983). It was reported that, feeding frequency had no effect on CH_4 production in dairy cows (Crompton et al., 2011). Cows did tend to consume more feed when it was freshly delivered (DeVries and Von Keyserlingk, 2005), but in another study (Crompton et al., 2011), in the once per day feeding treatment, CH_4 production peaked at 140 minutes after feeding and steadily decreased (at a rate of -0.0007 min^{-1}) thereafter.

The literature on the effect of feeding frequency on animal production is also scarce. For example Dhiman et al. (2002) did not report any production advantage of feeding lactating dairy cows once or four times daily. In some cases, milk production of dairy cows was reduced with frequent feeding, and this was attributed by the authors to more frequent handling (Phillips and Rind, 2001).

Concentrates Inclusion: Some experiments with lactating dairy cows and beef cattle have shown linear decreases in CH_4 emissions with an increase in the proportion of concentrate in the diet (Mc Geough et al., 2010; Aguerre et al., 2011).

However, concentrates generally provide more digestible nutrients (per unit feed) than roughage, which could increase animal productivity. For example, Huhtanen and Hetta (2012) in a meta-analysis of 986 dietary treatments reported a highly significant and positive relationship between dietary concentrate intake and production of milk, energy-corrected milk, and milk fat and milk protein. Hence, CH_4 expressed per unit product (i.e., Ei) is likely to decrease. In a European study (Eugène et al., 2011), the addition of starch and lipid combination to the diet of feedlot bulls reduced GHG emissions per unit of feed intake and BW gain. In another study by the same group, higher grain inclusion in the diet of Blond d'Aquitaine bulls (70 vs 21 to 41% grain; the type of forages also differed) resulted in a dramatic increase in Ym, from 3.2 to 6.9%, respectively (Doreau et al., 2011).

Which are belong to this term the inclusion of more concentrates as reported to decrease CH_4 but it carries more cost, as it well known that the concentrates has more higher prices, so it cannot be suggested for mitigation at all conditions, for which relates to TMR, also it cannot be generalized as one of mitigates in whole countries, some countries depend on grazing system to feed their animals.

3. Conclusions

Nowadays there are many methods for reducing CH_4 production in ruminants, through the information mentioned above there are some strategies for feed additives and feed management. For the additives, Ionophores were one of the common additives as a growth promotants and had studied as CH_4 inhibitor, in addition to that there were no any negative effects on Dry matter intake (DMI) and digestibility, and because of the suspensions around its resistance because of some rare cases are detected, Ionophore still need some long term studies for proving or denying this worries. Application of essential oils needs more in vivo studies to determine the influential amounts without any decline for DMI or digestion, and also the cost of these products need to be taken into considerations. Long term supplementation experiments also are needed, especially for Saponins and tannins, they have good CH_4 mitigation, but certain kinds of tannins there is no effect, for feed management its reported that concentrates of up to 35-40% or more is effective for inhibiting CH_4 production but in some regions it cannot be economically in the countries which have low prices of milk, then the cost of production will not meet with the prices. Lipids showed good inhibition indications, but DMI declinations were appeared during many experiments which reflected on the productivity of the animals, so this kind of strategies cannot be admitted as one of the inhibitors TMR also has good values reflect on the productivity and ruminal processes, but as mitigation method still needs ore researches to prove this exactly, so in this topic its recommended that still more researches are required and the suggested methods differ from country to country and from region to region.

References

Aaerstrup, F., F, Bager, NE Jensen, M. Madsen, A. Meyling, and HC Wegener, 1998. Surveillance of antimicrobial resistance in bacteria isolated from food animals to antimicrobial growth promoters and related therapeutic agents in Denmark. APMIS. 106, 606.

Aarestrup, F.M., Y. Agerso, P. Gerner-Smidt, M. Madsen, and L.B. Jensen, 2000. Comparison of antimicrobial resistance phenotypes and resistance genes in Enterococcus faecalis and Enterococcus faecium from humans in the community, broilers, and pigs in Denmark. Diagnostic Microbiology and Infectious Disease. 37 (2), 127–137.

Abo-Donia, F., A. El-Aziz, and T. Diraz, 2009. Impact of using sodium or calcium salts of fatty acids as sources of energy in buffalo rations during late pregnancy. Egyptian Journal of Nutrition and Feeds. 12 (1), 35–52.

Aguerre, M., M. Wattiaux, J.M. Powell, G. Broderick, and C. Arndt, 2011. Effect of forage-to-concentrate ratio in dairy cow diets on emission of methane, carbon dioxide, and ammonia, lactation performance, and manure excretion. Journal of Dairy Science. 94 (6), 3081–3093.

Appuhamy, J.R.N., A. Strathe, S. Jayasundara, C. Wagner-Riddle, J. Dijkstra, J. France, and E. Kebreab, 2013. Anti-methanogenic effects of monensin in dairy and beef cattle: A meta-analysis. Journal of Dairy Science. 96 (8), 5161–5173.

Asakawa, Y., A. Ludwiczuk, and F. Nagashima, 2012. Chemical Constituents of Bryophytes: Bio-and Chemical Diversity, Biological Activity and Chemosystematics. In *Progress in the Chemistry of Organic Natural Products.* Vienna: Springer. 93, 1–760.

Bargo, F., L. Muller, J. Delahoy, and T. Cassidy, 2002. Performance of high producing dairy cows with three different feeding systems combining pasture and total mixed rations. Journal of Dairy Science. 85 (11), 2948–2963.

Beauchemin, K., and S. McGinn, 2006. Methane emissions from beef cattle: effects of fumaric acid, essential oil, and canola oil. Journal of Animal Science. 84 (6), 1489–1496.

Beauchemin, K., M. Kreuzer, F. O'mara, and T. McAllister, 2008. Nutritional management for enteric methane abatement: a review. Animal Production Science. 48 (2), 21–27.

Beauchemin, K.A., S.M. McGinn, and H.V. Petit, 2007. Methane abatement strategies for cattle: Lipid supplementation of diets. Canadian Journal of Animal Science. 87 (3), 431–440.

Benchaar, C., and H. Greathead, 2011. Essential oils and opportunities to mitigate enteric methane emissions from ruminants. Animal Feed Science and Technology. 166, 338–355.

Blümmel, M., H. Makkar, and K. Becker, 1997. In vitro gas production: a technique revisited. Journal of Animal Physiology and Animal Nutrition. 77 (1-5), 24–34.

Boadi, D., C. Benchaar, J. Chiquette, and D. Massé, 2004. Mitigation strategies to reduce enteric methane emissions from dairy cows: update review. Canadian Journal of Animal Science. 84 (3), 319–335.

Busquet, M., S. Calsamiglia, A. Ferret, M. Carro, and C. Kamel, 2005.Effect of garlic oil and four of its compounds on rumen microbial fermentation. Journal of Dairy Science. 88 (12), 4393–4404.

Busquet, M., S. Calsamiglia, A. Ferret, and C. Kamel, 2006. Plant extracts affect in vitro rumen microbial fermentation. Journal of Dairy Science. 89 (2), 761–771.

Calsamiglia, S., M. Busquet, P. Cardozo, L. Castillejos, and A. Ferret, 2007.Invited review: essential oils as modifiers of rumen microbial fermentation. Journal of Dairy Science. 90 (6), 2580–2595.

Cao, Y., T. Takahashi, K.-i. Horiguchi, N. Yoshida, and Y. Cai, 2010. Methane emissions from sheep fed fermented or non-fermented total mixed ration containing whole-crop rice and rice bran. Animal Feed Science and Technology. 157 (1), 72–78.

Capper, J.L., R. Cady, and D. Bauman, 2009. The environmental impact of dairy production: 1944 compared with 2007. Journal of Animal Science. 87 (6), 2160–2167.

Chiquette, J., and C. Benchaar, 2005. Effects of different dose levels of essential oils compounds on in vitro methane production by mixed ruminal bacteria. Journal of dairy science. 306–306.

Conner, D.E., 1993. Naturally occurring compounds. Marcel Dekker, New York. Antimicrobials in Foods. Eds., Davidson, P. M., and A. L. Branen. 441–468.

Crompton, L., J. Mills, C. Reynolds, and J. France, 2011. Fluctuations in methane emission in response to feeding pattern in lactating dairy cows. In: *Modelling nutrient digestion and utilisation in farm animals*: Springer. 176–180.

Delgado, C.L., 2003. Rising consumption of meat and milk in developing countries has created a new food revolution. The Journal of Nutrition. 133 (11), 3907S–3910S.

DeVries, T., and M. Von Keyserlingk, 2005. Time of feed delivery affects the feeding and lying patterns of dairy cows. Journal of Dairy Science. 88 (2), 625–631.

Dhiman, T., M. Zaman, I. MacQueen, and R. Boman, 2002. Influence of corn processing and frequency of feeding on cow performance. Journal of Dairy Science. 85 (1), 217–226.

Dohme, F., A. Machmüller, A. Wasserfallen, and M. Kreuzer, 2000. Comparative efficiency of various fats rich in medium-chain fatty acids to suppress ruminal methanogenesis as measured with RUSITEC. Canadian Journal of Animal Science. 80 (3), 473–484.

Doreau, M., H. Van Der Werf, D. Micol, H. Dubroeucq, J. Agabriel, Y. Rochette, and C. Martin, 2011. Enteric methane production and greenhouse gases balance of diets differing in concentrate in the fattening phase of a beef production system. Journal of Animal Science. 89 (8), 2518–2528.

Duffield, T., A. Rabiee, and I. Lean, 2008. A meta-analysis of the impact of monensin in lactating dairy cattle. Part 2. Production effects. Journal of Dairy Science. 91 (4), 1347–1360.

Ellis, J., J. Dijkstra, A. Bannink, E. Kebreab, S. Hook, S. Archibeque, and J. France, 2012. Quantifying the effect of monensin dose on the rumen volatile fatty acid profile in high-grain-fed beef cattle. Journal of Animal Science. 90 (8), 2717–2726.

Eugène, M., D. Massé, J. Chiquette, and C. Benchaar, 2008. Meta-analysis on the effects of lipid supplementation on methane production in lactating dairy cows. Canadian Journal of Animal Science. 88 (2), 331–337.

Eugène, M., C. Martin, M. Mialon, D. Krauss, G. Renand, and M. Doreau, 2011. Dietary linseed and starch supplementation decreases methane production of fattening bulls. Animal Feed Science and Technology. 166, 330–337.

FAO, 2011. World livestock 2011 - livestock in food security. Food and Agriculture Organization of the United Nations (FAO), Rome.

Gerber, P., T. Vellinga, C. Opio, and H. Steinfeld, 2011. Productivity gains and greenhouse gas emissions intensity in dairy systems. Livestock Science. 139 (1), 100–108.

Goodrich, R., J. Garrett, D. Gast, M. Kirick, D. Larson, and J. Meiske, 1984. Influence of monensin on the performance of cattle. Journal of Animal Science. 58 (6), 1484–1498.

Grainger, C., and K. Beauchemin, 2011. Can enteric methane emissions from ruminants be lowered without lowering their production? Animal Feed Science and Technology. 166, 308–320.

Guo, Y., J.X. Liu, Y. Lu, W. Zhu, S. Denman, and C. McSweeney, 2008. Effect of tea saponin on methanogenesis, microbial community structure and expression of mcrA gene, in cultures of rumen micro-organisms. Letters in Applied Microbiology. 47 (5), 421–426.

Hollmann, M., and . Beede, 2012. Comparison of effects of dietary coconut oil and animal fat blend on lactational performance of Holstein cows fed a high-starch diet. Journal of Dairy Science. 95 (3), 1484–1499.

Hristov, A., K. Grandeen, J. Ropp, and M. McGuire, 2004. Effect of sodium laurate on ruminal fermentation and utilization of ruminal ammonia nitrogen for milk protein synthesis in dairy cows. Journal of Dairy Science. 87 (6), 1820–1831.

Hristov, A., C. Lee, T. Cassidy, M. Long, K. Heyler, B. Corl, and R. Forster, 2011. Effects of lauric and myristic acids on ruminal fermentation, production, and milk fatty acid composition in lactating dairy cows. Journal of Dairy Science. 94 (1), 382–395.

Hristov, A., J. Oh, J. Firkins, J. Dijkstra, E. Kebreab, G. Waghorn, H. Makkar, A. Adesogan, W. Yang, and C. Lee, 2013. Special topics—Mitigation of methane and nitrous oxide emissions from animal operations: I. A review of enteric methane mitigation options. Journal of Animal Science. 91 (11), 5045–5069.

Huhtanen, P., and M. Hetta, 2012. Comparison of feed intake and milk production responses in continuous and change-over design dairy cow experiments. Livestock Science. 143 (2), 184–194.

Kamra, D., 2005. Rumen microbial ecosystem. Current Science. 89 (1), 124–135.

Kamra, D., A. Patra, P. Chatterjee, R. Kumar, N. Agarwal, and L. Chaudhary, 2008. Effect of plant extracts on methanogenesis and microbial profile of the rumen of buffalo: a brief overview. Animal Production Science. 48 (2), 175–178.

Lee, C., A. Hristov, K. Heyler, T. Cassidy, M. Long, B. Corl, and S. Karnati, 2011. Effects of dietary protein concentration and coconut oil supplementation on nitrogen utilization and production in dairy cows. Journal of Dairy Science. 94 (11), 5544–5557.

Leng, R., 2008. The potential of feeding nitrate to reduce enteric methane production in ruminants. A Report to the Department of Climate Change Commonwealth Government of Australia. ACT Canberra Australia http://www.penambulbooks.com.

Li, W., and W. Powers, 2012. Effects of saponin extracts on air emissions from steers. Journal of Animal Science. 90 (11), 4001–4013.

Lin, B., Y. Lu, J. Wang, Q. Liang, and J. Liu, 2012. The effects of combined essential oils along with fumarate on rumen fermentation and methane production in vitro. Journal of Animal and Feed Sciences.575, 54.

Liu, J., W. Yuan, J. Ye, and Y. Wu, 2003. Effect of tea (Camellia sinensis) saponin addition on rumen fermentation in vitro. In *Matching Herbivore Nutrition to Ecosystems Biodiversity. Tropical and Subtropical Agrosystems. Proceedings of the Sixth International Symposium on the Nutrition of Herbivore.* Merida, Mexico.

Machmüller, A., and M. Kreuzer, 1999. Methane suppression by coconut oil and associated effects on nutrient and energy balance in sheep. Canadian Journal of Animal Science. 79 (1), 65–72.

Machmüller, A., 2006. Medium-chain fatty acids and their potential to reduce methanogenesis in domestic ruminants.Agriculture, Ecosystems &Environment. 112 (2), 107–114.

Makkar, H.P., S. Sen, M. Blümmel, and K. Becker, 1998. Effects of fractions containing saponins from Yucca schidigera, Quillaja saponaria, and Acacia auriculoformis on rumen fermentation. Journal of Agricultural and Food Chemistry. 46 (10), 4324–4328.

Manh, N., M. Wanapat, S. Uriyapongson, P. Khejornsart, and V. Chanthakhoun, 2012. Effect of eucalyptus (Camaldulensis) leaf meal powder on rumen fermentation characteristics in cattle fed on rice straw. African Journal of Agricultural Research. 7 (13), 1997–2003.

Mc Geough, E., P. O'Kiely, K. Hart, A. Moloney, T. Boland, and D. Kenny, 2010. Methane emissions, feed intake, performance, digestibility, and rumen fermentation of finishing beef cattle offered whole-crop wheat silages differing in grain content. Journal of Animal Science. 88 (8), 2703–2716.

McAllister, T., S. Oosting, J. Popp, Z. Mir, L. Yanke, A. Hristov, R. Treacher, and K.-J. Cheng, 1999. Effect of exogenous enzymes on digestibility of barley silage and growth performance of feedlot cattle. Canadian Journal of Animal Science. 79 (3), 353–360.

McGinn, S., K. Beauchemin, T. Coates, and D. Colombatto, 2004. Methane emissions from beef cattle: effects of monensin, sunflower oil, enzymes, yeast, and fumaric acid. Journal of Animal Science. 82 (11), 3346–3356.

McIntosh, F., P. Williams, R. Losa, R. Wallace, D. Beever, and C. Newbold, 2003. Effects of essential oils on ruminal microorganisms and their protein metabolism. Applied and Environmental Microbiology. 69 (8), 5011–5014.

Muller, H., J. Sax, M. Kirchgessner, 1980. Effect of frequency of feeding on energy losses in faeces, urine and methane in nonlactating and lactating cows. Zeitschrift fur Tierphysiologie, Tierernahrung und Futtermittelkunde. 44 (4/5), 181–189.

Oberthür, S., and C.R. Kelly, 2008. EU leadership in international climate policy: achievements and challenges. The International Spectator. 43 (3), 35–50.

Odongo, N., R. Bagg, G. Vessie, P. Dick, M. Or-Rashid, S. Hook, J. Gray, E. Kebreab, J. France, and B. McBride, 2007. Long-term effects of feeding monensin on methane production in lactating dairy cows. Journal of Dairy Science. 90 (4), 1781–1788.

Patra, A., D. Kamra, and N. Agarwal, 2006. Effect of plant extracts on in vitro methanogenesis, enzyme activities and fermentation of feed in rumen liquor of buffalo. Animal Feed Science and Technology. 128 (3), 276–291.

Patra, A., and J. Saxena, 2009. The effect and mode of action of saponins on the microbial populations and fermentation in the rumen and ruminant production. Nutrition Research Reviews. 22 (2), 204–219.

Patra, A.K., and J. Saxena, 2010. A new perspective on the use of plant secondary metabolites to inhibit methanogenesis in the rumen. Phytochemistry. 71 (11), 1198–1222.

Patra, A.K., and Z. Yu, 2013. Effective reduction of enteric methane production by a combination of nitrate and saponin without adverse effect on feed degradability, fermentation, or bacterial and archaeal communities of the rumen. Bioresource Technology. 148, 352–360.

Phillips, C., and M. Rind, 2001. The effects of frequency of feeding a total mixed ration on the production and behavior of dairy cows. Journal of Dairy Science. 84 (9), 1979–1987.

Poos, M., T. Hanson, and T. Klopfenstein, 1979. Monensin effects on diet digestibility, ruminal protein bypass and microbial protein synthesis. Journal of Animal Science. 48 (6), 1516–1524.

Potter, E., R. Muller, M. Wray, L. Carroll, and R. Meyer, 1986. Effect of monensin on the performance of cattle on pasture or fed harvested forages in confinement. Journal of Animal Science. 62 (3), 583–592.

Roehrmoser, G., H.L. Mueller, and M. Kirchgessner, 1983. Energy balance and energy utilization of lactating cows under restricted protein supply and subsequent refeeding. Zeitschrift fuer Tierphysiologie, Tierernaehrung und Futtermittelkunde (Germany, FR).

Rogers, K.H., and R. Bestbier, 1997. Development of a protocol for the definition of the desired state of riverine systems in South Africa: Department of Environmental Affairs and Tourism.

Russell, J.B., and A.J. Houlihan, 2003. Ionophore resistance of ruminal bacteria and its potential impact on human health. FEMS Microbiology Reviews. 27 (1), 65–74.

Sallam, S., I. Bueno, P. Brigide, P. Godoy, D. Vitti, and A. Abdalla, 2009.Efficacy of eucalyptus oil on in vitro ruminal fermentation and methane production. Options Mediterraneennes. 85 (85), 267–272.

Sallam, S.M., I.C. Bueno, M.E. Nasser, and A.L. Abdalla, 2010. Effect of eucalyptus (Eucalyptus citriodora) fresh or residue leaves on methane emission in vitro. Italian Journal of Animal Science. 9 (3), 58.

Steinfeld, H., P. Gerber, T. Wassenaar, V. Castel, M. Rosales, and C.d. Haan, 2006. *Livestock's long shadow: environmental issues and options*: Food and Agriculture Organization of the United Nations (FAO).

Ungerfeld, E., R. Kohn, R. Wallace, and C. Newbold, 2007. A meta-analysis of fumarate effects on methane production in ruminal batch cultures. Journal of Animal Science. 85 (10), 2556–2563.

Van Nevel, C., and D. Demeyer, 1992. Influence of antibiotics and a deaminase inhibitor on volatile fatty acids and methane production from detergent washed hay and soluble starch by rumen microbes in vitro. Animal Feed Science and Technology. 37 (1), 21–31.

Vincken, J.-P., L. Heng, A. de Groot, and H. Gruppen, 2007. Saponins, classification and occurrence in the plant kingdom. Phytochemistry. 68 (3), 275–297.

Waghorn, G., 2008. Beneficial and detrimental effects of dietary condensed tannins for sustainable sheep and goat production—Progress and challenges. Animal Feed Science and Technology. 147 (1), 116–139.

Wallace, R.J., 2004. Antimicrobial properties of plant secondary metabolites. Proceedings of the Nutrition Society. 63 (04), 621–629.

Wang, Y., T. McAllister, C. Newbold, L. Rode, P. Cheeke, and K. Cheng, 1998. Effects of Yucca schidigera extract on fermentation and degradation of steroidal saponins in the rumen simulation technique (RUSITEC). Animal Feed Science and Technology. 74 (2), 143–153.

Wang, Y., T.A. McAllister, L.J. Yanke, Z.J. Xu, P.R. Cheeke, and K.J. Cheng, 2000. In vitro effects of steroidal saponins from Yucca schidigera extract on rumen microbial protein synthesis and ruminal fermentation. Journal of the Science of Food and Agriculture. 80 (14), 2114–2122.

Wei-lian, H., W. Yue-ming, L. Jian-xin, G. Yan-qiu, and Y. Jun-an, 2005. Tea saponins affect in vitro fermentation and methanogenesis in faunated and defaunated rumen fluid. Journal of Zhejiang University Science B. 6 (8), 787–792.

Woodward, S., G. Waghorn, M. Ulyatt, K. Lassey, 2001. Early indications that feeding Lotus will reduce methane emissions from ruminants. In: *Proceedings-New Zealand Society of Animal Production*: New Zealand Society of Animal Production. 61, 23–26.

Woodward, S., 2006. Supplementing dairy cows with oils to improve performance and reduce methane-does it work? In *Proceedings-New Zealand Society of Animal Production*: New Zealand Society of Animal Production. 66, 176–181.

Yurtseven, S., and I. Ozturk, 2009. Influence of two sources of cereals (corn or barley), in free choice feeding on diet selection, milk production indices and gaseous products (CH_4 and CO_2) in lactating sheep. Asian Journal of Animal and Veterinary Advances. 4 (2), 76–85.

Zhou, Y., H. Mao, F. Jiang, J. Wang, J. Liu, and C. McSweeney, 2011. Inhibition of rumen methanogenesis by tea saponins with reference to fermentation pattern and microbial communities in Hu sheep. Animal Feed Science and Technology. 166, 93–100.

Reducing the Concentrations of Airborne Pollutants in Different Livestock Buildings

Thomas Banhazi
National Centre for Engineering in Agriculture (NCEA), University of Southern Queensland, West St, Toowoomba, QLD, 4530 Australia
Email: Thomas.banhazi@usq.edu.au

Abstract

The negative effects of high concentration of airborne pollutants on animal health, welfare and productivity are well documented. Reducing the concentration of airborne pollutants in livestock buildings is therefore an important task and could also help to reduce the occupational health and safety risk associated with farm work. The main objective of this research was to evaluate the effects of oil treatment on the concentration of airborne pollutants inside different livestock facilities. Air quality parameters were recorded in (1) a number of partially slatted, mechanically or naturally ventilated pig facilities and in (2) two poultry buildings. Airborne pollutant concentrations were measured and compared between the treatment and control facilities. The concentrations of both inhalable and respirable airborne particles were significantly ($P < 0.001$) reduced in the experimental livestock facilities. This technique would enable livestock producers to improve the environmental quality in livestock buildings at a relatively low cost.

Keywords: Air quality, reduction, emission, ammonia, dust, airborne particles

1. Introduction

Dust is one of the major airborne pollutants associated with intensive livestock production and determines the quality of the environment within livestock buildings (Banhazi et al., 2009a; Banhazi et al., 2009b). The negative effects of high concentration of bioaerosol on human and animal health, as well as on animal welfare and productivity are well documented (Donham and Leininger, 1984; Donham et al., 1984; Donham, 1991). Suspended airborne particles can also absorb toxic and noxious gases as well as bacteria components and act as vectors for these pollutants (Donaldson, 1977). High concentrations of airborne particles may contain bacterial toxins and appears to enhance both the prevalence and severity of respiratory diseases in pigs (Lee et al., 2005). An improvement in air quality within poultry buildings should enhance production efficiency and health of birds (Al Homidan et al., 1998). The litter is a major source of particles in poultry houses and its characteristics would affect airborne particle concentrations (Banhazi et al., 2008a). Reducing the concentration of airborne particles in livestock buildings is therefore an important component of good management and can improve production efficiency and reduce the potentially harmful effects of long term exposure to humans (Donham et al., 1989). Therefore, simple, low-cost and practical techniques, which will have the potential to deliver a significant reduction of dust, ammonia and other pollutant emissions cost effectively, need to be investigated, developed and evaluated (Takai and Pedersen, 2000; Banhazi et al., 2011). Spraying the floor of pig sheds with a mixture of oil and water (Takai et al., 1995) is a potentially beneficial technique.

Thus, the overall objective of this study was to evaluate the effects of oil spraying and oil impregnation primarily on the concentration of airborne particles inside livestock facilities, but the effects of the treatments on the concentration of other airborne pollutants were also investigated.

2. Materials and Methods

An automated oil spraying system was installed in a number of piggery buildings. General design concepts of oil spraying systems have been published previously (Lemay et al., 1999;

Takai and Pedersen, 1999; Banhazi, 2005) and the actual spray system was assembled from commercially available components. The spray system was positioned on the top of the pen walls by utilising a wire cable spanning between the pen walls.

Air quality parameters were recorded for 25 days in two partially-slatted, mechanically ventilated weaner rooms housing 89 pigs (approximate mean live weight 18 kg) and for 10 days in two partially-slatted, naturally ventilated grower rooms housing 91 pigs (approximate mean live weight 42 kg). The floor of one of the rooms (experimental facility) was sprayed daily with a canola oil, water and surfactant mixture at a 4:5:1 ratio and at the rate of 3 g pig^{-1} (6.3 g m^{-2}), using an automatic spraying system. The other room was not treated and served as a control facility. Air quality parameters were measured throughout the trials.

A classical comparative experiment was conducted at a South Australian poultry farm. Two identical and environmentally controlled broiler buildings (approximate size of 370 m^2) on the same poultry farm were selected for the experiment and chopped straw was used as litter material in both buildings. The bedding material in one of the buildings was treated with the oil/water mixture, while the other building stocked at the same rate was used as control building. Male meat birds were placed in the buildings at the same time and were stocked at 15.9 birds m^{-2}. The quantity of oil used represented approximately 7.5% of the weight of the bedding material. Canola oil, water and surfactant (emulsifier) were mixed in a drum at the ratio of 8:4:1. The mix was poured into a backpack-spraying unit containing 16 litres of the mixture and sprayed directly onto the litter that was then raked to homogeneously spread the mix.

During the studies in the piggery buildings, the sensors were attached by wire cable to the ceiling or a beam and were lowered to pig level (approximately 1.1 to 1.3 m) above the selected pens. In the poultry building, wire cages were positioned in the middle of both poultry buildings to protect the measuring devices, which were deployed within these cages in both the experimental and control buildings. This arrangement ensured that the concentrations of airborne pollutants and environmental parameters were measured at animal level.

Airborne particles concentrations inside the buildings was measured utilising the standard gravimetric method, as described previously (Banhazi et al., 2008b). Total inhalable and respirable particle concentrations were measured using GilAir air pumps (Gilian Instrument Corp., West Caldwell, N.J.) connected to cyclone filter heads (for respirable particles) and Seven Hole Sampler (SHS) filter heads (for inhalable dust) (Casella, Ltd., Kempston, U.K.) and operated over a 6 or 8-hour period at 1.9 and 2.0 l min^{-1} flow rate, respectively. After sampling, the filter heads were taken back to the laboratory and weighed to the nearest 0.001 milligram using certified microbalances and then the inhalable and respirable dust levels were calculated.

Ammonia NH_3 and carbon dioxide CO_2 were monitored continuously using a multi-gas monitoring machine in each building, as described previously (Banhazi et al., 2008b). Electrochemical (Bionics TX-FM/FN, Bionics Instrument Co., Tokyo, Japan) and infrared sensors (GMM12, Vaisala Oy, Helsinki, Finland) were used to detect the concentrations of NH_3 and CO_2, respectively. The gas sensors were enclosed in a shock-resistant electrical box and an air delivery system was used to deliver air samples from the sampling points within and outside the buildings to the actual gas monitoring heads.

Total viable airborne bacteria were measured using an Anderson viable six-stage bacterial impactor (Clarke and Madelin, 1987) filled with horse blood agar plates (HBA, Medvet Science Pty. Ltd., Stepney, Australia). The airspace was sampled for five minutes at a flow rate of 1.9 litre min^{-1} (Figure 2). The bacteria plates were incubated for 48 hours at 37°C and the number of colony forming units (CFUs) were counted manually to express the concentration of airborne microorganisms as CFUs m^{-3}.

A General Linear Model (GLM) was developed to determine the effects of the oil treatment on airborne pollutant concentrations, considering experimental effects and covariates such as CO_2 concentrations and age of birds $animal^{-1}$ (SAS, 1989; StatSoft, 2001).

3. Results and Discussion

3.1. Piggery buildings

The concentration of both inhalable and respirable airborne particles as well as airborne bacteria was significantly reduced in the experiment facilities (Table 1). Only the experimental effect proved to be significant; no other effects were identified by the statistical analysis to be significantly influencing airborne pollutant levels in the control and experimental rooms.

Table 1. Concentrations of respirable and inhalable airborne particles, viable bacteria and ammonia for the control and treatment rooms (Pigs).

Treatment	Respirable Particles (mg m^{-3})	Inhalable Particles (mg m^{-3})	Total Bacteria (× 1000 CFUs m^{-3})	Ammonia (ppm)
Weaner (Control)	0.212^{a}	4.118^{a}	71^{a}	10.1^{a}
Weaner (Treatment)	0.138^{b}	2.022^{b}	32^{b}	9.0^{a}
Grower (Control)	0.116^{a}	1.451^{a}	68^{a}	8.1^{a}
Grower (Treatment)	0.101^{a}	0.682^{b}	109^{b}	9.2^{a}

[a, b] Values in the same column and within the same age group (i.e., weaner vs. grower pigs) with different superscripts differ significantly ($P < 0.05$).

The experiment demonstrated a significant reduction in the concentrations of both inhalable and respirable airborne particles in the airspace following the direct spraying of an oil and water mixture onto the floor. This study confirmed previously published data (Takai et al., 1995) and the technique used in the experiment could be used by producers to effectively reduce dust levels in piggery building. This study demonstrated that oil spraying have the potential of significantly reducing pollutant concentrations cost effectively (Takai et al., 1995).

3.2. Poultry buildings

Table 2 shows the mean concentrations of inhalable and respirable airborne particles in the treatment and control rooms. The oil treatment significantly ($P < 0.001$) reduced the concentrations of both inhalable and respirable particles in the treatment room, which is consistent with the effect of oil treatment demonstrated in previous studies (Feddes et al., 1995; Banhazi et al., 2011).

The age of birds also had a significant effect on inhalable particles concentration ($P < 0.05$) but not on respirable particles. The inhalable particles concentration increased with the age of birds, which agreed with previous studies (Madelin and Wathes, 1989; Hinz and Linke, 1998). The internal temperature significantly ($P < 0.05$) affected the respirable particles concentration although it was not statistically significant for the inhalable particles concentration.

Table 2. Concentrations of respirable and inhalable airborne particles and ammonia for the control and treatment rooms (Poultry).

Treatment	Respirable Particles (mg m^{-3})	Inhalable Particles (mg m^{-3})	Ammonia (ppm)
Poultry (Control)	0.586^{a}	3.550^{a}	11.9^{a}
Poultry (Treatment)	0.329^{b}	2.560^{b}	4.9^{a}

[a, b] Values in the same column with different superscripts differ significantly ($P < 0.05$).

A significant reduction in ammonia concentration ($P < 0.001$) was also demonstrated in the treatment room (Table 2). A reduction of ammonia concentrations with the same type of oil treatment was reported in piggery buildings (Jacobson et al., 1998). However, the effect of the oil treatment on ammonia was not demonstrated in previous studies in poultry buildings (Feddes et al., 1995). One possible explanation for the positive effect demonstrated in this study is that the oil treatment might interfere with the bacteria flora in bedding responsible for ammonia generation from nitrogenous compounds, thus decreasing the ability of bacteria to generate ammonia (Banhazi et al., 2007). Despite the fact that the reduction of ammonia concentrations has not been fully explained, this finding was an important result because high ammonia levels are not advantageous for poultry production.

4. Conclusions

Overall, these studies have demonstrated that airborne particle (and potentially ammonia) concentrations can be significantly reduced in livestock buildings by either impregnating the bedding material with a relatively small amount of oil, or by directly spaying oil on the floor of the farm buildings.

However, oil application has to be made more practical via associated engineering developments, as the experimentally used manual spraying and raking (applied in the poultry facilities) is not practical under commercial conditions. In the future the oil should be directly incorporated into the bedding material before spreading. The spreading of the bedding material is normally associated with high airborne particle concentrations and the reduction of airborne particles during that time is most likely to have beneficial effect on worker safety and respiratory health.

In addition, the reduction of particle levels indoors will also reduce particle emissions, assuming constant ventilation rates. The future adoption of particle reduction strategies in the intensive livestock industries is important, due to the increasing environmental and occupational health and safety requirements. The oil application methods, utilised during these experiments, appeared to be useful and practical. However, further experiments are needed to assess the potentially beneficial effects of particle reduction on production efficiency that will further encourage producers to utilise these techniques.

Acknowledgements

The author wishes to acknowledge the financial support of the Australian Pork Limited (APL) and the Rural Industries Research and Development Corporation (RIRDC). In addition the author wishes to recognise the professional support of Dr C. Cargill (South Australian Research and Development Institute, SARDI), Dr J. Seedorf (Osnabrück University of Applied Sciences), M. Laffrique (Ecole Nationale Supérieure Agronomique de Rennes), Prof. J. Hartung, Hannover University, Germany, and the technical support of Mr K. Hillyard and Mr J. Wegiel of SARDI. An extended version of this paper was published previously in "Livestock Housing", Wageningen Academic Publishers (WAP) and has been partially reprinted with the kind permission of WAP.

References

Aarnink, A.J.A., E.N.J. van Ouwerkerk, M.W.A. Verstegen, 1992. A mathematical model for estimating the amount and composition of slurry from fattening pigs. Livestock Production Science. 31 (1–2), 133–147.

Al Homidan, A., J.F. Robertson, A.M. Petchey, 1998. Effect of environmental factors on ammonia and dust production and broiler performance. British Poultry Science. 39 (Supplement), S10.

Banhazi, T., 2005. Oil spraying systems for piggeries to control dust. In: *Proceedings of AAPV Conference*. Gold Coast, QLD, Australia, 15–19 May, 2005. Gold Coast, QLD, Australia: AVA. Ed., Fahy, T. 76–80.

Banhazi, T., M. Laffrique, J. Seedorf, 2007. Controlling the concentrations of airborne pollutants in poultry buildings. In: *XIII International Congress on Animal Hygiene*. Tartu, Estonia, June, 2007. Tartu, Estonia: Estonian University of Life Sciences. Ed., Aland, A. 302–307.

Banhazi, T.M., J. Seedorf, M. Laffrique, D.L. Rutley, 2008a. Identification of the risk factors for high airborne particle concentrations in broiler buildings using statistical modelling. Biosystems Engineering. 101 (1), 100–110.

Banhazi, T.M., J. Seedorf, D.L. Rutley, W.S. Pitchford, 2008b. Identification of risk factors for sub-optimal housing conditions in Australian piggeries - Part II: Airborne pollutants. Journal of Agricultural Safety and Health. 14 (1), 21–39.

Banhazi, T.M., E. Currie, M. Quartararo, A.J.A. Aarnink, 2009a. Controlling the concentrations of airborne pollutants in broiler buildings. In: *Sustainable animal production: The challenges and potential developments for professional farming*. Wageningen, The Netherlands: Wageningen Academic Publishers. Eds., Aland, A., F. Madec. 347–364.

Banhazi, T.M., E. Currie, S. Reed, I.-B. Lee, A.J.A. Aarnink, 2009b. Controlling the concentrations of airborne pollutants in piggery buildings. In: *Sustainable animal production: The challenges and potential developments for professional farming*. Wageningen, The Netherlands: Wageningen Academic Publishers. Eds., Aland, A., F. Madec. 285–311.

Banhazi, T.M., C. Saunders, N. Nieuwe, V. Lu, A. Banhazi, 2011. Oil spraying as a air quality improvement technique in livestock buildings: Development and utilisation of a testing device. Australian Journal of Multi-disciplinary Engineering. 8 (2), 169–180.

Clarke, A.F., T. Madelin, 1987. Technique for assessing respiratory health hazards from hay and other source materials. Equine Veterinary Journal. 19 (5), 442–447.

Donaldson, A.I., 1977. Factors influencing the dispersal, survival and deposition of airborne pathogens of farm animals. Veterinary Bulletin. 48 (2), 83–94.

Donham, K.J., J.R. Leininger, 1984. Animal studies of potential chronic lung disease of workers in swine confinement buildings. American Journal of Veterinary Research. 45 (5), 926–931.

Donham, K.J., D.C. Zavala, J.A. Merchant, 1984. Respiratory symptoms and lung function among workers in swine confinement buildings: a cross-sectional epidemiological study. Archives of Environmental Health. 39 (2), 96–101.

Donham, K.J., P. Haglind, Y. Peterson, R. Rylander, L. Belin, 1989. Environmental and health studies of farm workers in Swedish swine confinement buildings. British Journal of Industrial Medicine. 46, 31–37.

Donham, K.J., 1991. Association of environmental air contaminants with disease and productivity in swine. American Journal of Veterinary Research. 52 (10), 1723–1730.

Feddes, J.J.R., K. Taschuk, F.E. Robinson, C. Riddell, 1995. Effect of litter oiling and ventilation rate on air quality, health and performance of turkeys. Canadian Agricultural Engineering. 37 (1), 57–62.

Hinz, T., S. Linke, 1998. A comprehensive experimental study of aerial pollutants in and emissions from livestock buildings. Part 2: Results. Journal of Agricultural Engineering Research. 70 (1), 119–129.

Jacobson, L.D., B. Hetchler, K.A. Janni, L.J. Johmston, 1998. Odor and gas reduction from sprinkling soybean oil in a pig nursery. In: *ASAE Annual International Meeting*. Orlando, Florida, 12–16 July 1998. Orlando, Florida. 1–9.

Lee, C., L.R. Giles, W.L. Bryden, J.L. Downing, P.C. Owens, A.C. Kirby, P.C. Wynn, 2005. Performance and endocrine responses of group housed weaner pigs exposed to the air quality of a commercial environment. Livestock Production Science. 93 (3), 255–262.

Lemay, S.P., E.M. Barber, M. Bantle, D. Marcotte, 1999. Development of a sprinkling system using undiluted canola oil for dust control in pig buildings. In: *Dust Control in Animal Production Facilities*. Scandinavian Congress Center, Aarhus, 30 May–2 June. Scandinavian Congress Center, Aarhus: Danish Institute of Agricultural Science. Ed., Pedersen, S. 215–222.

Madelin, T.A., C.M. Wathes, 1989. Air hygiene in a broiler house: comparison of deep litter with raised netting floors. British Poultry Science. 30, 23–37.

SAS, 1989. SAS/STAT® User's Guide. Version 6, Fourth Edition, Volume 2,, Cary, NC: SAS Institute.

StatSoft, I., 2001. STATISTICA. StatSoft, Inc., Tulsa, OK, USA, data analysis software system p.

Takai, H., F. Moller, M. Iversen, S.E. Jorsal, V. Bille-Hansen, 1995. Dust control in pig houses by spraying rapeseed oil. Transactions of the ASAE. 38 (5), 1513–1518.

Takai, H., S. Pedersen, 1999. Design concept of oil sprayer for dust control in pig buildings. In: *Dust Control in Animal Production Facilities*. Scandinavian Congress Center, Aarhus, 30 May–2 June. Scandinavian Congress Center, Aarhus: Danish Institute of Agricultural Science. Ed., Pedersen, S. 279–286.

Takai, H., S. Pedersen, 2000. A comparison study of different dust control methods in pig buildings. Applied Engineering in Agriculture. 16 (3), 269–277.

Evaluation of Methods to Assess the Effectiveness of Different Ventilation Systems in Livestock Buildings: A Review

Yu Wang [a], Dapeng Li [a], Xiong Shen [b], Zhengxiang Shi [a,*]

[a] Key Laboratory of Agricultural Engineering in Structure and Environment, Ministry of Agriculture
College of Water Resources and Civil Engineering, China Agricultural University
P.O. Box 67, Beijing 100083, PR China

[b] Key Laboratory of Indoor Air Quality Control of Tianjin City,
School of Environment Science and Engineering, Tianjin University
Building 24, Tianjin 300072, PR China

* Corresponding author. Email: shizhx@cau.edu.cn

Abstract

Ventilation is believed to be an effective way of regulating indoor environment. It is used in buildings to create thermally comfortable environment with acceptable indoor air quality. In this article, state-of-the-art ventilation systems in animal feeding operations are introduced, and the indices of evaluating the effectiveness of different ventilation systems in animal feeding operations (AFOs) are revealed to judge the performance of a ventilation system quantitatively. In addition, the applied researches to predict airflow pattern and ventilation performance of different ventilation patterns are revised. Numerical simulation, laboratory experiment in test chamber and field experiment was adopted. According the papers reviewed, this paper put forward several major challenges that researchers will meet and directions that researchers should make endeavor to in the future study of predicting and optimization design of ventilation performance in animal feeding environment.

Keywords: Ventilation, evaluation, index, effectiveness

1. Introduction

Since mid-December 2014, there have been several ongoing highly pathogenic avian influenza (HPAI) H5 incidents along the Pacific, Central and Mississippi Flyways. Approximately 49.6 million commercial birds are affected and have been depopulated or are pending depopulation (USDA, 2015). The United States Department of Agriculture's Animal and Plant Health Inspection Service (APHIS) conducted investigations on over 80 commercial poultry farms and found that air samples collected outside of infected poultry houses contain virus particles, indicating that the virus could be transmitted by air (APHIS, 2015).

Air pollutants from animal feeding operations (AFOs) pose a potential threat not only for animal health, but also for the ambient environment. Researches have revealed that air pollutants have an important effect on hen welfare, health, and performance in commercial poultry and egg production (Wang et al., 2014). For example, high in-house NH3 concentrations can degrade bird health (Miles et al., 2004). PM particles may carry gaseous pollutants, bacteria, and viruses, and transport potentially harmful materials to the ambient environment of animal as well as person (Meng et al., 2012; Wang-Li et al., 2013).

Ventilation is believed to be an effective way of regulating indoor environment. It is used in buildings to create thermally comfortable environment with acceptable indoor air quality by regulating indoor air parameters, such as air temperature, relative humidity, air velocity, and chemical species concentrations in the air (Chen, 2009; Calvet et al., 2010).

Hence guarantee of indoor air quality is also a critical method to control the airborne diseases transmission and to reduce the infection risk in the indoor environment. Ni (2015) summarized that variations in indoor air quality could be correlated to several factors, among which were barn structures, manure handling and storage methods, animal densities, feed regimes, building ventilation, and farm management practices. Although large amount of studies have been focused on mitigating pollutant emissions from animal buildings and manure storage for air quality improvement, knowledge gaps still exist regarding the effect of ventilation on emissions and distributions of pollutants, the latter of which can influence uniformity of animal health and performance (Wang et al., 2014).

The objective of this paper is to provide an overview on the latest researches and their recent applications related to evaluating and predicting ventilation performance and indoor air quality in livestock building.

2. State-of-the-Art Ventilation Systems in Animal Feeding Operations

Ventilation in animal feeding facilities is a process of controlling several environmental factors by diluting inside air with fresh outside air, which can remove excess heat and moisture, minimise dust and odors, limit air pollutant concentration such as ammonia and carbon dioxide, and provide oxygen for respiration.

2.1. Natural ventilation

According to its energy saving effect, natural ventilated animal houses are widely used in temperate climates, e.g., most dairy houses are naturally ventilated with side wall and ridge openings in Europe (Pereira et al., 2010). Wind or buoyancy driven (Norton et al., 2009; Norton et al., 2010a, b; Shen et al., 2012) ventilation complied with the principle of reasonable vent opening area and positioning (Seo et al., 2009; Norton et al., 2010a; De Paepe et al., 2012) and the flow rate (Norton et al., 2009) provides comfortable thermal environment and air environment for agricultural structure and animal production.

2.2. Tunnel ventilation

The prevalence of intensive modern AFOs requires more precise environmental control for livestock and poultry building. Mechanical ventilation enables us to better control indoor thermal conditions, concentration of air pollutants, and air distribution by ventilation fans operating year-round (Hong et al., 2013).

Especially nearly the entire modern poultry industry around the world (both tropical and temperate climates) is using negative pressure tunnel ventilation with evaporative cooling misting or pad system in order to control the thermal environment inside the poultry building and minimize heat stress during the warmest seasons and hottest hours of the day (Wang-Li et al., 2013). In typical tunnel ventilation, ventilation air enters the house through air inlets on side walls (or one end wall) of the house. Each end wall (or another end wall) is normally equipped with several exhaust fans which located at different height and are ventilated in different stages. The minimum ventilation fans are also known as the primary representative exhaust fans (PREFs).

2.3. Cross ventilation

With regard to the type of mechanical ventilation, tunnel ventilation has yet to become widely used in Southern Europe where the cross ventilation systems are still used in most broiler houses (Blanes-Vidal et al., 2008; Calvet et al., 2010). This type of poultry house equipped for conventional cross ventilation is provided with a certain number of side-wall inlets and fans located at the opposite wall. Fresh air enters through wall inlets which are controlled by an opening system consisting of the bottom-hinged flaps.

2.4. Hybrid ventilation

Although natural ventilation has been applied for centuries, to overcome its drawbacks and meet its challenge, e.g., cleaning the exhaust air and maintaining appropriate thermal conditions in naturally ventilated building under cold weather (Rong et al., 2015), are still of much importance. In consequence the hybrid ventilation which combined natural ventilation with partial pit mechanical ventilation has been applied in swine or dairy cow houses (Rong et al., 2014; Rong et al., 2015).

For instance a modern swine confinement which is divided into two zones: the room where the pigs reside and the pit that stores the waste. With the purpose of guarantee of the indoor air quality (IAQ) within the confinement, fans located in the side wall of the room and in the end wall of the pit are used (Ecim-Djuric and Topisirovic, 2010; Zong et al., 2014a).

In some cases of poultry building, laying hen houses are equipped with local exhaust ventilation systems in deep pit or manure belt ventilation systems with forced air drying (Fabbri et al., 2007). In addition duct ventilation systems have been used in broiler houses to increase air temperature distribution uniformity and to prevent cold incoming air from directly reaching the bird zone in cold seasons when low air flow conditions are required for animal rearing (Mostafa et al., 2012).

3. Overview of Indices for Assessing and Predicting Ventilation Performance

Different kinds of barn, requirements of different environmental parameters, as well as complex air distribution conditions, may determine different ventilation performances. However it is generally identified that not only to maintain indoor thermal comfort, to remove indoor contaminants but also to maximize the utilization of energy are primary tasks of livestock building ventilation systems.

Indices for the ventilation efficiency were investigated actively in the 1980s and 1990s, but direct and continuous measurement of building ventilation is still a technical challenge especially for naturally-ventilated buildings and mechanically-ventilated buildings with large number of fans (Ni, 2015). Cao et al. (2014) reviewed current indices for assessing ventilation performance in built environment from five aspects including effectiveness of contamination removal, air exchange effectiveness, efficiency of heat removal, exposure effectiveness and effectiveness of an air distribution system. Besides that other concepts of the risk of airborne infection (Riley et al., 1978), the age of air (Sandberg, 1981), Purging Flow Rate (PFR) (Sandberg and Sjoberg, 1983) the visitation frequency (VF) (Csanady, 1983), ventilation effectiveness factor (VEF) (Zhang et al., 2001), the response coefficient (Li and Zhu, 2009), and net escape velocity (NEV) (Lim et al., 2013) have been used in researches of building ventilation.

3.1. Air exchange effectiveness

The air exchange effectiveness is the effectiveness of different ventilation systems in delivering the supply air to a particular point in a room (for example, how quickly the air in the ventilated space is changed by fresh air). Local air change effectiveness is defined as (David Etheridge, 1996):

$$\varepsilon_a = \frac{\tau_n}{2\overline{\tau_p}} \tag{1}$$

where τ_n is the nominal time constant for the room, which is the reciprocal of the air change rate. τ_n can be determined from the air supply rate Q and the room volume V, and $\overline{\tau_p}$ is the local mean age of air.

$$\tau_n = \frac{V}{Q} \tag{2}$$

The local mean age of air is calculated using the following equation:

$$\tau_p = \frac{1}{C_{(0)}}\int_0^\infty C_p(t)dt \tag{3}$$

where $C_{(0)}$ is the initial concentration of the tracer gas and $C_p(t)$ is the concentration at a certain point in the room at time t.

3.2. Effectiveness of contaminant removal

The contaminant concentration and greenhouse gases emission rate have been measured in more and more researches in order to obtain knowledge of release characteristics and to reduce the emission, which have a strong relationship with ventilation strategies. Ventilation effectiveness for the removal of contamination in the occupied zone is expressed as (Sandberg, 1981):

$$\varepsilon = \frac{C_e - C_s}{\bar{C} - C_s} \tag{4}$$

where C_e is the pollutant concentration in exhaust air, C_s the pollutant concentration in supply air and $\bar{C}$ is the mean contaminant concentration in the occupied zone. If $\bar{C}$ is the concentration at animal rearing level, then the ventilation effectiveness can be used to assess the efficiency of contaminant removal from the animal rearing zone. This index is indefinitely large when C_s equals to $\bar{C}$. However, to evaluate the ability of the ventilation system in reducing the concentrations of pollutants at a point, the absolute ventilation efficiency, which is expressed as given below, has been used.

$$\varepsilon_p = \frac{C_e - C_s}{C_p - C_s} \tag{5}$$

where C_p is the concentration at a certain point.

3.3. The risk of airborne infection

A concept of quantal infection was introduced by (Wells, 1957) for calculating the infection risk of airborne diseases. A quantum is the dose necessary to cause infection to a new susceptible and may be one or more airborne pathogens, which can attach to the droplet nuclei. Using the concept of quantal infection, Riley et al. (1978) developed the Wells–Riley equation for predicting the airborne infection risk. The equation can be written as follows.

$$P = \frac{C}{S} = 1 - e^{-Iqpt/Q} \tag{6}$$

where P is the probability of infection (risk), C is the number of cases to develop infection, S is the number of susceptibles, I is the number of source patients (infector), p is the pulmonary ventilation rate of each susceptible per minute (m^3/min), Q is the room ventilation rate (m^3/min), q is the quanta produced by one infector (quanta/min), and t is the duration of exposure (min).

Equation (6) has the following assumptions: As induced by air turbulence, there is a binomial probability property on the respiratory deposition of air-borne pathogens, which is not being well adjusted in current risk models.

4. Results and Discussion

Differs from that of residential environment for human beings, (the demand for comfort and indoor air quality is becoming higher and higher, many requirements are recommended in standards on ventilation effect by ASHRAE), it had been found that in many animal production facilities, the hygienic threshold values for indoor pollution are often exceeded (Wachenfelt, 1994). In modern intensive AFOs, to meet the requirements of ventilation rate and control the concentration of harmful gases and dust will directly affect the animal production performance. Therefore choosing the appropriate indices of assessing ventilation effect is extremely important for the control of indoor environment of animal production. A short summary of evaluation of ventilation performance of different AFOs is presented in Table 1. Ventilation performance or airflow measurement in the livestock buildings was identified as the main object of 14 studies in table1 and ammonia emission was the main objective of 9 studies. Given that quantification of ventilation effectiveness has been widely used to evaluate ventilation systems for air quality control in livestock buildings, researchers were interested in evaluating/improving assessment methods, and then especially the measurement accuracy of ventilation rate. Most of the studies adopted the ventilation effectiveness factor of ventilation rate (VR), and half of those investigated airflow pattern (AP) and air exchange rate (AR).

Table 1. A short summary of evaluation of ventilation performance of different AFOs.

Building	System[1]	Method	Indices[2]					Ref.[3]
			AP	VR	AR	CR	ER	
Fattening pig	NV	1:50 lab experiment	*					1
Livestock	MV	Simulation, 1:1 lab exp.		*				2
Livestock	NV	Simulation, 1:1 lab exp.	*		*			3
Farrowing pig	MV	Field experiment		*				4
Dairy cow	MV	Field experiment		*		*	*	5
Dairy cattle	NV	Field experiment		*		*	*	6
Livestock	NV	Simulation with CFD	*	*	*			7
Pig	MV	1:1 lab experiment	*	*	*	*		8
Broiler	MV	Field experiment		*				9
Livestock	HV	Simulation with CFD	*	*				10
Calf	NV	Simulation with CFD	*	*	*	*		11
Calf	NV	Simulation with CFD	*	*	*			12
Livestock	NV	Simulation with CFD	*					13
Fattening pig	HV	1:1 lab experiment		*	*	*	*	14
Pig	MV	Simulation with CFD	*					15
Dairy cattle	MV	1:2 lab experiment		*	*	*	*	16
Dairy cattle	NV	Simulation with CFD	*		*			17
Livestock	NV	Simulation with CFD	*	*	*	*	*	18
Dairy cow	HV	Field experiment		*	*	*	*	19
Fattening pig	HV	1:1 lab experiment		*		*	*	20
Dairy cow	HV	Simulation with CFD	*		*			21

[1] System: NV = naturally ventilated; MV = mechanically ventilated; HV = hybrid ventilated. [2] The indices are applied in research to evaluate the performance of ventilation system. AP = airflow pattern; VR = ventilation rate; AR = air exchange rate; CR = contaminant removal; ER = NH_3 / CO_2 emission rate. [3] References: 1. (Daskalov et al., 2005); 2. (Stables and Taylor, 2006); 3. (Bartzanas et al., 2007); 4. (Guarino et al., 2008); 5. (Bluteau et al., 2009); 6. (Ngwabie et al., 2009); 7. (Norton et al., 2009); 8. (Ye et al., 2009); 9. (Calvet et al., 2010); 10. (Ecim-Djuric and Topisirovic, 2010); 11. (Norton et al., 2010b); 12. (Norton et al., 2010c); 13. (Norton et al., 2010d); 14. (Saha et al., 2010); 15. (Seo et al., 2012); 16. (Wu et al., 2012a); 17. (Wu et al., 2012b); 18. (Bjerg et al., 2013a; 2013b; 2013c); 19. (Rong et al., 2014); 20. (Zong et al., 2014b); 21. (Rong et al., 2015).

However some difficulties and limitations do exist in the field application, for instance the factor of air exchange effectiveness and infection risk is related to structure configuration of indoor space and applied based on the assumption that the air is fully mixed which is unlikely to be true in typical livestock buildings. To estimate the AR, ventilation rate and mean age of air should be quantified at first while the uncertainty in the calculation of natural ventilation rates might induce misleading. Unlike the mechanically ventilated livestock buildings, it is more difficult to determine the naturally ventilation rates and the airflow in a real livestock building where the airflow is usually irregular. According these researches methods of volume flow rates and tracer gas decay were widely used in order to accurately quantify the natural ventilation rate, and Takai et al. (2013) found that in recent studies researchers did a lot of jobs on evaluating/improve known assessment methods and their measurement accuracy. To investigate and document the effectiveness of contaminant removal (CR) pollutant concentration in exhaust air is needed. The average concentration of all the sampling positions is generally used to represent the outlet gas concentration in the exit air (Zhang et al., 2005). Since the gas concentrations may vary significantly at different domains, the sampling positions for representative gas concentrations is crucial to improve the estimation accuracy of ventilation effectiveness and contaminant emissions. Airflow pattern (AP) have much influence on assessment of ventilation performance and the detailed airflow inside ventilated livestock buildings is increasingly described and predicted by Computational Fluid Dynamics (CFD) (Norton et al., 2010a, 2010b) as well as evaluating various methods of measurement.

Ventilation effectiveness of indoor animal production environment is affected by several factors, such as the natural climate conditions, sealing of corral building construction, management strategies, animal activities, etc. (Vranken et al., 2004). According to the high rearing density and uncertainty of animal activity, the flow field of animal indoor environment appeared to be more complicated. In addition, extremely high concentration of harmful gas or being placed in animal occupied zone may cause destruction of the experimental apparatus and sensors. A bad animal production environment brings difficulty to the ventilation index measurement and data acquisition. Especially in the field experiment, the obtaining of indices may bring a certain amount of stress to animals, which will affect the measurement results. It becomes more complicated when evaluating the performance of ventilation systems. In the design of air distribution, animal feeding device and the animals themselves as a source of buoyancy should be taken into account for the thermal plume from them influencing the airflow.

Although those researchers have given their opinions of proposed optimization measures of ventilation systems of AFOs, there are still several major challenges we will meet in the future study. Compared with the numerical simulation, field experiment is more costly and time consuming, results of experiment research are often thought to be more reliable. However, real and complex indoor environment is a great challenge for the accuracy and rationality of the experimental data due to the relatively large space, etc. The combination of field experiment and numerical simulation is a reliable research method and CFD simulations with finite field experiments are proved to be effective, reliable and economical. Field experiments make the CFD simulation results more convincing as validation of CFD simulation results.

To evaluate the efficiency of ventilation system of livestock and poultry building, reasonable indicators should be considered or introduced depending on the task of the system. Primary tasks have been identified including thermal comfort control, air exchange and contaminant removal. To fulfill these tasks, the combination of different types of ventilation, might have a better performance than applying one method. Methods to precisely measure air pollutant release and to acquire its data result from AFOs should be further studied. Energy saving potential of different ventilation systems applied in AFOs should be evaluated by taking into account conditions like animal thermal comfort and indoor air quality.

Acknowledgements

This work was supported by the National "Twelfth-Five Year" Research Program of China under Grant Number 2012BAD39B02.

References

APHIS, 2015. APHIS Releases Partial Epidemiology Report on Highly Pathogenic Avian Influenza.http://www.aphis.usda.gov/animal_health/animal_dis_spec/poultry/downloads/Epidemiologic-Analysis-June-15-2015.pdf.

Bartzanas, T., C. Kittas, A.A. Sapounas, C. Nikita-Martzopoulou, 2007. Analysis of airflow through experimental rural buildings: Sensitivity to turbulence models. Biosystems Engineering. 97 (2), 229–239. http://dx.doi.org/10.1016/j.biosystemseng.2007.02.009.

Bjerg, B., G. Cascone, I.-B. Lee, T. Bartzanas, T. Norton, S.-W. Hong, I.-H. Seo, T. Banhazi, P. Liberati, A. Marucci, G. Zhang, 2013a. Modelling of ammonia emissions from naturally ventilated livestock buildings. Part 3: CFD modelling. Biosystems Engineering. 116 (3), 259–275. http://dx.doi.org/10.1016/j.biosystemseng.2013.06.012.

Bjerg, B., P. Liberati, A. Marucci, G. Zhang, T. Banhazi, T. Bartzanas, G. Cascone, I.-B. Lee, T. Norton, 2013b. Modelling of ammonia emissions from naturally ventilated livestock buildings: Part 2, air change modelling. Biosystems Engineering. 116 (3), 246–258. http://dx.doi.org/10.1016/j.biosystemseng.2013.01.010.

Bjerg, B., T. Norton, T. Banhazi, G. Zhang, T. Bartzanas, P. Liberati, G. Cascone, I.B. Lee, A. Marucci, 2013c. Modelling of ammonia emissions from naturally ventilated livestock buildings. Part 1: Ammonia release modelling. Biosystems Engineering. 116 (3), 232–245. http://dx.doi.org/10.1016/j.biosystemseng.2013.08.001.

Blanes-Vidal, V., E. Guijarro, S. Balasch, A.G. Torres, 2008. Application of computational fluid dynamics to the prediction of airflow in a mechanically ventilated commercial poultry building. Biosystems Engineering. 100 (1), 105–116. http://dx.doi.org/10.1016/j.biosystemseng.2008.02.004.

Bluteau, C.V., D.I. Massé, R. Leduc, 2009. Ammonia emission rates from dairy livestock buildings in Eastern Canada. Biosystems Engineering. 103 (4), 480–488. http://dx.doi.org/10.1016/j.biosystemseng.2009.04.016.

Calvet, S., M. Cambra-López, V. Blanes-Vidal, F. Estellés, A.G. Torres, 2010. Ventilation rates in mechanically-ventilated commercial poultry buildings in Southern Europe: Measurement system development and uncertainty analysis. Biosystems Engineering. 106 (4), 423–432. http://dx.doi.org/10.1016/j.biosystemseng.2010.05.006.

Cao, G., H. Awbi, R. Yao, Y. Fan, K. Sirén, R. Kosonen, J. Zhang, 2014. A review of the performance of different ventilation and airflow distribution systems in buildings. Building and Environment. 73, 171–186. http://dx.doi.org/10.1016/j.buildenv.2013.12.009.

Chen, Q., 2009. Ventilation performance prediction for buildings: A method overview and recent applications. Building and Environment. 44 (4), 848–858. http://dx.doi.org/10.1016/j.buildenv.2008.05.025.

Chung, K.-C., S.-P. Hsu, 2001. Effect of ventilation pattern on room air and contaminant distribution. Building and Environment. 36 (9), 989–998. http://dx.doi.org/http://dx.doi.org/10.1016/S0360-1323(00)00051-2.

Csanady, G.T., 1983. Dispersal by Randomly Varying Currents. Journal of Fluid Mechanics. 132 (Jul), 375–394. http://dx.doi.org/Doi 10.1017/S0022112083001664.

Daskalov, P., K. Arvanitis, N. Sigrimis, J. Pitsilis, 2005. Development of an advanced microclimate controller for naturally ventilated pig building. Computers and Electronics in Agriculture. 49 (3), 377–391. http://dx.doi.org/10.1016/j.compag.2005.08.010.

David Etheridge, M.S., 1996. Building ventilation: theory and measurement: John Wiley & Sons.

De Paepe, M., J.G. Pieters, W.M. Cornelis, D. Gabriels, B. Merci, P. Demeyer, 2012. Airflow measurements in and around scale model cattle barns in a wind tunnel: Effect of ventilation opening height. Biosystems Engineering. 113 (1), 22-32. http://dx.doi.org/10.1016/j.biosystemseng.2012.06.003.

Ecim-Djuric, O., G. Topisirovic, 2010. Energy efficiency optimization of combined ventilation systems in livestock buildings. Energy and Buildings. 42 (8), 1165–1171. http://dx.doi.org/10.1016/j.enbuild.2009.10.035.

Fabbri, C., L. Valli, M. Guarino, A. Costa, V. Mazzotta, 2007. Ammonia, methane, nitrous oxide and particulate matter emissions from two different buildings for laying hens. Biosystems Engineering. 97 (4), 441–455. http://dx.doi.org/10.1016/j.biosystemseng.2007.03.036.

Guarino, M., A. Costa, M. Porro, 2008. Photocatalytic TiO2 coating to reduce ammonia and greenhouse gases concentration and emission from animal husbandries. Bioresource Technology. 99 (7), 2650–2658. http://dx.doi.org/10.1016/j.biortech.2007.04.025.

Hong, S.W., I.B. Lee, I.H. Seo, K.S. Kwon, 2013. The design and testing of a small-scale wind turbine fitted to the ventilation fan for a livestock building. Computers and Electronics in Agriculture. 99, 65–76. http://dx.doi.org/10.1016/j.compag.2013.08.020.

Li, X.T., F.F. Zhu, 2009. Response Coefficient: A New Concept to Evaluate Ventilation Performance with "Pulse" Boundary Conditions. Indoor and Built Environment. 18 (3), 189–204. http://dx.doi.org/10.1177/1420326X09104345.

Lim, E., K. Ito, M. Sandberg, 2013. New ventilation index for evaluating imperfect mixing conditions – Analysis of Net Escape Velocity based on RANS approach. Building and Environment. 61, 45–56. http://dx.doi.org/10.1016/j.buildenv.2012.11.022.

Meng, Q.Y., D. Svendsgaard, D.J. Kotchmar, J.P. Pinto, 2012. Associations between personal exposures and ambient concentrations of nitrogen dioxide: A quantitative research synthesis. Atmospheric Environment. 57, 322–329. http://dx.doi.org/10.1016/j.atmosenv.2012.04.035.

Miles, D.M., S.L. Branton, B.D. Lott, 2004. Atmospheric ammonia is detrimental to the performance of modern commercial broilers. Poultry Science. 83 (10), 1650–1654.

Mostafa, E., I.-B. Lee, S.-H. Song, K.-S. Kwon, I.-H. Seo, S.-W. Hong, H.-S. Hwang, J.P. Bitog, H.-T. Han, 2012. Computational fluid dynamics simulation of air temperature distribution inside broiler building fitted with duct ventilation system. Biosystems Engineering. 112 (4), 293–303. http://dx.doi.org/10.1016/j.biosystemseng.2012.05.001.

Ngwabie, N.M., K.H. Jeppsson, S. Nimmermark, C. Swensson, G. Gustafsson, 2009. Multi-location measurements of greenhouse gases and emission rates of methane and ammonia from a naturally-ventilated barn for dairy cows. Biosystems Engineering. 103 (1), 68–77. http://dx.doi.org/10.1016/j.biosystemseng.2009.02.004.

Ni, J.Q., 2015. Research and demonstration to improve air quality for the U.S. animal feeding operations in the 21st century - a critical review. Environmental Pollution. 200, 105–119. http://dx.doi.org/10.1016/j.envpol.2015.02.003.

Norton, T., J. Grant, R. Fallon, D.-W. Sun, 2009. Assessing the ventilation effectiveness of naturally ventilated livestock buildings under wind dominated conditions using computational fluid dynamics. Biosystems Engineering. 103 (1), 78–99. http://dx.doi.org/10.1016/j.biosystemseng.2009.02.007.

Norton, T., J. Grant, R. Fallon, D.-W. Sun, 2010a. Assessing the ventilation performance of a naturally ventilated livestock building with different eave opening conditions. Computers and Electronics in Agriculture. 71 (1), 7–21. http://dx.doi.org/10.1016/j.compag.2009.11.003.

Norton, T., J. Grant, R. Fallon, D.-W. Sun, 2010b. A computational fluid dynamics study of air mixing in a naturally ventilated livestock building with different porous eave opening conditions. Biosystems Engineering. 106 (2), 125–137. http://dx.doi.org/10.1016/j.biosystemseng.2010.02.006.

Norton, T., J. Grant, R. Fallon, D.-W. Sun, 2010c. Improving the representation of thermal boundary conditions of livestock during CFD modelling of the indoor environment. Computers and Electronics in Agriculture. 73 (1), 17–36. http://dx.doi.org/10.1016/j.compag.2010.04.002.

Norton, T., J. Grant, R. Fallon, D.-W. Sun, 2010d. Optimising the ventilation configuration of naturally ventilated livestock buildings for improved indoor environmental homogeneity. Building and Environment. 45 (4), 983–995. http://dx.doi.org/10.1016/j.buildenv.2009.10.005.

Pereira, J., T.H. Misselbrook, D.R. Chadwick, J. Coutinho, H. Trindade, 2010. Ammonia emissions from naturally ventilated dairy cattle buildings and outdoor concrete yards in Portugal. Atmospheric Environment. 44 (28), 3413–3421. http://dx.doi.org/10.1016/j.atmosenv.2010.06.008.

Riley, E.C., G. Murphy, R.L. Riley, 1978. Airborne spread of measles in a suburban elementary-school. American Journal of Epidemiology. 107 (5), 421–432.

Rong, L., D. Liu, E.F. Pedersen, G. Zhang, 2014. Effect of climate parameters on air exchange rate and ammonia and methane emissions from a hybrid ventilated dairy cow building. Energy and Buildings. 82, 632–643. http://dx.doi.org/10.1016/j.enbuild.2014.07.089.

Rong, L., D. Liu, E.F. Pedersen, G. Zhang, 2015. The effect of wind speed and direction and surrounding maize on hybrid ventilation in a dairy cow building in Denmark. Energy and Buildings. 86, 25–34. http://dx.doi.org/10.1016/j.enbuild.2014.10.016.

Saha, C.K., G. Zhang, P. Kai, B. Bjerg, 2010. Effects of a partial pit ventilation system on indoor air quality and ammonia emission from a fattening pig room. Biosystems Engineering. 105 (3), 279–287. http://dx.doi.org/10.1016/j.biosystemseng.2009.11.006.

Sandberg, M., 1981. What Is Ventilation Efficiency. Building and Environment. 16 (2), 123–135. http://dx.doi.org/Doi 10.1016/0360-1323(81)90028-7.

Sandberg, M., M. Sjoberg, 1983. The use of moment for assessing air quality in ventilated rooms. Building and Environment. 18 (4), 181–197. http://dx.doi.org/10.1016/0360-1323(83)90026-4.

Seo, I.-h., I.-b. Lee, O.-k. Moon, S.-w. Hong, H.-s. Hwang, J.P. Bitog, K.-s. Kwon, Z. Ye, J.-w. Lee, 2012. Modelling of internal environmental conditions in a full-scale commercial pig house containing animals. Biosystems Engineering. 111 (1), 91–106. http://dx.doi.org/10.1016/j.biosystemseng.2011.10.012.

Seo, I.H., I.B. Lee, O.K. Moon, H.T. Kim, H.S. Hwang, S.W. Hong, J.P. Bitog, J.I. Yoo, K.S. Kwon, Y.H. Kim, J.W. Han, 2009. Improvement of the ventilation system of a naturally ventilated broiler house in the cold season using computational simulations. Biosystems Engineering. 104 (1), 106–117. http://dx.doi.org/10.1016/j.biosystemseng.2009.05.007.

Shen, X., G. Zhang, B. Bjerg, 2012. Comparison of different methods for estimating ventilation rates through wind driven ventilated buildings. Energy and Buildings. 54, 297–306. http://dx.doi.org/10.1016/j.enbuild.2012.07.017.

Stables, M.A., C.J. Taylor, 2006. Non-linear Control of Ventilation Rate using State-dependent Parameter Models. Biosystems Engineering. 95 (1), 7–18. http://dx.doi.org/10.1016/j.biosystemseng.2006.05.015.

Takai, H., S. Nimmermark, T. Banhazi, T. Norton, L.D. Jacobson, S. Calvet, M. Hassouna, B. Bjerg, G.-Q. Zhang, S. Pedersen, P. Kai, K. Wang, D. Berckmans, 2013. Airborne pollutant emissions from naturally ventilated buildings: Proposed research directions. Biosystems Engineering. 116 (3), 214–220. http://dx.doi.org/10.1016/j.biosystemseng.2012.12.015.

USDA, 2015. Avian Influenza, http://www.usda.gov/wps/portal/usda/usdahome.

Villafruela, J.M., F. Castro, J.F. San José, J. Saint-Martin, 2013. Comparison of air change efficiency, contaminant removal effectiveness and infection risk as IAQ indices in isolation rooms. Energy and Buildings. 57, 210–219. http://dx.doi.org/10.1016/j.enbuild.2012.10.053.

Vranken, E., S. Claes, J. Hendriks, P. Darius, D. Berckmans, 2004. Intermittent Measurements to determine Ammonia Emissions from Livestock Buildings. Biosystems Engineering. 88 (3), 351–358. http://dx.doi.org/10.1016/j.biosystemseng.2004.03.011.

Wachenfelt, E.v., 1994. Modern technique gives less air pollutions in broiler houses. XII World Congress on Agricultural Engineering: Volume 1. Proceedings of a conference held in Milan, Italy, August 29 - September 1 1994. 590–595.

Wang-Li, L., Z. Cao, Q. Li, Z. Liu, D.B. Beasley, 2013. Concentration and particle size distribution of particulate matter inside tunnel-ventilated high-rise layer operation houses. Atmospheric Environment. 66, 8–16. http://dx.doi.org/10.1016/j.atmosenv.2012.03.064.

Wang, Z., T. Gao, Z. Jiang, Y. Min, J. Mo, Y. Gao, 2014. Effect of ventilation on distributions, concentrations, and emissions of air pollutants in a manure-belt layer house. The Journal of Applied Poultry Research. 23 (4), 763–772. http://dx.doi.org/10.3382/japr.2014-01000.

Wells, W.F., 1957. Airborne Contagion and Air Hygiene: An Ecological Study of Droplet Infection. London: Harvard University Press. 65 p.

Wu, W., P. Kai, G. Zhang, 2012a. An assessment of a partial pit ventilation system to reduce emission under slatted floor – Part 1: Scale model study. Computers and Electronics in Agriculture. 83, 127–133. http://dx.doi.org/10.1016/j.compag.2012.01.008.

Wu, W., J. Zhai, G. Zhang, P.V. Nielsen, 2012b. Evaluation of methods for determining air exchange rate in a naturally ventilated dairy cattle building with large openings using computational fluid dynamics (CFD). Atmospheric Environment. 63, 179–188. http://dx.doi.org/10.1016/j.atmosenv.2012.09.042.

Ye, Z., C.K. Saha, B. Li, G. Tong, C. Wang, S. Zhu, G. Zhang, 2009. Effect of environmental deflector and curtain on air exchange rate in slurry pit in a model pig house. Biosystems Engineering. 104 (4), 522–533. http://dx.doi.org/10.1016/j.biosystemseng.2009.09.015.

Zhang, Y.H., X. Wang, G.L. Riskowski, L.L. Christianson, 2001. Quantifying ventilation effectiveness for air quality control. Transactions of the ASAE. 44 (2), 385–390.

Zhang, G., J.S. Strom, B. Li, H.B. Rom, S. Morsing, P. Dahl, C. Wang, 2005. Emission of ammonia and other contaminant gases from naturally ventilated dairy cattle buildings. Biosystems Engineering. 92 (3), 355–364. http://dx.doi.org/10.1016/j.biosystemseng.2005.08.002.

Zong, C., Y. Feng, G. Zhang, M.J. Hansen, 2014a. Effects of different air inlets on indoor air quality and ammonia emission from two experimental fattening pig rooms with partial pit ventilation system – Summer condition. Biosystems Engineering. 122, 163–173. http://dx.doi.org/10.1016/j.biosystemseng.2014.04.005.

Zong, C., G. Zhang, Y. Feng, J.-Q. Ni, 2014b. Carbon dioxide production from a fattening pig building with partial pit ventilation system. Biosystems Engineering. 126, 56–68. http://dx.doi.org/10.1016/j.biosystemseng.2014.07.011.

Re-Design of Ventilation System of a Newly-Converted Grouped Sow Housing Facility Using Computer Simulation

Bernardo Predicala [*], Alvin Alvarado

Prairie Swine Centre Inc., Saskatoon, Saskatchewan S7H 5N9, Canada

* Corresponding author. Email: bernardo.predicala@usask.ca

Abstract

In response to increasing public demand for improved animal welfare in pig production, the industry is currently shifting from housing pregnant sows in traditional stalls to the more welfare-friendly group housing system. In addition, in the recently-revised Canadian code of practice, more floor space is allocated per sow than the current allocation in traditional stall systems, thus converting a sow barn facility to group housing system means reduced total number of animals can be housed in the same building footprint. This conversion leads to considerable changes in the environmental conditions of the room and thus requires re-consideration of the existing ventilation system. This study aimed to re-design and evaluate the mechanical ventilation system of a newly-converted group sow housing facility using computer simulation techniques. A computer simulation package which utilizes computational fluid dynamics (CFD) principles to numerically simulate fluid flow, heat and mass transfer, and mechanical movement, was employed. Various ventilation design configurations which included different sizes and locations of exhaust fans, and design and location of air inlets, were assessed based on thermal ventilation effectiveness. Results showed that the pre-conversion room with stall system and the converted group housing room with unmodified ventilation design were not very effective in removing heat from the animal occupied zone. Among all the design configurations investigated, tunnel ventilation system with the exhaust fans on one side and inlets on the opposite side had the highest heat removal effectiveness for both summer and winter conditions, thus this configuration was selected for actual in-barn evaluation.

Keywords: Swine, barn conversion, ventilation, computational fluid dynamics, group housing

1. Introduction

Traditionally, pregnant sows are kept in stall systems during their gestation period, but in response to increasing public demand for improved animal welfare in pig production, the recent revision to the Canadian Code of Practice for the Care and Handling of Pigs has mandated that sows should be reared using the more welfare-friendly group housing system (NFACC, 2014). In the process of converting from stalls to group housing system, most pork producers focus mainly on remodeling the penning, floor layout, and feeding system, while the ventilation system is frequently neglected or overlooked (Harmon, 2013). Because the new code requires more floor space allocated per sow (19-24 square feet per sow) than the current space allocation in traditional stall systems (typically less than 18 sq. ft. per sow), converting a sow barn facility to group housing system means that generally less total number of animals can be reared in the same building footprint.

Ventilation affects many aspects of the animal environment as well as barn operating costs (in particular, energy costs). Retaining the existing ventilation system as-is in a converted sow barn leads to over-ventilation during winter because the existing minimum ventilation fans are designed for higher animal density. According to Harmon et al. (2010), when ventilation is continued at the pre-remodeling level (prior to conversion to group housing), the building would be over-ventilated by about 33% higher than required. An estimate of energy use for an over-ventilated facility indicated that over-ventilating by 30% can raise heating energy consumption by 75%. During summer, the impacts are less pronounced but over-ventilation uses extra electricity which translates to higher electrical cost (Harmon, 2013). In addition, the transitioning of the ventilation system designed for stall system to group housing also requires careful reconfiguration to ensure proper air distribution throughout the room to eliminate dead spots (unventilated areas) and unwanted drafts. Air exchange is critical to providing a healthy environment that fosters efficient pig growth by reducing humidity and noxious gases like ammonia and carbon dioxide. Since under-ventilation creates an unhealthy environment and over-ventilation wastes valuable heating and electrical energy, finding the right balance is the key to a healthy environment for both animals and workers as well as to energy savings and efficiency (Harmon et al., 2010).

Assessing the effectiveness of various potential ventilation system design configurations in an actual barn set-up is a big challenge because of the costs associated with setting up the experiment to be able to collect data from each individual design option being tested. Hence, in this work, numerical computer simulation technique which utilizes computational fluid dynamics (CFD) principles to numerically simulate fluid flow, heat and mass transfer, and mechanical movement, was used as a tool to examine various design configurations and to determine the most effective design of the ventilation system for a converted group sow housing facility. The CFD modeling has been widely used in the study of airflow and air quality in greenhouses, animal buildings, and human-occupied structures (Jackson et al., 1999; Svidt et al., 1998; Gan, 1995). Adrion et al. (2013) used CFD techniques for airflow calculations in two pig fattening compartments and found that numerical simulation can offer a deeper insight into airflow in pig barns and into potential strategies for optimization. Sun et al. (2002) reported that CFD simulation may provide precise solutions and visualizations of velocity, pressure, temperature, turbulence, and gas concentrations throughout the entire building airspace without extensive testing of a full-scale model.

The overall goal of this project is to re-design the ventilation system of a gestation barn converted from traditional stalls to a group sow housing system, to reduce energy cost, and improve overall air quality and sow performance. Specifically, this study aimed to develop and evaluate different ventilation system design configurations using computer simulation techniques.

2. Materials and Methods

In this study, the mechanical ventilation system of a newly-converted group sow housing facility at the Prairie Swine Centre in Saskatchewan, Canada, was evaluated to determine the changes that are necessary when converting a gestation room from stalls to group housing system. The room has inside dimensions of 6.97 m (width) x 19.83 m (length) x 2.9 m (height).

The purpose of the barn ventilation system is to bring fresh air into the building through inlet openings, thoroughly mix it with stale inside air, and exhaust the moist contaminated air from the building. Exhaust fans and air inlets drive this air movement and are considered as critical components in mechanical ventilation systems. Thus, in this work, the ventilation system design parameters investigated include: (1) capacity and location of exhaust fans, and (2) size and location of air inlets. These two parameters were configured in such a way that the resulting ventilation system design followed the principles of either an upward airflow, downward airflow or tunnel ventilation. Summary of the different design configurations being investigated is shown in Table 1.

Each design configuration was assessed based on ventilation effectiveness which is an indicator for uniform mixing and elimination of dead zones and unwanted drafts (van Wagenberg and Smolders, 2002; Breum et al., 1990). Ventilation effectiveness was evaluated based on the effectiveness of removal of heat (HRE) from the animal occupied zone (AOZ) calculated as follows (van Wagenberg and Smolders, 2002):

$$\mathrm{HRE} = (\mathrm{T_{outlet}} - \mathrm{T_{inlet}}) / (\mathrm{T_p} - \mathrm{T_{inlet}}) \qquad (1)$$

where *HRE* is the dimensionless heat removal effectiveness at point *p*; T_{inlet} is temperature of the inlet air, °C; T_{outlet} is temperature of the outlet air, °C; T_p is temperature at *p* in the room, °C.

Table 1. Ventilation system design configurations investigated in this simulation work.

Housing type	Ventilation rate, m^3/s Summer	Ventilation rate, m^3/s Winter	Flow direction	Exhaust fan location	Inlet location	Code
Stall*	3.78	0.32	Downward	Wall	Ceiling	SDWC
Group**	3.78	0.32	Downward	Wall	Ceiling	GrDWC
Group***	2.94	0.25	Downward	Wall	Ceiling	GDWC
				Wall	Ceiling + recirculation duct	GDWC-RD
				Sidewall	Ceiling	GDSC
				Pit	Ceiling	GDPC
			Upward	Wall	Sidewall - close to floor	GUWSF
				Chimney (ceiling)	Sidewall - close to floor	GUCSF
				Chimney	Sidewall - close to ceiling	GUCSC
				Chimney	Ceiling	GUCC
			Tunnel	Wall	Wall - longitudinal	GTWW
				Right sidewall	Left sidewall	GTSS

*Pre-conversion barn with 54 sows in stall system and using the current ventilation system; **Converted barn with the new group housing lay-out and new number of sows (42) but using the existing (unmodified) ventilation system; ***Converted barn with ventilation system modified using various configuration options.

The value of HRE can be above or below 1. According to van Wagenberg and Smolders (2002), HRE values above 1 indicate that fresh air enters the AOZ first and then passes the contaminant sources on its way to the outlet, which indicates effective air displacement in the AOZ. HRE values below 1 indicate that the temperature level in the AOZ exceeds the temperature in the outlet air. Low values of HRE translate to high contaminant levels at the AOZ that are not being efficiently removed by the ventilation system.

Based on the above considerations, the following steps were conducted in this study:

1. A computer model of a barn before conversion to group housing was generated, using existing components of the current ventilation system

2. Using the model barn, computer simulation was conducted on both the pre-conversion barn as well as the converted barn with the new group housing lay-out and new number of sows but using the existing (unmodified) ventilation system to determine its current performance (baseline case)

3. Using the model of the converted barn, a series of simulations was conducted with various configurations of the ventilation system applied to the barn, based on the design parameters described above to assess their performance relative to the baseline case

4. From the results of the series of simulations, the various ventilation system configurations were ranked based on ventilation effectiveness. The ventilation configuration which showed the best performance was selected for actual in-barn evaluation in subsequent work.

The computer simulation was carried out using ANSYS Fluent 15.0 (ANSYS Inc., Canonsburg, PA, USA). The setting-up of models and mesh as well as the evaluation of results were done through the application of DesignModeler, Meshing and CFD-Post in the ANSYS Academic Research CFD Package (ANSYS Inc., Canonsburg, PA, USA). A standard κ-ε model with scalable wall functions was used. A pressure-based solver with SIMPLE algorithm was employed for the calculations.

3. Results and Discussion

3.1. Model of gestation rooms and boundary conditions

As shown in Figure 1, computer models of two sow gestation rooms with different geometries were generated in this simulation work. Figure 1A is the model of a pre-conversion barn with sows housed in stall system. A total of 54 sows were in the room, occupying all the available stalls. For this simulation, all sows were assumed to be standing at a height of 0.9 m. Figure 1B is the model of a converted barn with group housing layout. A total of 42 sows were housed in the room; 32 were assumed to be lying down (shown in rectangular blocks) and the rest were standing (shown in cylinders). Sows lying down were assumed to have a height of 0.425 m. For simulating some of the ventilation system configurations that involved pit ventilation, the converted barn geometry was enhanced by including the underfloor manure storage and slatted flooring in the model; this was not considered in the other models for two reasons: 1. to simplify the geometry of the model and to lessen the number of grid points in the generated mesh (which impacts the required computing power and time to run the simulations), and 2. It was assumed that there was minimal interaction between the room airspace and the underfloor space for some of these configurations, thus the simplified room geometry was used.

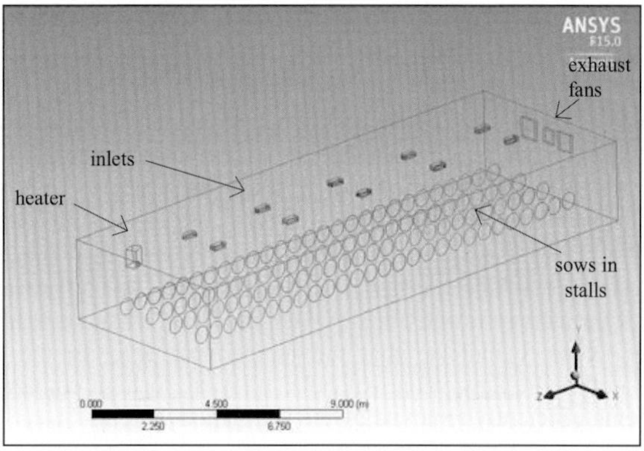

(A)

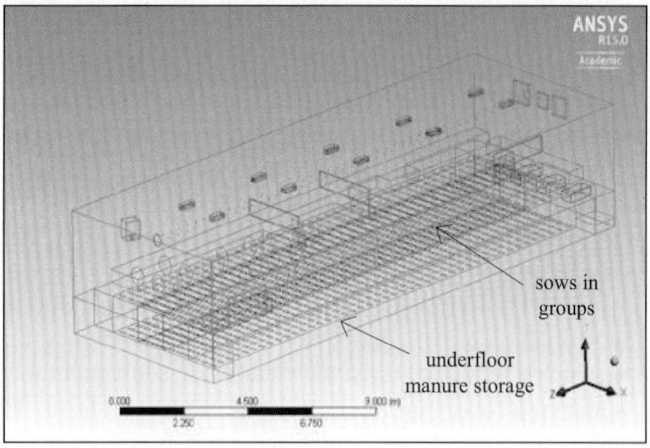

(B)

Figure 1. Two types of gestation room lay-out that were modeled for computer simulation: A) pre-conversion with stall system, and B). newly-converted with group housing system.

In this study, the developed model was used in simulations under winter and summer conditions. Table 2 shows the boundary conditions used in the simulation work. Inlet temperature and the corresponding moisture concentration were based from the average minimum winter and maximum summer temperatures from climatological records for Saskatchewan, Canada. Ventilation rates for the stall and group housing systems were based from 0.07 m^3/s-sow (150 cfm/sow) for summer and 0.006 m^3/s-sow (12 cfm/sow) for winter recommended by the Midwest Plan Service (MWPS, 1990). The total ventilation rate of the two housing systems were based on the number of sows housed in the room; 54 sows for the stall system and 42 sows for the newly-converted group system. The sows were assumed to be 190 kg on average, which impacts the total heat and moisture production of all the sows in the room. In addition, the heat transfer coefficient of the walls and ceilings were based on the thermal resistance (R-value) of the building components and insulation material.

Table 2. Boundary conditions used in the CFD simulations.

Parameters		Winter	Summer
Inlet			
	temperature, °C	-32	28
	moisture concentration, kg/m^3	0.0003	0.022
Outlet			
	ventilation rate, m^3/s (stall system)	0.31	3.82
	ventilation rate, m^3/s (group system)	0.24	2.97
Pigs			
	heat generation, W	288.42	243.47
	moisture production, mg/s	34.3	41.2
Heat transfer coefficient (insulation)			
	ceiling, W/m^2-K	0.19	0.19
	walls, W/m^2-K	0.28	0.28

3.2. Ventilation effectiveness

Table 3 shows the heat removal effectiveness (HRE) of the different ventilation system configurations during periods with high and low ventilation. The values represented nine different points in the room with each point located at the animal occupied zone (AOZ) approximately 1 m above the pen floor. According to van Wagenberg and Smolders (2002), the average HRE values in the AOZ provide a better indication of the effectiveness of air displacement. In general, HRE values at high ventilation rates were not influenced by the general air flow direction. The pre-conversion room with stall system (SDWC) had an average HRE value of 1.18 ± 0.49 which generally indicates effective air displacement in the AOZ. However, the resulting standard deviation (0.49) implies that the variation among the different points in the AOZ was high, indicating that the air was not homogeneously mixed. It was observed that certain points in the AOZ had HRE value less than 1 (lowest HRE was 0.61), indicating that the temperature at these points were higher than the temperature at the exhaust.

When the room was converted to group housing system but the existing ventilation design was retained (GrDWC), the average HRE value was almost the same (1.17) but the variability among the different points was reduced slightly from SD of 0.49 to 0.36. When the ventilation rate was adjusted in accordance with lesser number of sows after conversion (GDWC), a slight improvement in the HRE value and further reduction in variability were observed (1.20 ± 0.22).

In general, with the group housing layout and new ventilation design, HRE value increased particularly when the air inlets were located on the opposite side of the exhaust fans following the principle of tunnel ventilation system (GTWW). GTWW had an average HRE value of 1.32 ± 0.32, which was the highest among all the design configurations investigated. Also, for this configuration, all nine monitoring points in the AOZ had HRE values greater than 1 (lowest

HRE was 1.08) which indicate that the air was homogeneously mixed. Figure 2 shows the streamlines and air temperature distribution in the room at high ventilation rates during summer season. Effective removal of heat from the AOZ (HRE > 1) is desirable at high ventilation rates during summer season when ventilation is mainly for temperature control at the AOZ (van Wagenberg and Smolders, 2002).

Table 3. Mean (±SD) values of heat removal effectiveness (HRE) of the different ventilation system design configurations during summer and winter periods (N = 9).

Design code	Flow direction	Heat Removal Effectiveness	
		Summer	Winter
SDWC	Downward	1.18 (0.49)	0.85 (0.10)
GrDWC	Downward	1.17 (0.36)	0.83 (0.12)
GDWC	Downward	1.20 (0.22)	0.87 (0.10)
GDWC-RD	Downward	1.03 (0.23)	0.93 (0.09)
GDSC	Downward	1.18 (0.30)	1.06 (0.12)
GDPC	Downward	0.95 (0.07)	0.95 (0.04)
GUWSF	Upward	0.91 (0.24)	0.94 (0.06)
GUCSF	Upward	1.08 (0.32)	1.10 (0.14)
GUCSC	Upward	0.90 (0.23)	1.11 (0.07)
GUCC	Upward	1.19 (0.31)	1.10 (0.29)
GTWW	Tunnel	1.32 (0.23)	1.11 (0.12)
GTSS	Tunnel	1.01 (0.43)	1.06 (0.10)

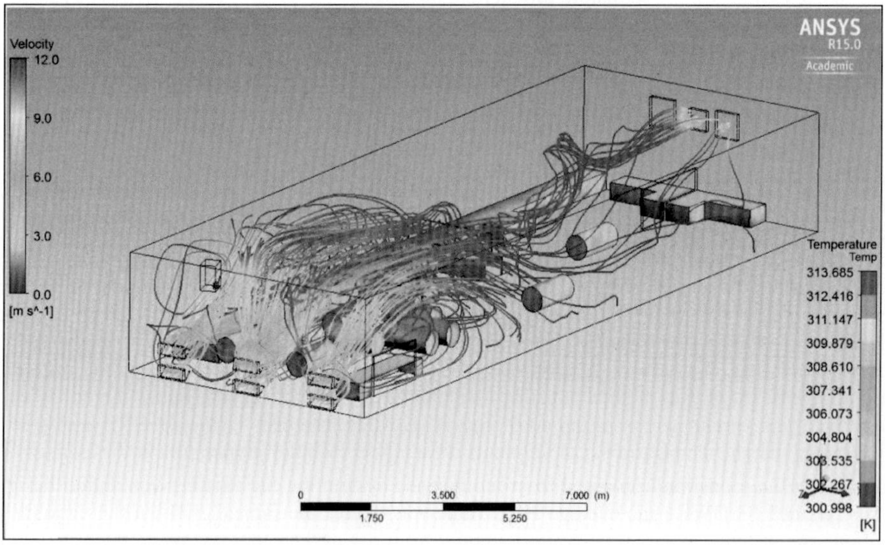

Figure 2. Ventilation air streamlines and temperature distribution in the room with tunnel ventilation system (GTWW) during summer period.

During winter period, all HRE values decreased which could be attributed mainly to the lower ventilation rates maintained in the rooms during the cold season. However, GTWW still had HRE values greater than 1 in all 9 monitoring points. On average, GTWW had an HRE value of 1.11 ± 0.12, which was the highest among all the designs tested for winter. Furthermore, the room with the stall system (SDWC) and the converted room with unmodified ventilation system (GrDWC) had average HRE values less than 1, indicating that part of the fresh air coming from the inlets was directly removed from the room without mixing and without causing air displacement in the AOZ. This may result to accumulation of high contaminant levels at the AOZ because stale air is not being efficiently removed by the ventilation system.

4. Conclusions

Results from the computer simulation work have confirmed the need to re-design the ventilation system of a newly-converted group sow housing facility. The pre-conversion room with the stall system as well as the newly-converted room with group sow housing layout and existing (unmodified) ventilation system had HRE values of less than 1, which indicated that air was not homogeneously mixed in the room. Among all the design configurations tested, tunnel ventilation system was the most effective in removing heat from the animal occupied zone in the room during both summer and winter seasons. This particular design will be implemented in actual in-barn tests to validate the results from computer simulation and to provide a more definitive investigation of the impact of the new ventilation design configuration on air quality, animal welfare, and overall production costs.

Acknowledgements

The research team would like to acknowledge the financial support for this research project from the Agriculture Council of Saskatchewan Inc. through the Advancing Canadian Agriculture and Agri-Food Saskatchewan (ACAAFS) Program. The authors also acknowledge the strategic program funding provided by Sask Pork, Alberta Pork, Ontario Pork, the Manitoba Pork Council and the Saskatchewan Agriculture Development Fund to the Prairie Swine Centre. In addition, the authors also wish to express their appreciation for the support of the production and research staff at Prairie Swine Centre in carrying out this study.

References

Adrion, F., J. Threm, E. Gallmann, W. Pflanz, and T. Jungbluth, 2013. Simulation of airflow in pig fattening houses with different air supply systems. Landtechnik. 68(2), 89–94.

Breum, N.O., H. Takai, and H.B. Rom, 1990. Upward vs. downward ventilation air flow in a swine house. Transactions of the ASAE. 33(5), 1693–1699.

Gan, G., 1995. Evaluation of room air distribution systems using computational fluid dynamics. Energy and Buildings. 23(2), 83–93.

Harmon J., 2013. Group housing systems: New and conversion construction. National Pork Board, Des Moines, IA.

Harmon, J, M. Hanna, and D. Petersen, 2010. Sizing minimum ventilation to save heating energy in swine housing. Iowa State University. PM2089J. https://store.extension.iastate.edu/Product/Sizing-Minimum-Ventilation-to-Save-Heating-Energy-in-Swine-Housing-Farm-Energy. Accessed May 15, 2015.

Jackson, C., J. Rehg, C. Rock, R. Henning and S. Reynolds, 1999. Computer simulation optimizes airflow design in non-rectangular animal room configurations. www.fluent.com/applicat/HVAC/articles/jude.htm. Accessed May 15, 2015.

MWPS, 1990. Mechanical ventilating systems for livestock housing. Midwest Plan Service, Ames, IA.

NFACC, 2014. Code of Practice for the Care and Handling of Pigs. Canada: National Farm Animal Care Council. https://www.nfacc.ca/pdfs/codes/pig_code_of_practice.pdf. Accessed May 15, 2015.

Sun, H., R. Stowell, H. Keener and F. Michel, Jr., 2002. Two–dimensional computational fluid dynamics (CFD) modeling of air velocity and ammonia distribution in a High–RiseTM hog building. Transaction of the ASAE. 25(5), 1559–1568.

Svidt, K., D. Zhang and B. Bjerg, 1998. CFD simulation of air velocity distribution in occupied livestock building. In Roomvent '98: 6th International Conference on Air Distribution in Rooms. Stockholm, Sweden: Royal Institute of Technology. 491–496.

Van Wagenberg, A. V., and M. A. H. H. Smolders, 2002. Contaminant and heat removal effectiveness of three ventilation systems in nursery rooms for pigs. Transactions of the ASAE. 45(6), 1985–1992.

Using Smart Ventilation Approaches for Complex Environmental Issues in Farm Animal Housing

Guoqiang Zhang

Department of Engineering, Aarhus University, Aarhus, DK-8000, Denmark

Email: Guoqiang.Zhang@eng.au.dk

Abstract

Ventilation is one of the most important approaches to achieve a desired indoor air quality and thermal environment for occupants in confined structures. In farm animal housing, intensive production, cost efficiency, working condition and environmental impacts make the issue complex, depending on many factors, such as the region climate, animal welfare and environmental regulation, etc. This paper presents some of the innovative approaches to face the challenges of low carbon and low emission ventilation as well as ventilation in hot climate regions and in continent climate regions. Two cases of new ventilation approaches are provided with on-site field measurement results. The research perspectives on complex livestock housing ventilation are included. The example of partial pit air exhaust ventilation showed that 58% and 51% ammonia emissions could be removed via the pit exhaust in the two runs, respectively. For the system using combined Natural ventilation (NV) and partial pit air exhaust (PPAE), the measured results revealed that 64%–83% of ammonia emissions were via partial pit ventilation. Both systems have potential to effectively reduce ammonia emissions by connecting the PPAE with an air purification unit.

Keywords: Ventilation design and control, livestock housing, indoor climate and air quality, Animal wellbeing, low emission

1. Introduction

Ventilation is essential for farm animal housing to remove the surplus heat and moisture from indoor space and maintain desired thermal environment and air quality for both animals and workers. Practices in design and control of ventilation in intensive confined livestock production vary a lot in different regions due to the factors such as tradition and experiences, climates and economy, etc.

By driving forces, ventilation can be classified as mechanical ventilation and natural ventilation. In applications of mechanical ventilation, a system can be classified / configured as mixing ventilation, vertical replacement ventilation, cross ventilation and tunnel ventilation. The characteristics of these ventilation forms are different and one can be superior to another under certain climate conditions and requirements on the housing system. Under complicated climatic regions with large seasonal climate variation and various requirements on regulations of indoor climate and air quality, choosing or configuring system as well as control strategy can be complex.

This paper will start with basic forms of ventilations currently applied or could be applied in livestock housing, and address the needs for a complex ventilation system that combines different ventilation forms into a housing system due to climate condition and performance request on the ventilation system.

2. Ventilation Methods

2.1. Ventilation system in livestock building ventilation

Ventilation forms as mechanical ventilation application in farm animal housing can be defined: (1) as negative pressure ventilation, positive pressure ventilation and equal pressure ventilation according to the system operation pressures; (2) as mixing ventilation with either jet or diffusion air supply, and replacement ventilation according to the supply air entry trajectories and interactive forms with room air; and (3) as a system with partial zone ventilation, e.g., precision zone air exhaust, local partial air in-taking by locations of ventilation openings.

2.2. Mixing ventilation

The so called mixing ventilation often refers to a system, in which the supply air is mixed with the room air before it reaches the occupant zoon.

2.2.1. Mixing ventilation with wall jet inlet

Mixing ventilation with wall jet inlets is often configured in a negative pressure ventilation system with sidewall inlets at one or two sides of the ventilated room depending on the building layout and internal partitions. The idea is to get sufficient mixing of the supply air and room air to avoid cold draft in winter period and high air speed in animal occupied zone. The exhaust air channel can be installed as ceiling top chimney, side wall outlets or under floor ducts.

Figure 1. A photo of housing using mixing ventilation with wall jet inlets.

2.2.2. Mixing ventilation with ceiling jet inlet

An alternative for ventilation air supply can be ceiling jet inlets, which are similar to the wall jet inlet but are installed at the ceiling of the ventilated room and take the supply air via ceiling room. The initial air jet direction can be vertical, horizontal or in an initial angle referring to the ceiling, depending on the desired airflow conditions in animal occupant zone under varied weather conditions.

Figure 2. A photo of mixing ventilation system with ceiling jet inlets under smoke testing.

2.2.3. Mixing ventilation with diffusion ceiling inlets

Another alternative of supply air into the ventilated room space is via the ceiling isolation/porous materials driven by negative pressure. Since the supply air has to go through the porous materials, the initial air speed into the room is generally low and also warmed up a little in cold weather period. Consequently, the system has advantage for avoiding draft in animal occupied zone (AOZ). However, the airflow speed in AOZ may be too low in warm weather, even ventilation rate reached the maximum.

2.3. Replacement ventilation

Replacement ventilation is a system that uses the fresh supply air to replace the room air to avoid it mixes with room air and to provide better air exchange efficiency. In general, the inlets and outlets of the system are placed in opposite ends of the ventilated room space.

2.3.1. Vertical replacement ventilation

In vertical replacement ventilation, the main airflow is moving either upward or downward vertically. Since the surplus heat produced by animals are at floor level and the air borne contaminants are form the sources at floor or under slatted floor, a vertical replacement ventilation with downward air motion would be efficient to remove the surplus heat and contaminant air. This can be done by placing the inlet at ceiling level and the exhaust opening at floor level or under floor. Disadvantages can be that the supply air temperature can be too low without mixing/preheating before it reaches AOZ and that the high ventilation airflow through the emission surface can induce high contaminant emissions, resulting in extra environmental negative impact.

2.3.2. Cross ventilation

Cross ventilation is a system that has air supply at a side of the room and air exhaust at another side of the room to create airflow across the ventilated room space. Cross ventilation can be used in case that higher air velocity is needed in AOZ, often in hot climate season or regions. However, considerations should be taken to avoid too high air speed in cold season in system design. In case that building width is large, temperature gradient and air contaminants at downwind side should be considered.

2.3.3. Tunnel ventilation

Tunnel ventilation is designed with a group of exhaust fans at one end of the building and air inlets at another end of the building. Similar to cross ventilation, tunnel ventilation is often used for creating high air speed in the ventilated room space to enhance convective heat transfer in AOZ to reduce heat stress of animals. The disadvantage of tunnel ventilation is similar as cross ventilation when the required ventilation rate is low at non-hot season when the air temperature gradient in room and pollutant air concentration near exhaust end can be very high.

2.4. Ventilation with cooling aids

Application of air conditioning system is rare in livestock production housing. Instead of that, using water evaporation in air for cooling down supply air or increase speed of the airflow at AOZ has been the major methods for reducing the heat stress of animals in hot climate season or regions.

2.4.1. Wet pad

Wet pad is often placed associated with ventilation air inlet where incoming supply air has to travel through the porous space of the pad panel. And the air passing the wet surface will generate water evaporation and consequently loss part of heat during the process and reducing the temperature of the ventilation air, Figure 3.

2.4.2. High pressure fogging

The method of high pressure fogging can be applied either in the room space directly or in the inlet air jets. With high operation pressure and desired nozzles, the water can be broken down to very fine particles as fog and can be easily absorbed in the room air or supply air; and consequently, the fogging can get the air temperature down.

2.4.3. Water sprinkler in room

Application of water sprinkler is similar to the fogging system. However, since the water particles are often larger than the fogging system, parts of them may drop on the floor surface or on animals in the room directly. The sprinkler is therefore often placed in room above the dunging area for pig buildings.

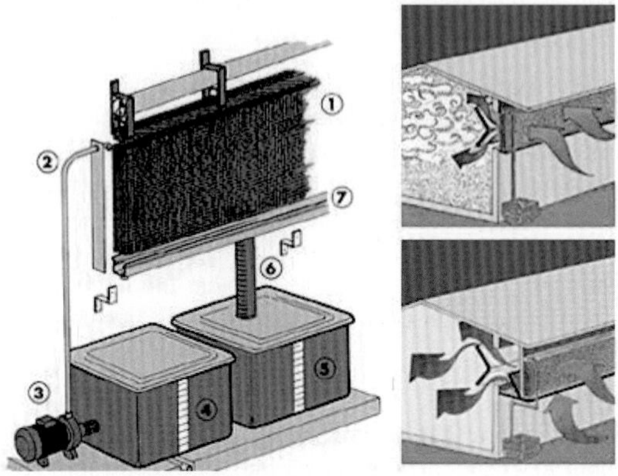

Figure 3. Schematic of wet-pad cooling for supplying air. Left: (1) pad; (2) water supply tube; (3) Pump; (4) supply water tank; (5) return water tank; (6) return water tank; (7) frame and water collection. Right: up, wet-pad active; down, wet-pad inactive.

2.4.4. Heat exchangers – passive and active

One of the passive heat exchange method is utilizing the earth heat potential. In this method, the ventilation supply air is driven through some underground air ducts so that it can be heated in winter and cooled in summer due to the temperature differences between the air above the earth and soil of 1.5 - 2m under the earth surface.

An active heat exchanger often consists of source heat/cool elements that allow the passing-through ventilation supply air to be heated/cooled. The source heat/cool elements are connected via a pump, a control valve to a heat reservoir using water circulation (e.g., via a ground heat source).

2.5. Complex ventilation

In the case that a single ventilation form cannot fulfil the desired objectives, complex ventilation with integrated functions based on more than two ventilation forms may be needed.

2.5.1. Partial Pit Air Exhaust ventilation

Partial pit air exhaust (PPAE) ventilation is an add-on exhaust in the main ventilation exhaust, which is located locally near to the main source zone of airborne contaminant, aiming at the most contaminated air is removed from the room directly before it is mixed with room air. In that way, the indoor air quality will be improved. By cleaning treatment of this partial exhaust air only, the capacity required for air cleaning unit can be smaller compared with the treatment of all exhaust air.

2.5.2. Combination of natural and mechanical ventilation

Natural ventilation (NV) has advantages of low noise and energy saving. Automatically-controlled natural ventilation system can provide reasonable good indoor climate conditions for most livestock production housing. However, for some animal species and in some climate regions the winter cold air draft could be a problem, since the pressure created by buoyant force is not strong enough to generate a strong inlet air jet to ensure a good mixing of incoming air with room air before it reach AOZ. The control ability for inlet air momentum is generally poor in a NV system. Also, it is almost impossible to control and clean the exhaust air from a naturally ventilated building. In this context, mechanical ventilation as a partial exhaust air driver can be applied to both control the inlet air jet momentums and remove the most contaminated air from the source location and connected with an air purification unit.

2.5.3. Combined mixing ventilation and replacement ventilation

Replacement ventilation such as wind tunnel ventilation or cross ventilation is an effective alternative in hot climate ventilation. However, in case that the season is not very hot, other ventilation forms can be superior to these two forms. In such cases, a complex ventilation by combining mixing ventilation and replacement ventilation may be considered.

3. Two examples of complex ventilations and Results

In this section two examples of complex ventilations in livestock production housing with field measurement results will be presented.

3.1. Partial Pit Air Exhaust ventilation in pig housing

Partial pit air exhaust ventilation (PPAE) is based on a hypothesis that the most pollutant air can be removed by an extra exhaust opening near the pollution sources for cleaning treatments, Figure 4. The PPAE airflow rate is designed and controlled as only a small portion of the total designed ventilation capacity for the room, which was often defined as 10% of the designed ventilation capacity of the building (DVCB) in most experiments reviewed in this paper. By applying an effective air purification unit only to this pit exhaust air channel, the required airflow capacity (consequently the power) of the cleaning unit will be much reduced. Several investigations has been reported since the concept is introduced in a Danish research project, Reduction of Odour and Ammonia Emission in and from Swine Housing (ROSES) (Zhang 2006a; 2006b; Saha et al., 2010; Bjerg and Zhang, 2013; Zong et al., 2014a, 2014b). The following example is part of the work reported by Zhang et al. (2013).

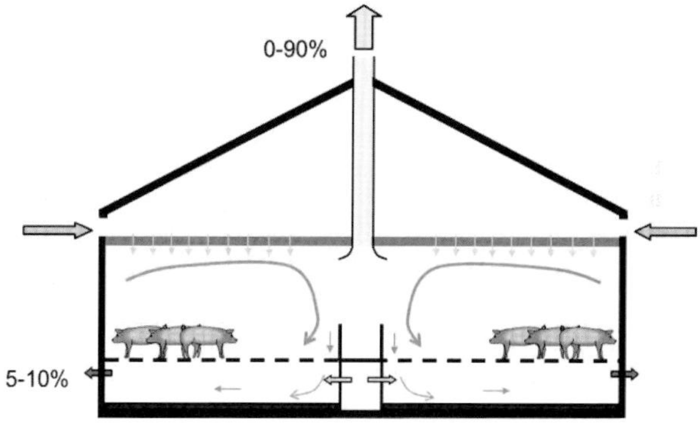

Figure 4. Schematic of a partial pit air exhaust in a system with ceiling inlet and a ceiling top air exhaust.

3.1.1. System description

The room used for the experiments had two pens and each with two third fully-slatted floor and one third drain floor. The opening area for the slatted floor was 16.5% and that for the drain floor was 8.5%. The slurry gutter under the pen floor was 0.9m deep, with draining pipes of 0.25m of internal diameter and a central valve for the removal of slurry. The pen partition on the floor was up to 1.2m high. In front of the pig pens there was inspection alley of 1.2m width. The outdoor air came through the attic and then entered via the diffuse ceiling into the production room, driven by negative operation pressure. The room exhaust opening was located at the ceiling top (ceiling exhaust) and the PPAE opening was located under the drain floor closed to the lying area. The capacity of the PPAE was 10% of maximum ventilation capacity of the system.

The ceiling exhaust unit was active as the main exhaust unit during the experiments and its varied ventilation rate was controlled by the climate computer according to the indoor thermal conditions measured by a temperature sensor and the reference indoor air temperature settings. The indoor air temperature sensor used for control system was placed at 1.6m above the pig lying area.

There were 32 pigs in the room for experiments and the pigs were equally divided and put into the two pens in the room, started at about 30 kg pig^{-1} and ended at about 100 kg pig^{-1}. Feed and drinking water were available all the time in ad labium. The type of feed for the pigs was ''DLG Finale Plus U Fuldfoder til slagtesvin'' (Eng. DLG Finale Plus U Complete Feed for finishing pigs) (DLG a.m.b.a., Copenhagen, Denmark) containing 40% wheat, 30% barley, 12% rapeseed, 7.45% wheat bran, 4.85% soya bean, 2.40% beet molasses, plus vitamins and minerals. The diet contained 15.5% raw protein.

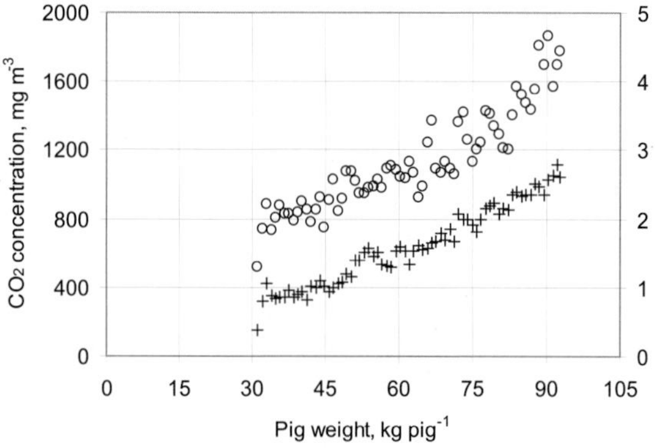

Figure 5. Indoor air quality represented in concentrations CO_2 and NH_3 measured at ceiling exhaust, o and +, respectively.

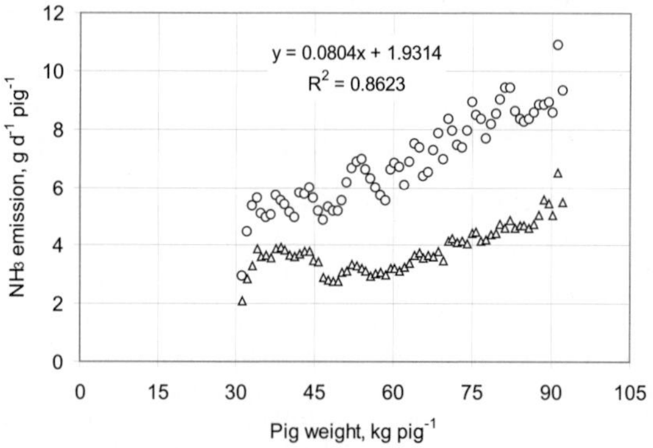

Figure 6. Total ammonia emission, o and emission via PPAE, Δ.

3.1.2. Results of PPAE

An example of recorded carbon dioxide (CO_2) and ammonia (NH_3) concentrations during an experimental run using PPAE is showed in Figure 5. An example of ammonia emission from the

PPAE system is given in Figure 6. The summarised results comparing with other experiments with and without PPAE are given in Table 1.

Table 1. Ammonia concentrations and emissions comparing with other experiments

References	Ventilation system	Exhaust position	Ventilation rate, $m^3\ h^{-1}\ pig^{-1}$	NH_3 concentration, ppm	NH_3 emission, $g\ d^{-1}\ pig^{-1}$	% of emission via pit exhaust
Koerkamp et al. (1998)	Conventional	ceiling	-	14.9	7.7	-
Saha et al. (2010)	Conventional	Ceiling	48.7	6.5	4.1	
	PPAE	Ceiling	61.4	3.8	2.5	43.2%
		Pit	11.6	10.8	1.9	
		Total	80.0	-	4.4	
Pedersen and Jensen (2010)	Conventional	ceiling	52	9.3	7.5	
	PPAE	Ceiling	49	2.6	1.8	64.7%
		Pit	10	20	3.3	
		Total	59	-	5.1	
Zhang et al. (2014)	PPAE R1	ceiling	75.5	1.8	2.4	55.6%
		pit	13.3	12.5	3.0	
		total	88.8		5.4	
	PPAE R2	ceiling	80.6	2.3	3.0	55.1%
		pit	12	18.6	3.8	
		total	92.6	-	6.9	
Zong et al. (2014)	PPAE S	Ceiling	80.6	2.1	2.9	48.2%
		Pit	12	16.6	2.7	
		Total	92.6		5.6	
	PPAE W	Ceiling	68.1	3.4	3.3	46.5%
		Pit	9.3	21.3	3.8	
		Total	77.4	-	7.1	

3.2. Integrated ACNV and PPAE

The idea of this type of complex ventilation is to combine automatically controlled natural ventilation (ACNV) and mechanical PPAE in livestock housing (Zhang, 2008; Wu et al., 2012a, 2012b; Rong et al., 2014). Together with the potential of natural ventilation for energy savings, the pit exhaust air in small quantities but relatively high contaminant concentrations can be cleansed by using an air purification system in order to effectively reduce gaseous emissions. The PPAE operation in winter period can also enhance the inlet air jet momentum control to create desired air mixing before the cold in-coming air entry to AOZ (animal occupant zone). In such a system, airflow near slatted the floor openings and exhaust locations is important and should be considered in system configurations and control. These considerations may include inlet design and locations, wind effects when ventilation openings are relative large. The following example is a part of the investigation work newly reported by Rong et al. (2014, 2015).

3.2.1. System description

The experiments with the integrated ACNV and PPAE were conducted in a new constructed dairy cattle building in mid-west Jutland, Denmark (Rong *et al.* 2014). The dimensions of the building were shown in Figures 7 and 8. The length of the building was 74.0m and the width was 45.0m. In house, there were 360 cows, averaged weight of 650 kg. The milk production was 9500–10,000 kg yr^{-1} cow^{-1}. The cows were milked twice per day (4:00–7:00 and 15:00–18:00). The feeding time for the cows was around 5:30 in the morning. The daily feeding consumption per cow was 54.67 kg d^{-1} and the contents of the feeding includes 63.5% corn silage, 25.9% grass silage, 6.1% HP soya, 3.76% Rapeseed cake, 0.25% saturated fat and 0.46% straw.

In this study, the mechanical PPAE system (fans power in total of 2.2 kW) ran through a year. In winter, the PPAE ventilation rate was controlled by indoor air temperature and CO_2 concentration until it arrives at minimum ventilation rate. In summer, the pit ventilation rate was 25% of the designed maximum ventilation rate (450 $m^3\ h^{-1}\ cow^{-1}$).

Figure 7. A photo of the dairy cattle building with hybrid ventilation.

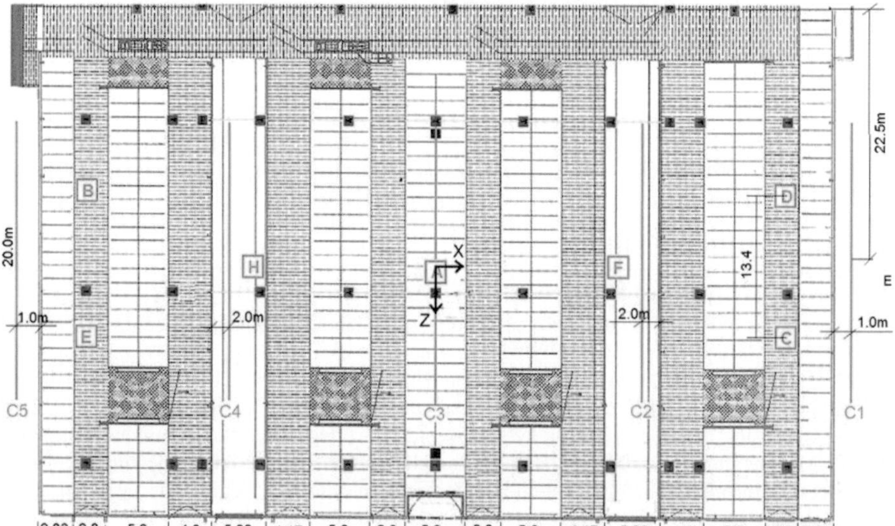

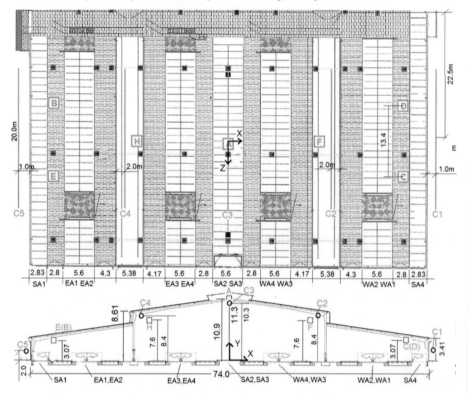

Figure 8. A diagram of the dairy cattle building with hybrid ventilation.

3.2.2. Field measurement results

A set of summarized data is given in Table 2, which presents the average emission rates in three units, per HPU, per LU and per m^2 of slatted floor. The mean emission rate of NH_3 via NV during the whole measuring period was 4.5 g HPU^{-1} d^{-1} with indoor average air temperature of 7.8°C in winter and 17.7 g HPU^{-1} d^{-1} with indoor average air temperature of 19.9°C in summer.

The average emission rate of methane (CH_4) through NV was 129.3 g HPU^{-1} d^{-1} and 247.0 g HPU^{-1} d^{-1} in winter and summer, respectively. The average NH_3 emission rate via the pit ventilation was 35.1 g HPU^{-1} d^{-1} in summer and 21.8 g HPU^{-1} d^{-1} in winter. Through pit ventilation, in average 83% of NH_3 emissions was in winter. In summer, in average 64% of NH_3 emissions was via the PPAE system. Without the air purification unit, the total NH_3 emission was at the similar level compared with a conventional naturally ventilated dairy cattle building.

Table 2. Ammonia and methane emission of the NV-PPAE cattle house.

Period	location	gas	Emission ($g\ HPU^{-1}\ d^{-1}$)		Emission ($g\ LU^{-1}\ d^{-1}$)		Emission ($g\ m^{-2}\ d^{-1}$)*	
			mean	SD	mean	SD	mean	SD
02/20/2013	Room	NH_3	4.53	1.49	4.95	1.63	1.58	0.52
		CH_4	129.25	34.11	141.34	37.3	45.24	11.94
03/13/2013	Pit**	NH_3	22.12	1.24	24.17	1.35	7.71	0.43
		CH_4	129.63	35.38	141.76	38.69	45.37	12.38
07/15/2013	Room	NH_3	17.72	8.47	19.38	9.26	6.2	2.96
		CH_4	246.97	73.6	270.07	80.48	86.44	25.76
08/16/2013	Pit	NH_3	32.23	4.19	35.06	4.62	11.22	1.48
		CH_4	28.31	5.65	30.95	6.12	9.91	1.96

* The emission was summarized as gram per day per square meter of slatted floor, which was 1421.5 m^2 in the cattle building

** During the winter experiments, the concentration in the pit was not measured continuously. The average value was calculated according to the data obtained from 20/2, 23/2, 27/2, 5/3, 8/3 and 13/3.

4. Conclusions and Future Research Perspectives

Ventilation is essential for farm animal housing to remove the surplus heat and moisture from indoor space and maintain desired thermal environment and air quality for both animals and workers. The characteristics of varied ventilation system designs and controls are different and one can be superior in performance than another under certain weather conditions and climate regions. In complicated climatic regions with large climate variation in seasons and varied requirements on regulations of indoor climate and air qualities, choosing or configuring system as well as control strategy can be complex. Cooling of the indoor air and supply air, control of the airflow speed in animal occupant zone can be required for an optimal performance of a ventilation system. And in some of the conditions, a complex system design and control will be necessary.

In the reported example of PPAE mechanical ventilation system, the 58% and 51% emitted NH_3 was removed via the pit exhaust channel in the two production periods. The partial pit exhaust ventilation has a potential to reduce NH_3 emissions by connecting it to an effective air purification unit. Since only a limited portion of the ventilation air is needed to be cleaned, the capacity required for the cleaning system will be reduced.

In the system using combined NV and PPAE, the measured results revealed that 64%–83% of NH_3 emissions were via partial pit ventilation. About 50% of CH_4 in winter but only 10% of CH_4 in summer were emitted via partial pit ventilation. The results showed that NH_3 emissions through natural ventilation were around 60% lower than the value found in the literature.

Complex ventilation with smart design and control can provide promising engineering solutions in livestock housing under complex conditions. Further research efforts aiming at both system modelling and application are needed.

References

Bjerg B; Zhang G. (2013). CFD Analyses of the influence of ventilation system on the effectiveness of a partial pit exhaust. *Acta Horticulturae.* 1008: 143–150. http://www.actahort.org/books/1008/1008_18.htm

Rong L; Liu D; Pedersen EF; Zhang G. 2014. Effect of climate parameters on air exchange rate and ammonia and methane emissions from a hybrid ventilated dairy cow building. *Energy and Buildings*. 82: 632–643. http://dx.doi.org/10.1016/j.enbuild.2014.07.089

Rong L; Liu D; Pedersen EF; Zhang G. 2015. The effect of wind speed and direction and surrounding maize on hybrid ventilation in a dairy cow building in Denmark. *Energy and Buildings*. 86: 25–34. http://dx.doi.org/10.1016/j.enbuild.2014.10.016

Saha C K; Zhang G; Kai P; Bjerg B. 2010. Effects of a partial pit ventilation system on indoor air quality and gaseous emissions from a growing/finishing pig building. *Biosystems Engineering*. 105(3): 279–287.

Wu W; Kai P; Zhang G. 2012. An assessment of a partial pit ventilation system to reduce emission under slatted floor — Part 1: Scale model study. *Computer and Electronics in Agriculture*. 83: 127–133.

Wu W; Zhang G; Bjerg B; Nielsen PV. 2012. An assessment of a partial pit ventilation system to reduce emission under slatted floor - Part 2: Feasibility of CFD prediction using RANS turbulence models. *Computer and Electronics in Agriculture*. 83: 134–142

Zhang G. 2006a. Reduction of odour source in and emissions from swine buildings – ROSES. DaNet News Letter, Dec. 2006. Research Centre Bygholm, Danish Institute of Animal Sciences.

Zhang G. 2006b. Descriptions of research project ROSES (Reduction of odour source in and emissions from swine buildings). http://web.agrsci.dk/jbt/roses/

Zhang G. 2008. Description of research project LEK "Low Emission Cattle Buildings (LavEmission Kvægstald)". http://web.agrsci.dk/jbt/LEK/beskriv.htm

Zhang G; Kai P; Zong C. 2013. Precision ventilation for optimal indoor air quality and reduction ammonia emission from pig production housing. In Precision Livestock Farming'13: the proceedings of the EC-PLF Conferences, p. 883–892, Leuven, Belgium, 2013 September 10–12

Zong C; Feng Y; Zhang G; Hansen MJ. 2014a. Effects of different air inlets on indoor air quality and ammonia emission from two experimental fattening pig rooms with partial pit ventilation system — summer condition. *Biosystems Engineering*. 122: 163–173 http://dx.doi.org/10.1016/j.biosystemseng.2014.04.005

Zong C; Zhang G; Feng Y; Ni J-Q. 2014b. Carbon dioxide production from a fattening pig building with partial pit ventilation system. *Biosystems Engineering*. 126: 56–68 http://dx.doi.org/10.1016/j.biosystemseng.2014.07.011.

Study on Pig Manure Odor Removal Efficacy by Wormcast Packing

Shihua Pu [a,b,c], Zuohua Liu [a,b], Baozhong Lin [a,b,c], Feiyun Yang [a,b], Dingbiao Long [a,b,*]

[a] Chongqing Academy of Animal Sciences, Rongchang, Chongqing 402460, China

[b] Key Laboratory of Pig Industry Sciences, Ministry of Agriculture, Rongchang, Chongqing 402460, China

[c] Chongqing Key Laboratory of Pig Industry Sciences,
National Scientific Experiment Station on Breeding facilities engineering,
Ministry of Agriculture, Rongchang, Chongqing 402460, China

*Corresponding author. Email: longjuan880@163.com

Abstract

In order to improve the removal efficacy on pig manure odor, a biological filter was developed with the packing of wormcast, sawdust, perlite and activated carbon and its effect was evaluated. The effect on pig manure odor removal using three different packing, which had C/N ratios from 25 to 40, was compared under different environmental conditions. Results showed that the removal efficacy on hydrogen sulfide (H_2S) from anaerobic fermentation was up to 92.96% when using the packing of "wormcast + sawdust + activated carbon". Compared with the other two types of packing, it had better removal efficacy and longer stability, but prone to be influenced by environmental factors including temperature, humidity and packing height. The packing cost can be reduced by appropriately supplementing with certain amount of absorption materials, perlite for example, without significantly affecting the removal efficacy.

Keywords: Pig manure, odor, sawdust, removal, hydrogen sulfide (H_2S)

1. Introduction

Nowadays, odor emission from livestock industry is a severe environmental issue, increasingly attracting the attention of the society. Aiming at improving the livestock environment and reducing odor emission, several deodorization methods, such as adsorbent, spraying, natural ventilation, oxygen ion group et al., with effect on deodorization to some extent were adopted by the livestock industry. However, because of the disadvantages of poor persistence and high operation cost, those methods are not practical on farms, and some may also lead to secondary pollution. Thus, biological filter processing with a new integrated material being able to effectively control odor diffusion, reduce the cost, and lower energy consumption without a secondary pollution through odor is popularly needed by farm enterprises.

In this study, a packing made of wormcast, activated carbon and wood chips, abundant in a variety of microbial compost, were adopted to develop a biological filter. The removal processing of hydrogen sulfide (H_2S) from animal housing by using the biological filter with the new packing was tested and its efficacy was analyzed. The design of the biological filter, its optimal operation parameters and packing combinations were also determined.

2. Materials and Methods

2.1. Design parameters and performance indicators of the biological filter

A similar design of a biological filter referred to the experiment methods provided by Schmidt et al. (2004) at University of Minnesota, USA shown in Table 1 was used in this study.

2.2. Filter and packing

The device consisted of a bioreactor, a bottom gas filling system, a nutrient solution spraying circulation system, a ventilation system, and a temperature and humidity testing auxiliary system. Its pillar is made of PVC pipes, with diameter of 33 cm, height of 125.5 cm, filled with activated carbon (Hommy, Jiangshu, China), wood chips, wormcast (excretion of No.

2 Taiping worm) and perlite, with a total height of 90 cm. Each layer of a filter cylinder was equipped with a net PVC clapboard to support the packing. A layer of gravel was placed at the bottom of the filter to make the gas uniformly distributed inside the filter pillar. A water distribution system with mixed solution (Nutrient Agar, Pepton, NaCl 5g Agar, distilled water , etc.) was installed at the top of the filter to satisfy the growth need of the microbes attaching on the packing. It adjusted the humidity of the packing, and contributed to the outlet of the metabolites attached at the packing surface through the drainage system at the bottom.

There were three groups of packing in the filter: packing A with activated carbon + sawdust + perlite, packing B with wormcast + sawdust + activated carbon, and packing C with wormcast + sawdust + compost.

Table 1. Design parameters of the biological filter

Design parameters	Formula	Control range
Filter packing volume V_m, m^3	$V_m = Q \times EBCT$	0.02–0.06
Empty bed contact time, min	$EBCT = k \times D^2 \times H \times Q^{-1}$	1–3
Surface load UAR, m h^{-1}	$UAR = Q\ A_m^{-1}$	60–17.1
Packing area A_m, m^2	$A_m = V_m H_m^{-1}$	0.02–0.07

2.3. Sample collection and data analysis

The inlet flowrate of filters was 1.2 $m^3\ h^{-1}$, empty bed contact time was 120–180s and the initial inlet concentration of H_2S was 15.13–19.28mg m^{-3}. Since the beginning of the trial, the H_2S samples were collected from the bottom (30 cm), middle (60 cm) and the top (90 cm) of the device every 2 days, with one circle of 15 days. The H_2S concentrations in the samples were determined by means of gas detection tubes (RAE, the accuracy is 0.1 ppmv, USA); the humidity of packing was recorded by a temperature and humidity auto-measurement system (RS-13 Temperature and Humidity Sensor, ESPEC, Japan). The data were processed and analyzed by means of software Origin 8.0, Excel and Spss 19.0.

3. Results and Discussion

3.1. Influence of different packing combination on removal efficacy of H_2S

As shown in Figures 1–6, the H_2S removal rate of packing A was the highest during the period of d 4–6 reaching up to 90.12%; and it declined after d 6 with the decreasing of the packing pH. The removal rate of packing B rose slowly in the period of d 1–6, and reached the peak on d 7 at 92.96%. Similarly, removal rate of packing C increased relatively slower and reached the highest on d 8 at 87.21%. Compared with the other two groups, packing A was better at the early stage due to physical absorption, and reached the highest on d 6; but the removal efficiency greatly decreased at a later period due to the absorption saturation with pH decreasing. The removal efficiency of packing B increased slowly at the early stage; but the removal effect was considerable, with the highest of 92.96% because microbial activity in the packing gradually enhanced by the metabolism of S^{2-} and SO_4^{2-}. At the later stage, the removal efficiency declined due to the increase of S^{2-} and SO_4^{2-}, and the decrease of pH. As a result of physical absorption of activated carbon, the H_2S removal efficiency presented strong sustainability. The H_2S removal efficiency of packing C was better than B because it had more microbial content than B. At the later stage, due to the gas short-cut caused by the increase of humidity, the remove efficiency and continuity of packing C were lower and poorer than packing B, respectively.

3.2. The influence of different filler height on H_2S removal rate

As shown in Figure 7, the H_2S removal rate increased with the increase of filler layer height. While the filler layer height increased, the path of H_2S through the filler layer was prolonged;

the retention and the contact reaction time between gas and microbial membrane increased, therefore, the removal rate was accelerated.

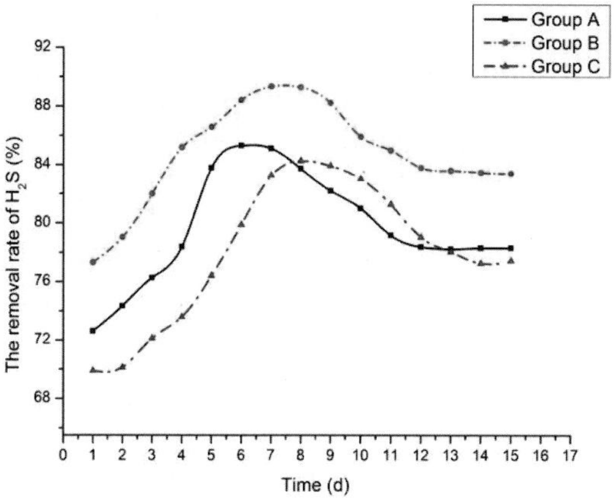

Figure 1. Removal rate of H_2S at the bottom of the device.

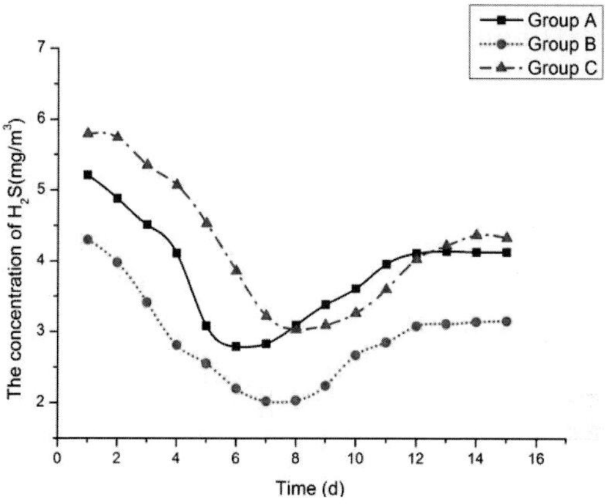

Figure 2. Concentration of H_2S at the bottom of the device.

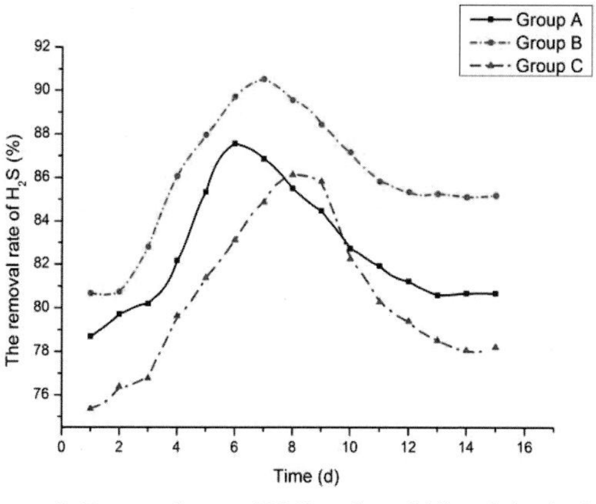

Figure 3. Removal rate of H_2S at the middle of the device.

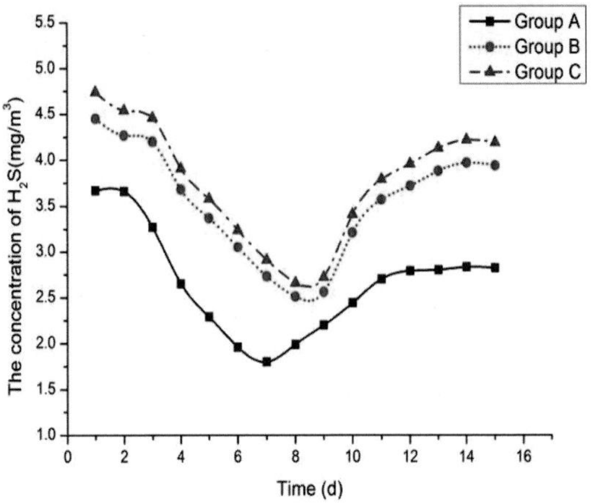

Figure 4. Concentration of H_2S at the middle of the device.

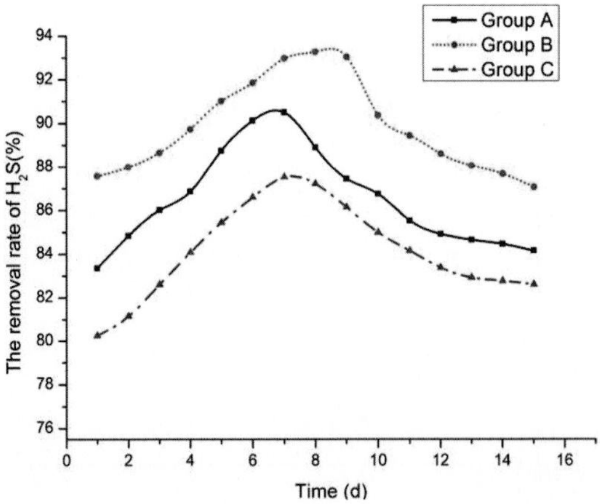

Figure 5. Removal rate of H_2S at the top of the device.

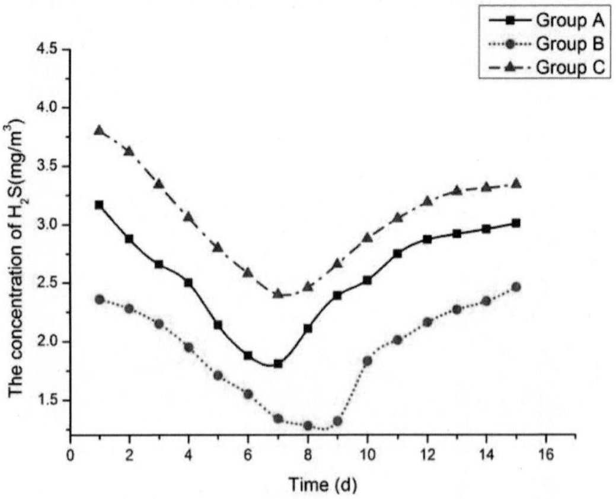

Figure 6. Concentration of H_2S at the top of the device.

Similarly, the increase of the filler layer height can improve the removal efficiency. In practice, the pressure loss of the gas through the biological filter will be increased, and the removal efficiency will be reduced accordingly. However, the constant increase of the height of packing will increase the operating costs; but it is negligible to improve the efficiency.

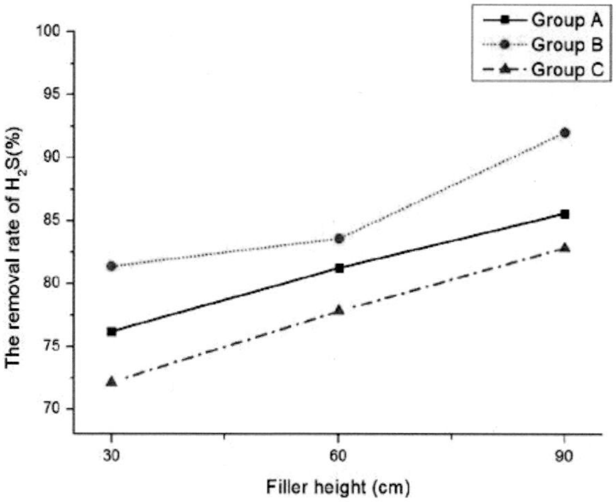

Figure 7. Influence of filler height on removal rate of H_2S.

3.3. The influence of different humidity on H_2S removal rate

As shown in Figure 8, the removal rate increased with the humidity increase when the filler humidity was at lower level (30–60%) and declined when humidity was up to 60%. This was because of the presence of a few microorganisms in the packings, and their activity was low at a lower humidity. The viscosity of filler particles was increased under higher humidity, which resulted in the reduction of "gas short cut", and the contact area of H_2S gas and filler, the H_2S removal efficiency was reduced. It can be seen that the most optimal humidity for the removal of H_2S was 50–65%.

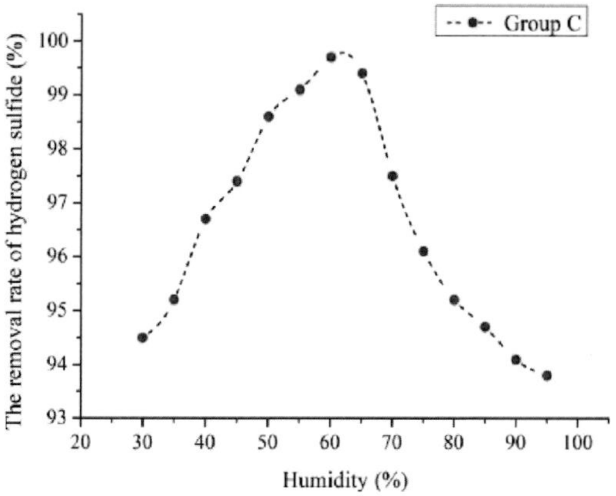

Figure 8. Influence of humidity on removal rate of H_2S.

4. Conclusions

1. Comparing the removal efficiencies of the three groups, packing A mainly belonged to physical absorption with low cost, and resistance to the influence caused by environmental

factors while its removal efficiency was dependent on the absorption quantity and sustainability; both packing B and C combined the physical absorption and biological removal with long sustainability and good efficiency, but they were prone to be influenced by environmental factors such as temperature, humidity and packing height, especially the essential impact of humidity and nutrients content on microorganisms, the operation cost of later period was slightly higher than that of packing A. It is concluded that the best removal efficiency can be achieved based on packing C supplemented with inorganic material such as perlite at a rational ratio, for the purpose of reducing the operation costs and energy consumption, while improving the removal efficiency.

2. Regarding the impact of filling height on removal efficiency, the removal effect of packing C had a positive linear relationship with filling height, but in practice, the pressure loss of gas through the biological filter would be increased with the increase of filling height, which result in a lower removal efficiency, so the appropriate filling height can improve the H_2S removal efficiency; and the removal effect achieved the best when the humidity range of group C was 50 –65%.

References

Li, L., J.X. Liu, 2001. The selection of biological techniques and processes for treatment of VOC and odours. Techniques and Equipment for Environmental Pollution Control. 2(5), 41–17.

Ni, M.J., S.R. Cui, 2005. The development and prospect in application of the biofilter in treating agriculture odor. Journal of Agricultural Mechanization Research. 5(3), 221–223.

Schmidt, D., L. Jacobson, and R. Nicolai. 2004. Biofilter design information. http://www1.extension.umn.edu/agriculture/manure-management-and-air-quality/air-quality/biofilter-design-information/. Accessed September 26, 2015.

Tu, D.Y., H.M. Dong, Z.K. Zhang, 2005. Application of biofilter on odor control. Journal of Anhui Agricultural University. 32(3), 397–401.

Zhang, P.Y., G. Yu, Z.P. Jiang, 2000. Treatment of volatile organic compounds and odour by biofiltration. Techniques and Equipment for Environmental Pollution Control. 1(1), 1–7.

Theme III:

Manure Management and Utilization to Reduce Environmental Impact

Analysis of Nitrogen Cycle in Jiangxi Yufeng Beef Cattle Farm

Run-hang Li [a], Shi-tang Yang [b], Yu-jie Lou [a,*]

[a] College of Animal Science and Technology, Jilin Agricultural University, Changchun, Jilin Province, China

[b] Yufeng Beef Cattle Farm, Gao'an, Jiangxi Province, China

* Corresponding author. Email: lyjjlau@163.com

Abstract

The Yufeng Beef Cattle Farm, the largest grass-cattle farm in Jiangxi Province, has an area of about 200 ha, including a pasture of 100.67 ha and an orchard of 10 ha, with an annual silage herbage production of 31,700 tons. The herd kept at the farm was around 2200 heads and the number of slaughtered cattle was up to 3000 heads per year. Through the two main models including "Manure – Herbage" and "Manure – Fruit", nitrogen from cattle manure has been fully recycled at the farm. When excluding the external factors such as atmospheric nitrogen deposition, the total nitrogen surplus of the farm was -44,095 kg; the nitrogen surplus of the soil in pasture and orchard was -43,930 kg. The fact that nitrogen surplus was negative indicated a continuous decline of the total amount of soil-surface nitrogen. Thus, the farm needed a certain amount of nitrogen fertilizer input, in order to maintain the nitrogen balance for grass planting.

Keywords: Nitrogen cycle, nitrogen balance, nitrogen surplus, cattle, manure treatment

1. Introduction

The rapid development of intensive cattle farming could significantly improve the productivity and the profits in this field (Lv, 2012). The high feeding rates in the intensive system lead to more manure production than the land capacity, which can result in a large quantity of nutrient lost to the environment. The cattle farms under this system are facing many challenges related to the environment, society and some other aspects (Feng, 2014). The excessive application of nitrogen (N) does not only reduce the N use efficiency on cattle farms, but also exacerbate the N load in the environment. A cattle farm produces N output that ranges only from 15%–35% (Powell, 2010). The N cycle refers that the N comes into the ecosystem through various ways, after many successive transformations, and leaves the system in varying degrees. The loop is open so that it has a complex exchange with the environment (Lu, 1992 and Wu, 2013). In some areas of South China, environmental pollution problems, which are faced by large-scale intensive cattle farms, are more serious. The main problem is that the N production in the environment has exceeded its absorption capacity. Although there are a variety of manure treatment technologies, they only played a limited role in reduction without solving the ultimate problem of nutrient losses radically. This study is to analyze the N cycle in Yufeng Beef Cattle Farm (YBCF) in Jiangxi Province by on-site data collection and samples testing for three years.

2. Overview of the YBCF

The YBCF is the largest herbage – cattle farm in Jiangxi Province with a total area pf about 200 ha, including 100.67 ha pasture and 10 ha orchard (Guo, 2011). The output of herbage to the surrounding was 31,700 tons per year. The annual cattle population was around 2200 heads and the annual number of slaughter was up to 3000 heads. The recycling of cattle farming model YBCF adopted the southern by using the way that feeding herbage to cattle and backing the manure to the pasture in order to reduce the pollution (Dai, 2014). This model transformed the shortage of herbage in the southern to the view of supplying high-quality herbage around the year. This provided a feasible way to carry out the large-scale cattle breeding in Jiangxi Province. The N balance was conducive to study N cycle in a cattle farm by quantifying the types of its total N input and output. It has achieved effective applications in some ecological farming cycle models, including "Manure – Herbage" and "Manure – Fruit" for YBCF at present. The N cycle model for the YBCF was summarized in Figure 1.

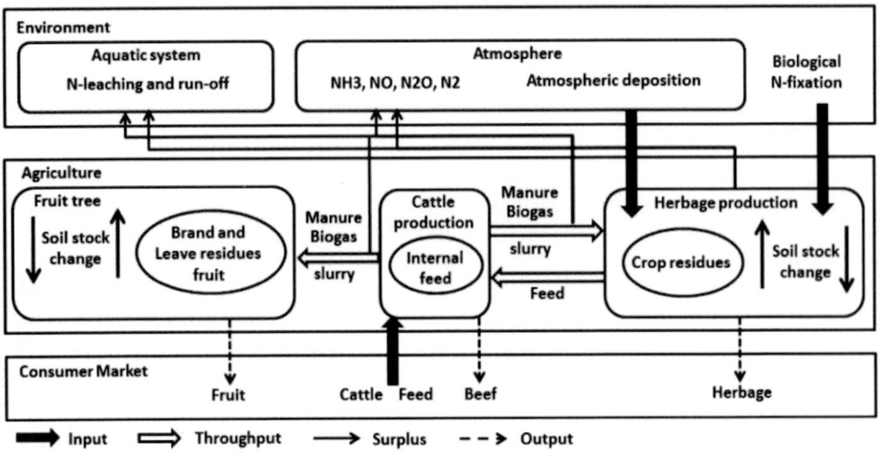

Figure 1. Schematic diagram of N cycle on YBCF.

2.1. N Input and output on YBCF

The YBCF had more than one breed of cattle. *Simmental* was the main breed. Its main fine material was mixed according to the nutritional needs of cattle with premix feed, energy feed, and protein feed. The main source of concentrate was brewer's grains (wet, BGS), and the main green fodder was *Pennisetum Purpureum* (PP) and *Perennial Ryegrass* (PR). The N input of YBCF is shown in Table 1.

Table 1. Nitrogen input of YBCF.

Input item	Average input (kg/year)	N rate (% FM)	Average N input (kg/year)
Corn	3,120,000	1.1968	37,340
Bran	1,030,000	0.2400	2472
Soybean meal	460,000	6.1632	28,351
Cottonseed meal	310,000	6.4768	20,078
Premix feed	210,000	0	0
BGS	7,300,000	1.0895	79,534

Note: FM means fresh matter.

Through adopting the acquisition of cattle to increase the cattle population, YBCF completed the production process after fattening and selling. The average of annual population was 2000 heads and the number of slaughtered animals was 2750 heads. Fattening cattle entered to the farm at 300 kg average weight. The selling weight at average slaughter was 600 kg. The pasture area was about 100.67 ha, including PP area of 80 ha and PR area of 26.7 ha, with the total annual production of fresh herbage of 39,000 tons. In addition to satisfy the breeding use of own farm, the sale of silage packages was about 31,700 tons every year. The orchard area was about 10 ha, including citrus tree area of 6.67 ha and pear tree area of 2.67 ha, with the total annual fruit output of 333 tons. The N output of YBCF is shown in Table 2.

2.2. Feces and urine production on YBCF

The clearance of feces in the barn was mainly manually done on YBCF. The feces were put into the fermentation pool (with O_2) and the urine flowed into the biogas slurry pool (without O_2) through underground pipes (Zhu, 2013). The biogas slurry (or residue) and manure were applied to pasture and orchard after completion of the fermentation. Biogas as a fuel was transported to the houses for nearby residents through a pipeline (Hou, 2014). It was measured that the average daily production of feces (wet) was 22 kg and urine was 10 kg per cattle under normal conditions (Gao, 2014). The N production of the feces and urine (Fan, 2006) on YBCF is shown in Table 3.

Table 2. Nitrogen output of YBCF.

Output item	Average output (kg/year)	N rate (% FM)	Average N output (kg/year)
Beef	825,000	3.560	29,370
PP	31,700,000	0.570	180,690
Citrus	237,619	0.576	1369
Pear	95,119	0.464	441

Table 3. Notrogen production of feces and urine on YBCF.

Item	Average production (kg/piece)	Average population (piece)	Average total production (kg/year)	N rate (% FM)	Average N production (kg/year)
Feces	8030	2000	16,060,000	1.0370	166,542
Urine	3650		7,300,000	0.1807	13,191

3. Nitrogen Cycling and Utilization on YBCF

If no practical solution is found for the large amount of feces and urine, they will lead to serious pollution problems and destruction of the surrounding environment. Furthermore, they might affect the health of animals and people. At present, there are two main models of N cycling and utilization for YBCF, i.e., "Manure – Herbage" and "Manure – Fruit", being used to solve the manure utilization problem.

3.1. Model of Manure – Herbage

In the south of China, there are three planting seasons, which are different from the north of China. So there is much more N transferred into plants in the south of China. This means that more N fertilizers are needed in the pasture. In this ecological chain cycle, the organic N in the feces and biogas slurry (or residue) was transformed into inorganic N by degradation of the soil organisms. Then, the reused organic N is transformed from inorganic N by plants absorption and synthesis. There were 34,200 tons of PP (Liu, 2006) and 4800 tons of PR (Zhang, 2008) produced every year on YBCF. Among them, the N input sources (without gas consideration) were organic fertilizers and N output sources were PP and PR. The N production of pasture on YBCF was show in Table 4.

Table 4. Nitrogen production of pasture on YBCF.

Item	Planting area (ha)	Average production (kg/ha)	Average total production (kg/year)	N rate (% FM)	Average total N production (kg/year)
PR	26.7	180,000	4,806,000	0.56	26,914
PP	80.0	427,500	34,200,000	0.57	194,940

3.2. Model of Manure – Fruit

The completion of the fermentation of the feces and slurry (or residue) was supplied to the orchard regularly. The reused organic N was synthesized the same way as the plants in pasture (Cao, 2002). Ripe fruits were eventually harvested and taken to the market through unified picking and packing system. Sales were the same as production basically. Among them, N input source (no considering the gas) was organic fertilizer and N output sources were citruses and pears (Yang, 2008). The N production of orchard on YBCF is show in Table 5.

Table 5. Nitrogen production of orchard in YBCF.

Item	Planting area (ha)	Average production (kg/ha)	Average total production (kg/year)	N rate (% FM)	Average total N production (kg/year)
Citrus tree	6.67	35,625	237,619	0.576	1369
Pear tree	2.67	35,625	95,119	0.464	441

4. Results and Discussion

The N Surplus can be determined through the N Balance (N Input and output). The importance of knowing the N surplus leads to some contribution for the N management not only for the farm but also for the environmental legislations. The N balance is easy to quantify and its application attracts researchers, farm operators, and policy makers.

The most successful application of N surplus is the design of Mineral Accounting System (MINAS) in De Marke Cow Farm system in Netherlands. The concept of N surplus (excluding atmospheric N factors) was introduced in the Legislation (Langeveld, 2007). The calculation method of N surplus was according to the following equation:

$$NS = N_{input} - N_{output} \tag{1}$$

where NS is the N surplus by kg; N_{input} is the total N input, kg; N_{output} is the total N input, kg.

The main calculation methods of N balance including farm-gate balance and soil-surface balance, both of which can calculate N balance in different types and scales of agricultural systems, and are important tools for monitoring the effectiveness of agri-environmental policy (Wang, 2006).

4.1. Soil-surface N balance on YBCF

In accordance with Eq. (1), excluding the external factors such as atmospheric N deposition, we obtained the equation of the soil-surface N balance on YBCF:

$$NS_S = N_{OF} - \left(N_{PR} + N_{PP} + N_{Citrus} + N_{Pear}\right) \tag{2}$$

where NS_S is the N surplus of soil-surface, kg; N_{OF} is the total N of organic fertilizer, kg; N_{PR}, N_{PP}, N_{Citrus} and N_{Pear} are, respectively, the total N production of PR, PP, citrus and pear, production, kg.

Applying the data from the YBCF into Eq. (2), NS_S = -43,930 kg was obtained. The N surplus is negative, indicating that the total amount of annual soil-surface N continued to decline. Based on the principle of soil-surface N balance, 43,930 kg of N should be applied to pasture and orchard each year so that the problem of lack of soil N can be solved.

4.2. Farm-gate N balance on YBCF

In accordance with Eq. (1), excluding the external factors such as atmospheric N deposition, we obtained the equation of farm-gate N balance on YBCF:

$$NS_F = \left(N_C + N_B + N_{SM} + N_{CM} + N_{BGS}\right) - \left(N_{PR} + N_{PP} + N_{Citrus} + N_{Pear}\right) \tag{3}$$

where NS_F is the N surplus of farm-gate, kg; N_C, N_B, N_{SM}, N_{CM}, N_{BGS} are respectively the total input N of corn, bran, soybean meal, cottonseed meal and BGS, kg; N_{Beef}, N_{PP}, N_{Citrus} and N_{Pear} are respectively the total output N of beef, PP, citrus and pear, kg.

Applying the data into Eq. (3), NS_F = -44,095 kg was obtained. The N surplus is negative, indicating that the total amount of annual farm-gate N balance continued to decline. However, there is a difference between NS_F and NS_S because of N loss existing in cattle body production process as the amount of 165 kg. Based on the principle of farm-gate N balance, 44,095 kg of N should be applied to the farm each year so that N balance on YBCF can be maintained.

Table 5. Nitrogen production of orchard on YBCF.

Item	Planting area (ha)	Average production (kg/ha)	Average total production (kg/year)	N rate (% FM)	Average total N production (kg/year)
Citrus tree	6.67	35,625	237,619	0.576	1369
Pear tree	2.67	35,625	95,119	0.464	441

4.3. Improvement and development of YBCF

In accordance with the NS_F and NS_S, the development schemes fall into two types, namely:

4.3.1. Direct purchase of inorganic N fertilizer

According to NS_S, by removing other factors, 43,930 kg of N should be required to direct input to the pasture and orchard. If all the inorganic N fertilizer used is 100% of urea (N rate is 46.6%), the amount of urea fertilizer needed is 94,271 kg. At this time, NS_F will be -165 kg. The YBCF can maintain N balance by reducing the PP (N rate is 0.57% in FM) output to 28,869 kg.

4.3.2. Expand scale of cattle breeding and use the increasing manure as organic fertilizer

It seems more reasonable and effective to complement the annual output of N by the way of ecological cycle. According to NS_S and by removing the other factors, 43,930 kg of N should be required for direct input into the pasture and orchard. If all of the organic N fertilizers are produced from the feces and urine of the cattle, the number of cattle population needs to be increased to 488.84 heads based on the calculated data from Table 3.

Each cattle needs 27 kg feed, including 7 kg of fine materials, 10 kg of BSG and 10 kg of PP. If the number of cattle population increases by one head, the farm will need to input 7 kg of fine material and 10 kg of BSG. At the same time the number of PP output will reduce by 10 kg. Table 6 shows the composition of fine material on YBCF.

Table 6. Composition of fine material on YBCF.

Item	Composition rate (%)	DM rate (%, FM)	CP rate (%, DM)	N rate (%, FM)
Corn	61.0	88.0	8.50	1.197
Bran	20.0	89.3	1.68	0.240
Soybean meal	9.0	90.0	42.80	6.163
Cottonseed meal	6.0	92.0	44.00	6.477
Premix feed	4.0	100.0	0	0
Total rate	100.0			
Total N rate				1.7214

In addition, according to existing cattle population (2000 heads) and slaughter number (2750 heads), there is a result that cattle population increasing of each herd will bring 1.375 heads of slaughter.

Totally, through calculation, it is shown that if the population number increases to 488.84 heads, the YBCF needs to input 1,248,984 kg of fine material and 1,784,263 kg of BSG with the amount of N 21,498 kg and 19,440 kg, respectively. At the same time, YBCF needs to reduce 1,784,263 kg of PP output with the reducing of output N of 10,170 kg and increase 201,646 kg of beef (Niu, 2011) output with increasing N output of 7179 kg. At this time, NS_S changes to 0, but the NS_F changes to -165 kg. So YBCF needs to additionally reduce 28,869 kg of PP output to make NS_F change to 0 in order to achieve N balance finally.

4.3.3. Relationship between increasing of cattle population and inorganic n fertilizer input

Let us consider X as the increasing number of cattle population and Y as the input of inorganic N fertilizer (100% urea); according to the soil-surface N balance on YBCF, the equation can be as following:

$$X \cdot N_{FU} + Y \cdot R_U = |NS_S| \tag{4}$$

where N_{FU} is the N production of feces and urine for each cattle, kg; R_U is the N rate of urea, %; NS_S is the N surplus of soil-surface, kg.

Applying the data into Eq. (4), we got: $Y = 94271.11 - 192.85X, X \leq 488.84$.

When $X = 0$, $Y = 94271.11$, namely all the fertilizer is urea; when $X = 488.84$, $Y = 0$, namely all the fertilizer comes from feces and urine.

4.3.4. Relationship between increasing of cattle population and the reducing of PP output

Let X be the increasing number of cattle population and Z be the reducing number of PP output; according to farm-gate N balance on YBCF, the equation can be drawn as follows:

$$(X \cdot N_{FU} + Y \cdot R_U) - (X \cdot N_{Beef} - Z \cdot R_{PP}) = |NS_F| \tag{5}$$

where N_{FU} is the N production of feces and urine for each cattle, kg; R_U is the N rate of urea, %; N_{Beef} is the N production of beef for each cattle, kg; R_{PP} is the N rate of PP, %, wet; NS_F is the N surplus of farm-gate, kg.

Applying the data into the Eq. (5), we got: $Z = 3650.00\,X + 28869.31, X \leq 488.84$.

When $X = 0$, $Z = 28869.31$, namely all the fertilizer is urea, YBCF need to reduce the PP output by 28,869.31 kg so that NS_F can change to 0; When $X = 488.84$, $Z = 1813132.14$, namely when the number of cattle population increased by 488.84 heads, YBCF needs to reduce PP output by 1,813,132.14 kg in order to make NS_F change to 0.

5. Conclusions

The Yufeng Beef Cattle Farm could solve the problem of feces and urine through utilizing the two main models: Cattle – Feces, Biogas Slurry (or Residue) – Pasture – Herbage; and Cattle – Feces, Biogas Slurry (or Residue) – Orchard – Fruit, so that nitrogen could be fully recycled in this cattle farm. By excluding the external factors such as atmospheric nitrogen deposition, the total nitrogen surplus of the farm was -44,095 kg and the nitrogen surplus of the soil in pasture and orchard was -43,930 kg. The fact that nitrogen surplus was negative indicated that the total amount of annual soil-surface N started to decline. The YBCF needed a certain amount of nitrogen fertilizers input, including organic nitrogen and inorganic nitrogen fertilizers, in order to maintain the nitrogen balance on the cattle farm. The relationship between the increasing number of cattle population (X) and inorganic nitrogen fertilizer input (Y) was $Y = 94271.11 - 192.85X, X \leq 488.84$. The relationship between the increasing number of cattle population (X) and the reducing number of PP output (Z) was $Z = 3650.00\,X + 28869.31, X \leq 488.84$.

Acknowledgements

Thanks to Yufeng Beef Cattle Farm for providing the investigation site and the data used in this paper. Thanks to the authors whose references are quoted in this paper. The authors are grateful to China Agriculture Research System for the funding (Code: nycytx-38).

References

Cao, C.L., 2002. Effect of Nitrogen and Nitrogen Form on Physiological Characteristics and Yields of Crop. Ph.D. diss., Northwest A&F University, Xi'an, China.

Dai, Z.H., X.H. Gan, G.H. Xu, R.M. Wang, X.G. Yu, 2014. Observation for silage effect of *Pennisetum Purpureum* in cattle breeding. Jiangxi Journal of Animal Husbandry & Veterinary Medicine. 1, 40–41.

Fan, X., L.J. Han, C.J. Huang, G.Q. Huang, 2006. Determination of nutrient contents in beef manure with near infrared reflectance spectroscopy. Transactions of the Chinese Society for Agricultural Machinery. 37(3), 76–79.

Feng, F., Y.S. Li, Y.T. Zhuang, Y.Y. Liu, W. Chen, S.C. Xing, Y. Wang, X.D. Liao, 2014. Status analysis on pollution on large-scale model farms in Haikou. Acta Ecologiae Animalis Domastici. 35(11), 45–52.

Gao, Y.Y., 2014. Difference in Different Growth Stages of Simmental Hybrid Beef Cattle Emission N, P and Methane. M.S. thesis, Henan Agricultural University, Zhengzhou, China.

Guo, A.Z., M.R. Qu, B.Z. Huang, Z.Q. Peng, 2012. Report on the development of cattle standardized breeding in southern areas. Animals Breeding and Feed. 11, 1–7.

Hou, R.L., 2014. Research on the Excrement and Biogas Rreatment of Scale Pig Farms. M.S. thesis, Shihezi University, Shihezi, China.

Langeveld, J., A. Verhagen, J. Neeteson, 2007. Evaluating farm performance using agri-environmental indicators: recent experiences for nitrogen management in the Netherlands. Journal of Environmental Management. 82, 363–376.

Liu, X.F., 2006. Effects of Nitrogenous Fertilizer on Guimu-1 Hybrid Pennisetum Urpureum Production Characteristics and Nutrition. M.S. thesis, Hunan Agricultural University, Changsha, China.

Lu, B.Y., 1992. General situation of research on nitrogen cycling in agroecosyscem. Journal of Shandong Agricultural University. 23(4), 457–460.

Lv, C., 2012. Effect of Roughage Combinations on N, P Emission and Study on Farmland Carrying Capacity of Dairy Cow. M.S. thesis, Henan Agricultural University, Zhengzhou, China.

Niu, L., 2011. Study on Quantity Evaluation of Chinese Simmental Cattle and Rapid Determination of Beef Quantity by Near-infrared Spectroscopy. M.S. thesis, Agricultural University of Hebei, Baoding, China.

Powell, J.M., C.J. Gourley, C.A. Rotz, 2010. Nitrogen use efficiency: a measurable performance indicator for dairy farms. Environmental Science and Policy. 13, 217-228.

Wang, B.H., 2006. Nitrogen Cycling and Balance in the Farmer Household's Agro-animal Husbandry Ecosystem of Yaozhou Region, Shanxi Province. M.S. thesis, Northwest A&F University, Xi'an, China.

Wu, L.F., Z. Ou'yang, X.L. Xie, 2013. Nitrogen and phosphorus balance of cropland at regional scale for integrated crop-livestock farming system in two different areas. Journal of Natural Resources, 26(6), 943–954.

Yang, S.Q., 2008. Studies on the Influence of Soil and Leaf Nutrient on Citrus Fruit's Output and Quality-taking Citrus Orchards in Zhonxian County of Chongqing as an Example. M.S. thesis, Southwest University, Chongqing, China.

Zhang, L., D.Y. Liu, T. Shao, 2008. The feeding value utilization prospect of lolium multiflorum. Pratacultural Sience. 25(4), 64–69.

Zhu, H.S., F.Y. Zuo, Y.F. Gao, Z.P. Bai, 2013. Collection and composition analysis of solid manure on a large-scale beef cattle farm in Chongqing. Heilongjiang Animal Science and Veterinary Medicine. 11(1), 70–71.

Application of Livestock Manure to Corn Fields in Comparison with Commercial Fertilizer Application

Robert Spajić [a,*], Robert T. Burns [b], Davor Kralik [a], Đurđica Kovačić [a], Katarina Kundih [c], Daria Jovičić [a]

[a] Department for Animal Husbandry, Faculty of Agriculture, University of J.J., Strossmyer Osijek, 31000 Osijek, Croatia

[b] The University of Tennessee Extension, Knoxwille, USA

[c] Department for Swine Production, Belje d.d., 31326 Darda, Croatia

* Corresponding author. Email: rspajic@pfos.hr

Abstract

This paper will present the results of the field trials of manure application to the corn crop, with comparison of corn yields on the field where artificial fertilizer is applied. The field trials were conducted with emphasis on replacement of the whole nutrient requirements (N, P, and K) from the commercial fertilizer to the corn plant, with nutrients sourced from livestock manure. The paper will show the analysis of swine farm manures applied to the fields and their results compared to the usage of commercial fertilizer. Application was made by a new application model of usage the injection equipment for farm manures in the mentioned region. The trial was done on one 200 ha field, divided into four plots (about 50 ha each plot), where two plots were applied with swine manure and two other plots with artificial fertilizer. Base for quantities of manure used per hectare was N content in manure and it was calculated on 200 kg of N applied per ha. Each manure sample was analyzed on total N and ammonium N, and applied based on corn requirements, as well as artificial fertilizer. Significant differences on corn yields occurred where manure was applied to the fields. With the comparison to the artificial fertilizers, application of manure resulted in 15% higher yields. The fields that were treated with commercial fertilizers have corn yield of 5.20 and 6.27 t ha^{-1}, while the fields that were treated with manure have 7.30 and 6.60 t ha^{-1} of corn yield.

Keywords: Commercial fertilizer, manure application, corn production, nitrogen content

1. Introduction

Application to soil is one of the most widely accepted and best economic and agronomic ways of manure usage (Sawyer et al., 2003) which aroused interest in new manure management technologies with an emphasis on minimal negative effects to the environment. Animal manure is an excellent nutrient source because it contains most of the plant essential elements in spite the fact that their concentrations tend to be low. Many studies have shown that land application of manure will produce crop yields equivalent or higher than that obtained with commercial fertilizers (Risse et al., 2006). Fertilization with animal manure has a positive effect on soil nutrient availability, soil physical properties, crop yields, and pasture carrying capacity (Costa et al., 2014; Lamb and Schmitt, 2001). However, the use of manure as fertilizer can lead to significant losses of nitrogen which can be lost through leaching, denitrification or ammonia volatilization. Ammonia volatilization represents the primary nitrogen loss concern when fertilizing with manure. When considering ways of manure application to soil, it is already well documented that manure injection methods are far better than application of manure over the soil surface (Sutton et al., 1982; Laboski et al., 2013; Groot et al., 2007). When manure is applied over the soil surface, it can rapidly lose nitrogen through ammonia volatilization. But direct manure injection into the soil can greatly decrease losses sometimes to only a few percent (Costa et al., 2014; Maguire et al., 2009). In field trials Smith et al. (2009) reported about 59% decrease in nitrogen losses when injecting manure into the soil compared to surface application. This reduction ammonia losses management provides other benefits too, like controlling odor and incorporating phosphorus, which limits runoff losses (Jokela and Meisinger, 2008).

Manure is typically applied to land for corn production before planting (in the spring) or after harvest (in the fall). Application is spring is recommended to increase efficiency of nitrogen use and decrease potential for leaching and runoff losses. Fall application of manure is usually done for economic and practical reasons like limited storage, field conditions, and time constraints in spring etc. (Williams et al., 2012; Jokela et al., 1999).

The aim of this paper was to present trials of swine manure application to the soil prior to corn (Zea Mais) seeding and it's comparison to commercial fertilizers application. It was heavy sandy soil type region where trials were conducted. Usage of swine manure as a replacement for the commercial fertilizer will be shown additionally through the financial analysis of the trials.

2. Materials and methods

Swine manure from large swine operation unit (1400 sows, 6000 gilts, and 26,000 piglets) was collected in the swine farm under the animals, held on fully slatted floor into the all three production segments (sows, gilts, and piglets). Swine manure was collected from the shallow pits (60 cm deep). It was collected on a monthly base into the central pumping system and directly pumped into the closed steel tanks (figure 1). Each tank was sized for six month of storage capacity.

Swine manure and commercial fertilizer were applied on the soil prior to corn seeding. The trial was held on a 200.60-ha surface plot, divided into four smaller plots (plot 1 = 46.07 ha, plot 2 = 48.13 ha, plot 3 = 50.92 ha, plot 4 = 55.48 ha). Plots 1 and 3 were covered with swine manure only, and plots 2 and 4 were covered with standard commercial fertilizer. Basic focus of manure quantity calculation was based on N level in swine manure and commercial fertilizer. The level of P and K were recorded for both, swine manure and commercial fertilizer additionally. Quantities of applied manure and commercial fertilizer to the soil were compared on N level. Basic amount requirements, for both swine manure and commercial fertilizer, to the corn, were defined on 200 kg of available N.

Figure 1. Closed steel manure tanks.

Application rate of swine manure was based on the lab analysis of the manure samples, taken from the stainless steel covered tanks, prior to application. Analyzed manure samples results were held in September 2011 prior to autumn application and in March 2012, prior to spring application. Each manure sample was collected in triplicate and mixed as one representative sample. Variation between the samples was not monitored during the trials.

Table 1 shows the amount of manure applied on each plot (plots 1 and 3). Each plot was covered with manure application twice, in September 2011 first time and in April 2012 the second time. Different amount of manure was applied in each pass on the plots.

Every sample of manure was analyzed on chemical composition of the nutrients presented in Table 3. Manure application was placed with manure application umbilical system produced by PCE (Puck Custom Enterprises) (Figure 2), set up with tool bar 7.5 m wide, with injection sweeps - Dietrich type (depth = 25 cm under the soil surface).

Figure 2. PCE manure application system.

System was supplied with the swine manure from the steel tanks by pumping unit Cornell, powered by John Deere motor 375 HP, with application capacity of 180 m^3 per hour. Plot 1 was applied in autumn (79 m^3 ha^{-1}) and spring (39 m^3 ha^{-1}). Plot 3 was applied in autumn (66 m^3 ha^{-1}) and spring (28 m^3 ha^{-1}). Manure was applied on both plots on the same day, in autumn and spring. Commercial fertilizer was applied with Amazon granule fertilizer spreader. Both application systems were controlled by Ag Leader GIS navigation systems (Figure 3).

Figure 3. Ag Leader GIS navigation system.

Manure application in autumn was held prior to early winter plowing of the soil. The spring application of manure was held on a plowed ground, before closing the soil surface and prior to corn (Zea Mais) seeding process.

Manure samples were analyzed by Croatian Department for organic fertilizers and soil analysis. Each sample was freshly prepared and analyzed for total solids (TS), pH, total N, ammonium N, total P, total K, total Ca, and total Mg. The TS was determined by gravimetric method. The pH level was measured at a room temperature by pH in H_2O - HRN EN 13037:1999 method. Total N was determined by Kjeldahl block digestion and steam distillation unit. Bremnner's method was used for determination of ammonium N. Total P was determined by macro-destruction spectrophotometric method. Photometric method was used for determination of total K while total Ca and Mg were determined using atomic absorption spectrophotometer.

Commercial fertilizer was applied to the soil in following forms of fertilizer MAP (mono-ammonium phosphate), UREA (urea) and KAN (potassium-ammonium nitrate). The KAN commercial fertilizer was not monitored by AG Leader GPS system.

Corn (Zea Mais) was sowed in the end of April 2012 on the same day on all four plots. The study was conducted in Brestovac – Karanac (Baranja County, eastern Croatia).

3. Results and Discussion

Table 1 represents corn yields obtained by different rates and amounts of manure application based on N content, while Table 2 represents corn yields obtained by different amounts of applied commercial fertilizers. Corn yields collected on all four plots are presented as total yield per plot and as yield per ha. Plots 1 and 3 have significantly higher yields per ha basis (7.30 t ha^{-1} and 6.60 t ha^{-1}, respectively) than plots 2 and 4 (5.20 t ha^{-1} and 6.27 t ha^{-1}, respectively).

Table 1. Corn yields (grain) resulted by N usage through the application of manure expressed for each round of application.

	Plot 1 application		Plot 3 application	
	1st Sept. 2011	2nd April 2012	1st Sept. 2011	2nd April 2012
Manure applied (m^3 ha^{-1})	79	39	66	28
Manure applied (m^3 plot^{-1} round^{-1})	3639.53	1796.73	3366	1428
Total N (kg t^{-1})	1.7	3.1	2.5	3.1
Usable N (kg per ha)	107	109	132	78
Corn yield, grain (t ha^{-1})	7.30		6.60	

Table 2. Application rates of artificial fertilizer, total N amount from fertilizers, with results of the corn yields, obtained by application of commercial fertilizers.

	Plot 2	Plot 4
MAP 12-52-0 (kg ha^{-1})	203	203
UREA 46% (kg ha^{-1})	293	293
KAN 27% (kg ha^{-1})	150	150
Total N applied (kg ha^{-1})*	200	200
Corn yield (t ha^{-1})	5.20	6.27

* Calculated from the artificial fertilizer

Manure application system that was used in this trial and that was presented through this paper has significantly lower loses of N during and after the application than standard application systems that are applying manure on the soil surface. It was calculated that loses of total N from swine manure applied in autumn are 20% and loses of total N from swine manure applied in spring are 10%. These calculations are based on estimated losses during manure application (technical evaporation during operation, nitrogen loses in the soil during winter period etc.) and presented through the numbers expressed as usable N. In comparison with the manure application system presented in this trial, standard application systems that are applying manure on the soil surface, has significantly higher losses of N (up to 80% of N losses).

Manure that was applied to the field was analyzed prior to fertilizing and the results of the analysis are shown in Table 3. Even though the costs per ha were higher in case of manure application the yields were significantly higher. Additionally, through the manure application case it was strongly improved nutrient management plan and negative environmental influence caused by non-adequate managing of the manure was reduced to a minimum

Value of the manure was calculated based on N level applied in each application round based on manure analysis. Component of N was calculated based on value of N expressed through the UREA N, MAP and KAN artificial fertilizer values. Based on the price of the artificial fertilizer, the value of 1kg of N from manure was calculated as 0.61 \$ kg^{-1} of total N. In our case the price of artificial fertilizers was sourced as an average price from period 2011 – 2012. In Table 4 it is calculated the amount of N per m^3 and multiplied with price of 0.61 \$ kg^{-1}.

Table 3. Average chemical composition of nutrients (kg t^{-1}) of applied manure to the soil.

	Manure with 1,7 total N	Manure with 2,5 total N	Manure with 3,1 total N
Total solids (kg t^{-1})	8.90	20.00	20.70
pH	7.64	7.66	7.31
Total N (kg t^{-1})	1.70	2.50	3.10
Ammonium N (kg t^{-1})	1.40	1.70	0.21
Total P (kg t^{-1})	0.23	0.53	1.60
Total K (kg t^{-1})	0.88	1.44	7.60
Total Ca (kg t^{-1})	0.41	0.96	4.63
Total Mg (kg t^{-1})	0.19	0.39	1.23

If we compare financial results with the costs of production, it is visible that plots 1 and 3 had better yields than plots 2 and 4. Costs of production per kg of corn are presented as follow in brackets (plot 1 = 0.054 $ per kg; plot 3 = 0.052 $ per kg) in comparison with plots 2 and 4 (plot 2 = 0.063 $ per kg; plot 4 = 0.052 $ per kg) as shown in Tables 4 and 5. As we can see there was significant difference in yields and cost production between plot 1 and 2. On the other hand, even though the yields were higher in plot 3 comparing it with plot 4, there was not any differences in cost production values.

Table 4. Financial analysis of manure application in corn production.

Item	Plot 1 application		Plot 3 application	
	1st	2nd	1st	2nd
Manure per ha (m^3)	79.00	39.00	66.00	28.00
Value of manure based on N ($ m^{-3})	1.04	1.89	1.53	1.89
Total manure per plot (m^3)	3,639.26	1,796.73	3,366.00	1,428.12
Total manure value per application ($)	3,784.83	3,395.82	5,149.98	2,699.15
Total ha per plot	46.07		50.92	
Total value of manure per plot ($)	7,180.65		7,849.13	
Value of manure per ha ($)	155.86		154.15	
Costs of manure application per m^3 ($)	2.00		2.00	
Costs of manure application per ha ($)	236.00		188.00	
Total costs per ha ($)	391.86		342.15	
Total of corn per ha (kg)	7,300.00		6,600.00	
Total costs per kg of produced corn	0.0537		0.0518	

Table 5. Financial analysis of commercial fertilizers application in corn production.

Item	Plot 2	Plot 4
MAP 12-52-0 (kg ha^{-1})	203.00	203.00
UREA 46% (kg ha^{-1})	293.00	293.00
KAN 27% (kg ha^{-1})	150.00	150.00
Total ha per plot	48.13	55.48
Total value of commercial fertilizer per plot ($)	14,636.33	16,871.47
Value of commercial fertilizer per ha ($)	304.10	304.10
Costs of fertilizer application per ha ($)	20.83	20.83
Total costs per ha ($)	324.93	324.93
Total of corn per ha (kg)	5,200.00	6,270.00
Total costs per kg of produced corn	0.0625	0.0518

4. Conclusions

Integrating livestock manure as a fertilizer for crop production is a well-accepted way for its disposal and has value for farmers because it contains most of the plant essential elements and can improve soil quality. The application of swine manure increased corn yields in comparison with commercial fertilizers. It is evident that swine manure can be efficient as organic fertilizer when applied properly with adequate equipment and adequate agricultural time frames.

Significant reduction of cost in production of corn can be implemented if we treat the manure as something that is already here, something that is already paid through swine production costs. Our analysis in this paper is comparison of two technologies' costs, with strong accent on reduction of potential negative influence on environment. We will say additionally that it is important to remember the potential negative environmental impact is turned into significant benefit for the producers and farmers.

References

Costa, M., Shigaki, F., Alves, B., Kleinman, P. and Pereira, M. (2014). Swine manure application methods effects on ammonia volatilization, forage quality, and yield in the Pre-Amazon Region of Brazil. Chilean Journal of Agricultural Research 74(3), 311–318.

Groot, J.C.J., Van Der Ploeg, J.D., Verhoeven, F.P.M. and Lantinga, E.A. (2007). Interpretation of results from on-farm experiments: Manure-nitrogen recovery on grassland as affected by manure quality and application technique. 1. An agronomic analysis. Netherlands Journal of Agricultural Science 54, 235–254.

Jokela, B., Bosworth, S. and Tricou J. (1999). Direct incorporation of liquid diary manure for fall and spring application of corn. Lower Missisquoi Water Quality Project. University of Vermont, Burlington, Vermont. Duration of project: 1995–1997.

Jokela, B. and Meisinger, J. (2008). Ammonia emissions from field-applied manure: management for environmental and economic benefits. In *Proceedings of the 2008 Wisconsin Fertilizer, Aglime & Pest Management Conference Vol. 47.* Madison, Wisconsin, January 15–17, University of Wisconsin, 199–208.

Laboski, C.A.M., Jokela, W. and Andraski, T.W. (2013). Dairy manure application methods: N credits, gaseous N losses, and corn yield. In *Proceedings of the 2013 Wisconsin Crop Management Conference Vol. 52.* Madison, Wisconsin, January 15-17, University of Wisconsin, 20–31.

Lamb, J.A. and Schmitt, M.A. (2001). Management of turkey and swine manure derived nitrogen in a sugarbeet cropping system. Soil Management 2001, 32, 125–134.

Maguire, R., Heckendorn, S.E. and Jones, B. (2009). Fertilizing with manures. Virginia Cooperative Extension, publication 452-705. https://pubs.ext.vt.edu/452/452-705/452-705.html Accessed June 19, 2015.

Risse, M., Cabrera, M.L., Franzluebbers, A.J., Gaskin, J.W., Gilley, J.E., Killorn, R., Radcliffe, D.E., Tollner, E.W. and Zhang, H. (2006). Land application of manure for beneficial reuse. National Center for Manure & Animal Waste Management. USDA – Fund for Rural American Grant.

Sawyer, J.E., Lundvall, J.P., Rakshit, S. and Mallarino, A.P. (2003). Liquid swine manure nitrogen utilization for crop production. In *Proceedings of the 15th Annual Integrated Crop Management Conference*. Ames, Iowa, December 3-4, Iowa State University, IA. 95–112.

Smith, E., Gordon, R., Bourque, C., Campbell, A., Génermont, S., Rochette, P. And Mkhabela, M. (2009). Simulated management effects on ammonia emissions from field applied manure. Journal of Environmental Management 90, 2531–2536.

Sutton, A.L., Nelson, D.W., Hoff, J.D. and Mayrose, V.B. (1982): Effects of injection and surface applications of liquid swine manure on corn yield and soil composition. Journal of Environmental Quality 11(3), 468-472.

Williams, M.R., Feyereisen, G.W., Beegle, D.B. and Shannon, R.D. (2012): Soil temperature regulates phosphorus loss from lysimeters following fall and winter manure application. Transactions of the ASABE 55(3), 871-880.

Co-digestion of *Enteromorpha* and Chicken Manure for Enhancing Methane Production

Ruirui Li [a], Na Duan [a,*], Yuanhui Zhang [a,b], Baoming Li [a], Zhidan Liu [a], Haifeng Lu [a]

[a] China Agricultural University, Beijing 100083, China

[b] University of Illinois at Urbana-Champaign, Urbana, IL 61801, USA

* Corresponding author. Email: duanna@cau.edu.cn

Abstract

Anaerobic digestion (AD) of *Enteromorpha* and chicken manure (CM) was conducted in triplicate to determine the biogas production rates of algae for 25 d, with a constant total VS content (10g for each bottle) and *Enteromorpha*/CM ratios (0:10, 2:8, 3:7, 5:5, 7:3, 8:2, 10:0, blank) at 35°C. The objective of this wok was to determine the feasibility of co-digestion of *Enteromorpha* and CM, and optimum *Enteromorpha*/CM proportion on methane production and macronutrient degradation in laboratory scale. Results showed that the co-digestion of *Enteromorpha* and CM led to the highest CH_4 production of 35.48 mL/g VS at the ratio of 7:3, demonstrating 185 times improvement comparing with that from the solo-digestion of *Enteromorpha*. But it deteriorated as the ratio increased due to the increased ammonia inhibition. The concentration of total nitrogen (TN), total phosphorus (TP), volatile fatty acids (VFAs) and pH values showed no significant difference. However, removal rate of total organic carbon (TOC) in co-digestion was much higher than that of solo-digestion, indicating that co-digestion of *Enteromorpha* and CM was a promising technology for both waste treatment and clean energy production.

Keywords: Anaerobic digestion, TP, TAN, TOC removal, energy production

1. Introduction

Eutrophication causing the bloom of *Enteromorpha* in rivers and lakes has caused great distress to China for years. Jiaozhou Bay, located in the south of Shandong Peninsula, which is an excellent natural harbour, is suffering the most pressing problem of eutrophication and harmful algal blooms (commonly referred to as green tide). The strong reproductive capacity of floating *Enteromorpha* might be the main reason of green tide occurrence. *Enteromorpha* itself is non-toxicity. Whereas if the blooms mushroom at an alarming rate, it will block the sun and inhibit the growth of benthic algae, causing unbalance of species in ocean system. Not only this, it will consume amounts of oxygen in sea and some chemical substances it produces may have bad effect on other sea creatures. It will seriously interfere with the progress of water activities as well as tourisms. Qingdao International Sailboat Center for the 2008 Olympic Games was invaded by *Enteromorpha* blooms and heroic efforts were paid to remove this toxic biomass from the rowing course. As refloatation is considered to be the quickest way to reduce green tide, thousands of tons of *Enteromorpha* were refloated every day. It is one of the feasible sewage revitalizing ways to recover and reuse nitrogen and phosphorus by using farmland as the treatment system to eliminate pollution in foul water. However, if it is not disposed effectively and completely, it will easily cause secondary environmental pollution.

The algal biomass, especially the oil-rich species, has been considered as bioenergy (generally biodiesel or ethanol) source for several years (Zhong et al., 2013). *Enteromorpha* is not rich in oil; so it is less attractive for biodiesel. For those species, anaerobic digestion (AD) is the simplest and most efficient way for energy recovery. *Enteromorpha* is rich in proteins, carbohydrates, fibers, fatty acids and several minerals, but low in fats. This made *Enteromorpha* an ideal feedstock for biogas or methane production. As to the possible ammonia inhibition due to its high protein levels and hard degradability due to its cell wall, physico-chemical pretreatment, co-digestion, or control of gross composition are strategies that can significantly and efficiently increase the conversion yield of the algal organic matter into methane (Sialve et al., 2009). AD is a key process that can solve the environmental issues as well as the economical and energetic balance of such a significant technology. AD demands for less energy and produces more renewable ones in forms of biogas and waste treatment. It does not need advanced equipment to pump oxygen, meaning a much lower cost for electric energy than aerobic digestion. It was reported that the energy consumed by AD was only 1/10 of that of aerobic digestion. Moreover, in developed countries, energy transformed by the biogas it produced can help generate 33-100% electricity for sewage treatment plant in turn.

The world seems to be raising its energy needs owing to an expanding population and people's desire for higher living standards. Diversification biofuel sources have become an important energy issue in recent times. Among the various resources, algal biomass has received much attention in the recent years due to its relatively high growth rate, its vast potential to reduce greenhouse gas emissions and climate change, and their ability to store high amounts of lipids and carbohydrates. These versatile organisms can also be used for the production of biofuel (Wu et al., 2014). Biofuels derived from microalgae and biofuels derived from macroalgae are the two principal components of people's interests in bioenergy. Microalgae has drawn so much attention as a source for biofuel. While the production of macroalgae for biofuels was first tried and tested in the late 1960s (Hughes et al., 2012), energy converted via AD presented an excellent result due to its biochemical components making it an ideal feedstock. Macroalgae as a source of bioenergy first received intensive scrutiny as part of the US Ocean Food and Energy Farm project as proposed by Wilcox (Bird et al., 1987), initiated in 1973 and lasting over a decade (Bird et al., 1987). The biological gasification of macroalgae was well proven in the later decades of the 20th century and AD technology has sufficiently matured to offer a range of possibilities to further optimize methane yields. Added value could be achieved by processing part of the crop for human and animal foodstuffs, and food supplements, for its mineral content for animal feeds, as an organic slow release fertilizer, and potential bio-active compounds (Imhof et al., 2004). Compared to first generation biofuels, macroalgae have inherent advantages that make them environmentally sustainable. It is prudent to develop the technology required to obtain significant quantities of biofuel from marine biomass in time to help meet Europe's energy needs and climate change targets.

Anaerobic digestion of algal bloom biomass can be traced back to the first oil crisis, and was thought to be the simplest and most effective way for energy recovery and sewage treatment. This paper highlights the co-digestion of *Enteromorpha* and chicken manure (CM), evaluating at different CM *Enteromorpha* ratios. The black controller was inoculated with granular sludge only. The objective of this wok was to determine the feasibility of co-digestion of *Enteromorpha* and CM, and optimum CM proportion on methane production and macronutrient degradation in laboratory scale.

2. Materials and Methods

2.1. Substrates and inoculant

The substrate *Enteromorpha* was obtained from China Ocean University Organism Project Development Company, and was crushed with a pulverizer (Minye Yungkang Industry Co., Ltd., Zhejiang, China) and sieved to 60 mesh to achieve a uniform sample. Chicken manure was a mixture of chicken stool and urine freshly collected from Shandong Minhe Co., Ltd. Effluent from an anaerobic digester fed with waste water solids (Xiaohongmen Sewage Treatment Plant, Beijing, China) providing the microbial consortia was used as an inoculant.

2.2. Batch laboratory AD tests

The bench-scale AD tests for determining the anaerobic biodegradability and ultimate CH_4 yield of *Enteromorpha* were carried out by using Automatic Methane Potential Test System (AMPTS) II (Bioprocess, Sweden) with software of AMPTS v5.0 (Badshah et al., 2012). The AMPTS II has three main units. A water-bath incubation unit which allows up to 24 glass bottles of 250 mL containing the substrate and/or inoculum which is incubated at 35°C.The biogas produced and CH_4 content was measured online periodically using the automated data-acquisition system. Batch experiments were conducted in triplicate to determine the biogas production rates of algae for 25 d, with a constant inoculum and 8 *Enteromorpha*/CM ratios (10:0, 8:2, 7:3, 5:5, 3:7, 2:8, 0:10, and 0:0) at 35°C.

2.3. Analytical methods

2.3.1. Methane

Methane concentration in the biogas was analyzed using a gas chromatograph (GC 910, Kechuang, China) equipped with a thermal conductivity detector (TCD) and argon as the carrier gas. The injector, oven and detector temperatures were 150, 120 and 150°C, respectively.

2.3.2. Physicochemical parameters

The total solids (TS), volatile solids (VS), total carbon (TC), total phosphate (TP) and total ammonia nitrogen (TAN) were measured by using the APHA standard methods (Eaton, A.D., 2005). The protein content was determined based on the total Kjeldahl nitrogen (TKN) measurement using the correction factor 6.25 (Msuya, 2008). The total lipid content was analyzed gravimetrically from the extract obtained with diethyl ether in a Soxtec System HT (HT2 1045, Tecator, Sweden) (Barje, 2008). Carbohydrate was estimated as the remaining fraction of VS after the determination of protein and lipid.

3. Results and Discussion

3.1. Characterization of substrates

Table 1 shows the chemical properties of *Enteromorpha* in terms of TS, VS, proteins, lipids, cellulose, hemicellulose, lignin and other carbohydrates. Results revealed the organic fraction contributed to the major part of the fresh biomass, representing a protein content of 17.73±0.39% TS, cellulose content of 12.35±0.20% TS and carbohydrate content of 31.82±1.01% TS, respectively. As for C/N ratios, fresh biomass had the value of 7.67, which is lower than the most suitable value for optimum AD operation of 15–30 (Xu, 2012). Thus, in batch experiments of the AD study, the inoculum and substrate concentration of 10 g VS was applied to avoid methanogenesis inhibition by NH_3 (Liu, 2012). The characteristics of the substances used in this anaerobic co-digestion were listed in Table 1.

Table 1. Main characteristics of the substances used in this anaerobic co-digestion.

	Enteromorpha	Chicken manure
TS, %	98.91±2.22	19.83±0.56
VS,%	70.40±1.72	81.0±1.92
Proteins,% TS	17.73±0.39	NA
Lipids,% TS	1.67±0.06	NA
Cellulose,% TS	12.35±0.20	NA
Hemicellulose,% TS	30.7±0.16	NA
Lignin,% TS	5.73±0.14	NA
Other carbohydrates,% TS	31.82±1.01	NA
C/N	7.67	9.28

3.2. TP, TAN, pH value and TOC removal

From Figures 1–4, we can see that TP and pH value did not show significant difference. The pH value has been almost constant at around 7.0. For each group, the final TP content changed less than 100mg/L compared with the initial. That was because the predominant bacteria, methanogens, did not absorb phosphorus during its metabolism. There was a balance of each microorganism population, maintaining a stable and suitable pH value in the system. The TAN varied during the hydrolysis, acetoxylation and methane-producing process. With the production of methane, TOC removed certainly, varying from 0.93% to 57.57%. The best group for TOC removal is NO.6, with the value of 57.57±6.90%.

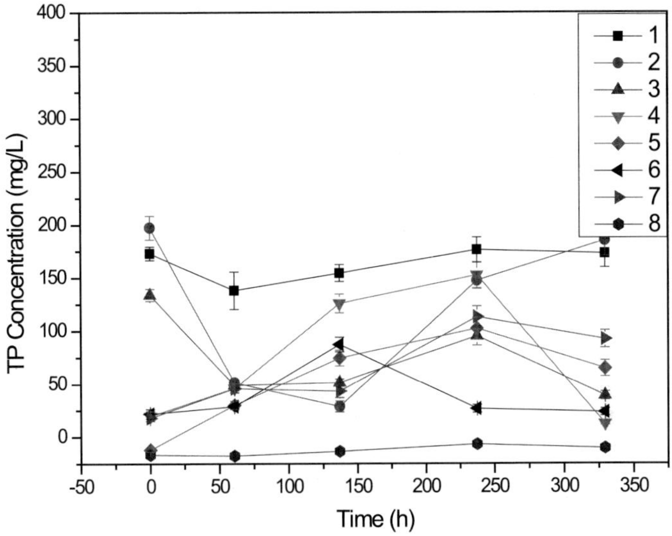

Figure 1. Total phosphorus (TP) during the process.

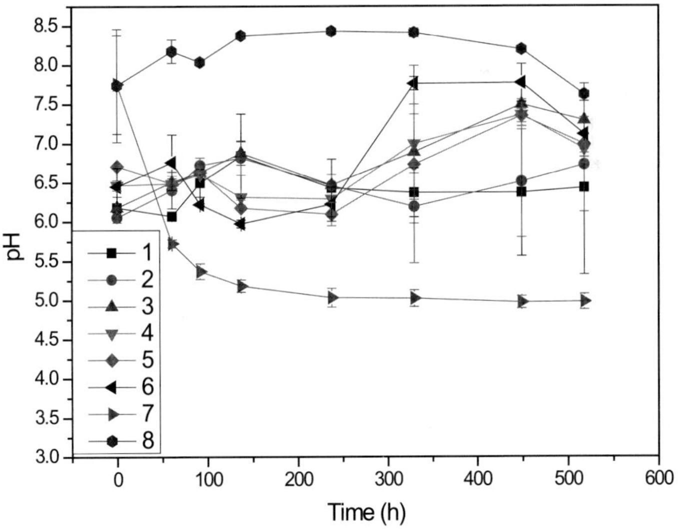

Figure 2. pH value during the process.

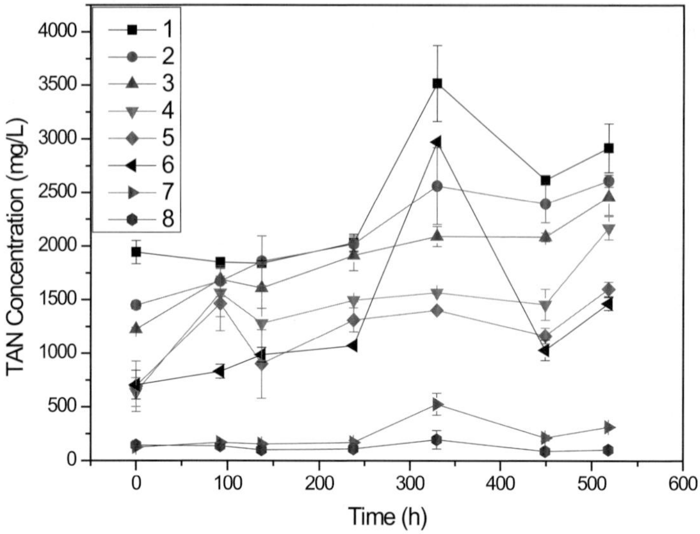

Figure 3. Total ammonia (TAN) during the process

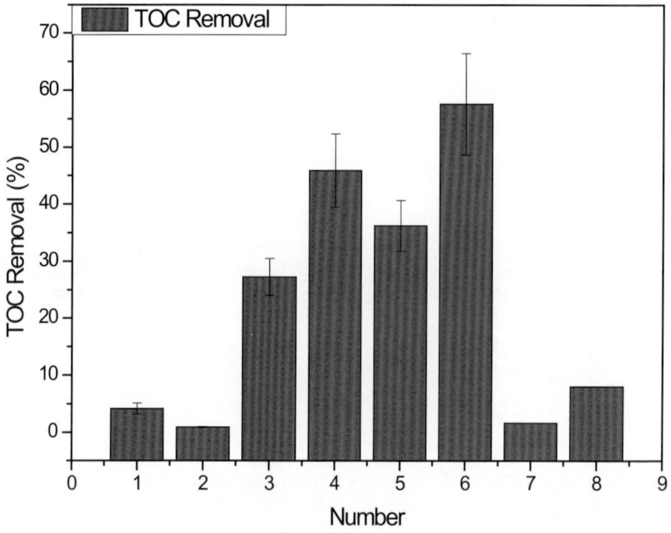

Figure 4. TOC removal rate during the process

1-*Enteromorpha*/CM = 0:10; 2-*Enteromorpha*/CM = 2:8; 3-*Enteromorpha*/CM = 3:7; 4-*Enteromorpha*/CM = 5:5; 5-*Enteromorpha*/CM = 7:3; 6-*Enteromorpha*/CM = 8:2; 7-*Enteromorpha*/ CM = 10:0; 8-Blanck

3.3. *Enteromorpha* biomethane production potential

Figure 5 shows the cumulative CH_4 yield as a function of time from AD of different substrates. The lowest cumulative CH_4 yield was 0.192 mL/g VS when the substrate ratio was 7-*Enteromorpha*/CM = 10:0. While the cumulative CH_4 yield was highest for AD of NO.3 (*Enteromorpha*/CM = 3:7), with the value of 354.82±19.45mL, that is 35.48mL/g VS, and followed by NO. 6 (*Enteromorpha*/ CM = 8:2), with the value of 34.48 mL/g VS. Results reveal that only co-digestion can improve the efficiency of methane production. We can infer that microorganisms in CM help to disrupt the cell wall of the biomass and subsequent VS release. Considering the TOC removal rate, we choose NO. 6 as the ideal group.

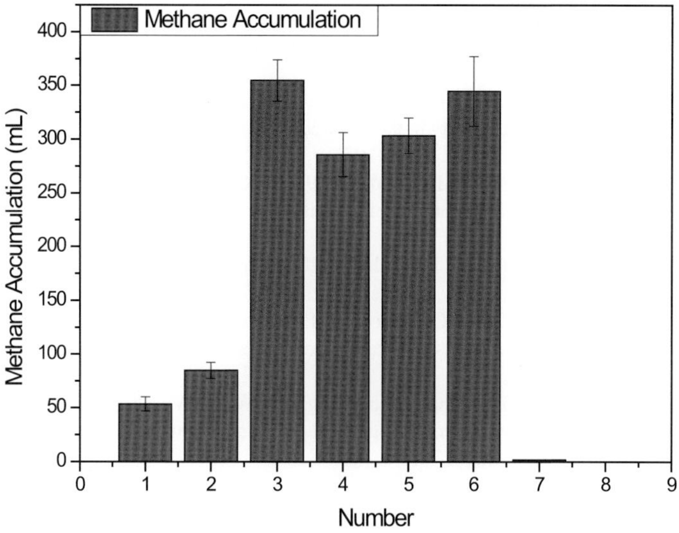

Figure 5. Methane accumulation during the process

4. Conclusions

Co-digestion of *Enteromorpha* and CM resulted in improved CH_4 yield of 35.48 mL/g VS at the *Enteromorpha*/CM ratio of 3:7, while digestion of *Enteromorpha* simply led to optimized CH_4 production of 0.19 mL/g VS. Besides, removal rate of total organic carbon (TOC) is about 60 times more than that of solo-digestion. During the co-digestion and digestion, pH, TAN and TP corroborated the appropriate stability of the anaerobic digestion. Co-digestion of *Enteromorpha* and CM exhibited a better performance at the aspect of TOC removal rate and methane production than solo-digestion of *Enteromorpha* only, indicating that co-digestion of *Enteromorpha* and CM was a promising technology for both waste treatment and clean energy production.

References

Bird K.T., and P.H. Benson, 1987. Oceanic farming of Macrocystis, the problems and non-problems. Seaweed Cultivation for Renewable Resources. 16, 39-68.

Bird K.T., and P.H. Benson, 1987. Oceanic farming of Macrocystis, the problems and non-problems. Seaweed Cultivation for Renewable Resources. 16, 102-109.

Badshah, M., D.M. Lam, J. Liu, and B. Mattiasson, 2012. Use of an automatic methane potential test system for evaluating the biomethane potential of sugarcane bagasse after different treatments. Bioresource Technology. 114, 262–269.

Barje, F., S. Amir, P. Winterton, and E. Pinelli, 2008. Phospholipid fatty acid analysis to monitor the co-composting process of olive oil mill wastes and organic household refuse. Journal of Hazardous Materials. 154(1-3), 682–687.

Hughes, A.D., S.K. Maeve, K.D. Black, and M.S. Stanley, 2012. Biogas from Macroalgae: is it time to revisit the idea? Biotechnology for Biofuels. 5, 1–7.

Eaton, A.D., L.S. Clesceri, and A.E. Greenberg, 2005. *Standard Methods for the Examination of Water and Wastewater*, 21st ed. American Public Health Association/American Water Works Association/Water Environment Federation, Washington DC, USA.

Imhof M., and L. Bounoua, 2011. Human consumption of primary productivity. Human consumption of primary productivity. 7, 55-70.

Liu, X., W. Wang, and Y. Shi, 2012. Pilot-scale anaerobic co-digestion of municipal biomass waste and waste activated sludge in China: effect of organic loading rate. Waste Management. 32, 2056–2060.

Msuya, F., and A. Neori, 2008. Effect of water aeration and nutrient load level on biomass yield, N uptake and protein content of the seaweed Ulva lactuca cultured in seawater tanks. Journal of Applied Phycology. 20(6), 1021–1031.

Sialve B., N. Bernet, and O. Bernard, 2009. Anaerobic digestion of microalgae as a necessary step to make microalgal biodiesel sustainable. Biotechnology Advances. 27(4), 409-416.

Wu Y.H,, H.Y. Hu, Y. Yu, and T.Y. Zhang, 2014. Potential of the micro and macro algae for biofuel production: a brief review. BioResources. 9(1), 1606–1633.

Xu, F., J. Shi, and W. Lv, 2012. Comparison of different liquid anaerobic digestion effluents as inocula and nitrogen sources for solid-state batch anaerobic digestion of corn stover. Waste Management. 3, 26–32.

Zhong W., L. Chi, Y. Luo, Z. Zhang, Z. Zhang, and W.M. Wu, 2013. Enhanced production from Taihu lake blue algae by anaerobic digestion with con straw in continuous feed digesters. Bioresource Technology. 134, 264–270.

Combination of Electrolysis and Microalgae Cultivation to Treat Effluent from Anaerobic Digestion of Poultry Manure

Xinfeng Wang [a,b], Haifeng Lu [a,b,*], Li Zhang [a], Mengzi Wang [a,b], Yu Zhao [a], Baoming Li [a,b,*]

[a] Key Laboratory of Agricultural Engineering in Structure and Environment, Ministry of Agriculture; College of Water Resources and Civil Engineering, China Agricultural University, Beijing 100083, China

[b] Center for Environment Research of Animal Health Culture, Beijing 100083, China

*Correspondence author: Email: libm@cau.edu.cn, hflucau@163.com

Abstract

For further usage of poultry manure anaerobic digestion effluent (PMADE), it is a wise chose to make PMADE a fertilizer. However, wastewater containing high ammonia nitrogen (NH_3-N) is generated during the manufacturing process of liquid fertilizer. In this study, the wastewater was pretreated by electrolysis to degrade NH_3-N concentration to a low contain, and then used for cultivation of two types of microalgae, *S. platensis* and *C. pyrenoidosa*. The NH_3-N, total organic carbon (TOC) and inorganic carbon (IC) were removed by electrolysis greatly. The removal ratio increased with time. The addition of sodium chloride (NaCl) had an effect on carbon and nitrogen removal after 2 h electrolysis. Here, wastewater treated by electrolysis was used to cultivate *C. pyrenoidosa* and *S. platensis* with the NH_3-N concentration of 500 mg/L and 100 mg/L respectively. The growth rate of *C. pyrenoidosa* in electrolysis treatment wastewater was 0.019g/d higher than that of *C. pyrenoidosa* in BG-11. *S. platensis* was suitable to grow in the wastewater pretreated by electrolysis with NaCl dosage. The cost of electrolysis treatment was below 0.25 Yuan/ L wastewater.

Keywords: Nutrients removal, animal waste, electrolysis treatment, poultry manure, digestate

1. Introduction

In 2009, with the total of poultry products of 130.6 million tons in China (FAO, 2009), there is a great amount of poultry manure needs to be treated. Poultry manure contains a plenty variety of elements, such as nitrogen, carbon and phosphorous, which could be recovered. Anaerobic digestion is an alternative way to reuse carbon resource in poultry manure. Moreover, harmful bacteria and virus in poultry waste could be killed by anaerobic fermentation (IEA Bioenergy, 2010). Nevertheless, anaerobic digestion effluent without proper treatment can cause eutrophication (Holm-Nielsen et al., 2009). To prevent eutrophication, poultry manure anaerobic digestion effluent from biogas projects is mostly likely used as liquid fertilizer to cultivate crops (Liedl et al., 2006), turning waste into value-added products. However, the concentrations of nitrogen (N) and phosphorus (P) in poultry manure anaerobic digestion effluent (PMADE) are high, which are not suitable for crop growth. Membrane has been chosen to turn the digested wastewater into fertilizer (Cathie et al., 2014). Still, during production of fertilizer by membrane, a special effluent is generated. This effluent contains high ammonia and phosphors concentration but low carbon concentration, which is hard to be disposed of by traditional activated sludge methods. Hence, to develop a suitable treatment method is emergent.

Electrochemical treatment has been used to treat high ammonia wastewater. Lei and Maekawa (2007) evaluated electric current, NaCl dosages and initial pH with Ti/Pt–IrO_2 electrode and found that the electric current and NaCl dosage had a considerably larger effect on the oxidization of ammonia. The results also indicated that a 1% NaCl dosage was almost saturated. Therefore, this dosage could be regarded as sufficient for ammonia removal (Ihara et al., 2006).Additionally, the color of anaerobic digested poultry manure wastewater could be cleared because of oxidation during the electrolysis treatment (Yetilmezsoy and Sakar, 2008).

As a main force to treat slightly polluted wastewater, microalgae can absorb the nutrients in the wastewater for their growth and multiplication. With a proper nutrition concentration in the wastewater, microalgae can take advantage of carbon, phosphors, ammonium nitrogen (NH_3-N) from the feedstock. Lots of researches have been done to cultivate microalgae with wastewater. Singh et al. (2011) employed microalgal system to treat effluent from poultry litter anaerobic digestion, with 6% (v/v) concentration of the effluent in deionized water, and found three types of algal strains that are suitable candidates for the wastewater treatment. Wang et al. (2015) used PMADE dilution pretreated by electrolysis to cultivate Chlorella vulgaris, and obtained a maximal algal biomass accumulation of 0.53 g/L. Thus moderate electrochemical treatment coupled with microalgae cultivation could enhance the pollutants removal for wastewater.

In this work, electrolysis treatment combining microalgae cultivation was used to treat PMADE to realize wastewater treatment and useful biomass production. The electrolysis treatment as the pretreatment was used to degrade the pollutants to a certain level for the next stage of microalgae cultivation.

2. Materials and Methods

2.1. Poultry manure anaerobic digestion effluent and microalgae strains

The PMADE was obtained from the biogas plant of Minhe Bio-technology Limited Company, Shandong Province, China. The wastewater (Table 1) from liquid biofertilizer making process was generated by ultrafiltration (UF) membrane and was called UFW.

Table 1. Characteristics of UFW.

Parameter	Value	Parameter	Value	Parameter	Value
TP(mg/L)	85	TOC(mg/L)	1,889	Ca (mg/L)	20.4
TN(mg/L)	4,500	IC(mg/L)	3,100	K (mg/L)	2,740.1
NH_3-N (mg/L)	2,590	pH	8.08	Na(mg/L)	678.2

The *S. platensis* (FACHB-314) and *C. vulgaris* (FACHB1067) were obtained from the Institute of Hydrobiology of the Chinese Academy of Sciences (Wuhan, China) and were cultivated in standard culture medium Zarrouk (Yoshitomo and David, 1995) and BG-11 (Rippka et al., 1979), respectively.

2.2. Experimental setup

2.2.1. Electrode board and electrochemical treatment

The material of electrode board was Pt/Ti- IrO_2, and the size was 15cm × 9.6cm. Batch experiments were conducted in a 1500 mL electrolysis tank and the wastewater volume was 1000mL.The distance of the two electrode material was 6cm. The direct currant voltage was set as 15V. The total effective electrode area was 96cm^2. The electrodes were connected to a DC power supply. The temperature was not controlled in this study. NaCl was chose as an alternative addition in the experiment, making the initial weight ratio of NaCl to NH_3-N was 4:1. The UF1 was UFW with no NaCl addition; and UF2 was UFW with 15.2 g/L NaCl addition. The NH_3-N concentration of the water was measured every 0.5 h. During electricity treatment, 10 mL water was taken every 0.5 h to determine the NH_3-N concentration of the water. When the concentration of NH_3-N was around 100 mg/L or 500 mg/L, the electrolysis pretreatment was stop and the water was used for microalgae cultivation.

2.2.2. Microalgae cultivation

Batch experiments were conducted in 250 mL flasks. The total volume of cultivation broth was 160mL. The volume ratio of wastewater to microalgae broth was 1:1. The *S. platensis* was cultivated in the wastewater with the NH_3-N concentration was around 100 mg/L, and *C. pyrenoidosa* was cultivated in the wastewater with the NH_3-N was approximately 500 mg/L. The control experiment was with Zarrouk and BG11 cultivate medium, respectively. The microalgae were cultivated under 28°C with a daily lighting schedule of 12 h on and off. The light intensity was 170 μmol photons m^{-2} s^{-1}. The pH value of water culturing *C. pyrenoidosa* was adjusted to 7.1 ± 0.1 by 1 mol/L HCl and 1 mol/L NaOH every day. TP, NH_3-N, TOC, and dry cell biomass were detected every three days.

2.3. Analysis methods

The TP and NH_3-N were tested according to the Chinese State Environmental Protection Agency Standard Methods (SEPA, 2002). Dry cell weight of microalgae was evaluated by 0.45 μm glass microfiber filters at 105 °C for 24 h. Total organic carbon (TOC) values for molecular weight distribution were tested by a Total Organic-Carbon Analyzer TOC-VCPN (Shimadzu Co. Company, Japan). The pH and light intensity was monitored using a PSH-3 pH meter (Shanghai Precision and Scientific Inc., China) and a LI-250A light meter (LI-COR Inc., Canada), respectively. The whole experiment was taken in duplicate, and the results were averaged.

3. Results and Discussion

3.1. Influence of NaCl dosage on current and voltage change

The current and voltage change of DC power supply during electrolysis treatment without NaCl and with NaCl were shown in Figure 1.

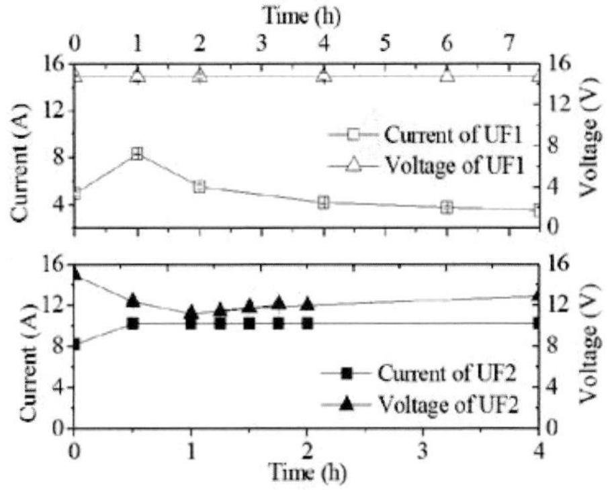

Figure 1. The current and voltage change of DC power supply during electrolysis treatment.

As mentioned above in section 2.2 experiment set-up, the DC voltage of electrolysis process was set as 15v. During the electrolysis treatment of UF1, the voltage was constant and the current had an increase during the first hour and dropped thereafter. During the electrolysis treatment, the voltage was not the constant as settled in some situation. When NaCl was added in UFW, the voltage of the electrolysis process changed automatically from 15V turning into smaller, and the currant was bigger than UF1. That might be the NaCl dosage existed in the UF2 made the electric current disorder. From Figure 1, the electric current of UF1 was 3–5 A for most of the time and 8A for 1 h. The voltage was 15V and the power was in Eq (1).

$$W \approx 15V \times 8A \times 1h + 15V \times 4A \times (T-1)h \approx 60Wh + 60Wh \times T \quad (1)$$

When it came to UF2, the electric current was 10.2A for most of the time and 8A for about 0.5 h. The voltage was 11–15V and the power was in Eq (2).

$$W \approx 15V \times 8A \times 0.5h + 12V \times 10.2A \times (T-0.5)h \approx 120Wh \times T \quad (2)$$

So, when the electrolysis time was 1h, the electricity consumption was 120 Wh regardless whether NaCl was added or not; when the electrolysis time was over 1 h, NaCl addition required more electricity than when no NaCl was added. When treating high NH_3-N concentration wastewater, over an hour was needed. So NaCl addition would increase electricity consumption. Here, the power of UF2-4, UF1-7.5, UF2-3 and UF1-4 cost was about 510, 480, 300, and 360 Wh, respectively. The price of electricity is about 0.5 RMB Yuan/1000 Wh. Then electrolysis treatment costs no more than 0.25 Yuan.

3.2. Effect of electrolysis process on nutrients removal

The water characteristics after the electrolysis treatment of UF1for 4 h (UF1-4) and for 7.5 h (UF1-7.5), as well as the electrolysis treatment of UF2 for 3 h (UF2-3) and for 4 h (UF2-4) were as shown in Table 2.

Table 2. Characteristics of the treated water.

Treatment	UF2-4	UF1-7.5	UF2-3	UF1-4
Final volume (mL)	640	630	770	755
NH_3-N (mg/L)	91	107	525	528
TP (mg/L)	218	241	201	207
TOC (mg/L)	1399	1723	1374	1863
IC (mg/L)	731	752	866	959
pH value	9.48	9.85	9.12	9.15

The temperature of the electrolyte solution increased along with the electrolysis process and reached up to 78°C. When the temperature rose to a certain degree, NH_3-N and Cl_2 overflew from the water. The volume of the wastewater decreased greatly, and about 60% (v/v) of the initial wastewater was left. This mainly due to high temperature during electrolysis process, part of H_2O was converted to vapor and part of H_2O broke down into H_2 and O_2.

When the electrolysis process ended, the pH value of the water increased to 9.15 - 9.85, higher than the initial pH value of the wastewater. One reason for the pH value of the water increasing was that H_2O was consumed and OH^- was produced by electrolysis anode. The reaction equation of H_2O as oxide material was in Eqs. (3 and 4) (Panizza and Cerisola, 2001).

$$O_2 + H_2O + 2e^- \rightarrow HO_2^- + OH^- \text{ (in alkaline solution)} \tag{3}$$

$$O_2 + 2H^+ + 2e^- \rightarrow H_2O_2 \text{ (in acidic solution)} \tag{4}$$

Here, the reaction equation of oxide material was Eq. (3). Another reason for pH value increasing might be CO_2 stripping from wastewater followed Eq. (5), which caused a shift of pH to the alkaline side.

$$HCO_3^- \leftrightarrow CO_2 + OH^- \tag{5}$$

When the electrolysis process began without NaCl, bubbles were generating at the surface of electrodes and they were bigger than those small gas bubbles with diameters ranging from 22 to 50μm in the research of Ketkar et al. (1991). That partly due to TOC/IC turned into carbon dioxide, and NH_3 turned into nitrogen gas. During the first hour, as the electrolysis time increased the bubbles were bigger. The bubbles began to degrade 1.5 h later. Yet the bubbles generated faster with NaCl addition in UF2 during the electrolysis process than in NaCl absence. The biggest bubbles in UF2 occurred after 0.5 h, while that of UF1 was after 1.5 h. So NaCl addition would accelerate the process of electrolysis.

During the electrolysis, IC decreased dramatically and TOC decreased slightly, so under electrolysis, IC was easier to react than TOC. The IC removal efficiency was high (69%-76%), mainly because of CO_2 stripping from wastewater. Lower TOC removal efficiency (28%-47%) was caused by the low Cl_2 in the free chlorine in the pH range (8.08-9.48). The result was comparable to that obtained by Lei et al. (Lei and Maekawa, 2007). The reactions of carbon were given as follow in Eqs. (6) to (9) (Szpyrkowicz et al., 2001; Lei and Maekawa, 2007).

$$CO_3^{-2} + 2H^+ \rightarrow HCO_3^- + H^+ \rightarrow CO_2 + H_2O \text{ (Inorganic carbon)} \tag{6}$$

$$R(CHO) + MO_X[\cdot OH] \rightarrow MO_X + CO_2 + H^+ + e^- \tag{7}$$

$$(CHO)_n + O_2 \rightarrow CO_2 + H_2O_2 \tag{8}$$

$$(CHO)_n + Cl_2 \rightarrow CCl_4 + HCl + H_2O \text{ (Organic carbon)} \tag{9}$$

$$NH_3 + H_2O \leftrightarrow NH_4^+ + OH^- \tag{10}$$

For a given pH and temperature, two forms of ammonia established an equilibrium following the equation Eq. (10) (Emerson et al., 1975). The NH_3-N concentration of the water decreased with the time of the electrolysis process. NH_3-N elimination could be attributed to electro-oxidation process (Lin and Wu, 1996), and the main products of NH_3-N were nitrogen gas and nitrate (Kapałka et al., 2010). Li and Liu (2009) found that approximately 88% of NH_3-N was removed and part of NH_3-N was released as N_2. The percentage of ammonia existed in ionized form as a function of pH and temperature, it was apparent that as the pH (below 10) increases above 7 ammonia existing in its un-ionized form (NH_3) also increases (Lin and Wu, 1996). Of these two forms of ammonia, the un-ionized one was much easier to oxidize (Benefield et al., 1982). Here, the pH value of the water was higher than 9 and lower than 10, so the ammonia nitrogen oxidizes mainly followed Eq. (11) (Chen et al., 2007). In addition, ammonia could also be removed by ammonia stripping in the pH rage of 9.0 and 11.6.

$$2NH_3 + 6OH^- \leftrightarrow N_2 + 6H_2O + 6e^- \quad (11)$$

The temperature was not controlled during the electrolysis process and the final volume decreased nearly by 1/3; the whole phosphonium ion in the electrolysis tank decreased, while the phosphonium ion in per liter increased. The TP concentration of water increased during the electrolysis process, which might because the amount of total phosphorus and the water-soluble phosphorus in the electrolysis tank did not change greatly during the electrolysis process, while the volume of the wastewater was decreasing. Additionally, the phosphate ions adhere to the electrode could also decrease the removal rate of NH_3-N concentration in the water (Chen, 2008).

3.3. Influence of NaCl dosage on NH_3-N removal

As it shown in Figure 2, the NH_3-N concentrations of the electrolysis process changed greatly. The difference between the NH_3-N concentrations of UF1 and UF2 was not big during the first 2.5 h; after this period, the difference was evident. It took 7.5 h for the NH_3-N concentration of UF1 below 100 mg/L and 3.5h for UF2. The pH value of the water was higher than 9 and lower than 10, so the ammonia nitrogen oxidizes mainly followed equation Eq. (11). Some Cl^- might already exist in UFW, NaCl addition could not increase Eq. (12) reaction at the beginning, until 2.5 h later the additional Cl^- did have some effect on the NH_3-N removal. The additional NaCl would accelerate the speed of NH_3-N removal. The electrolysis reaction with chlorine as oxide material was list in Eqs. (13) and (14) (Lin and Wu, 1996).

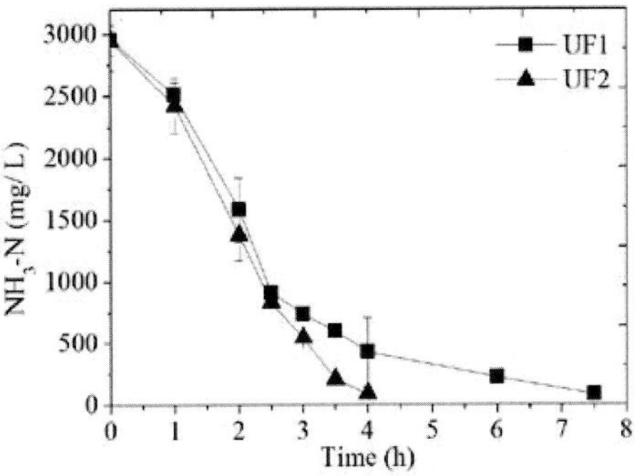

Figure 2. The NH_3-N concentrations of wastewater during the electrolysis process.

$$2Cl^- + H_2O\text{-}e^- \rightarrow Cl_2(aq) + H_2O \rightarrow HClO + Cl^-\text{-}e^- \rightarrow OCl^- + H^+ \quad (13)$$

$$2NH_4^+ + 3HClO \rightarrow N_2 + 3H_2O + 5H^+ + 3Cl + 2e^- \quad (14)$$

So adding NaCl or was determined by the pH value and Cl- concentration of water body.

3.4. Effect of wastewater pretreated by electrolysis on microalgae cultivation

The UF1-4 was used for cultivating *C. pyrenoidosa* and was called UF1-4-C; UF1-7.5 was used for cultivating *S. platensis* and was called UF1-7.5-S; UF1-7.5 was used for cultivating *C. pyrenoidosa* and was called UF2-3-C; UF2-4 was used for cultivating *S. platensis* and was called UF2-4-S. The biomass growth and the nutrients (TP, NH_3-N, TOC and IC) concentration change in culture media for *S. platensis* was shown in Figure 3.

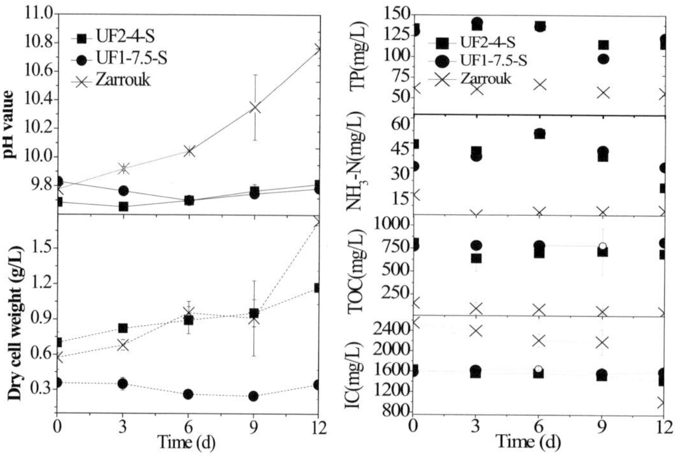

Figure 3. Biomass growth of *S. platensis* and the nutrients change in culture media. Left is the pH value and DCW of *S. platensis*; right is TP, NH_3-N, TOC and IC in culture media.

The pH value and dry cell weight (DCW) of Zarrouk had a great increase from 10.0 to 10.5 and 0.3 g/L to 1.6 g/L respectively. The pH value and DCW of UF2-4-S increased a little and increased from 0.3 g/L to 1.2 g/L respectively. This might due to the lack of IC and enrich of TP, and the initial pH value was higher than cultivate medium. The pH value and DCW of UF1-7.5-S almost did not change. That may be the osmotic pressure of UF1-7.5 was not suitable for *S. platensis* to grow. So cultivating *S. platensis* not only need a low NH_3-N concentration, but also salinity density for a suitable osmotic pressure (Çelekli and Yavuzatmaca, 2009). With the growth of *S. platensis*, the IC was consumed and followed Eq. (15) (Ruiz-Martinez A, 2012). So with the consume of $CO_3^{2-}$and HCO_3^-, the concentration of H^+ decreased and the pH value of the water became higher, leading to the biomass weight becoming heavier.

$$CO_3^{-2} + 2H^+ \leftrightarrow HCO_3^- + H^+ \leftrightarrow CO_2 + H_2O \rightarrow (CH_2O) + O_2 \tag{15}$$

The TP concentration of UF2-4-S and UF1-7.5-S had a little increase in the first 6 days and decreased later. The NH_3-N removal rate of the UF2-4-S and UF1-7.5-S was about 20%. The NH_3-N of the UF2-4-S and UF1-7.5-S increase a high concentration at 6th day and decreased later, the NH_3-N removal rate of the UF2-4-S and UF1-7.5-S is about 30%. The TOC of Zarrouk decreased little. While the IC of Zarrouk decreased greatly, the IC removal efficiency is about 60%. The IC, TOC of UF2-4-S and UF1-7.5-S did not change so much, and the biomass UF2-4-S and UF1-7.5-S.

The biomass growth and the nutrients (TP, NH_3-N, TOC and IC) concentration change in culture media for *C. pyrenoidosa* was shown in Figure 4. The pH value of culture medium for *C. pyrenoidosa* was adjusted by 1 mol/L HCl and 1 mol/L NaOH, BG11 was easier than UF2-3-C and UF1-4-C to adjust to neutral state, which might be the chemical composition of BG11 was inorganic, while UF2-3-C and UF1-4-C were organic which had a strong ability to buffer.

The TP and NH_3-N concentration of UF2-3-C, UF1-4-C and BG11 had a little increase, the IC of UF2-3-C and UF1-4-C had a great decrease, and the IC removal rate of UF2-3-C and UF1-4-C was over 60%.The dry cell weights of UF2-3-C and UF1-4-C was increasing all along the experiment period. The growth rate of UF2-3-C and UF1-4-C was 0.029g/d and 0.025g/d respectively and higher than that of BG11 with 0.01g/d. The algae from wastewater were grown better than BG11, and this was because mix-trophic cultivation was better than photographic. And many studies about microalgae cultivation have been done to using high TOC contained wastewater by mix-trophic cultivation (Lee, 2004).

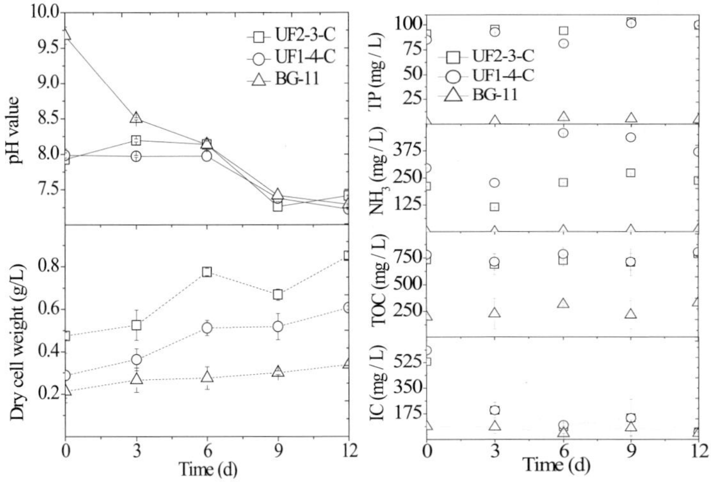

Figure 4. Biomass growth of *C. pyrenoidosa* and the nutrients change in culture media. Left is the pH value and DCW; right is TP, NH_3-N, TOC and IC of cultivation medium.

4. Conclusions

Electrolysis treatment is a feasible way to treat polluted wastewater, since after electrolysis treatment the TC, NH_3-N, and TP of the wastewater have a decrease. The NaCl dosage would have an effect on the electrolysis, while its effect is not so substantial when there were oxidation irons in water. The cost that per liter wastewater needed in electrolysis treatment is not higher than 0.25 Yuan. It is possible to cultivate *C. pyrenoidosa* and *S. platensis* in electrolysis pretreated wastewater. The growth rate of *C. pyrenoidosa* in electrolysis treatment wastewater was 0.019g/d higher than in BG-11.

Acknowledgements

This work was supported by the Program of Outstanding Talents and Innovative Research Teams in Agriculture (2011-049), National Natural Science Foundation of China (51308535), and Chinese Universities Scientific Fund (2015SYL002). The authors want to give thanks to the technical sponsor and wastewater from Minhe Animal Husbandry Incorporated Company. Many thanks to Jiqin Ni (Purdue University), Raquel Lamartin (China Agricultural University) and Gibson for improving the paper.

References

Benefield, L. D., Judkins, J. F., Weand, B. L., 1982. *Process Chemistry for Water and Wastewater Treatment*. Englewood Cliffs, NJ: Prentice-Hall.

Çelekli, A. Yavuzatmaca, M., 2009. Predictive modeling of biomass production by Spirulina platensis as function of nitrate and NaCl concentrations. Bioresource Technol., 100, 1847–1851.

Chen, J., Shi, H., and Lu, J., 2007. Electrochemical treatment of ammonia in wastewater by RuO2–IrO2–TiO2/Ti electrodes. J. Appl. Electrochem., 37 (10), 1137–1144.

Chen, J., 2008. Study and Application of Electrochemical Oxidation Technology for Ammonia Removal. Doctoral Thesis, Beijing: Tsinghua University, Department of Environmental Science and Engineering.

Emerson, K., Russo, R.C., Lund, R.E., Thurston, R.V., 1975. Aqueous ammonia equilibrium calculations: effect of pH and temperature. Journal of the Fisheries Board of Canada, 32 (12), 2379–2383.

FAO, 2009. Food and Agricultural Commodities Production: 2009-2010. FAOSTAT. Rome: Food and Agriculture Organization (FAO) of the United Nations.

Holm-Nielsen, J.B., Al Seadi, T., and Oleskowicz-Popiel, P., 2009. The future of anaerobic digestion and biogas utilization. Bioresource Technol., 100 (22), 5478-5484.

IEA Bioenergy, 2010. Utilisation of digestate from biogas plants as biofertiliser. www.biogas.org.nz/Publications/Resources/utilisation-of-digestate-biogas-to-biofertiliser.pdf. Accessed September 1, 2015.

Ihara, I., Umetsu, K., Kanamura, K., Watanabe, T., 2006. Electrochemical oxidation of the effluent from anaerobic digestion of dairy manure. Bioresource Technol., 97 (12), 1360-1364.

Kapałka, A., Cally, A., Neodo, S., Comninellis, C., Wächter, M., and Udert, K.M., 2010. Electrochemical behavior of ammonia at Ni/Ni (OH)2 electrode. Electrochem. Commun., 12 (1), 18 –21.

Ketkar, D.R., Mallikarjunan R., Venkatachalam S., 1991. Electroflotation of quartz fines. Int. J. Miner. Process. 31, 127–138.

Lee, Y.K., 2004. Algal nutrition heterotrophic carbon nutrition. *Handbook of Microalgal Culture: Biotechnology and Applied Phycology*. Ed. A. Richmond. Oxford: Blackwell Science Ltd. 116–124.

Lei, X. Maekawa, T., 2007. Electrochemical treatment of anaerobic digestion effluent using a Ti/Pt–IrO2 electrode. Bioresource Technol., 98 (18), 3521–3525.

Li, L. Liu, Y., 2009. Ammonia removal in electrochemical oxidation: mechanism and pseudo-kinetics. J. Hazard. Mater., 161 (2), 1010–1016.

Liedl, B.E., Bombardiere, J., Chaffield, J.M., 2006. Fertilizer potential of liquid and solid effluent from thermophilic anaerobic digestion of poultry waste. Water Sci. Technol., (53), 69–79.

Lin, S.H. Wu, C.L.,1996. Electrochemical removal of nitrite and ammonia for aquaculture. Water Res., 30 (3), 715–721.

Panizza, M. Cerisola, G., 2001. Removal of organic pollutants from industrial wastewater by electrogenerated Fenton's reagent. Water Res., 35 (16), 3987–3992.

Rippka, R., Deruelles, J., Waterbury, J.B., Herdman, M., Stanier, R.Y., 1979. Generic assignments, strain histories and properties of pure cultures of cyanobacteria. Journal of General Microbiology, 111, 1–61.

Ruiz-Martinez, A., Garcia, N.M., Romero, I., Seco, A., Ferrer, J., 2012. Microalgae cultivation in wastewater: nutrient removal from anaerobic membrane bioreactor effluent. Bioresource Technol., 126, 247–253.

SEPA, C., 2002. Water and Wastewater Analyzing Methods, 4th ed. Beijing: China Environmental Science Press.

Singh, M., Reynolds, D.L., Das, K.C., 2011. Microalgal system for treatment of effluent from poultry litter anaerobic digestion. Bioresource Technol., 102 (23), 10841-10848.

Szpyrkowicz, L., Kelsall, G.H., Kaul, S.N., Faveri, M.D., 2001. Performance of electrochemical reactor for treatment of tannery wastewaters. Chem. Eng. Sci., 56 (4), 1579–1586.

Wang, M., Wu, Y., Li, B., Dong, R., Lu, H., Zhou, D., Cao, W., 2015. Pretreatment of poultry manure anaerobic-digested effluents by electrolysis, centrifugation and autoclaving process for Chlorella vulgaris growth and pollutants removal. Environ. Technol., 36 (7), 837–843.

Yetilmezsoy, K. Sakar, S., 2008. Improvement of COD and color removal from UASB treated poultry manure wastewater using Fenton's oxidation. J. Hazard. Mater., 151 (2), 547–558.

Yoshitomo, W., David, O.H., 1995. A new strategy for lipid production by mix cultivation of Spirulina platensis and Rhodotorula glutinis. Energy Conversion, 36, 721–724.

Two-stage Microalgae Cultivation Using Diluted Effluent of Anaerobic Digestion to Remove Ammonia Nitrogen and Phosphorus

Xinfeng Wang [a,b], Haifeng Lu [a,b,*], Li Zhang [a], Yu Zhao [a], Baoming Li [a,b,*]

[a] Laboratory of Environment-Enhancing Energy (E2E), Key Laboratory of Agricultural Engineering in Structure and Environment, Ministry of Agriculture;
College of Water Resources and Civil Engineering, China Agricultural University, Beijing 100083, China

[b] Center for Environment Research of Animal Health Culture, Beijing 100083, China

* Corresponding author. Email: limb@cau.edu.cn, hflucau@163.com

Abstract

Microalgae can be used not only to produce many products for people, such as food, bio-oil, and health care products, but also to purify water by absorbing nutrients from the waterbody. In this study, two kinds of microalgae were used to clean ammonia nitrogen and phosphorus existing in wastewater through a two-stage cultivation process. In the first-stage cultivation, *Chlorella vulgaris* (*C. vulgaris*) was cultivated under a light intensity of 170 μmol m^{-2} s^{-1} and cycle of 12 h light and 12 h darkness in a diluted effluent from anaerobic digestion of poultry manure at 28°C. The *C. vulgaris* was then harvested by ultrafiltration (UF) membrane. In the second-stage cultivation, the water harvested from UF membrane was used to cultivate *Spirulina platensis* (*S. platensis*) by adding Zarrouk. With a small volume of algae broth, harvesting algae with membrane costs less energy, but with a large amount of microalgae loss. At a lower ammonia concentration (8–16 dilution ratio), *S. platensis* could grow well and remove ammonia completely. The two-stage cultivation of microalgae not only produced more algae biomass, but also reduced total phosphorus (TP) and total nitrogen (TN) in wastewater.

Keywords: Microalgae harvest, ultrafiltration membrane, anaerobic digestion effluent, nutrient removal, *C. vulgaris*, *S. platensis*

1. Introduction

Consumption of meat and eggs has been increasing with improved living standards (FAO, 2006). From 1980 to 2000, the annual rate for meat and egg productions increased for about 7.7% and 10.9%, respectively (FAO, 2005). The total number of chickens and annual chicken manure generated in China was 7658.7 million and 279.5 Mt, 8047.2 million and 293.7 Mt, 8435.8 million and 307.9 Mt in 1999, 2005, 2010, respectively (Li et al., 2005). Due to the large amount of poultry manure, anaerobic fermentation has been utilized to treat the manure. Besides methane generation, the anaerobic fermentation process results in effluents containing organic matter, nitrogen and phosphorus elements that can be reused by living organisms (Cheung and Wong, 1981).

Due to its high growth rates, protein and fat contents, microalgae can be used for both healthy products (Miguel et al., 2006; Spoalore et al., 2006) as well as biofuel (Mata et al., 2010). Currently, researchers are focusing on microalgae growth and wastewater treatment, in a win-win strategy to resolve the energy and environmental challenges. Wastewater contains a lot of nutrients such as nitrogen, carbon, phosphorous for microalgae growth. Zhou et al. (2012) developed a sequential two-stage microalgae cultivation process for economically viable and environmentally friendly production of both renewable biofuel and high-value microalgae used as animal feed and simultaneous animal wastewater treatment. After the algal biomass was harvested, the residual supernatant was reused as media to culture other kinds of microalgae.

With a proper nutrient concentration in the wastewater, microalgae can grow and multiply quickly. Usually, the effluent from the anaerobic digestion needs to be diluted for microalgae cultivation, thus more water is consumed during the dilution process. However, different types of microalgae vary in the ability to withstand the wastewater, e.g., *Spirulina* has a faster growth rate and is often chosen to treat slightly polluted wastewater (Chang et al., 2013) or highly polluted wastewater with high dilution ratio. On the other hand, Chlorella has a slower growth rate, but strong stability in wastewater, hence, it is often chosen to treat wastewater containing more nutrients or with low dilution ratio.

In this work, two stages of microalgae cultivation were used to treat effluent from anaerobic digestion of poultry manure (PMADE) for wastewater treatment and produce useful biomass. In the first stage, *C. vulgaris* cultivation was used to degrade the pollutants present in wastewater to a lower level. During this stage, the algae harvesting method of UF membrane was also investigated. In the second stage, *S. platensis* cultivation utilized the nutrients in the recycled water to gain more biomass.

2. Materials and Methods

2.1. Wastewater

The PMADE was obtained from a biogas plant of Minhe Bio-technology Limited Company, Shandong Province, China. The wastewater (Table 1) from liquid bio-fertilizer production process using PMADE was generated by ultrafiltration (UF) membrane and is hereby referred to as UFW.

Table 1. Characteristics of UFW.

Item	Value	Item	Value	Item	Value
TP (mg/L)	85	Ca (mg/L)	20.35	Cu (μg/L)	415.2
TN (mg/L)	4,500	B (mg/L)	5.33	Mn (μg/L)	83.1
NH_3-N (mg/L)	2,590	Co (mg/L)	2.99	Mo (μg/L)	43.7
TOC (mg/L)	1,889	Fe (mg/L)	2.97	Cr (μg/L)	109.1
IC (mg/L)	3,100	K (mg/L)	2740	As (μg/L)	130.5
TC (mg/L)	4,989	Mg (mg/L)	3.84	Se (μg/L)	10.1
pH value	8.08	Na(mg/L)	678	Cd (μg/L)	25.0
		Zn (mg/L)	1.9	Pb (μg/L)	4.7
		Al (mg/L)	4.34	Hg (μg/L)	0.38

2.2. Microalgae

The *S. platensis* (FACHB-314) and *C. vulgaris* (FACHB1067) were obtained from the Institute of Hydrobiology of the Chinese Academy of Sciences in Wuhan, China and were cultivated in standard culture medium Zarrouk (Yoshitomo and David, 1995) and BG-11 (Rippka et al., 1979), respectively.

2.3. Experimental setup

There were two stages in the algal cultivation system. In the first stage, *C. vulgaris* was cultivated in wastewater with high NH_3-N concentration. In the second stage, *S. platensis* was cultivated in wastewater with low NH_3-N concentration. A schematic diagram of algal cultivation system with membrane is shown in Figure 1.

During the first step of *C. vulgaris* cultivation, the cultivation tank was first filled with diluted wastewater, and then inoculated with 0.08 g/L *C. vulgaris*. The microalgae were cultivated for 14 days under 28°C. The light intensity was 170 μmol m^{-2} s^{-1} and a daily lighting schedule was 12 hours on/off. Batch experiments were taken out in 2000 mL flasks with 160 mL microalgae broth, 200 mL UFW and 1240 mL distilled water. The pH value was adjusted to 7.1 ± 0.1 by 1 mol/L NaOH and 1 mol/L HCl. When the microalgae were in the steady growth phase, control valve (2) was opened. The mixture of *C. vulgaris* and wastewater was flowed through pump (3), flow meter (4), pressure meter (5) and into UF membrane (6), before the filtration began.

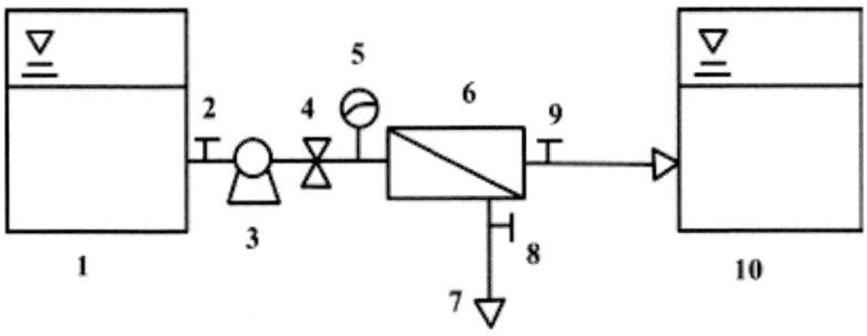

Figure 1. Schematic diagram of algal cultivation system with membrane.

1. Feed solution for *C. vulgaris* cultivation; 2, 8, and 9. Control valve; 3. Pump, 4. Flow meter; 5. Pressure meter; 6. UF membrane; 7. Algae; 10. Dilution for *S. platensis* cultivation.

During the first stage, microalgae were harvested by the UF membrane (6) and the water was recycled. The UF membrane was made of Hydrophilic PES (Chaoyu Company, Guangzhou, China). The molecular cutoff weight as specified by the manufacturer was 10,000 Dalton. The initial flux of the UF membrane system was 10 L/h. Initial pressure was 0.08 Mpa and the working pressure was controlled below 0.3 Mpa. After the microalgae were all harvested, back wash was conducted using distilled water. The flux of the backwash was 30L/H.

In the second stage, *S. platensi* was cultivated. The recycled water was diluted and used for *S. platensi* cultivation. The effect of different NH_3-N concentrations and dilution ratios on *S. platensis* growth was as presented in Table 2. The NH_3-N×4, NH_3-N×8, NH_3-N×10, NH_3-N×16 was used as the inoculum and was inoculated with 50mL algae broth. The wastewater volume to total volume ratio were 1:4, 1:8, 1:10, 1:16, respectively. Blank was inoculated with 110mL culture medium and 50mL algae broth. The cultivation conditions of microalgae were the same as in the first stage of microalgae cultivation. Batch experiments were conducted with a volume of 160mL in 250 mL flasks, and the experiment design was as shown in Table 2. The total phosphorus (TP), ammonia nitrogen (NH_3-N), total nitrogen (TN), and dry cell biomass were measured every three days.

Table 2. *S. platensis* cultivation by different NH_3-N concentration

Item	Wastewater volume (mL)	Culture medium volume (mL)
Blank	0	110
NH_3-N ×4	40	70
NH_3-N ×8	20	90
NH_3-N ×10	16	94
NH_3-N ×16	10	100

2.4. Analysis method

The TP and NH_3-N were tested using the Chinese State Environmental Protection Agency Standard Methods (SEPA, 2002). Dry cell weight of microalgae was evaluated by 0.45 μm glass microfiber filters at 105°C for 16 hours. Total organic carbon (TOC) values for molecular weight distribution were tested by a Total Organic-Carbon Analyzer TOC-VCPN (Shimadzu Corporation Company, Japan). The pH and light intensity were monitored using a PSH-3 pH meter (Shanghai Precision and Scientific Inc., China) and a LI-250A light meter (LI-COR Inc., Canada), respectively. The whole experiment was conducted in duplicate, and the results are the average values.

3. Results and Discussion

3.1. First stage of *C. vulgaris* cultivation

The initial (d 0) and final (d 14) characteristics of the culture water (NH_3-N, TOC, IC and TP) and dry cell weight of microalgae in the first stage were as shown in Figure 2.

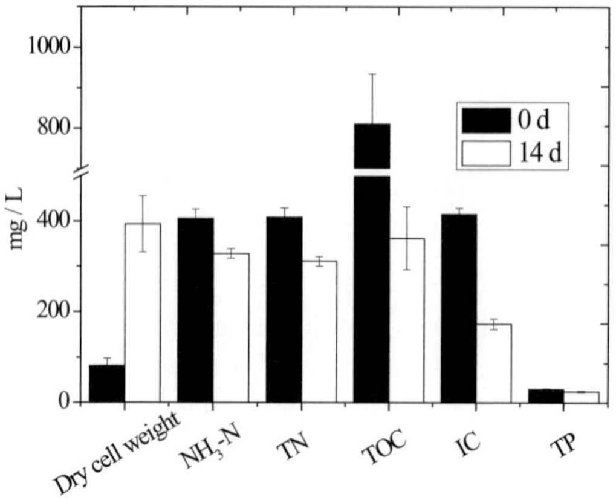

Figure 2. The initial (0 d) and final (14 d) characteristics of culture water (NH_3-N, TOC, IC and TP) and dry cell weight of microalgae. The error bars represent the standard deviation.

The pH value was decreased from 8.5 to 7.6. The removal rates of TP, NH_3-N, and TN were 17%, 19% and 24%, respectively. The dry cell weight was increased 3.8 times, from 81 to 394 mg/L. The TOC and IC removal rates were 55% and 58%. Both TOC and IC were removed. This implies that the microalgae were cultured mixotrophically. For *C. vulgaris*, the photosynthetic and heterotrophic maximum specific growth rates were comparable (Lee, 2004), and mixotrophic maximum specific growth rate was almost equal to the sum of the former two. Therefore, using wastewater as culture medium for *C. vulgaris* cultivation was faster.

3.2. *C. vulgaris* collection through UF membrane

When the microalgae were in the plateau growth stage, the microalgae were harvested by UF membrane introduced in section 2.2. The microalgae backwash used 1500mL water, 0.06 Mpa, 110s. The results of microalgae harvested and water recycled by UF membrane were as shown in Figure 3. The efficiency of the microalgae harvest was 81%. About 96.8% water was recycled from the UF membrane.

The pore size of UF membrane was 10,000 Dalton. Part of phosphorus iron and ammonia nitrogen was trapped. The NH_3-N concentrations before UF membrane, after UF membrane and backwash were 328 mg/L, 227 mg/L, and 29 mg/L, respectively. The TN concentration before UF membrane, after UF membrane and backwash was 311 mg/L, 224 mg/L, and 90 mg/L, respectively. The TP concentrations before UF membrane, after UF membrane and backwash was 24.6 mg/L, 22.8 mg/L, and 9 mg/L, respectively.

Mackay and Salusbury (1988) thought membrane was a better choice when the algae broth volume was below 20,000 L. The UF membrane enriched *C. vulgaris* for 3 times and the final concentration of the algae reached 1.2 g/L in this study. It was very low compared to the results of Zhang et al. (2010), who enriched Scenedesmus quadricauda 150 times by UF membrane, and the final concentration of the algae reached 154.85 g/L. This difference might be due to the fact that the initial concentration of dry cell weight and the harvest operation were different.

Besides membrane, chemical coagulation has often been used for microalgae harvest. Although chemical coagulation is cheap, addition of chemical material would increase ash content by up to 7 times (Hu et al., 2013; Tian et al., 2015) and cause algae cells to break (Gonzalez-Torres et al, 2014). High ash content or a broken algae cell would have negative effects on biocrude oil production in hydrothermal liquefaction (Kuo, 2010). In particular, microalgae have higher ash content, which results in lower yields of biocrude oil (Anastasaki and Ross, 2011; Li et al., 2012). Using membrane in microalgae harvest could avoid chemical material addition in algae, so that the ash of algae cell could not be added or the cell of the algae would not be broken.

During the membrane operation, microalgae often attached to the membrane and caused flux decline. To clean a fouled membrane, there was a need for backwash which consumes more energy. Usually, more water was needed to wash the algae from the membrane. When the membrane was fouling heavily, chemical cleaning was often applied. To avoid fresh water usage, we recommend the use of algae broth to wash the membrane.

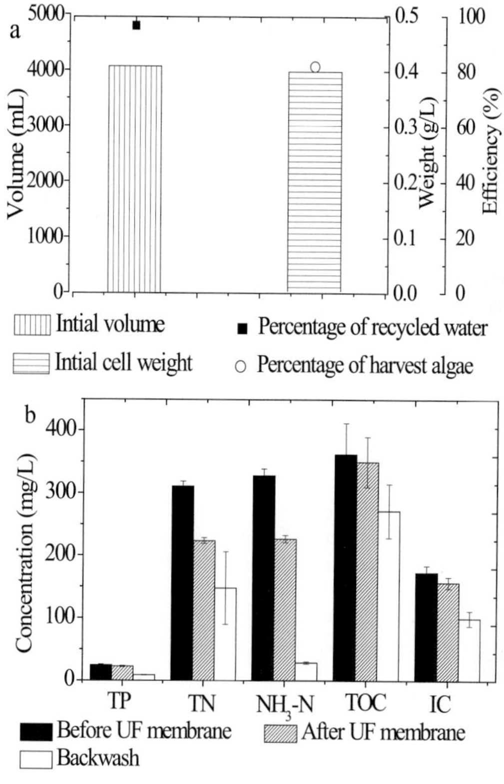

Figure 3. The results of microalgae harvest and water recycled with UF membrane. a. Efficiency of the harvested algae and recycled water; b. water characteristics before and after UF membrane and backwash

3.3. Influence of NH_3-N concentration on *S. platensis* growth in second stage

After the *C. vulgaris* was harvested by UF membrane, the water was recycled to cultivate *S. platensis* with different dilution ratios. The influence of NH_3-N concentration on *S. platensis* growth was shown as in Figure 4.

The initial dry cell weight of *S. platensis* in the water was 0.16 g/L. After the 15-day cultivation, the dry weight of *S. platensis* was 5, 3.45, 3.15, 2.95 and 0.18 g/L for the blank, NH_3-N×16, NH_3-N×10, NH_3-N×8 and NH_3-N×4, respectively. The growth rates of *S. platensis* from high to low ranked as 0.22, 0.20, 0.19, and 0.001 g/d/L in the wastewater with ammonia nitrogen concentration of NH_3-N×16, NH_3-N×10, NH_3-N×8, and NH_3-N×4, respectively. The volumetric productivity were comparable to that of Yuan, et al. (2011) which was (0.16-0.28) g/d/L with a temperature of 20°C and approximately 400 μmol m^{-2} s^{-1} light intensity, and with 61.8 mg/L ammonia.

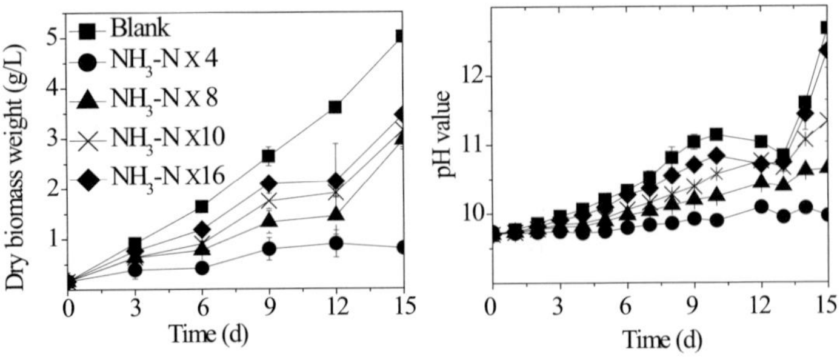

Figure 4. The dry cell weight and pH value of *S. platensis*

All the culture mediums experienced increase in pH value. Additionally, as the biomass obtained in the water increased, the pH value of the water was also elevated. The IC was consumed as a result of *S. platensis* growth (Abelson and Hoering, 1961; Coleman and Colman, 1981). The $CO_3^{2-}$and HCO_3^- was also utilized. This led to a decrease in the concentration of H^+ decreased; hence, the pH value of the water increased and the biomass weight increased. The chemical reaction followed Eq. (1) (Ruiz-Martinez et al., 2012).

$$CO_3^{-2} + 2H^+ \leftrightarrow HCO_3^- + H^+ \leftrightarrow CO_2 + H_2O \rightarrow (CH_2O) + O_2 \quad (1)$$

The NH_3-N concentrations of NH_3-N×16, NH_3-N×10, NH_3-N×8, and NH_3-N×4 were lowest on the day 6 when almost 100% of the NH_3-N was removed. *S. platensis* grew well and eliminated 100% of the ammonia in the NH_3-N×8 culture medium. This volume was higher than that of Canizares and Dominguez (1993), who reported a maximal removal of nitrogen and phosphorus by *Spirulina* occurred in 50% diluted swine waste stabilized by aeration, corresponding to about 42 mg of ammonia nitrogen per liter of nitrogen (Figure 5).

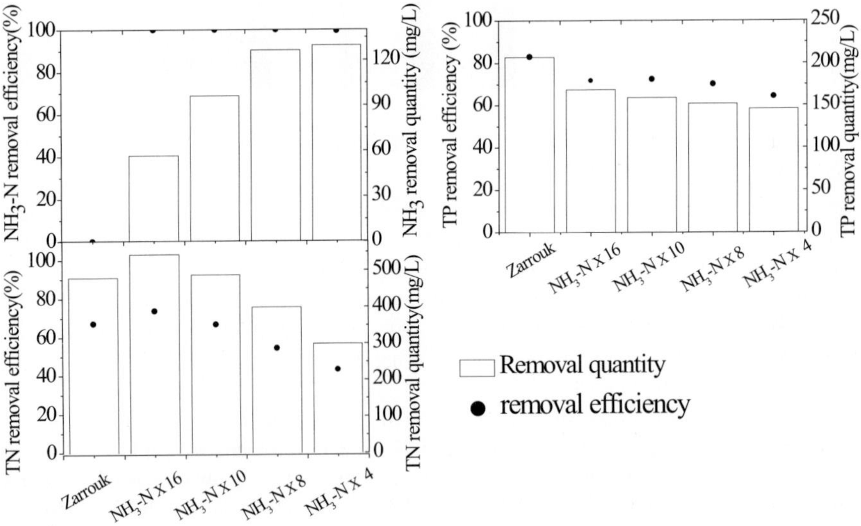

Figure 5. NH_3-N, TN and TP removal efficiency and removal quantity of the water culturing *S. platensis.*

In NH_3-N×4, the growth of *S. platensis* was inhibited, but the NH_3-N was removed. This could be due to the fact that the culture conditions were high temperature and high pH values, which favored the stripping of ammonia.

The TN removal efficiencies of Blank, NH_3-N×4, NH_3-N×8, NH_3-N×10, and NH_3-N×16 were 81%, 61%, 75%, 83%, and 84%, respectively. The TP removal efficiencies of Blank, NH_3-N×4, NH_3-N×8, NH_3-N×10, and NH_3-N×16 were 10%, 18%, 34%, 49%, and 79%, respectively. The NH_3-N×20 had the highest removal efficiency of NH_3-N, TN and TP. Based on an assumption that the approximate molecular formula for microalgae was CO0.48H1.83N0.11P0.01 (Chisti, 2007; Grobbelaar, 2004), the biomass of NH3-N×16 was the highest, and *S. platensis* in NH_3-N×16 took up more nutrients to composite the organics in algae body.

The removal rate of TP, NH_3-N, and TN was 17% (from 29.7 to 24.6 mg/L), 19% (from 406 to 328 mg/L) and 24% (from 409 to 311 mg/L), respectively, in the first-stage microalgae cultivation. The removal rate of NH_3-N, TP, and TN was 99–100% (from 20–30 to 0 mg/L), 18–49% (from 218 to 115 mg/L), and 61–83% (from 697 to 220 mg/L), respectively, in the second-stage microalgae cultivation. The first-stage of biomass accumulation was about 0.3 g/L in 12 days and the second-stage of biomass accumulation reported 3 g/L in 15 days. For different microalgae, there might be a relationship between NH_3-N concentration and microalgae biomass: high NH_3-N results in low productivity while low NH_3-N results in high productivity.

3.4. Cost evaluation of the two-stage microalgae harvest system

There were two components of cost during microalgae harvesting in this system: infrastructure investment and operation cost. For the shorter service life, the membrane system investment and operation cost was relatively cheap. The membrane system contained a pump cost of 300 RMB Yuan and a UF membrane of 200 Yuan. However, the service life of the UF membrane was short (only 2–3 years). As mentioned above in section 2.2, the power of the pump was set at 19.2 W, and 1000 mL algae broth took 5 min; so the electricity power input was about 1.6 Wh. Because the electricity cost is about 0.58 Yuan/kWh, the cost translated to 9.28×10^{-4} Yuan/L water. When the membrane is fouled, it needs to be back washed. This would cost more energy.

Recycling harvest water could obviously save water, during the algae culture process. When membrane was used in water treatment, not only algae but also bacteria could be removed from the wastewater. This could reduce the possibility of bacterial infection. Additionally, taking advantage of membrane to harvest algae could clean almost 100% algae from the culture medium and prevent different kind of microalgae mixture in water reuse. What's more, microalgae harvest by membrane could not increase ash content of algae.

4. Conclusions

Two kinds of microalgae were used to clean ammonia nitrogen and phosphorus existing in wastewater through a two-stage cultivation process. With a small volume of algae broth, harvesting algae with membrane costs less energy, but with a large amount of microalgae loss. With a lower ammonia concentration (8–16 dilution rate), Spirulina platensis could grow well and remove ammonia completely. The two-stage cultivation of microalgae could not only reduce TP and TN existing in wastewater, but also recycled wastewater for algae cultivation.

Acknowledgements

This work was supported by the Program of Outstanding Talents and Innovative Research Teams in Agriculture (2011-049), National Natural Science Foundation of China (51308535) and Chinese Universities Scientific Fund (2015SYL002). The authors want to give thanks to the technical sponsor and wastewater from Minhe Animal Husbandry Incorporated Company. Many thanks to Lingying Zhao (Ohio State University), Jiqin Ni (Purdue University), Raquel Lamartin (China Agricultural University) and Gibson for improving the paper.

References

Anastasakis, K., Ross, A.B., 2011. Hydrothermal liquefaction of the brown macro-alga Laminaria saccharina: effect of reaction conditions on product distribution and composition. Bioresource Tech. 102, 4876–83.

Chang, Y., Wu, Z., Bian, L., Feng, D., Leung, Y.C., 2013. Cultivation of Spirulina platensis for biomass production and nutrient removal from synthetic human urine. Appli. Energy. 102, 427–431.

Cheung, Y.H., Wong, M.H., 1981. Properties of animal manures and sewage sludges and their utilisation for algal growth. Agri. Waste. 3, 109–122.

FAO, 2005. FAOSTAT data. Rome: Food and Agriculture Organization (FAO) of the UN.

FAO, 2006. Livestock Sector Brief. China. Rome: Food and Agriculture Organization (FAO) of the UN, Livestock Information, Sector Analysis.

FAO, 2015. A glass of milk for every child – Asia's dairy farmers aim to boost production, livelihoods and nutrition. Rome: Food and Agriculture Organization (FAO) of the United Nations. www.fao.org/asiapacific/news/detail-events/en/c/281449/. Accessed Sept. 7, 2015.

Hu, Z., Zheng, Y., Yan, F., Xiao, B., Liu, S., 2013. Bio-oil production through pyrolysis of blue-green algae blooms (BGAB): product distribution and bio-oil characterization. Energy. 52, 119–125.

Gonzalez-Torres, A., Putnam, J., Jefferson, B., Stuetz, R.M., Henderson, R.K., 2014. Examination of the physical properties of Microcystis aeruginosa flocs produced on coagulation with metal salts. Water Res. 60, 197-209.

Kuo, C.T, 2010. Harvesting natural algal blooms for concurrent biofuel productionand hypoxia mitigation. MS Thesis, Illinois, USA: University of Illinois at Urbana-Champaign.

Li, J., Hu, R., Song, Y., Shi, J., Bhattacharya, S.C., Salam, A.P., 2005. Assessment of sustainable energy potential of non-plantation biomass resources in China. Biomass and Bioenergy. 29:167–177.

Mackay, D., Salusbury, T., 1988. Choosing between centrifugation and crossflow microfiltration. Chemical Engineer. 477, 45–50.

Teresa, M.M., António, M.A., Nidia, C.S., 2010. Microalgae for biodiesel production and other applications: A review. Renewable & Sustainable Energy Re. 14, 217–232.

Miguel, H., Alejandro, C., and Elena, I., 2006. Sub- and supercritical fluid extraction of functional ingredients from different natural sources: Plants, food-by-products, algae and microalgae A review. Food Chem. 98, 136–148.

Rippka, R., Deruelles, J., Waterbury, J. B., Herdman, M., Stanier, R.Y., 1979. Generic assignments, strain histories and properties of pure cultures of cyanobacteria. J. of General Microbio. 111, 1–61.

SEPA, C., 2002. Water and Wastewater Analyzing Methods, 4th ed. China Environmental Science Press, Beijing.

Spoalore, P., Joannis-Cassan, C., Duran, E., Isambert, A., 2006. Commercial application of microalgae. J. of Biosci. & Bioengi. 101, 87–96.

Tian, C., Liu, Z., Zhan, Y., Li, B., Cao, W., Lu, H., Duan, N., Zhang, L., Zhang, T., 2015. Hydrothermal liquefaction of harvested high-ash low-lipid algal biomass from Dianchi Lake: Effects of operational parameters and relations of products. Bioresource Tech.184, 336–343.

Yoshitomo, W., David, O.H., 1995. A new strategy for lipid production by mix cultivation of Spirulina platensis and Rhodotorula glutinis. Energy Convers. 36, 721–724.

Zhang, X., Hua, Q., Sommerfeld, M., Puruhito, E., Chen, Y., 2010. Harvesting algal biomass for biofuels using ultrafiltration membranes, Bioresource Tech. 101, 5297–5304.

Zhou, W., Hu, B., Li, Y., Min, M., Mohr, M., Du, Z., Chen, P., Ruan, R., 2012. Mass cultivation of microalgae on animal wastewater: a sequential two-stage cultivation process for energy crop and omega-3-rich animal feed production. Appli. Biochem. & Biotech. 68, 348–363.

Overview of United States and European Union Manure Management and Application Regulations

Robert T. Burns [a,*], Robert Spajić [b], Davor Kralik [b], Đurđica Kovačić [b], Katarina Kundih [c]

[a] The University of Tennessee Extension, Knoxville, TN 37796, USA

[b] Department for Animal Husbandry, Faculty of Agriculture, University of J.J., Strossmyer Osijek, 31000 Osijek, Croatia

[c] Department for Swine Production, Belje d.d., 31326 Darda, Croatia

* Corresponding author. Email: rburns@utk.edu

Abstract

This paper presents an overview of current manure management guidelines and manure application regulations in the United States and the European Union. A review of current United States and European Union regulations with discussion on the challenges faced regarding the development and implementation of these regulations can provide valuable insight in regards to developing regulations in other locations that protect the environment while allowing for continued growth of the poultry and livestock sectors. This paper provides a review of the existing regulatory frameworks currently in place in both the United States and the European Union as well as comparisons of manure storage guidelines as well as manure nutrient regulations. Regulation related to land application in regards to manure application setbacks, manure application methods, application timing as well as nitrogen and phosphorus management and application limits are discussed. A review of regulatory approaches that have been successful as well as those that have proven to be cost prohibitive or challenging to implement will be discussed.

Keywords: Manure management, nutrient management land application, regulations, permitting

1. Introduction

The United States (US) and the European Union (EU) have both developed comprehensive animal manure management and utilization regulations. Regulations in both the US and the EU consider manure storage, nutrient management, and land application. The US and EU are also similar in that fact that no single set of regulations governs all of the US or all of the EU respectively. In the US, state level manure management regulations are required to enforce the national level federal regulations, but they may be stricter than the national regulations. Similarly, member countries within the EU may choose to implement stricter regulations than those imposed by the EU. United States and EU manure management regulations are reviewed and individual sections provide an overview of permitting, manure storage, nutrient management planning and land application regulations.

2. Permit Requirements

2.1. United States permit requirements

Animal producers in the US are subject to federal regulations as well as the state regulations where their operation is located. Most producers deal directly with their state regulatory agency rather than the US government regarding manure management regulations. This is due to the fact that regulatory authority of the US Clean Water Act is delegated to 46 of the 50 US states. The national manure management regulations in the US have undergone multiple litigation driven changes over the past decade. Continued changes to the national regulations has resulted in a complex set of manure management and usage rules across the country that vary by state. While the ability of US states to regulate animal production systems at differing levels has provided state specific flexibility, it has also led to a high level of frustration for many animal producers who have to deal with regulations that both vary by location and change over time.

Animal production systems are typically categorized as Animal Feeding Operations (AFOs) and Concentrated Animal Feeding Operations (CAFOs). Facilities that stable, confine, feed or maintain animals for a total of 45 days or more in any 12-month period, and that do not sustain vegetation of some type over the area where animals are confined are defined as AFOs. Facilities that qualify as AFOs and stable or confine specific numbers of animals are classified as either Medium CAFOs or Large CAFOs by the US Environmental Protection Agency (US EPA). In the 46 US states with delegated regulatory authority from the US Environmental Protection Agency (EPA), CAFO National Pollution Discharge Elimination (NPDES) permits are issued and regulated by state governments. Many states currently require Large CAFOs to obtain a NPDES permit. From 2003 to 2012, the US EPA required facilities defined as Large CAFOs to obtain an individual NPDES permit. The legality of the federal governments requirement that a facility obtain a NPDES permit based on animal numbers and a presumed potential to discharge was challenged in a court case. The resulting ruling removed the federal requirement that animal facilities must apply for a NPDES permit based on the assumed potential to discharge. Under the current rule, any CAFO facility that has a manure discharge and does not have a NPDES permit in place is subject to much higher level of fines compared to the same situation where the facility has a NPDES permit in place. Even though the federal CAFO rule underwent this change regarding the duty to apply for a NPDES permit, most states did not change their CAFO rule requirements. Because states are allowed to implement rules more stringent than federal regulations, most states maintained the requirement that facilities have an individual NPDES permit if they were defined as Large CAFOs based on the threshold animal numbers shown in Table 1. An individual NPDES permit is unique to each facility. As such the limitations and requirements in an individual permit are based on the specific details of each animal production facility. Because of this uniqueness, a public hearing is required to be held prior to the issuance of an individual NPDES permit.

Table 1. US CAFO size threshold animal numbers.

Animal sector	Size Thresholds (number of animals)	
	Large CAFOs	Medium CAFPs
Beef Cattle - heifers, steers, bulls and cow/calf pairs	≥ 1,000	300 - 999
Mature Dairy cattle, whether milked or dry	≥ 700	200 - 699
Veal Calves	≥ 1,000	300 - 999
Swine (weighing over 55 pounds)	≥ 2,500	750 - 2,499
Swine (weighing less than 55 pounds)	≥ 10,000	3,000 - 9,999
Horses	≥ 500	150 - 499
Sheep or Lambs	≥ 10,000	3,000 - 9,999
Turkeys	≥ 55,000	16,500 - 54,999
Laying Hens or Broilers (liquid manure handling systems)	≥ 30,000	9,000 - 29,999
Chickens other than Laying Hens (other than a liquid manure handling systems)	≥ 125,000	37,500 - 124,999
Laying Hens (other than a liquid manure handling systems)	≥ 82,000	25,000 - 81,999
Ducks (other than a liquid manure handling systems)	≥ 30,000	10,000 - 29,999
Ducks (liquid manure handling systems)	≥ 5,000	1,500 - 4,999

The majority of states also require facilities defined as Medium CAFOs to obtain a general NPDES permit. General permits have predetermined conditions, so no public hearing is required and it is typically faster and easier for a Medium CAFO to obtain a general permit than for a Large CAFO to obtain an individual NPDES permit. The US EPA considers two or more AFOs to be a single operation if they adjoin each other or if they use a common area or system for the land application of manure. If these conditions are met, the number of animals at each operation must be added together to determine if the facility meets the threshold numbers to be designated as either a Large or Medium CAFO facility.

2.2. European Union permit requirements

At the current time in the EU there is a permit obligation defined under the Integrated Pollution Prevention and Control Directive (IPPC Directive) that was established in 2007 and revised in 2009 and again in 2014. This last revision was made in November 2014 and a finalized version of the IPPC Directive is expected to be published in 2015. The IPPC Directive is focused on swine and poultry farming operations. Any producer, farmer or farm facility that has more than 750 sows per site, or 2000 finishers per finish cycle round (one cycle is 105 days) or, 40,000 bird spaces (ducks, chickens, turkey etc.) is required to have an IPPC license. The IPPC license is issued by the Ministry for Environment and Protection of each EU member state. The IPPC license is an operational license that is focused on the total influence of whole farm emissions to air, water and soil.

The IPPC license is separate from the environmental impact assessment (EIA) study requirement. An EIA study is a document that is required from the local and governmental authorities for existing or new livestock production where the farm unit is greater than 500 Animal Units (AU), where an AU is defined as 500 kg of animal mass. For each animal type there are different coefficients used for calculating the number of AU's on a given farm. Swine and poultry farms that are required to obtain an EIA license are also required to obtain an IPPC license. Dairy and beef cattle farms however, are not subject to the IPPC permitting process. Dairy and beef cattle farms are only subject to the EIA permitting process. Farms with less than 500 AU can be required the local authority to obtain an EIA permit due to the cumulative environmental influence when multiple farms are located in the same region. The IPPC and EIA permitting process is similar for all 28 EU member states.

3. Manure Storage

3.1. United States manure storage regulations

Under US regulations, the required volume of a manure storage structure usually defined by the farm facility's nutrient management plan. Nutrient management plans consider the crop type and rotation on the land where manure will be applied. Land application periods are identified within the nutrient management plan when land application of manure is appropriate based on the crop type and rotation on the farm. This information in-turn defines the storage interval between planned manure land application events. Some state regulations require AFOs and or CAFOs to maintain a minimum manure storage volume adequate to store at least six months of manure, process water and rainfall runoff from the facility. In addition is it common requirement in the majority of states that manure storage structures maintain a "freeboard" depth that must be held in reserve to ensure that manure storages do not overflow during storm or other unplanned events that generate wastewater. States regulations typically require that a minimum of 0.3 meter of freeboard be maintained in manure storage systems. Many states require that up to 0.6 meters of freeboard be maintained in manure storage systems.

It is common for state regulations to require that United States Department of Agriculture (USDA) Natural Resources Conservation Service (NRCS) state conservation practice standards be met regarding manure storage construction specifications. The NRCS maintains a Field Office Technical Guide (FOTG) that includes conservation practice standards that are designed to meet state and local criteria. While the NRCS is not a regulatory agency, state regulatory agencies routinely require animal facilities to meet NRCS conservation practice standards under their AFO and CAFO permit requirements.

Prevention of manure seepage from the storage structure is a primary aspect of manure storage structure design and storage volume requirements are commonly included in AFO and CAFO permits. Permit requirements may require that a manure storage structure meet a specified maximum seepage limit (such as the establishment of a maximum allowable specific discharge for a compacted clay liner) or define allowable material specifications (such as a minimum thickness impermeable liner material, or a minimum concrete thickness). Regulations in the US have become more stringent over time in regards to allowable seepage from manure storage structures. From the 1970's through the 1990's earthen holding ponds or lagoons with compacted clay liners were the predominate manure storage structure type used in the US. During this period regulations typically required that unit seepage from these facilities should not exceed 9400 liters per hectare per day (1000 gallons per acre per day). This seepage limit could be achieved in a 3 meter deep storage with a 0.3 meter thick compacted clay liner with a coefficient of permeability in the liner of 1×10–7 centimeters per second (NRCS, 2008). In the 2000's many states began requiring the use of synthetic liners made from materials such as high-density polyethylene (HDPE), linear low-density polyethylene (LLDPE) or ethylene propylene diene monomer (EPDM) in earthen manure structures. In addition some state regulations also began require on-going groundwater monitoring to provide confirmation that earthen manure structures were not seeping into the surrounding groundwater. The added cost of on-going groundwater monitoring for earthen structures required by these regulations resulted in many producers opting to use concrete or fiberglass coated steel tanks for manure storage rather than earthen structures.

The siting of manure storage structures is also regulated in the US. Each state determines set-back requirements that define the separation distance required between a manure storage structure and various water sources and public sites. At a minimum states require that manure storage structures be built no closer than 305 meters (1000 feet) from public water supply well or surface intake structure, 91 meters (300 feet) from surface water, drainage inlets or off-site wells and 30 meters (100 feet) from on-site wells, property lines and public roads.

3.2. European Union manure storage regulations

Manure storage facilities are subject to permitting buy both the EIA and IPPC processes under EU regulations and directives. Storage facilities should be of sufficient size to store all types of manure and process water until land application. Manure storage length for the different climatic regions of the EU are shown in Table 2 below.

Climatic difference is the primary parameter that determines the required storage period length for each EU country under the IPPC Directive. The Directive also requires that all manure storage facilities should be waterproof without any possibility for leakages or drainage into the soil and ground-water. Some countries, including Germany, Holland, Belgium and Norway, require that manure storage facilities have impermeable covers to reduce aerial emissions, but most EU Countries do not have this requirement and it is allowable to utilize permeable materials such as a natural solids crust, straw or LEKA as a manure storage cover.

4. Nutrient Management Planning

4.1. United States nutrient management plans

4.1.1. Permit nutrient management plan requirements

The fundamental focus of nutrient management planning is to ensure that manure is applied to cropland at the correct nutrient application rate, correct time and correct location. Both individual NPDES AFO/CAFO permits and general AFO/CAFO permits require the animal production facility to develop a nutrient management plan. The level of nutrient management planning required to obtain an individual NPDES AFO/CAFO permit is generally greater than that required for general permit. The core consideration of a nutrient management plan is the management of manure application such that the application rates, methods and timing of manure nutrients is sufficient to achieve optimum crop yields while not resulting in significant transport of nutrients due to leaching, surface runoff, or gaseous losses.

Table 2. Required manure storage capacity for EU climatic regions (IPPC, 2003).

EU Member State	External Manure Storage Capacity (Months)	Climate
Belgium	4-6	Atlantic/Continental
Luxembourg	5	Atlantic/Continental
Denmark	6-9	Atlantic
Finland	12	Boreal
Germany	6	Continental
Austria	4	Continental
Greece	4	Mediterranean
Ireland	6	Atlantic
Italy	3 for solid and 5 for slurry	Mediterranean
Portugal	3 - 4	Mediterranean
Spain	3 or more	Mediterranean
Sweden	8 – 10	Boreal
The Netherlands	6 for pig slurry	Atlantic
UK	4 -6	Atlantic

Because nutrient management plans (NMPs) include a large range of information, typically only specific parts of the nutrient management plan are included in the actual AFO/CAFO permit. Basic nutrient management plans focus on establishing manure application rates. Plans may be either nitrogen or phosphorus based. Manure is an un-balanced plant fertilizer and when applied at the correct rate to satisfy plant nitrogen needs, it results in the application of phosphorus at rates in excess of plant needs. Similarly, if manure is applied to meet crop phosphorus requirements, nitrogen will usually be under applied for crop requirements. Nitrogen based nutrient management planning allows for the most cost effective use of manure nutrients and is recommended when the over application of phosphorus does not represent a water quality issue. When the over application of phosphorus does represent a water quality issue, phosphorus based nutrient management planning should be utilized.

4.1.2. Comprehensive nutrient management planning

Where basic nutrient management plans deal only with the management of nutrient application, Comprehensive Nutrient Management Plans (CNMPs) are intended to address all resource concerns on the farm. The CNMP concept was developed by the U.S. Department of Agriculture (USDA) Natural Resources Conservation Service (NRCS) to address conservation planning for animal feeding operations. Additionally, the Environmental Protection Agency (EPA) recognizes that CNMPs address the requirements of the Nutrient Management Plan necessary for the maintenance of a National Pollutant Discharge Elimination System (NPDES) permit. National USDA policy states that animal feeding operations should have a CNMP to be eligible to receive certain cost-share funding, such as Environmental Quality Incentive Program (EQIP) assistance. In 2002, the Farm Security and Rural Investment Act (Farm Bill) increased the amount of conservation program funds available to animal feeding operations, and introduced the mechanism for using certified Technical Service Providers (TSP) as a source of technical assistance for producers. Producers can contact a TSP, request the development of a CNMP and then potentially be reimbursed for a TSP's services with conservation program funds, depending on fund availability in their state.

A CNMP is a conservation system developed in accordance with NRCS planning policy that addresses all of the conservation aspects of an animal feeding operation. The purpose of a CNMP is to identify resource concerns on the farm and to establish a Resource Management System (RMS) specific to that farm that addresses existing natural resource concerns. A Resource Management System is simply a set of Conservation Practices designed to address these concerns. While all CNMPs should share similar components, each one is unique to a particular livestock production operation. The six basic elements that may be addressed in a CNMP are; Manure and Wastewater Handling and Storage, Nutrient Management, Land Treatment Practices, Record Keeping, Feed Management and Other Utilization Activities. A CNMP should identify and address natural resource concerns in regards to soil, water, air, plants, animals and people on the farm. Addressing these six elements will assist the producer in meeting soil, water, and air conservation goals as well as reducing the potential and actual threats to water quality and public health from their operations. When implemented, a CNMP should help ensure that both production and natural resource protection goals are achieved. Because a CNMP utilizes conservation practices to beneficially use animal manures, it also assists animal feeding operations in meeting regional and US national water quality goals and regulations.

While all six elements should be considered by a producer, and implemented accordingly, they do not all have to appear within the CNMP. At a minimum, a nutrient plan should address actions pertaining to an animal operation's production area and the land on which the manure and organic by-products will be applied (USDA-NRCS, 2003). This means that if an operator applies manure to his crop production area, the plan should at least cover elements 1-4 as listed above (Manure and Wastewater Handling and Storage, Nutrient Management, Land Treatment Practices and Record Keeping). These four elements represent the minimum expectation of a nutrient plan required by federal environmental regulations, often referred to a Nutrient Management Plan (NMP). However, a review of the feed management program and other utilization options often reveals additional valuable options for environmental improvements for some operations.

4.1.3. Comprehensive nutrient management planning elements

Manure and Wastewater Handling and Storage. This element addresses issues concerning manure storage structures and other areas within a production facility used for manure transfer, treatment, and/or storage. Within this section, the CNMP should identify concerns and provide documentation of adequate manure collection, storage, and/or treatment to allow for land application of the material, as well as for dead animal disposal. In addressing these concerns, the CNMP should also take air quality and pathogens into consideration.

Nutrient Management. This element addresses issues concerning the land application of manures and other nutrient sources to production fields associated with the livestock operation. The CNMP should outline and implement land application practices that minimizes the potential for adverse impacts to the environment and public health. Consideration should also be given to air quality impacts, pathogen movement to water, and salt and heavy metal build-up.

Land Treatment Practices. This element addresses soil loss and runoff that may be present on fields receiving manure and other soil amending nutrients. Within this section, conservation practices are used as components of a Resource Management System to address the resource concerns that have been identified on the farm. A Resource Management System is a suite of Conservation Practices that will be implemented on the animal feeding operation to address any soil, water, air, plant, animals and human resource concerns that have been identified. There are hundreds of Conservation Practices that can be potentially included in a RMS. Some examples of Conservation Practices are buffer strips, diversions, minimum tillage, etc. Traditional Land Treatment Practices (LTP) include buffer strips, various conservation tillage methods, grassed waterways and terraces. Land Treatment Practices are typically practices that have been traditionally implemented on the land with assistance by NRCS. The LTP element should include a schedule of planned implementation dates as well as an explanation of the operation and maintenance activities required to keep the practices functional. This part of a plan should also recognize the value of organic matter in manure for improving soil quality and reducing soil erosion and runoff

Record Keeping. This element is essential to the implementation of a CNMP. Records document and demonstrate that activities associated with the CNMP have been implemented. It is the responsibility of the production facility owner/operators to maintain the required records. It is important for producers to understand that a CNMP is dynamic. The original CNMP document that is prepared is basically a list of practices and management actions that will be implemented on the farm. As such, a CNMP is not in place until these practices and management actions have been implemented on the farm. From this standpoint, the most important part of any CNMP is records that show how the plan was implemented. A producer does not truly "have" a CNMP until he/she has implemented the actions contained in the plan and has records to document this implementation. Records are important to provide information to modify a CNMP. For instance, the original CNMP will suggest manure application rates and timing, cropping systems, and fields selected for receiving manure. However, a producer may need to modify these management practices for the upcoming year because new information is available, such as soil tests, stalk nitrate tests, yields, etc.

Feed Management. The inclusion of a feed management plan can be a critical component for CNMPs. Feed program decisions directly impacts the quantity of manure nutrients excreted and available for use as a fertilizer resource, the land needed for apply the nutrients, and the magnitude of air emission (odor, ammonia, and others) from a livestock operation. A CNMP plan provider should work with a producer and the farm's nutritionist to determine how much a feed management plan could potentially assist in meeting the CNMP goals. If sufficient opportunity for benefit exists, then the producer should consider enlisting the help from a qualified animal nutritionist to develop a Feed Management Plan for the CNMP. In some situations the implementation of a Feed Management Plan may provide large reductions in excreted nutrients and reduce the land requirement from a nutrient balance standpoint. A professional animal nutritionist should be consulted if feed management alternatives are being considered. Example of strategies to consider may include phase feeding, split-sex feeding, amino acid supplemented low crude protein diets, the use of low phytin phosphorus grain, growth promotants, processing of feeds and specific enzymes.

Other Utilization Activities. When insufficient land area is available and application of manure nutrients are greatly in excess of crop requirements with the potential for causing an environmental risk or if air quality concerns exist for an operation, alternative utilization methods may be considered. For pork and poultry operations that buy the majority of the feed used, export of manure becomes a critical other utilization option to consider. Emerging technologies may be an option for reducing nutrients and minimizing odors. The best possible alternatives should be similar in cost to traditional methods. Many times, however, alternative utilization options cost more than traditional land application. Because these alternatives are not conventional, industry standards do not always exist, and specific NRCS conservation practice standards may not be available. Land grant universities or other research organizations may provide more recent information on emerging alternatives.

4.2. European Union nutrient management plans

Many of the elements that are included in the US nutrient management regulations are similar in European Union. Swine and poultry farmers are obligated to obtain a permitting based on IPPC Document and Nitrate directive. Dairy farmers are obligated to obtain a permit based only on the Nitrate directive at this time.

Manure and Wastewater Handling and Storage. The focus of the IPPC permitting process is based on the current IPPC Document. Manure storage facilities should be monitored periodically (typically 4 times per year) and monitoring and inspection is required to be retained by the farmer for a five years period. Any modification or damage to the manure structure, handling system or other component of manure management system triggers a requirement for additional inspection and reporting requirements.

Nutrient Management. Nutrient management plans in in Europe are still in the development process. In many cases in Europe, manure nutrients are not applied properly. A growing number of farmers have begun to implement nutrient management practices that utilize manure to replace artificial fertilizer applications of N, P and K. Additionally the IPPC Document requires calculating the whole farm nutrient generation at the farm location, and a permit cannot be approved without the demonstration that sufficient land for proper manure application is available. Different manure treatment methods have been investigated over the last decade, but nutrient management has been recognized are more important than treatment in European farm systems. The amount of manure nutrients produced per location is still the dominant factor in the EU permitting process.

Land Treatment Practices. In the EU, the primary land application focus is on manure application techniques. It should be noted that there is very little focus on soil erosion in the EU when compared to the US. Techniques such as manure injection and proper manure application timing are an imperative for majority of the farmers. Still however, many EU farmers are broadcasting manure on the land surface where 60-80% of the nutrients are lost through runoff, leaching and volatilization.

Record Keeping. The EU permit process requires record keeping as part of the management planning and data collection. All documentation needs to be keep for a five year period. All of the documents associated with the permit are subject to inspection and follow-up enforcement. The majority of the EU Countries typically inspect 5-10% of permitted farms per year.

5. Manure Land Application

5.1. United States manure land application regulations

Manure land application rates are defined within nutrient management plans. When specific manure application rates determined in the nutrient management plans are written into CAFO/AFO permits they become regulatory limits. These nutrient application rates are used in conjunction with manure nutrient analysis to set field specific manure application rates in NMPs and CNMPs. In addition to manure application rates, NMPs and CNMPs also define manure land application timing, method and location specifics and limitations.

Nutrient management plans typically require producers to apply manure nutrients to crops prior to or during active crop growth periods. Many state manure application regulations do not allow manure application when the ground is frozen or in periods where no actively growing crop is or will be available to utilize the applied nutrients. Allowable application timing varies by application method. For example, manure application via either surface broadcast or sub-surface injection may be allowed during cool season months while only sub-surface injection may be allowed in warm season months. In some cases regulations may also require that specific application methods be used. For example, surface broadcast of manure may not be allowed at all in some locations.

Manure application location is managed through recommended manure application setback distances. Table 3 gives typical US manure application setback distances. It is very common for states to develop their own manure application setback rules, but they are usually similar to the values shown in Table 3.

5.2. European Union manure land application regulations

The Main issue for the EIA and IPPC licenses is manure management and control. All of the EU countries use N as a basic manure evaluation unit, where 170 kg of N is the allowed limit for annual nitrogen application. Counties entering the EU are allowed a transition period where 210kg of N is the maximum allowed to be applied in a one year period. Current technology is providing the opportunity for crop rotations that make two harvests in one year possible. So it is likely that over the next 3-5 years provisions will be made to allow the application of more than 170 kg of N where two harvesting occur in one season, which is a common practice in mid and southern Europe Countries.

Table 3. Typical US non-application buffer widths.

Object, Site	Situation	Buffer Width
Well	Located up-slope of application site	46 m (150 ft)
Well	Located down-slope of application site provided conditions warrant application	92 m (300 ft)
Waterbody / Stream*	Predominate slope < 5% with good vegetation3/	9 m (30 ft)
Waterbody / Stream*	Predominate slope 5-8% with good vegetation3/	15 m (50 ft)
Waterbody / Stream*	Poor vegetative cover or Predominate slope > 8%3/	31 m (100 ft)
Waterbody / Stream*	Cultivated land, low erosion	9 m (30 ft)
Public Road	Irrigated wastewater	15 m (50 ft)
Public Road	Solids applied with spreader truck	15 m (50 ft)
Dwelling	Other than Producer	92 m (300 ft)
Public Use Area	All	92 m (300 ft)
Property Line	Located downslope of application site	9 m (30 ft)

*Waterbody includes pond, lake, wetland, or sinkhole. "Open" sinkholes should be protected the same as a well. Where sinkholes are not "open", a buffer width should be established in the flat area around the rim of the basin before the change in slope up out of the basin begins. Stream includes both perennial and intermittent streams. Good vegetation refers to a well-managed, dense stand which is not overgrazed.

6. Discussion

Manure management in the United States has been driven by US EPA environmental policy, and to a large degree this policy has been shaped by litigation from both environmental and agricultural interests. The result has been a perception by producers of an unstable regulatory environment. This perception has resulted in wait and see attitude by many producers in regards to the adoption of comprehensive nutrient management planning, unless required by regulation. There has been little advancement in regards to the development of economically viable manure treatment systems.

The First page of IPPC Directive document is supported with a sentence that explains that all the regulations that are presented in the document should be economically feasible for farmers to implement. This sentence is interpreted differently by many of the regulators, and the farmers also economic feasibility a different way as well. The IPPC Directive Document definitely offers a wide range of technical solutions to the existing and future problems in EU farming sector, but it still should be adopted to include technologies developed in the future. Manure handling still remains the biggest discussion point, where more developed countries such as Denmark, Holland, Germany, Norway, require higher standards for the manure storage and handling process than the less developed EU countries. Additionally air emissions from animal production will become a point of great discussion in the near future in the EU. The primary issue in all of these discussions is the economic feasibility of the regulations on farming in majority of the EU countries.

7. Conclusions

In the United States the land application of manure as a crop fertilizer is the manure management strategy that continued to be both environmentally sustainable and economically feasible. Producers and equipment manufacturers have both recognized that the best way to manage manure from both an economical and environmental standpoint is land application as a crop fertilizer. Over the last decade years great strides have been made in developing much higher efficiency manure application equipment. In the US sub-surface injection has been recognized as the most cost-efficient and environmentally friendly method to land apply manure slurries. Equipment manufacturers have steadily increased the capacity of slurry injection units. Currently manure injection systems that can inject manure slurries in excess of 1000 liters per minute are readily available in the US.

In the EU techniques such as manure injection currently provide the most economical way to manage manure management. This is due to the cost savings associated with replacing artificial fertilizers with manure nutrients in systems where livestock and arable farming systems are co-located. There continues to be a strong public demand for environmental protection, but the question of how to meet this demand while maintain profitable farming conditions for producers is the current challenge to be solved. One smart member of the Technical Working Group in IPPC Directive Committee said "We need to be aware that if all the hours spent in these meetings making conclusions and regulations don't kills us, we should still be cautious, because maybe the farmers will".

References

European Commission. 2003. Integrated Pollution Prevention and Control Reference Document on Best Available Technology for Intensive Rearing of Poultry and Pigs. European Commission

United States Department of Agriculture - Natural Resources Conservation Service (USDA-NRCS). 2008. Appendix 10D Design and Construction of Impoundments Lined with Clay or Amendment Treated Soil. Part 651 Agricultural Waste Management Field Handbook. United States Department of Agriculture.

United States Department of Agriculture - Natural Resources Conservation Service (USDA

NRCS). 2003. Part 600.5 – Comprehensive Nutrient Management Planning Technical Guidance. National Planning Procedures Handbook.

Theme IV:

Animal Behavior, Welfare, and Health

Farm Auditing of Animal Welfare and the Environment in the US —An Overview Provided Based on Experience

Lesa Vold

Egg Industry Center located at Iowa State University, Ames, IA 50011-3310, USA

Email: lvold@iastate.edu

Abstract

This paper presents an overview of the history and several programs and their processes that make up the assessment and audit industry that currently exists in the US. Understanding the foundation and history of where these programs started, the timelines in which they developed and how they were developed can help inform new policy and program development avoid the pitfalls that exist in the US.

Keywords: Assessments, audits, certification, management systems, consumer voice, brand awareness, extension

1. Introduction

Audits in the US used to be equated with one agency and one commodity: an individual's money and the Internal Revenue Service (IRS). These audits were never a positive experience and often resulted in fines for taxes that were still due in sums large enough that most people could never pay. Ultimately, this resulted in a fear of the term "audit" for most Americans. Deeply rooted in tradition, problem solving, independence and a will to survive, the American farmer was the last one who wished to have an audit. After all, farming is not consistently done every time, what standard can farming be judged by? This was less than thirty years ago. Today, the increased pressure for more, faster and with less with a lawsuit looming around every corner has every industry looking for solutions from anyone who has a sound background for advice. This includes the American farmer. It takes a tremendous amount of money for companies to gain market share for their products. Products must be not only produced as expected, but labeled with the correct niche phrases, packaged neatly, stocked in the correct locations at the store and priced right before they are even considered for purchase. To invest all the time, energy and effort it takes to carved out market share and have one bad social media post drive a company to bankruptcy (reference to Beef Products Inc., or BPI, and their lean fine textured beef product which was caused a media scandal and subsequent lawsuits in 2012) is something that shakes the American business model to the core. Not only are companies learning that it is no longer business as usual, but audits are quickly becoming a welcome opportunity to prove quality of many types and assure customers of their purchases. This is establishing a culture of total transparency in an ever-changing marketplace. As this philosophy permeates the boardrooms of America, it ends up affecting the entire food chain – and that starts on the farm.

2. Why Assessments and Audits?

2.1. Terminology

Auditee – Individual or operation being audited.

Inspection – A term developed by the US government for oversight activities they conducted over various facets of the food chain. On-farm inspections typically occurred in conjunction with a regulation that needed to be verified. In processing plants for beef, pork lamb, etc. these inspections looked at sanitation and co-mingling of products, temperature of facilities and hiring practices for personnel. From a farm standpoint, inspections come from various agencies of the government as well as national, state and local branches of the government. Topics for inspections are items such as milk storage bulk tank regulations regarding the temperature the milk is stored at the type of storage tank it is kept in to the nutrient management plan that must be developed and the associated records one must keep that are required for proof of manure and/or wastewater application from a facility. Inspections are often pass/fail and have consequences that include large fines or a business order to cease and desist operation until an issue is resolved.

On-Farm Assessment – A term that developed out of necessity because people feared the term audit. The term was developed by industry to provide the agriculture community a space for which an on-farm educational exercise could occur where individuals were not going to get fined or fear being told to stop operation. These experiences enabled producers to learn what was expected of them and the standards to which they were being held by regulatory agencies, the public and later their customers. These visits were also often accompanied with advice on how a producer could get from where they were to where they needed to be or resources for answering such questions.

On-Farm Audit – This is a repeatable pass/fail or scored verification of whether an operation is achieving a certain standard set forth by a government body, a voluntarily membership body (like an industry association), a customer or other standard creating body

In its truest form, there is no exchange of information on how something should be done or changed, only that is needs to. Because of most entities inability to conduct a true audit, levels of auditing have been developed:

- First Party
- Second Party
- Third Party

First Party Audit – A review of requirements performed by yourself or your own organization. This can also be called an internal audit. It provides no check and balance measure of trust worthiness since it is often to an individual's advantage to glance over items that can be of significant concern to others if it is associated with cost or met with a lack of technical expertise available to solve the issue. Often this audit system is plagued by a lack of documentation since it is time consuming to write a report for yourself. Lack of documentation provides another trustworthiness gap when it comes to proving that audits are occurring or that issues are being noticed and addressed.

Second Party Audit – A review of requirements conducted by a party that is related to you or your business but may or may not have a stake in the outcome of the audit results. The most widely used form of this audit is called a customer or supplier audit. If company B is buying materials from company A, then company B has individuals who go visit company A to ensure their product requirements are being met. While company B has a vested interest in ensuring what they are receiving is right, they also have a cost to bear if they decide to drop a supplier and locate another one. In this case, the do have a vested interest in the outcome of the audit or the actions that are taken as a result of the audit. Another example of why this type of audit relationship is tough is because by nature, the purchasing customer should be explaining what they expect – which often occurs during this audit interaction. In comparison to the definitions applied above, this becomes an assessment and not an audit. Another example of a second party audit would be a hired consultant. Again, unless hired only to conduct audits, they often work on the how and problem solving of issues at an operation. This person can definitely be qualified to perform audits, but at some point they are likely evaluating their own work. In this situation the creditability and trustworthiness of the audit can be questioned by outside stakeholders. Most often reports are provided, but the formality of the reports varies greatly.

Third-Party Audit – A review of requirements by a party that has no vested interest in the outcome of an audit. These audits are conducted by individuals who are totally un-related to the operation and are 100% independent of any repercussions the results of the audit could have. Entities who conduct these types of audits have adopted many names in the US such as: consultants, certifiers, certification bodies, special departments of government agencies, etc.

Ultimately they are the best at obtaining a true objective look at whether an operation is meeting a given requirement. These entities are skilled at their profession and well equipped to provide detailed reports back to the auditee. Often these reports are used with others in the food chain in the name of transparency.

2.2. Societal drivers

One question that any country needs to address is if this system fits their societal structure. Drivers for this system in the US have been mainly economical. America is a very litigious society. If a company that gets sued by a customer can prove that their supplier was really the one with an issue, it could mean the difference between operating a company for another day or closing its doors. On the other hand, sometimes companies are not being sued, but they need a way to prove the claims that they make about their products.

Sometimes an organization has a new or different process that enhances a product, or they take additional care in part of the production of a product. Because market share acquisition is so hard, every marketable angle of a product is exploited. Part of the market change that has come into play in this regard is a shift from marketing of products based on a product attribute, to marketing a product based on non-product attributes. For example, marketing a piece of pork loin based on the size of the loin is a product attribute, but marketing that same piece of pork as a sustainably raised piece of pork is a non-product attribute. This is something consumers can't see, so they expect proof to exist behind that claim if the organization wants them to remain loyal or pay more money.

As a result of these two drivers, assessments and audits have become an easy way to self-identify issues of concern and take steps to ensure problems do not result in lost business.

In the case of animal production, producers have economic incentives for a system of assessments and audit as well. Assessments and audits are a great educational tool. Assessors and auditors often gain a wealth of experience on practices to meet standards because they see so many different operations. They also have a wealth of knowledge in the interpretation of the standards or requirements to which they audit. This is why auditors are often under strict confidentiality contracts and often cannot even discuss the client they have visited, let alone what they have seen. However, in the world of agriculture, the vast body of knowledge is not proprietary. The vast body of knowledge revolves around something called best management practices and audits are a great evaluation tool for these items.

In closing this section on societal drivers, I will leave you with this thought. Remember that it was less than 30 years ago when the US did not operate on a system of audits. This is significantly less than the span of on person's career. Keep in mind that the societal driver may not yet exist for building a brand and protecting that brand, but rest assured that someday it will because a global consumer and a global economy is the world structure we have embraced.

3. Building a Structure for Success

3.1. Where we started

The US agriculture technical service providers of 30 years ago were called Agriculture Extension Agents. Part of the US history was the establishment of the Morrill Act of 1862. It created, through the sale of land, proceeds by which public colleges could focus on agriculture and mechanical arts. The "land grants" provided funding for 69 colleges across the U.S. In 1914 the Smith Lever Act took an idea circulated by a farm journal to visit and share about the advances on the farm and formalized it into the U.S. Agriculture Extension Service. USDA reports, "At that time, more than 50 percent of the U.S. population lived in rural areas, and 30 percent of the workforce was engaged in farming." The extension service, housed at each of the land-grant colleges, was created to help foster this transfer of knowledge on rural and agricultural issues.

This newly created body was funded by national, state and local government partnerships. As a result, the technology transfer was free to the average producer. However after World War I but prior to the 1960's, extension took on a new role looking for more involvement with the local people and in local situations. New areas of outreach were developed including community and rural development and family living. The farm crisis of the late 1980's created the initial downsizing of extension. As federal and state funding for education has declined so has the funding available for extension.

When technology transfer started in the U.S. it was due to meetings and farms visits organized by extension staff. One or two day meetings would be held where professors could talk about their areas of expertise and research. These meetings gave producer's time to talk with each other and gain valuable experience from the technical experts. If a farmer was struggling with something at the farm, the local county extension agent acted as a liaison to review the situation and take it back to those at the university for potential solutions. These on-farm visits became the initial assessment in agriculture. It was a time to determine what needed changed to improve a given situation and how best to accomplish that task.

3.2. Where we are today

The result in a declining funding system for extension caused consolidation across the countryside. The local extension agent became an area or regional agent and the time and resources available for the producer decreased. As generations of University-educated folks became available, the rise of the consultant occurred. These individuals were timelier than the local extension agent that now had a larger territory to cover. And a consultant often had extreme expertise in a given area instead of being a "generalist" like the extension agent had to be. While there were advantages to this system, it came at a price. The government did not subsidize consultants and they needed to charge for their time. This was a new concept in agriculture and many producers who received free extension services for years were not positioned, nor willing, to incur the costs of expertise on coming on their farms.

It was here that many of the breed organizations stepped in. The US commodity check-off system helped establish a system by which funds could be delegated to groups who were overseen by the USDA. These groups could use these set aside funds for specific purposes only, but it included commodity marketing and as an outreach of that, producer education and related programs. The breed associations could afford to able to bring collective expertise together from the universities and the consultants and develop robust programs to educate their producers. Many of the programs that exist today exist because there was a need to educate producers on a new regulation rule or best management practice. Once that education occurred the industry adopted the practice as the industry baseline for performance. As the legal and marketing climate in the US began to change, assessments and audits of baseline industry expectations were a natural place to start...and so they did.

3.3. The look and feel of an industry-driven program

Livestock commodity check-off organizations (livestock associations) could see the rapid shift away from available technical expertise at the extension service level and the need for private expertise to keep their producers successful. This started with the overwhelming need for help during the farming crisis of the 1980s. Those who remained in the industry were driven and more open than ever to better ways to do everything from manage the finances to till the fields to feed the hogs. The livestock associations were in a unique position to understand the regulatory issues the producers were going to have to face and have the ability to pull together groups of producers to educate them on the subject.

Much like the cooperative extension service beginning, the association found value for their members in providing one or two-day meetings where their members could learn about a variety of issues that were going to affect their operations or dive deeply into a complex issue where the industry needed expanded knowledge. As the government and public began calling for accountability for practices, the associations developed the two-prong program approach of classroom education and on-farm verification of technical application.

This process involved both consultants and extension service or other university personnel. It required that the experts come together on a given topic and establish the baseline standards to which individuals should be held. Items were often categorized into things that had to be done vs. things that would be good if they were done. The experts had to agree on practices that would help move producers toward a common goal. The key to these programs was the flexibility the industry had in administering them. When done prior to government regulation, they had the ability to focus the programs on incremental improvement. This was a steep contrast to the government inspection that could require immediate and extensive upgrades along with a penalty fee for noncompliance and daily noncompliance fees until your operation was brought back into compliance.

Most programs were set up with an expert panel that advised the standards that needed set. The associations then found members to help develop the program based on the expert's advice. Sometimes program development was outsourced to government or private companies, but program approval always rested with the industry members. Auditors for industry-led programs were often contracted companies or government employees.

3.4. Challenges that exist

As the need for assessment and audit services has grown in agriculture, it has failed to take into account a structure that will keep it successful. The following section outlines some of the pitfalls of the US and Global Audit Systems.

3.4.1. Good ideas don't always take off

I was amazed when I was brought into manage an assessment program that was free to producers and the company struggled to give the free service away. Sometimes even when the intent is right and the design is right, there are other forces at work that the market didn't consider. Many times these are the unspoken rules of a culture, a tradition, a challenge to old systems that individuals are not yet ready to let go. Sometimes these issues can be rooted in generational gaps or technology gaps. I often joked with my past employer, we were always right; we were just always five years ahead of the curve. It made us the experts at what we did, but it was often painful and expensive to be the leader. As the leader you were the ones who incurred all the research and development costs, you were the ones who got your ideas stolen, you were the ones that educated everyone on the paradigm shift that needed to occur and then someone else walked in and took the business. You just plain invested more sweat, blood and tears but you also gained the respect that money can't buy. If at first you don't succeed, don't give up. Keep working toward what you know will be the way of the future.

3.4.2. We are NOT all the same – we are human

Some people are driven…and others just aren't. I remember that one of the programs I was involved with provided information and answers but then it did two forms of evaluation. One was a traditional survey of service: did you feel this was helpful, do you feel it enabled you to change your operation, did you get useful suggestions, do you feel like you can take action to improve, etc. The second was an actual surprise verification of actions taken. Of course we would leave if a producer got really upset and didn't want to part-take, but we would show up a year later and see how many of the recommendations that had been made during the past visit had actually been implemented. If they hadn't been implemented why hadn't they been implemented? It seemed in an overwhelming majority of the cases that the easy items got handled, but rarely did the big items (no matter if they were going to cause a compliance fine if found or not) were dealt with. This was so disheartening to me. I believed in the good in people. I really believed that if you just point them in the right direction that they would always do the right thing. I was wrong. The path of least resistance will always prevail; it is human nature.

Remember that the people working for you are human. There is thing called human error that is hard to overcome – you need to factor it in to everything you do. You need to train and drill endlessly on matters that are important. TRUST, but verify in everything you do. Last but not least, spend the money to hire the right people. If you hire the driven ones, your operational costs will be lower in the long term.

And a caution about driven people, be aware of burnout. Good managers and good complex supervisors, they are hard to find. They are hard to find because they are hard-working, dedicated individuals who treat your business like it was their own. But just as with any machine, constant use with no maintenance will find you investing again in a new machine for several times the cost of maintaining the old one. Often times newer is not better in the livestock production environment. There are so many variables that need to be learned over time. High turnover is a public relations issue, it is an economic issues and it is a worker care issue. You must show you care, before employees will react as if they care.

3.4.3. A divided industry

I always thought that we had industry support when we worked on industry standards. We had producers in the room helping guide the process; we had producer pilot farms where we could train people on the process and what to expect while providing assessment and audit services. I just assumed the industry was united in its efforts; it rarely is. Some of this ties into Section *3.4.2* but sometimes it has to do with jealousy over another's situation or lack of decision making authority where it is needed. Every industry can be divided into individuals who are early-adopters and those who are not. The U.S. livestock producer mentality in regard to assessments and audits was strongly polarized in various ways. (1) Livestock species – remember the role of the livestock associations? Some livestock species very quickly adopted to an assessment and audit approach, others have been extremely late to the party and have done so to the detriment of the entire animal agriculture sector whose trustworthiness has been damaged in the eyes of the consumer. (2) Within my association – This ranges from things like I didn't get to sit on the organizing committee to, I don't have the ability to implement this on my farm and I am not willing to deal with that reality yet. Either way, putting an issue off doesn't make it go away. I often found if you could ask why enough times you would eventually get to the real reason someone was resisting change. The key to this is that you cannot have a conversation that digs that deep without developing extensive trust and trust takes time. Until then, you need to expect that there will be division.

3.4.4. This takes expertise

No one ever told the higher-level auditing bodies that auditors that audit other industries will not clear the hurdle that is called livestock agriculture. Auditing in this sector takes a special kind of person. This person is not easily found and depending on when you find them in their career their time actually able to audit can be limited. Auditing at a third-party level takes extensive travel. Many people need to be at the right time of their life to really want to do this for a period of time. It also takes a love of animals, a love for agriculture, and love of seeing others grow. It takes a special set of expertise that takes years to train. Good auditors are hard to keep because if they don't want to settle down for their own reasons, some other company is out there head-hunting their knowledge and will pay them more and promise them less work. In my experience it takes specific set of skills to make a successful auditor and sometimes even when you think it will work, it doesn't. Unfortunately in the US, we were able to get by on the deep and extensive expertise of our veteran auditors for over 10–15 years, but there was no auditor education program set up to feed new individuals through the correct education and training system to have them experienced and where they needed to be when the time comes. This training program can almost go all the way back to the University level and the college level courses that these individuals take. These individuals need to understand a broad range on things that happen on the farm and not a lot in one specific area like nutrition or reproduction. The auditor's needs are move diverse, across disciplines of animal sciences and agricultural and biosystems engineering, construction engineering, forestry, statistics, economics and human relations and management experience. This becomes a challenge in the current American university structure. Above all they have to be a process thinking that can see interactions in things before they even happen. This is not a trait that is easily taught at any level of education.

3.4.5. Activists

It is not if, it is a matter of when. Whether they attack your farm or the industry, it still hurts. How are you prepared to answer their call that animals are not fit for human consumption? How are you protecting your classrooms and educating your youth on the importance of agriculture.

4. Audit Areas and Opportunities

4.1. Who are the players/what are the programs

4.1.1. Animal welfare

- Industry-led Animal Welfare
    - United Egg Producers - UEP Certified
    - National Pork Board- Common Audit
    - National Milk Producers Federation – Farm Program
    - Milk and Dairy Beef Quality Assurance Program
    - National Chicken Council – Animal Welfare Guidelines
    - American Meat Institute –
    - National Turkey Federation – Best Management Practices for Turkey
    - National Cattleman's Beef Association – Beef Quality Assurance
    - American Sheep Industry Association
- Independent Sources (NGOs) Animal Welfare
    - American Humane Association - American Human Certified
    - Humane Farm Animal Care - Certified Humane
    - Animal Welfare Institute - Animal Welfare Approved
    - AnimalHandling.org
    - Global Animal Partnership – 5 Step Program
    - Food Alliance.org – Sustainably Produced Food
- Private Companies – Animal Welfare
    - Validus
    - Silliker
    - Frost, PLLC – FACTA – Animal Welfare Humane Certified

4.1.2. Environment

- Private Companies – Environment
    - Validus - Good Environmental Livestock Production Practices
        - General Site
        - Buildings
        - Sheds and Lots
        - Manure Storage
        - Nutrient Utilization
        - Mortality Management

4.1.3. Food safety

- Global Food Safety Initiative (GFSI)
  - SQF
  - Global Red Meat Standard

4.2. What challenges exist

- Defining the measure of success
- Hidden agenda's
- Confusion among the choices available for animal welfare
- Duplication, duplication, duplication.
- Training of a competent auditor pool

5. Conclusions

1. Social expectations are a strong economic market force. Companies need to know what their consumer's expectations are and strive to meet them.
2. Trust among consumers is a fleeting concept. Everything that is not transparent is suspect and technology makes it hard to stop communication of concerned citizens.
3. With prosperity comes adversaries', adversaries will always try and erode your stakeholder's confidence in your ability to do what you do best.
4. Audits and assessments make a great combination to improve education and provide an organization a proper defense of a given production system.
5. Set up a system that avoids the pitfalls of the systems already established.
6. Work with global certification groups to make things simpler and more cost effective for your producers.

References

Library of Congress. Morill Act of 1862. Web Guides – Primary Documents in American History. Statutes at Large 503 p. https://www.loc.gov/rr/program/bib/ourdocs/Morrill.html. Accessed September 24, 2015.

United States Department of Agriculture – National Institute of Food and Agriculture. Cooperative Extension History. http://nifa.usda.gov/cooperative-extension-history. Accessed September 24, 2015.

Purdue University. History of Cooperative Extension Service. January 2001. West Lafayette, Indiana.

United States Department of Agriculture National Agriculture Library. Farm Animals – Animal Welfare Audits sand Certification Programs. https://awic.nal.usda.gov/farm-animals/animal-welfare-audits-and-certification-programs/animal-welfare-audits-and-2. Accessed September 24, 2015.

Animal Sound...Talks!
Real-time Sound Analysis for Health Monitoring in Livestock

Dries Berckmans [a,*], Martijn Hemeryck [a,b], Daniel Berckmans [c], Erik Vranken [c], Toon van Waterschoot [b]

[a] SoundTalks, Kapeldreef 60, 3001 Leuven, Belgium

[b] KU Leuven, ESAT-ETC/STADIUS, Kasteelpark Arenberg 10, 3001 Leuven, Belgium

[c] KU Leuven, M3-BIORES, Kasteelpark Arenberg 30, bus 2456, 3001 Leuven, Belgium

* Corresponding author. Email: dries.berckmans@soundtalks.com

Abstract

Precision livestock farming (PLF) is a livestock management technology that puts the living organism in the centre and equipment around to measure the animal's response. Sound-based PLF techniques have significant advantages over other technologies such as cameras or accelerometers. Besides the fact that microphones are contactless and relatively cheap, there is no need for a direct line of sight, while large groups of animals can be monitored with a single sensor in a room. The objective of this paper is to present an example of a successful sound-based PLF product, in order to encourage the reader to consider performing sound-based PLF research in the future. The respiratory distress monitor, a tool that automatically monitors the respiratory health status of a group of pigs, was selected for discussion in this paper. Results of five different use cases are discussed to show the effectiveness of the respiratory distress monitor as an early warning tool for respiratory problems in a pig house. It is demonstrated that the tool works for the early detection of animal responses due to technical issues (ventilation problems) and health issues in a wide range of different conditions in commercial European pig houses.

Keywords: PLF, acoustic monitoring, respiratory distress monitor, early warning, cough

1. Introduction

In recent years, several factors have stressed traditional livestock farming. Firstly, the global meat demand has grown extremely as the world population continues to grow. Furthermore, income per capita is increasing, particularly in the upcoming industries like the BRIC-countries. This enables massive new groups of people to consume meat. Another trend specific to consumers in developed countries is the augmented concern towards ethical and environmentally friendly meat production. Additionally, there is a clear need for a reduction in the use of antibiotics in intensive livestock production (Aarestrup, 2012; Kimman et al., 2010). Finally, the global market has seen a shift towards fewer, internationally operating retailers. As the retail market uses price as a primary means of competition, livestock farmers face ever lower margins per animal, which in turn forces them to increase efficiency and exploit economies of scale. This leads to a lower number of farms with more animals per farm (Marquer et al., 2014).

To increase efficiency in livestock farming, Hanton and Leach proposed to combine the information of biological processes with principles and practices of modern engineering and technology (Hanton and Leach, 1981). In essence, livestock farming is described as a process control technology with the living organism at its centre. Berckmans expanded the idea of livestock farming as a process control technology and coined the term Precision Livestock Farming (PLF) (Berckmans, 2006). PLF is based on three guiding principles. Firstly, PLF does not aim to replace the farmer but intends to be a decision support tool. Secondly, the animal is to be considered the most crucial part in the biological production process. Lastly, three conditions are important for favourable monitoring and control: the animal variables need to be monitored continuously, the prediction (expectation) of the animal variable should be reliable with respect to environmental changes and the prediction needs to be integrated with on-line measurements into an analysing algorithm.

Different kinds of sensors are employed to facilitate PLF in a practical setup, e.g., cameras (CCTV, infra-red, 3D, thermal…), accelerometers, flow meters… A particularly interesting type of PLF techniques uses microphones to capture sound. Sound contains a lot of useful information about the animal and its environment (communication, health, welfare). Capturing sound with a microphone is contactless, does not depend on lighting conditions (these pose a real problem for many cameras in practical conditions), allows the monitoring of large groups of animals with a single sensor, is relatively cheap, does not need a direct line of sight, copes with a wide range of temperatures, can be used indoor and outdoor and has an acceptable lifetime.

Recently, a number of examples of sound-based PLF-technology have emerged in different species. Moura et al. showed how thermal (dis-) comfort can be monitored in broilers based on sound analysis (Moura et al., 2008), while Aydin et al. demonstrated the monitoring of broiler feed intake by pecking sounds (Aydin et al., 2014). Rhim et al. developed oestrus detection in sows based on vocalisations (Rhim et al., 2008); Hillman et al. used acoustic monitoring of pigs for the detection of thermal (dis-) comfort (Hillmann et al., 2004), while Manteuffel et al. developed a pig scream detection algorithm (Manteuffel and Schön, 2002).

The objective of this paper is to present an example of a successful PLF application based on sound, in order to encourage the reader to consider sound-based PLF research. The respiratory distress monitor for pigs was selected as the PLF example in this paper. The automatically generated respiratory distress index was compared to findings of the farmer, the veterinarian and a trained assessor.

2. Materials and Methods

2.1. EU-PLF setup

The data used for this paper were obtained in the EU-funded Collaborative Project EU-PLF (EU-PLF, 2012). Pig sounds were captured during 60 fattening rounds in 40 compartments divided over 10 fattening pig farms (4 compartments per farm) across Europe. The different farms had significantly different climatological conditions, housing layout, pig breeds and management styles. The farmer and local veterinarian kept track of their findings in logbooks, without access to use the respiratory distress monitor in real-time (the results were only discussed retrospectively when the fattening batches were finished). Moreover, trained animal experts performed welfare and health assessments of key indicators at discrete moments during the fattening period. The assessments were used as a reference in this paper and were the results of a 10-minute standardised manual cough count in the compartment.

2.2. The measurement system

The hardware used to capture the sounds was a SoundTalks' SOMO+ sound recording device. The phantom-powered (i.e., DC electric power is transmitted through the microphone cables to operate the microphones that contain active electronic circuitry) microphones were connected in a balanced way. This permits the use of long microphone cables, with limited susceptibility to noise. The microphones were typically centred with respect to the position of the animals that were monitored (e.g., pig pen). The microphones were fixed at a height of 2 metres to be close enough to animals, yet not too close for the animals to reach them. Recordings were continuous (24/24, 7d/w) in files of 5 minutes duration. The recording parameters were set to 16 signed integer bit resolution, with a sampling rate of 22.05 kHz (standard WAV file format). The embedded sound card was fanless and protected from the harsh environment by a sealed enclosure. The microphone was protected from the environment by a thin and flexible cover, designed to not interfere with the sound acquisition in the frequency range of interest, i.e., from 1 kHz to 5 kHz (Ferrari et al., 2008). The equipment, much like similar PLF technologies, was subject to a range of robustness-related issues typical in the farm environment (Banhazi et al., 2014).

2.3. Pig cough monitoring

The idea to use automated cough detection in pigs as an early warning tool for respiratory problems in a pig house is not new. Van Hirtum et al. first developed algorithms to discriminate pathological from non-pathological pig coughs (Van Hirtum, 2002). A physiological model to describe the various phases of cough was later developed and Guarino et al. translated the academic results to a field test on a commercial pig fattening farm (Guarino et al., 2008). Ferrari et al. did a characterisation of pig cough sounds using time and frequency-derived parameters (Ferrari et al., 2008), while Exadaktylos et al. proposed a first real-time algorithm for cough detection in controlled conditions (Exadaktylos et al., 2008). Further details on these studies are described in the paper by Vandermeulen et al. (2013).

The link between automatically measured cough and anomalies in respiratory porcine behaviour has been validated extensively in three clinical field trials (Finger et al., 2014; Genzow et al., 2014a, 2014b). The field trials took place on a commercial pig fattening farm, during three different fattening batches. Various clinical tests such as (discrete) serum drawings were used as diagnostics. The serum drawings were tested for *Swine Influenza Virus* (SIV), *Mycoplasma Hyopneumoniae* (M. Hyo) and *Porcine reproductive and respiratory syndrome virus* (PRRSV). The work showed that the automated cough measurement was an objective and unbiased indicator of porcine respiratory health for all three trials.

2.4. Respiratory distress monitor

In 2011, the scientific results of pig cough detection algorithms, as described above, were translated into a commercial product (the respiratory distress monitor). A number of challenges inherent to the process of commercialising scientific research have been faced. The objective of producing the commercial tool was not to count all individual pig coughs accurately in an automated way, but to have a robust tool that gives an early and reliable warning of respiratory problems in a commercial pig house. As pig cough sounds differ between seasons, between race of pigs, between types and stages of diseases (co-infections) and conditions (acoustical, climatological, management etc.), the algorithms that were developed in laboratory conditions must be modified to build a robust tool that works in real practice. All case studies in section 3 are described in terms of a *respiratory distress index* (RD-index). The RD-index is a measure for respiratory distress, relative to the number of animals present in the compartment. RD-values below 10 usually indicate moments with low respiratory distress, while higher RD-values indicate respiratory problems. In all the graphs, a single value of the RD-index is shown per day. It is shown in section 3 that an early warning based on the RD-index was often several days faster than the observation by the farmer / veterinarian, who sometimes even did not detect the respiratory problem.

3. Results and Discussion

Table 1 summarises the country, the total number of pig places and the assessment data of the different case studies. The total number of pig places in the compartment is given in the format a × b = c, with '*a*' the number of pens, '*b*' the number of animals per pen and '*c*' the total capacity of animals in the compartment. The last three columns describe the assessment data, i.e., the date of the assessment, the actual number of pigs in the compartment during the assessment and the number of coughs counted in 10 minutes by the assessor.

3.1. Case 1: Co-infection PRRSV, SIV and M. Hyo

Figure 1 shows the evolution of the respiratory distress as a function of time, measured on a farm in the Netherlands. On October 18th 2014, 79 piglets of 10 weeks old entered the compartment. Necropsy (PCR-analysis of lung section) was performed on two piglets that had died before reaching the age of 10 weeks. Both piglets were positive for Mycoplasma

Hyopneumoniae (*M. Hyo*) and negative for Influenza-A virus (*SIV*), PRRSV NA and PRRSV NA-HP. One piglet was further positive for PRRSV EU and weakly positive for PCV-2, while the other piglet was negative for both. All animals received vaccination against PCV-2, but not against PRRSV, M. Hyo and Influenza.

Table 1: Overview of the country, total number of pig places and assessment data per case.

Case	Country	Pig places	Date (yyyy-mm-dd)	# Pigs	10 min assessor cough count
1	The Netherlands	$8 \times 10 = 80$	2014-10-24	79	38
			2014-11-21	79	108
			2014-12-19	79	27
2	The Netherlands	$8 \times 10 = 80$	2015-03-13	72	76
			2015-04-15	72	4
3	Hungary	$20 \times 10 = 200$	2014-04-18	200	1
			2014-05-23	199	3
4	France	$16 \times 12 = 192$	2014-06-18	191	67
			2014-07-21	186	50
5	Spain	$4 \times 25 = 100$	2014-03-13	99	0

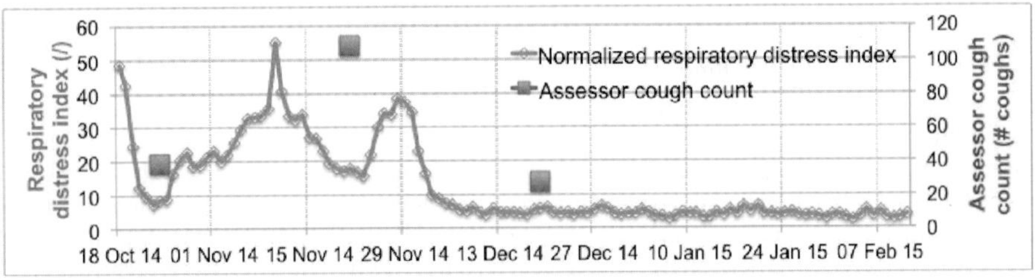

Figure 1: Case 1, the Netherlands, co-infection PRRSV, SIV and M. Hyo.

The veterinarian of the farm decided that treatment with doxycycline and sodium salicylate was necessary when the piglets entered the compartment on October 18th. This treatment was stopped after 4 days. It is clear that the respiratory distress index in this period is in good correlation with the observation of the veterinarian. The index starts at a high level (48), before the effect of the treatment becomes clear and the index drops to values below 10 after 5 days.

From October 23rd onwards, the respiratory distress was rising to reach a peak level of 55 on November 10th. The farmer and veterinarian, who could not consult the graph of respiratory distress in this two-direction blind test, were unaware of further respiratory problems in this period and no treatment was given to the animals. A routine diagnostic test (oral fluid multiplex PCR test) on November 12th revealed that the animals were positive for Mycoplasma Hyorhinis, weakly positive for Mycoplasma Hyopneumoniae and negative for PRCV, PRRSV, SIV, PCMV and PCV-2. In neighbouring compartments, positive cases of SIV were found. Because high coughing levels were noticed, an 8-day treatment with doxycycline and sodium salicylate was restarted on November 13th. This type of treatment usually does not last longer than 5 days on this farm, but the severity of the respiratory problem forced a longer treatment period. The effect of the treatment is clearly visible, as the respiratory index drops in this period. The added value of the respiratory distress monitor is clear in this case as the rise of distress could have been noticed more than 2 weeks earlier. Faster treatment of the animals would have resulted in less economic loss (higher average daily gain, lower feed conversion ratio, etc.).

A third increase in respiratory distress was visible around November 29th. Based on oral fluid samples from neighbouring compartments, the veterinarian believes that this increase was most probably caused by a combination of diseases, i.e., PRRSV, SIV and M. Hyo. After November 29th, the respiratory distress index returned to normal values, i.e., below 10.

The automatically generated respiratory distress index is in good correlation with the findings of the human assessor, as a high number of coughs were counted on 3 occasions. It is clear that the continuous automated measurement of respiratory distress gives a much clearer picture of the complex respiratory situation on the farm. In combination with diagnostics and the knowledge from farmer and veterinarian, the respiratory distress monitor proves to be a tool with added value and economic impact.

3.2. Case 2: SIV

Figure 2 shows the evolution of the respiratory distress in function of time, measured on a farm in the Netherlands. On February 24th 2015, 72 piglets of 10 weeks old were placed in the compartment.

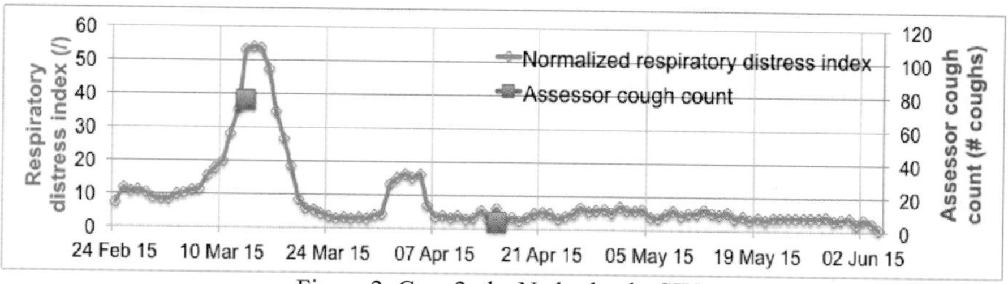

Figure 2: Case 2, the Netherlands, SIV.

The clear increase in the respiratory distress index from March 8th–14th corresponds very well with the high number of coughs (76) counted by the assessor on March 13th. Only on March 12th, i.e., about 4 days after the increase in the automatically generated respiratory distress index, the farmer noticed an increased level of coughs during his routine check. A 3-day group treatment with doxycycline and sodium salicylate was immediately started and from March 15th onwards, the respiratory index continued to drop. A level below 10 was reached again on March 19th. Based on oral fluid samples (from this compartment and other compartments in the farm), the veterinarian concludes that SIV is the probable cause of the respiratory distress.

A second, smaller peak in the respiratory distress index is visible between April 1st–7th. The farmer noticed an increase in coughing in a few animals on April 2nd, i.e., again one day later than the sudden increase in the respiratory distress index. A few animals were treated with doxycycline and sodium salicylate for 3 days. The respiratory distress graph correlates well with the observation of the farmer. The lower respiratory distress values in the second peak can be explained by a lower number of animals that were affected by the disease in this period. The hypothesis of the veterinarian is that the majority of the animals were already infected with SIV during the first peak, while the remaining animals (a smaller group) were only affected a few weeks later. Further diagnostics information to support this hypothesis was not available here.

The low value of the assessor cough count on April 15th (4) is in excellent correlation with the respiratory distress index, which had already dropped below 10 on April 6th.

3.3. Case 3: pleuritis / pneumonia

Figure 3 shows the evolution of the respiratory distress index for a batch of fattening pigs on a Hungarian farm. The farmer did not notice any respiratory problems during the round and no medical treatment was given to the animals. The findings of the assessor (1, resp. 3 coughs) support the idea that there were few respiratory problems on April 18th and May 23rd. Near the end of the batch though, the respiratory distress index starts to increase to reach a maximal value of 14 just before the end of the fattening round. Slaughterhouse data of 25 selected pigs revealed that 12/25 animals had pneumonia while 3/25 animals had pleuritis. No diagnostic data was available to retrieve the exact cause of the disease, but it is clear that there was a respiratory problem that remained unnoticed by the farmer. The hypothesis is that earlier medical intervention (triggered by the respiratory distress monitor) would have reduced the economic losses for the farmer.

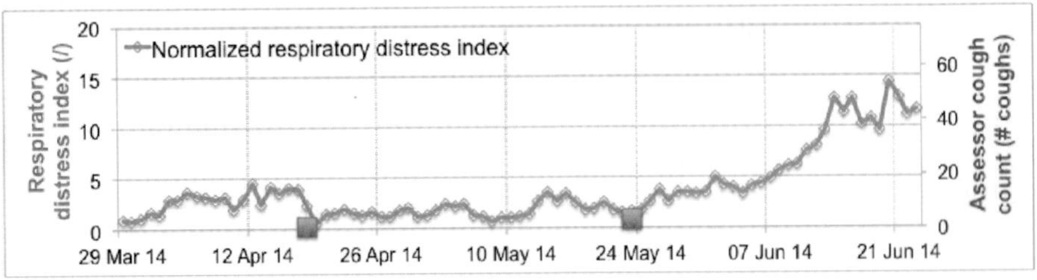

Figure 3: Case 3, Hungary, pleuritis / pneumonia.

3.4. Case 4: air washer problems

Case 4 presents the data of a fattening batch between the end of April till the end of August 2014 on a farm in France (Figure 4). Between May 30th and June 5th, there was some missing data due to humming (undesired noise caused by the electric circuit) in the microphone. Before entering the fattening compartment, the piglets were vaccinated against M. Hyo and PRCV; no further vaccines were given after April 21st.

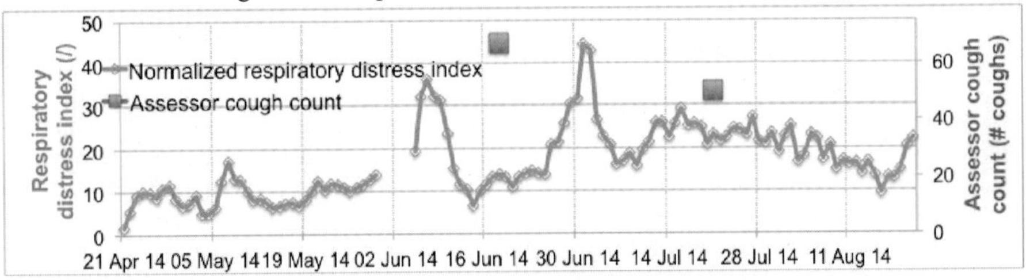

Figure 4: Case 4, France, problems with air washer.

On May 21st, the farmer first noticed that *something* was wrong on the farm. Unaware of the cause of the problem at that time, he reported that indoor temperatures were unusually high (31°C) although the ventilation was working at maximal capacity. The situation was monitored more precisely and on May 30th a blocked (by dust and bacteria) air washer was identified as the underlying cause of the problem. The farm was equipped with an air washer in order to reduce the dust and odour levels. As a first solution, 1 of the 3 layers of the air washer was removed on May 30th to increase the airflow through the pig house. The farmer reported that the pigs started to cough more in the next days and on June 3rd a 6-day treatment with aspirin and vitamin C was started. The veterinarian indicated that the pigs were suffering from pneumonia, although the exact underlying pathogens were unknown. The respiratory distress index clearly shows a good correlation with the findings of the farmer and the veterinarian in this period, with a peak up to 36 on June 7th.

Since the problems with the air washer persisted, the farmer removed the remaining 2 layers of the air washer on June 13th. Interestingly, the farmer reported that the cough levels dropped significantly after the treatment period of June 3rd–8th, but not to levels he would consider normal for his farm. The latter is reflected by the respiratory distress index, although there was a clear secondary peak around July 2nd that was not noticed by the famer (and therefore no treatment was given). Additional diagnostics and / or treatment seemed appropriate and would have resulted in less economic losses due to higher feed conversion rates and lower average daily weight gains. The manual coughs counted by the assessor (respectively 67 and 50 coughs in 10 minutes) confirmed that there were respiratory problems in this batch.

3.5. Case 5: thermal shock

Figure 5 shows the evolution of the respiratory distress index measured on a fattening farm in Spain. Except for the beginning and the end of the fattening round, the RD-index was below scale 10 indicating no significant respiratory problems. The farmer and the veterinarian did not notice any respiratory problems during this fattening batch; hence no medical treatment was given during this round. The only medical interventions were traditional vaccinations against Aujeszky on January 25th and February 16th. The assessment on March 13th (0 coughs) further supports the idea that the respiratory status of the pigs was good during the middle of the fattening round.

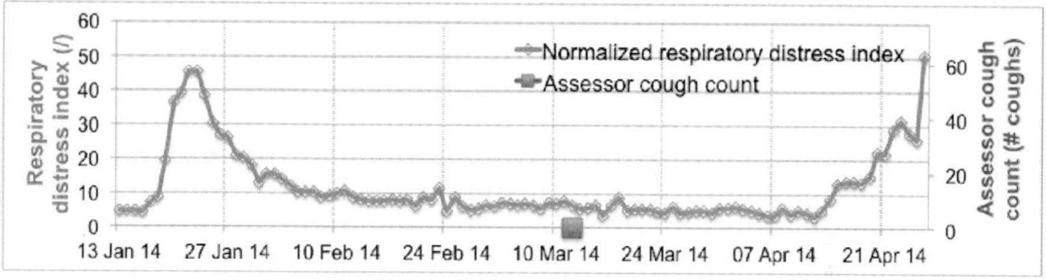

Figure 5: Case 5 Spain, thermal shock.

The peak in respiratory distress index in the beginning of the round is very characteristic and similar peaks were measured for other batches / compartments on this farm. The peaks were a lot higher during winter times and occurred typically within the first week after placing the animals in the compartment. As the building was very open (suited to the hot climate that occurs during most of the year) and the farmer did not pre-heat the compartment prior to the arrival of the pigs, the hypothesis was that the thermal shock between the nursery and the fattening compartment makes the pigs vulnerable to different respiratory infections. However, no further data was available to support this hypothesis, although it is well known that thermal shock can have these effects

From April 12th onwards, a clear increase in the respiratory index is visible again. The farmer did not notice any respiratory problems and no actions were taken. Based on previous examples, it was assumed that there was a respiratory problem and that the economic losses could have been reduced if the farmer would have taken appropriate measures near the beginning (pre-heating) and end of the batch (treatment).

In summary, this paper presented five different cases to analyse the reliability of the continuous, automated monitoring of respiratory distress in pigs.

4. Conclusions

This work demonstrated how the tool gives earlier warnings (up to 2 weeks earlier!) compared to a situation where the farmer and veterinarian rely on their own routine observations without the monitor. Different causes of respiratory problems ((co-) infections as well as technical problems with air washers and ventilation) have all been shown to result in an increase in automated respiratory distress measure.

Acknowledgements

The authors gratefully acknowledge the European Community for financial participation in Collaborative Project EU-PLF KBBE.2012.1.1-02-311825 under the Seventh Framework Programme. This work is further supported by funding of the IWT Baekeland (Agentschap voor Innovatie door Wetenschap en Technologie — Agency for Innovation by Science and Technology in Flanders) (No. IWT 140245).

References

Aarestrup, F. 2012. Sustainable farming: Get pigs off antibiotics. Nature 486, 465–466.

Aydin, A., Bahr, C., Viazzi, S., Exadaktylos, V., Buyse, J., and Berckmans, D. 2014. A novel method to automatically measure the feed intake of broiler chickens by sound technology. Comput. Electron. Agric. 101, 17–23.

Banhazi, T., Vranken, E., Berckmans, D., Rooijakkers, L., and Berckmans, D. 2014. Practical problems associated with large scale deployment of PLF technologies on commercial farms. In *Sessions of the 65th EAAP Annual Meeting*. Copenhagen, Denmark: Wageningen Academic Publishers. 105-112

Berckmans, D. 2006. Automatic on-line monitoring of animals by precision livestock farming. In *Livestock Production and Society*, R. Geers, and F. Madec, eds. Wageningen Academic Publishers. 51–54.

Exadaktylos, V., Silva, M., Aerts, J.M., Taylor, C.J., and Berckmans, D. 2008. Real-time recognition of sick pig cough sounds. Comput. Electron. Agric. 63, 207–214.

Ferrari, S., Silva, M., Guarino, M., and Berckmans, D. 2008. Analysis of cough sounds for diagnosis of respiratory infections in intensive pig farming. Trans. ASABE 51, 1051–1055.

Finger, G., Hemeryck, M., Gomez-Duran, O., and Genzow, M. 2014. Practical application of the Pig Cough Monitor in a German fattening pig herd with PRDC. In *Proceedings of the 23rd IPVS Congress*, Cancun, Mexico. 207–208.

Genzow, M., Gomez-Duran, O., Hemeryck, M., and Finger, G. 2014a. Course of cough in two batches of fattening pigs with different respiratory pathogen exposure. In *Proceedings of the 23rd IPVS Congress*, Cancun, Mexico. 212-213

Genzow, M., Gomez-Duran, O., Strutzberg-Minder, K., Finger, G., and Hemeryck, M. 2014b. Monitoring of a commercial fattening herd by means of the Pig Cough Monitor and oral fluid diagnostics. In *Proceedings of the 23rd IPVS Congress*, Cancun, Mexico. 205-206.

Guarino, M., Jans, P., Costa, A., Aerts, J.-M., and Berckmans, D. 2008. Field test of algorithm for automatic cough detection in pig houses. Comput. Electron. Agric. 62, 22–28.

Hanton, J.P., and Leach, H.A. 1981. Electronic livestock identification system. U.S. patent 4 262 632 A

Hillmann, E., Mayer, C., Schön, P.-C., Puppe, B., and Schrader, L. 2004. Vocalisation of domestic pigs (Sus scrofa domestica) as an indicator for their adaptation towards ambient temperatures. Appl. Anim. Behav. Sci. 89, 195–206.

Van Hirtum, A. 2002. The acoustics of coughing. PhD Thesis. KU Leuven. Leuven, Belgium.

Kimman, T., Smits, M., Kemp, B., Wever, P., and Verheijden, J. 2010. Banning Antibiotics, Reducing Resistance, Preventing and Fighting Infections: White Paper on Research Enabling an Antibiotic-free' Animal Husbandry. Wageningen Academic UR.

Manteuffel, G., and Schön, P.C. 2002. Measuring pig welfare by automatic monitoring of stress calls. Agrar. Berichte 29. 110-118

Marquer, P., Rabade, T., and Forti, R. 2014. Pig farming in the European Union: considerable variations from one Member State to another. Eurostat Stat. Focus 2014. 15/2014, 1-12

Moura, D.J. de, Nääs, I. de A., Alves, E.C. de S., Carvalho, T.M.R. de, do Vale, M.M., and Lima, K.A.O. de 2008. Noise analysis to evaluate chick thermal comfort. Sci. Agric. 65, 438–443.

Rhim, S.J., Kim, M.J., Lee, J.Y., Kim, N.R., and Kang, J.H. 2008. Characteristics of Estrus-related Vocalizations of Sows after Artificial Insemination. J. Anim. Sci. Technol. 50, 401–406.

Vandermeulen, J., Decré, W., Berckmans, D., Exadaktylos, V., Bahr, C., and Berckmans, D. 2013. The Pig Cough Monitor: from research topic to commercial product. In *Proceedings of Precision Livestock Farming '13*, Leuven, Belgium. Berckmans D., J. Vandermeulen eds. Rosseels Printing, 717–723.

Techniques for Measuring Animal Physiological and Behavioral Responses with Respect to the Environment

Tami M. Brown-Brandl

USDA Agricultural Research Service US Meat Animal Research Center, Clay Center, NE 68933, USA

Email: tami.brownbrandl@ars.usda.gov

Abstract

Environmental effects cause animal production inefficiencies and animal well-being issues. Thus, many experiments have been designed to understand thermal stress and to test different means to relieve it. There are multiple physiological responses and behavior/activities that can be measured to discern the level of response to environmental stressors. Additionally, choices of instrumentation and interpretation of the responses are dependent on understanding the heat balance of a homeotherm. This paper investigates the current understanding of heat balance concepts for livestock, and reviews various methods of measuring common physiological measurements (body temperature, respiration rate, and surface temperatures) are included as well as behavioral and activity level measurement methods

Keywords: Behavior, body temperature, image analysis, respiration rate, thermal images

1. Introduction

Regulatory measurements are valuable in the evaluation of physiological and behavioral responses to thermal challenges. A careful assessment of (1) objectives of the study, (2) environment conditions, and (3) an understanding of the thermal-regulatory mechanisms will help guide the choice of which measurements and methods should be used. This paper is a review of several different methods of measuring physiological and behavioral responses.

2. Physiological Measurements

2.1. Heat balance

The physiological and behavioral responses of animals are impacted by the environment. Understanding thermal balance of an animal helps explain the measured responses. Maintaining the balance in the homeostasis of animals, has both physiological and thermodynamic components. Equation 1 describes the overall process of homeostasis in an animal, where HP is heat production, HL is heat loss, ΔT_{body} = change in body temperature (°K), c_p = specific heat of the whole animal (J kg^{-1} °K^{-1}), and m = mass of the animal (kg).

$$HP - HL = \Delta T_{body} \times c_p \times m \tag{1}$$

The difference between heat production and heat loss results in an accumulation of heat in the body, which is manifested as a change in body temperature. Numerically, this accumulation is the change in body temperature multiplied by the specific heat of the whole animal and mass of the animal. The specific heat of the whole body has been reported to be 3.47 kJ kg^{-1} °K^{-1} (Blaxter, 1989). This number can vary depending on the exact make-up of the body.

Heat production is a by-product of the breakdown and utilization of feedstuffs. Classically, heat production has been divided into four components: basal metabolism, heat of digestion, heat of activity, and production metabolism (heat from the production of milk, egg, etc.). Basal metabolism is the heat produced from the maintenance of body's cells (no active digestion or movement). Basal metabolism is very difficult to measure, and is not typically completed on animals. A surrogate measurement for basal metabolism in animals is fasting heat production. Heat of digestion is the heat resulting from the intake and digestion of feedstuff. Heat of activity is the heat generated in muscles during physical activity. Production metabolism is the heat created during the physiological processes that yield products such as milk in the case of a dairy cow or eggs in the case of the laying hen. Heat production can be measured by indirect calorimetry (Nienaber and Maddy, 1985). In this procedure, heat production is calculated with Eq. (2) by measuring the consumption of oxygen (O_2, l) and the production of carbon dioxide (CO_2, l) and methane (CH_4, l, in the case of ruminants) to calculate the total heat production (HP, units) of the animal (Nienaber and Maddy, 1985).

$$HP = 16.18 \times O_2 + 5.02 \times CO_2 - 2.17 \times CH_4 \tag{2}$$

Heat lost from the body is accomplished by two physical processes: sensible and latent heat loss (Figure 1). Sensible heat loss is the process of losing heat by conduction (heat lost to another solid object), convection (heat lost from the body to a fluid, which can be air or water), and radiation (heat lost from the body to another body through radiant energy). Sensible heat can be gained or lost from an animal and is dependent on the temperature gradient between the animal's surface and its surroundings. Latent heat is the heat lost by evaporation of moisture from the surface of the skin or the respiratory tract of the animal. Latent heat can only be lost, never gained from the environment. While the concepts of heat transfer are relatively simple when applied to a static non-biological object, heat transfer from a dynamic living animal is quite complex.

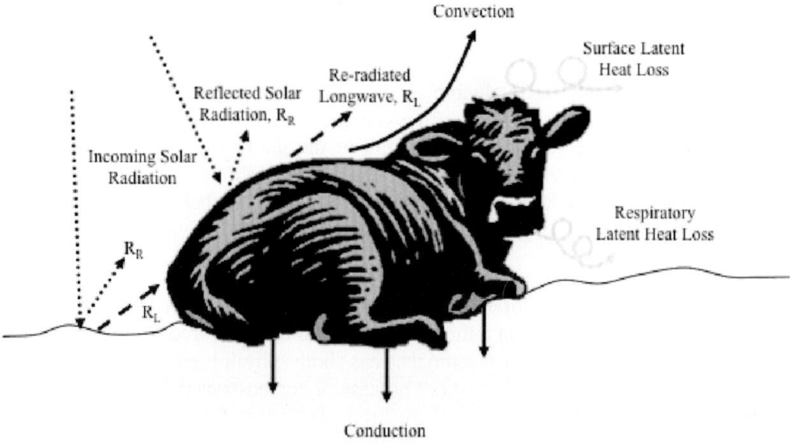

Figure 1. A steer lying in the open exposed to sun. All modes of sensible and latent heat loss are shown.

While it is difficult to accurately estimate heat loss from an animal using a series of simple equations, it is important to understand the components of each mode of heat transfer as this impacts the animal's behavior. Understanding the different components helps develop different management strategies. Conductive heat transfer is a function of temperature gradient between the animal and the surface in contact, the contact area and the resistance of the contact tissue and coat. When an animal is standing the contact area is minimal. Convective heat transfer is defined by temperature gradient between the surface and the air, the air velocity, and the exposed surface area of the animal. The important factors in the radiation component are temperature difference between the surface of the animal and its surroundings and the exposed surface area. A more detailed explanation and the equations used to calculate heat loss can be found in Ehrlemark and llvik (1996), Turnpenny et al. (2000) or McGovern and Bruce (2000).

All of these factors are very difficult to model, because an animal is very adept at changing their heat transfer mechanisms to balance the heat loss with the heat production under production conditions. An animal can change its conductive heat transfer simply by changing the amount of surface area that is in contact with the surface or is changing the area where they are lying. For example, the animal could choose to lay on a cool wet surface or on warm dry ground. The convective heat transfer is affected by hair coat characteristics, which can be changed over time. The density of the coat can change as the animal is exposed to an environment for a prolonged period (i.e., thick winter coat). The thickness of the coat can be changed instantly, by piloerection or fluffing up of the hair coat to add more "dead air space". The convective transfer and latent heat losses both change depending on how saturated the hair coat is. The animal can elect to stand facing the wind, or stand across from the wind or avoid the wind. Changing the position of the body to maximize or minimize the area exposed to the radiant source (for example, the sun or the night-time sky, or possibly a heater) thus changing the radiant heat transfer. An animal in an open outside pen during the summertime may seek shade when faced with extreme heat. Shade can take on many different forms including the shade of the fence line or shade created by the other animals. These behavioral changes are very dynamic and thus are a major factor in maintaining homeothermy.

Heat loss can be measured by direct calorimetry, which is the direct measurement of radiation, convection, conduction, and latent heat losses. Direct calorimetry measurements are accomplished by using either a heat sink calorimeter or a gradient layer calorimeter (Blaxter, 1989). Direct calorimetry is usually used to measure heat loss on small animals.

Heat stress is a complex interaction between physiology and behavior of the animal and the physics of heat transfer. While the physics of the heat transfer component can be described using a set of equations, the physiology can only be approximated with equations, and the behavior of the animals adds a dynamic component to heat transfer and heat balance equations. Therefore, to mathematically capture the complete and accurate interactions of these factors is very difficult. However, understanding these factors is invaluable in further study of the impact and management of thermal responses in livestock species.

When monitoring thermal regulatory responses consideration of the methods needs to be taken.

2.2. Body temperature

Monitoring body temperature can be broken into two categories: spot monitoring and dynamic monitoring. The disadvantage to spot monitoring is that the animals need to be restrained. The simple act of restraining the animal can significantly change the body temperature. The degree of change is dependent on the amount of physical exertion the animals expends.

Body temperature can be quickly taken using a mercury rectal thermometer. The Body temperature can be monitored on a dynamic basis. Two separate points need to be considered before monitoring body temperature: (1) the attributes of the measuring device including the logger and the temperature sensor, and (2) location of the sensor itself.

Specific characteristics of the measuring device include the size and the thermal mass of the sensor, the accuracy, the resolution, the total memory of the device, and the battery life. The size of the sensor is important for the logistical reasons, but in addition a large device has more thermal mass. The placement of the actual temperature sensor within the device is critical. The reaction time of the sensors can vary immensely depending on size and location of the sensor. Before choosing a sensor, be aware of the accuracy and the resolution of the sensor. The accuracy is how close the reading is to the actual temperature, and the number of significant digits being recorded determines the resolution. Total memory of a device defines the number of data points that can be stored before the logger needs to be downloaded. Currently, the total memory of the device seems to be the limiting factor on data collection with some devices.

Several different locations have been documented in the literature for monitoring body temperature including: tympanic temperature, ear canal temperature, rectal temperature, vaginal temperature, abdominal cavity temperature, and digestive track temperature. Every method has advantages and disadvantages associated with the method. Careful consideration needs to be taken when choosing the method.

Tympanic temperature methods have been developed for both cattle and pigs. The method described in Brown-Brandl et al. (1999) for cattle and Hanneman et al. (2004) for pigs. Tympanic temperature was considered a good measure of body temperature because of the close proximity of the tympanic membrane to the hypothalamus, the thermal regulatory center of the brain. One disadvantage of tympanic temperature is the need for frequent changes between left and right ears. Seven to 10 days is about the maximum to leave the probe sealed in the ear, after that the risk of ear infections is high. In addition, the loggers are either wrapped into the ear (making the ear heavy), or contained in a pocket on a head halter or harness or vest on the back of the animals. This can lead to broken leads and missing data.

Ear canal temperatures were developed based on the tympanic temperature method. Sometime ear canal temperatures are mistakenly referred to as tympanic temperatures; however sensors in the ear canal do not approach the tympanic membrane and, therefore, may not track accurately with tympanic temperature. This method typically uses an iButton®, which is a small logger and temperature sensor in one package. The advantage is the entire package (logger and sensor) is placed into the ear, eliminating the risk of broken leads and missing data. The requirement to change ears is the same as when tympanic temperatures are measured. In addition, the iButton® does not set deep in the ear canal; therefore, it can be influenced by outside temperatures during extreme weather (Meyer et al., 2015).

Rectal temperature probes can be used to collect dynamic body temperature data. Rectal probes can be easily installed, and held in place with elastic bands, to allow for the animal to move and stretch with ease. When measuring rectal temperature, the probe needs to be at a depth to ensure the sensors to be beyond the rectal sphincter to ensure accurate body temperature monitoring. A drawback of this system is that the maximum length of time for monitoring rectal temperature is 3–5 days without causing irritation to the surrounding tissue. Description of two different methods can be found in Brown-Brandl et al. (2003) and Reuter et al. (2010).

Vaginal temperature probes can be held in place with a specially designed device, so unlike rectal probes, there is no need to have an external device to secure the logger in place. A few different designs have been developed and used (Hillman et al., 2009; Burdick et al., 2012). Vaginal temperature can be used in group penned animals since there is no external securing device. Vaginal temperature can be left in place in heifers for up to 3 – 4 weeks.

Abdominal cavity temperature (Brown-Brandl et al., 2003) and digestive track temperatures (Rose-Dye et al., 2014) require a specialized sensor that can broadcast the information outside the animal. In addition, to gather data from the abdominal cavity requires surgery to install the sensors and a special processing facility to retrieve the sensors. The data from the abdominal cavity is good, but completely dependent of the system used to collect the data. The data from the digestive track, especially in cattle, is influenced by water intake, so data may be unusable for extended periods of time, and it can be difficult to know for certain if the data is "good data" or not. In swine, the digestive track data is not influenced by water intake, but the sensors pass through the digestive system in a matter of days. Battery power in either system is critical, because data will be collected for an extended period of time and a significant amount of energy is required to broadcast the data outside the animal.

2.3 Thermal images

Amount of blood flow to the surface of the skin is important in understanding the heat balance of an animal. An animal will shift blood flow to the surface of the skin/respiratory track to increase heat loss. When an animal is cold, the blood flow is shunted to the interior of the body to preserve heat.

Thermal cameras can help detect the surface temperature of objects, and thus have been determined to be useful in detecting changes in the surface temperature and thus blood flow to the surface of the skin (Bagavathiappan et al., 2009; Ring and Ammer, 2012), see Figure 2. These changes have been used to diagnosis some illnesses (Eddy et al., 2001), determine thermal comfort (Brown-Brandl et al., 2013), detect estrous (Sykes et al., 2012), and relationship with core body temperature (Tan et al., 2009).

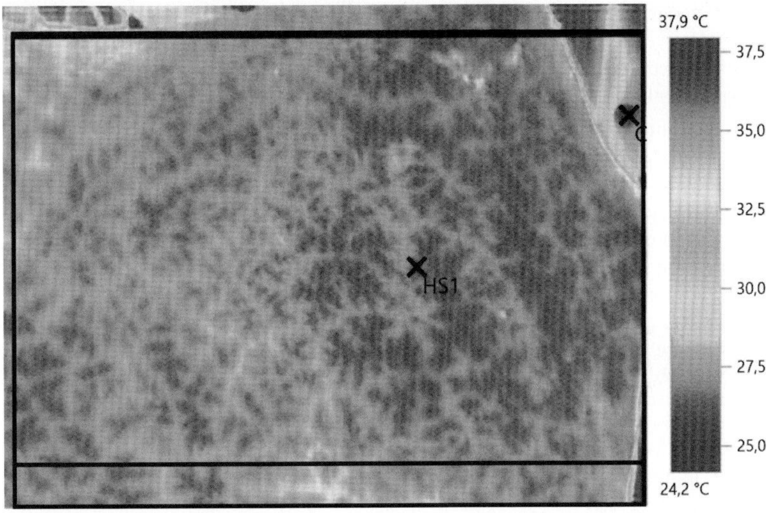

Figure 2. Thermal image of a *bos indicus* steer in summer time conditions (image courtesy of Dr. Luciane Silva Martello). In this image you can see the blood flow to the surface of the skin.

2.2.1. Cameras available

The cameras vary in the number of pixels, accuracy of the thermogram, and ability to zoom, focus, ability to save an image and the amount of time to save an image. Two example cameras: FLIR T400 and FLIR 640 (FLIR Systems, 27700 SW Parkway Ave, Wilsonville, OR 97070): FLIR T400 infrared camera, which measures surface temperatures and generates a thermogram in color (thermal sensitivity 0.05°C at 30°C; accuracy of ± 2% of reading; IR resolution of 320 × 240 pixels). The FLIR T640 camera has more resolution and slightly better thermal sensitivity (thermal sensitivity 0.035°C at 30°C; accuracy of ± 2% of reading; IR resolution of 640 × 480 pixels). There are an endless number of different cameras and different options.

2.2.2. Capturing images

The animal's skin temperature is very dynamic, extra care needs to be taken when capturing thermal images. Activity increases heat production, and therefore changes the animal's thermal state. Water/manure applied to the surface of the animals and the sun shining on the animals drastically change the surface temperature on the skin. Also, the thickness of the hair coat can sometimes make it difficult to detect changes in skin temperature.

2.2.3. Analyzing the images

Software packages are available for analyzing thermograms; FLIR has several levels of software packages. The most comprehensive package is ResearchIR, this allows for many different analyses of the image. There are also tools within Matlab (MathWorks, Inc., 3 Apple Hill Dr, Natick, MA 01760) for analysis of thermal images.

2.4 Respiration rate

Respiration rate is an important thermal regulatory mechanism especially for species that rely on panting for latent heat transfer. When the outside temperature begins to rise and approach the surface temperature of the animal, the animals rely more and more on latent heat loss to maintain homeostasis. Therefore, in panting species (cattle, pigs, sheep, poultry) their respiration rate increases.

Measuring of respiration rate can be done manually. The easiest method is to time 5 or 10 flank movements, then take the number of flank movements (5 or 10) and divide by the time in seconds and multiple by 60 to get breaths/min. To get accurate respiration rate counts animals need to be acclimated to the observers and any equipment they are carrying (stop watch, clip board, etc.). Observers need to be patient and make sure that the animals are acting normally before counts are recorded. The benefit to hand measurements is many animals can be recorded during an experiment; but the number of records is typically 1–4 measurements a day.

Several electronic systems for the recording of respiration rate have been developed. They rely on the physical movement of the flank (Eigenberg et al., 2000), the temperature change between inhaled and exhaled air (Nagl et al., 2003), and sound analysis (Eigenberg et al., 2002). The advantages of electronic devices are that dynamic data can be collected and there is no need to have a human present, which could change the animal's response. Careful selection of the methods of securing the device and the data logger is needed, as covering any portion of the skin changes the thermoregulatory response. However, the current electronic devices to monitor respiration rate, are time consuming to properly install and require daily management to ensure that the sensors are placed correctly. These time commitments limit the application of these sensors to only a few at a time.

3. Behavior and Activities

Many different behaviors and activities have been recorded and analyzed. For this paper, the discussion will focus only on those activities and behaviors that can be quantified electronically.

3.1. Feeding behaviors

Feeding behavior and feed intake are influenced by many factors including social, thermal, and health related. Feeding behavior in livestock species has been reported in different studies (Bach et al. 2004; Bigelow and Houpt 1988; Chapinal et al. 2007; Morgan et al. 2000; Nienaber et al. 1990, 1991). Each study includes different parameters of feeding behavior, which have included feed intake, meal (bout) length, meal (bout) interval, number of meals (bout) per day, total time spent eating, and rate of eating.

Systems are currently available to measure feed intake in association with feeding behavior for cattle (Basarab et al., 2003; Chapinal et al., 2007; Kelly et al., 2010), swine (Andree and Huegle, 2001; Chapinal et al., 2008; Hyun and Ellis, 2002; Nienaber et al., 1991), small livestock (Basarab et al., 2003; Gipson et al., 2006, 2007; Goetsch et al., 2010), and poultry (Puma et al., 2001). These systems provide the user with feed intake data (Basarab et al., 2003) in addition to feeding behavior but, they limit access to feeders by a single animal at a time; furthermore, the high cost of the units result in only a limited number of feeding stations being placed in a pen. While feed intake is a very important parameter in some studies (i.e., genetic evaluation, nutrition studies), it may not be necessary for others.

Limited access to the feeder can alter the animal's behavior and may not be representative of behavior in commercial sized pens. A system to monitor only feeding behavior can be designed so that feeding behavior is not influenced. Furthermore, this system has less equipment and no moving parts; therefore, is less costly and easier to maintain. Therefore, systems that allow recording of feeding behavior (without feed intake) may be more applicable to production animal facilities or for experiments, which hope to replicate industry conditions. The parameters measured in such a system include meal (bout) length, meal (bout) interval, number of meals (bout) per day, and total time spent eating. Feed intake (kg day^{-1}) and rate of eating (g min^{-1}) are two parameters that cannot be monitored in such a system.

Radio frequency identification systems have been used to record feeding behavior. The tags typically used in these systems are passive tags, meaning there is no battery in the tag itself. The tag is charged when placed in the vicinity of a specialize antenna. Once the tags are charged they transmit their number, which is received and read by the same antenna. The tags and the antennas are specialized and designed for a specific frequency. The antennas are placed over the feeding spaces and record animals that are at the feeder. Each frequency of RFID system has positives and negatives associated with it. An example of a low-frequency system is described in detail in Brown-Brandl et al. (2011) and an example of a high-frequency system is described in detail in Maselyne et al. (2014).

3.2. General activity

Activity level and selective behaviors significantly impact the thermal status of an animal. Therefore, it may be of interest to record behavior or activity changes in this at different environments (Brown-Brandl et al., 2006; Huynh et al., 2005). Behaviors can be recorded by live observation, using video or time-lapse photos. There are advantages and disadvantages to each method. However, care must be taken to ensure that activity level and/or behaviors are accurately recorded and that observers are not interfering with the animals. Also, it must be noted that video/time-lapse images can easily be recorded but it time consuming to analyze.

Automated analysis methods to detect behavior are being developed but are specific to specialized circumstances and generally limited to a single animal within an image (Berckmans et al., 2008; Cangar et al., 2008; Lao et al., 2015; Leroy et al., 2006). Activity level or huddling behavior has been used in groups of animals (Shao and Xin, 2000). Many analyzes methods include image analysis toolbox within Matlab. This allows investigators to use many different tools to select the region of interest (ROI) and differentiate it for the background. The key to image analysis is to have a consistent background color that is different from the ROI. Also, lighting becomes a critical factor in capturing images to be used in this type of analysis.

A camera that has gained some interest recently is the Kinect camera (Microsoft Corporation, 1 Microsoft Way, Redmond, WA). The Kinect camera captures a digital image along with a depth image. The depth image is stored as an array that is 640 × 480 pixels and gives the distance from the camera at each point. The advantage of the camera is that images can be analyzed without worrying about lighting. In addition, heights of animals can be determined from the depth image; thus, the difference between a standing, sitting, or laying animal can be monitored even when it would be difficult to discern in a digital image (Lao et al., 2015). The curvature of the spine can be detected (Viazzi et al., 2014) thus helping detect lameness in dairy cattle. The uses of this camera are quite varied and not fully utilized at this point.

4. Conclusions

Thermal stress causes animal production inefficiencies as well as serious animal well-being issues. Therefore, many experiments have been designed to understand thermal stress and to test different means to relieve it. Techniques for measuring animal physiological and behavioral responses are varied and are still developing. Careful selection of methods is needed to ensure that the data captured is reflective of the true animal's response to the environment and not to the method being applied. There is no single right method; careful selection of methods must be completed prior to each experiment. Considerations to include are (1) number of animals being monitored, (2) length of time the measurements need to be completed, (3) minimizing the amount of handling, and 4) determine if it is important to replicating industry standards.

References

Andree, H., T.U. Huegle, 2001. Effect of single animal feeding stations and group size on growing performance of group housed fattening pigs. ASAE, St. Joseph, MI.

Bach, A., C. Iglesias, I. Busto, 2004. Technical note: A computerized system for monitoring feeding behavior and individual feed intake of dairy cattle. Journal of Dairy Science. 87, 4207-4209.

Bagavathiappan, S., T. Saravanan, J. Philip, T. Jayakumar, B. Raj, R. Karunanithi, T.M.R. Panicker, M.P. Korath, K. Jagadeesan, 2009. Infrared thermal imaging for detection of peripheral vascular disorders. Journal of Medical Physics / Association of Medical Physicists of India. 34 (1), 43–47. http://dx.doi.org/10.4103/0971-6203.48720.

Basarab, J.A., M.A. Price, J.L. Aalhus, E.K. Okine, W.M. Snelling, K.L. Lyle, 2003. Residual feed intake and body composition in young growing cattle. Canadian Journal of Animal Science. 83 (2), 189–204.

Berckmans, D., C. Bahr, T. Leroy, X. Song, E. Vranken, W. Maertens, J. Vangeyte, A.V. Nuffel, B. Sonck, 2008. Automatic detection of lameness in dairy cattle - Analyzing image parameters related to lameness. 2008. 949–956.

Bigelow, J.A., T.R. Houpt, 1988. Feeding and Drinking patterns in young pigs. Physiology and Behavior. 43, 99-109.

Blaxter, K., 1989. *Energy Metabolism in Animals and Man.* Cambridge: Cambridge University Press.

Brown-Brandl, T.M., R.A. Eigenberg, 2011. Development of a livestock feeding behavior monitoring system. Transactions of ASABE. 54 (5), 1913–1920.

Brown-Brandl, T.M., R.A. Eigenberg, G.L. Hahn, J.A. Nienaber, 1999. Measurements of bioenergetic responses in livestock. ASAE.

Brown-Brandl, T.M., J.A. Nienaber, R.A. Eigenberg, G.L. Hahn, H.C. Freetly, 2003. Thermoregulatory responses of feeder cattle. Journal of Thermal Biology. 28, 149–157.

Brown-Brandl, T.M., J.A. Nienaber, R.A. Eigenberg, T.L. Mader, J.L. Morrow, J.W. Dailey, 2006. Comparison of heat tolerance of feedlot heifers of different breeds. Livestock Science. 105, 19–26.

Brown-Brandl, T.M., R.A. Eigenberg, J.L. Purswell, 2013. Using thermal imaging as a method of investigating thermal thresholds in finishing pigs. Biosystems Engineering. 114, 327–333.

Brown-Brandl, T.M., T. Yanagi, H. Xin, R.S. Gates, R. Bucklin, G.S. Ross, 2003. A new telemetry system for measuring core body temperature in livestock and poultry. Applied Engineering in Agriculture. 19 (5), 583-589.

Burdick, N.C., J.A. Carroll, J.W. Dailey, R.D. Randel, S.M. Falkenberg, T.B. Schmidt, 2012. Development of a self-contained, indwelling vaginal temperature probe for use in cattle research. Journal of Thermal Biology. 37, 339–343.

Cangar, T., Leroy, M. Guarino, E. Vranken, R. Fallon, J. Lenehan, J. Mee, D. Berckmans, 2008. Automatic real-time monitoring of locomotion and posture behaviour of pregnant cows prior to calving using online image analysis. Computers and Electronics in Agriculture. 64 (1), 53–60.

Chapinal, N., D.M. Veira, D.M. Weary, M.A.G. von Keyserlingk, 2007. Technical Note: Validation of a System for Monitoring Individual Feeding and Drinking Behavior and Intake in Group-Housed Cattle. Journal of Dairy Science. 90 (12), 5732-5736.

Chapinal, N., J.L. Ruiz-de-la-Torre, A. Cerisuelo, M.D. Baucells, J. Gasa, X. Manteca, 2008. Feeder use patterns in group-housed pregnant sows fed with an unprotected electronic sow feeder (Fitmix). Journal of Applied Animal Welfare Science. 11 (4), 319-336.

Eddy, A.L., L.M. Van Hoogmoed, J.R. Snyder, 2001. The Role of Thermography in the Management of Equine Lameness. The Veterinary Journal. 162 (3), 172–181. http://dx.doi.org/http://dx.doi.org/10.1053/tvjl.2001.0618.

Ehrlemark, A.G., K.G. llvik, 1996. A model of heat and moisture dissipation from cattle based on thermal properties. Transactions of the American Society of Agricultural Engineers. 39 (1), 187–194.

Eigenberg, R.A., G.L. Hahn, J.A. Nienaber, T.M. Brown-Brandl, D. Spiers, 2000. Development of a new respiration rate monitor for cattle. Transactions of the ASAE. 43 (3), 723–728.

Eigenberg, R.A., T.M. Brown-Brandl, J.A. Nienaber, 2002. Development of a Respiration Rate Monitor for Swine. Transactions of the ASAE. 45 (5), 1599–1603.

Gipson, T.A., A.L. Goetsch, G. Detweiler, R.C. Merkel, T. Sahlu, 2006. Effects of the number of yearling Boer crossbred wethers per automated feeding system unit on feed intake, feeding behavior and growth performance. Small Ruminant Research. 65 (1-2), 161-169.

Gipson, T.A., A.L. Goetsch, G. Detweiler, T. Sahlu, 2007. Effects of feeding method, diet nutritive value and physical form and genotype on feed intake, feeding behavior and growth performance by meat goats. Small Ruminant Research. 71 (1-3), 170-178.

Goetsch, A.L., T.A. Gipson, A.R. Askar, R. Puchala, 2010. Invited review: Feeding behavior of goats. Journal of Animal Science. 88 (1), 361-373.

Hanneman, S.K., J.T. Jerurum-Urbaitis, D.R. Bickel, 2004. Comparison of methods of temperature measurement in swine. Laboratory Animals. 38, 297–306.

Hillman, P.E., K.G. Gebremedhin, S.T. Willard, C.N. Lee, A.D. Kennedy, 2009. Continuous measurements of vaginal temperature of female cattle using a data logger encased in a plastic anchor. Applied Engineering in Agriculture. 25 (2), 291-296.

Huynh, T.T.T., A.J.A. Aarnink, W.J.J. Gerrits, M.J.H. Heetkamp, T.T. Cahn, H.A.M. Spoolder, B. Kemp, M.W.A. Verstegen, 2005. Thermal Behaviour of growing pigs in response to high temperature. Applied Animal Behaviour Science. 91, 1–16.

Hyun, Y., M. Ellis, 2002. Effect of group size and feeder type on growth performance and feeding patterns in finishing pigs. Journal of Animal Science. 80 (3), 568-574.

Kelly, A.K., M. McGee, D.H. Crews, Jr., A.G. Fahey, A.R. Wylie, D.A. Kenny, 2010. Effect of divergence in residual feed intake on feeding behavior, blood metabolic variables, and body composition traits in growing beef heifers. Journal of Animal Science. 88 (1), 109-123.

Lao, F., T.M. Brown-Brandl, J.P. Stinn, L. Kai, G. Teng, H. Xin, 2015. Automatic identification of lactating sow behaviors through depth image processing. Computers and Electronics in Agriculture. Submitted August 3, 2015.

Leroy, T., E. Vranken, A. Van Brecht, E. Struelens, B. Sonck, D. Berckmans, 2006. A computer vision method for on-line behavioral quanitications of individually caged poultry. Transactions of the ASABE. 49 (3), 795–802.

Maselyne, J., W. Saeys, B. De Ketelaere, K. Mertens, J. Vangeyte, E.F. Hessel, S. Millet, A. Van Nuffel, 2014. Validation of a High Frequency Radio Frequency Identification (HF RFID) system for registering feeding patterns of growing-finishing pigs. Computers and Electronics in Agriculture. 102, 10–18. http://dx.doi.org/http://dx.doi.org/10.1016/j.compag.2013.12.015.

Mayer, J.J., J.D. Davis, J.L. Purswell, E.J. Koury, N.H. Younan, J.E. Larson, and T.M. Brown-Brandl, 2015. Development and characterization of a continuous tympanic temperature logging (CTTL) probe for bovine animals. Transactions of the ASABE Submitted Aug 17, 2015.

McGovern, R.E., J.M. Bruce, 2000. AP—Animal Production Technology: A Model of the Thermal Balance for Cattle in Hot Conditions. Journal of Agricultural Engineering Research. 77 (1), 81–92.

Nagl, L., R. Schmitz, S. Warren, T. Hildreth, H. Erickson, D. Andresen, 2003. Wearable sensor system for wireless state-of-health determination in cattle. In: *Proceeding of the 25th Annual International Conference of the IEEE EMBS, Cancun, Mexico*. 3012–3015.

Nienaber, J.A., A.L. Maddy, 1985. Temperature controlled multiple chamber indirect calorimeter-design and operation. Transactions of the ASAE. 28 (2), 555–560.

Nienaber, J.A., T.P. McDonald, G.L. Hahn, Y.R. Chen, 1990. Eating dynamics of growing-finishing swine. Transactions of the ASAE. 33 (6), 2011-2018.

Nienaber, J.A., T.P. McDonald, G.L. Hahn, Y.R. Chen, 1991. Group feeding behavior of swine. Transactions of the ASAE. 34 (1), 289-294.

Puma, M.C., H. Xin, R.S. Gates, D.J. Burnham, 2001. An instrumentation system for studying feeding and drinking behavior of individual poultry. Applied Engineering in Agriculture. 17 (3), 365-374.

Reuter, R.R., J.A. Carroll, L.E. Hulbert, J.W. Daily, M.L. Galyean, 2010. Technical note: Development of a self contained, indwelling rectal temperature probe for cattle research. Journal of Animal Science. 88, 3291–3295.

Ring, E.F.J., K. Ammer, 2012. Infrared thermal imaging in medicine. Physiological Measurement. 33, R33–R46.

Rose-Dye, T.K., Burciaga-Robles, C.R. Krehbiel, D.L. Step, R.W. Fulton, A.W. Confer, C.J. Richards, 2014. Rumen temperature change monitored with remote rumen temperature boluses after challenges with bovine viral diarrhea virus and *Mannheimia haemolytica*. Journal of Animal Science. 89, 1193-1200.

Shao, B., H. Xin, 2000. Real time assessment and control of pig thermal comfort behavior by computer imaging. ASAE.

Sykes, D.J., J.S. Couvillion, A. Cromiak, S. Bowers, E. Schenck, M. Crenshaw, P.L. Ryan, 2012. The use of digital infrared thermal imaging to detect estrus in gilts. Theriogenology. 78, 147–152.

Tan, J.-H., E.Y.K. Ng, U. Rajendra Acharya, C. Chee, 2009. Infrared thermography on ocular surface temperature: A review. Infrared Physics & Technology. 52 (4), 97–108. http://dx.doi.org/http://dx.doi.org/10.1016/j.infrared.2009.05.002.

Turnpenny, J.R., J.A. Clark, A.J. McArthur, C.M. Wathes, 2000. Thermal balance of livestock. 1. A parsimonious model. Agricultural and Forest Meteorology. 101 (1), 15–27.

Viazzi, S., C. Bahr, T. Van Hertem, A. Schlageter-Tello, C.E.B. Romanini, I. Halachmi, C. Lokhorst, D. Berckmans, 2014. Comparison of a three-dimensional and two-dimensional camera system for automated measurement of back posture in dairy cows. Computers and Electronics in Agriculture. 100, 139–147. http://dx.doi.org/http://dx.doi.org/10.1016/j.compag.2013.11.005.

Influence of Group Size and Space Allowance on Production Performance and Mixing Behavior of Weaned Piglets

Bin Hu [a,b], Chaoyuan Wang [a,b,*], Zhengxiang Shi [a,b], Baoming Li [a,b], Zuohua Liu [c], Baozhong Lin [c]

[a] Department of Agricultural Structure and Bioenvironmental Engineering, College of Water Resources and Civil Engineering, China Agricultural University, Beijing 100083, China

[b] Key Laboratory of Agricultural Engineering in Structure and Environment, Ministry of Agriculture, Beijing 100083, China

[c] Chongqing Academy of Animal Science, Chongqing 402460, China

* Corresponding author. Email: gotowchy@cau.edu.cn

Abstract

Field experiment was conducted to evaluate effect of different group sizes (10, 20 and 40 pigs per group) and space allowances (0.3, 0.4 and 0.5 m^2 per pig) on production performance and mixing behaviors (feeding, drinking, excreting, lying and aggression) of weaned piglets. The piglets in large group (40 pigs per pen) illustrated a better production performance. Lying, feeding, and drinking behavior were not affected by the group sizes, however, the fighting behavior of the piglets in large group was greatly reduced. Furthermore, aggressive behavior was significantly lowered ($P < 0.05$) by the setting of playing toys inside the pens for the groups of 10, and 20 pigs/pen. Significant reductions of both duration and frequency of fighting behavior were found in the second 24 h after mixing ($P < 0.05$), indicated that the group was becoming stabilized. Production performance of piglets with the largest space allowance was significantly improved ($P < 0.05$), and the fighting behavior was effectively reduced as well ($P < 0.05$). Again with the toys, a considerable reduction on the aggression was observed, especially for the group with smallest space allowance ($P < 0.05$).

Keywords: Productivity, mixing behavior, stocking density, group size

1. Introduction

Group size and space allowance of housing can affect the pigs' behavior and production performance. Studies have revealed the effects of group size, stocking density, and space allowance on production performance and behaviors of pigs since 1960s (Bryant et al., 1972; Bryant et al., 1974; Gehlback et al., 1966; Jensen et al., 1962; Patterson et al., 1985). Most indicate that group sizes might have less influence on production performance (Nielsen et al., 1995; Schmolke et al., 2003; Spoolder et al., 1999; Randolph et al., 1981; Turner et al., 1999; Turner et al., 2002). However, Spicer and Ahernr (1987) detected appreciable impact on Average Daily Gain (ADG) of weaned piglets due to different group sizes and stocking density. Usually, inadequate stocking density leads to higher aggressive behavior, while the fights in small groups are fewer, since the individual is more familiar and the competition for resources is less (Forkman et al., 2004). Whereas, with larger group and less communication, pigs will be more likely to fight, and it is more difficult to establish a stable and sustainable social hierarchy. The fighting between pigs does not happen randomly (Dugatkin et al., 2003). When establishing the social hierarchy inside a group, severe fights may occur, especially at mixing. Food competition is another important reason. More fighting was found in larger groups, however, some researchers concluded contrary findings (Turner et al., 2001; Andersen et al., 2004).

The objectives of this paper were to assess the influence of stocking density and group size on behavior and production performance during the mixing process of weaned piglets, and to provide practical recommendations for pig barn design and management under typical Chinese production system.

2. Materials and Methods

2.1. Experiment setup

The study was carried out at a commercial pig farm of Chongqing Academy of Animal Sciences, Chongqing, China. The farm was built in 2010 and approximately 1,000 sows were housed during the experiment. In each weaned piglet barn, there were 6 independent compartments. Four continuous compartments (A, B, C, and D) in a same barn were chosen and modified for conducting the experiments. The excreting area of the pens was equipped with slatted floor, while other areas inside the pen remained unchanged with solid concrete floor.

In experiment 1, 280 weaned piglets were mixed into 12 pens inside the 4 compartments to test the influence of different group sizes and the setting of playing toys on the behavior and performance of the piglets. The three group sizes tested in the experiment were 40 (Large), 20 (Medium) and 10 piglets per pen (Small).The space allowance for each piglet of all the groups were kept at 0.4 m^2. Inside each compartment, three pens of piglets with different group sizes (Large, Medium and Small) were set. Additionally, compartments C and D were installed with playing toys as treatment group, while A and B as control. A feeding trough and two duck-billed drinkers were provided for every 10 pigs.

In experiment 2, 240 weaned piglets were assigned into 3 groups with different stocking densities. The space allowances were 0.5, 0.4, and 0.3 m^2/pig for Large, Medium, and Small pens, respectively; and the constant group size was 20 piglets per pen. Similarly, C_1 and D_1 compartments were treatment group, and A_1 and B_1 were control group.

Each experiment lasted 6 weeks. Due to commercial activities, the ages of piglets in the two experiments ranged from 33 to 36 days. Hence, the initial body weights of the piglets when mixing were variable from 11.20 kg to 13.64 kg.

2.2. Production performance measurement and behavior recording

Temperature and relative humidity of each compartment were automatically monitored by sensors to make sure that all groups have similar thermal environment.

During the experiments, initial and final body weights of the piglets were measured. The pigs were fed four times a day at 8:30 am, 11:30 am, 2:30 pm and 5:30 pm, and the feed intake of the group was recorded each time. In each pen, 5 piglets were randomly selected and marked with colors on the back for behavior analysis. Behaviors of the piglets, including lying, feeding, drinking, excreting and aggression (biting or fighting), were continuously recorded by infrared HD cameras (DS-2DF8223IW-A, HIKVISION, Beijing, China) for the entire experimental period, of which the effective behavior (drinking and aggressing) was defined as the action lasting for more than 3 sec. Duration and frequency of behaviors in the first 48 h after mixing were manually analyzed via the video by a same experienced technician.

Data were statistically treated using SPSS (Statistical Product and Service Solutions), version 20 (IBM, 2011, NY, USA). The influence of group sizes on pigs' average daily gain (ADG), average daily feed consumption (ADFC), feed consumption rate (FCR) and behaviors was assessed by analysis of variance.

3. Results and Discussion

3.1. Influence of group size on mixing behavior and production performance of weaned piglets

3.1.1. The influence of group size on production performance

The production performance of weaned piglets with different group sizes are shown in Table 1. Results indicated that the ADG of the smallest group (10 pigs/pen) was the least ($P < 0.05$) both in the control and treatment. No statistical difference on feed conversion rate (FCR) was observed, although the Large group had the highest FCR. The overall ADGs for the control and treatment groups averaged 0.41 kg/d and 0.40 kg/d, meanwhile FCRs were 43.5% and 44.6%, respectively, which suggested that the setting of playing toys had no effect on the growth performance, and may be beneficial to the FCR. Generally, bigger group size had better production performance and FCR in our study, which varies from some previous studies with the findings of having similar performance among different groups (Simon et al., 2003; Hyun et al., 2002; Birte et al., 1995). Stephanie et al. (2004) ever found the growing-finishing pigs in larger groups had lower production performance, which is also different from this study.

Table 1. Production performance of the piglets in different group sizes.

Group	Group Size	ADG, kg/d	ADFC, kg/pig	FCR, %
Control (A, B)	10	0.38±0.02a	0.88±0.10a	43.5
	20	0.43±0.01b	1.03±0.12b	42.2
	40	0.43±0.02b	0.98±0.10b	44.9
Treatment (C, D)	10	0.34±0.02a	0.81±0.09a	42.7
	20	0.40±0.02b	0.90±0.08b	45.0
	40	0.46±0.03c	1.00±0.07c	46.2

[a] Different letters indicate significant difference ($P < 0.05$) in the same column;
[b] ADG = Average Daily Gain, ADFC = Average Daily Feed Consumption, FCR = Feed Conversion Rate.

3.1.2. Influence of group size on mixing behaviors for the first 48 h

Average duration and frequency of different behaviors in the first 48 h after mixing are summarized in Table 2. For every 24 h, the average duration of lying behavior in control and treatment groups was 16.9 h and 17.6 h, which accounted for 70% and 73% of the day, respectively. In A and B compartments, the piglets in Large group spent shorter time in lying than those of medium and small groups ($P < 0.05$); while in the compartments C and D equipped with playing toys, there was no difference in lying duration in different groups. The piglets in 40 pigs/pen spent more time in eating and drinking, but no significant difference was detected from the other two groups. For every 24 h, excreting duration for the piglets in different groups ranged 0.05 h to 0.11 h, which accounted for 2.4%–4.2% of the total time.

Table 2. Duration and frequency of different behaviors of the piglets in different group sizes.

Behavior	Group size (pigs/pen)	Control Group		Treatment Group	
		Duration (h)	Frequency	Duration (h)	Frequency
Lying	10	17.21±1.12a	68.90±19.42a	17.17±1.35	53.75±17.43
	20	17.39±1.43a	57.70±14.38b	17.86±0.97	57.95±14.53
	40	15.98±1.21b	62.00±11.42a	17.64±1.42	63.60±10.87
Feeding	10	1.94±0.42	38.10±22.34a	1.45±0.52	25.15±14.52
	20	1.91±0.94	25.60±13.47b	1.64±0.47	26.15±12.45
	40	2.05±0.24	32.85±9.45ab	1.86±0.76	30.90±9.37
Drinking	10	0.27±0.10	53.00±22.38a	0.15±0.11	33.55±21.32
	20	0.25±0.15	33.30±14.35b	0.16±0.10	25.40±17.23
	40	0.26±0.14	33.10±9.43b	0.19±0.07	31.80±9.42
Excreting	10	0.05±0.02a	11.75±4.01a	0.06±0.17	11.25±3.78
	20	0.11±0.02b	16.00±5.64b	0.08±0.04	12.40±5.64
	40	0.06±0.02a	11.05±3.67a	0.08±0.21	13.60±2.74
Aggression	10	0.10±0.10a	12.50±11.20	0.08±0.01a	3.00±9.43a
	20	0.27±0.30b	9.25±9.25	0.18±0.06b	7.25±8.34ab
	40	0.13±0.14a	9.00±6.20	0.13±0.10ab	10.20±6.20b

[a] Different letters indicate significant difference ($P < 0.05$) in the same column.

The results also showed that on average, piglet in the Medium group spent 0.22 h per 24 h in fighting with others, which was more than the other two groups (vs. 0.09 h for Small group, and 0.13 h for Medium group). Generally, group sizes had no significant influence on feeding, drinking and excreting behavior of the weaned piglets after mixing in the first 48 h, while it did on aggressive behavior. Compared to the smallest groups, fighting behavior for the piglets in larger groups lasted longer. Piglets in medium groups of both the control and treatment performed the longest aggressive behavior. One of the possible reasons was that group of 10 pigs/pen was a natural group size which was beneficial to the reduction of fighting behavior; while weaker piglets in the Large group may easier find space to escape because of larger public area.

3.1.3. Influence of playing toys on behaviors of the piglets

For the controlled and treated piglets, the mean lying duration did not vary remarkably. In every 24 h, piglets in treatment group spent 0.3 h less in feeding, and its drinking time was also significantly less, suggesting that the piglets might be distracted to play the novel toys after mixing, and consequently spent much less time in exploring the feed and water. Excreting duration did not show significant difference between treatment and control groups. More importantly, the piglets of the treatment group performed much less fighting behavior during the first 48 h after mixing, and the frequency was lower as well.

Generally, duration and frequency of lying, feeding, drinking and excreting behavior of the piglets in the control and treatment groups were not affected by the setting of playing toys; however, the aggression was remarkably reduced. The toys may be attractive to piglets to explore, play and chew, especially in the beginning of mixing, hence distracted them from conducting aggression behavior in the treatment group.

3.1.4. Behaviors of piglets for the first and second 24 h after mixing with different group sizes

The mixing behavior expressed by duration and frequency for the first and second 24 h after mixing are summarized. In this paper, only the first 48 h behavior after mixing was analyzed because usually a stable and sustainable social hierarchy would have been established within the first two days for weaned piglets. The results showed that the average duration of lying, feeding, drinking and excreting behavior almost did not change for the first and second 24 h for both the control and treatment groups. While a big reduction on aggressive behavior was detected in the second 24 h, and the duration was significantly reduced by 0.08, 0.20, and 0.05h for the Small, Medium and Large groups in control, and 0.04, 0.16, and 0.09hin treatment, respectively. Furthermore, the frequency of the different behavior had a similar trend for both the treatment and control groups in the first 48 h after mixing.

In summary, the lying, feeding, drinking and excreting behavior in the first two days after mixing did not change significantly, while duration and frequency of aggression of the piglets reduced significantly in the second 24 h.

3.2. Influence of space allowance on mixing behavior and production performance of piglets

3.2.1. Influence of space allowance on production performance

Results indicated ADG of the piglets with 0.5 m^2/pig stocking density was significantly higher than the other two groups ($P < 0.05$), as shown in Table 3. For the pigs without toys, the FCR increased as the stocking densities decreasing from 0.3 m^2/piglet to 0.5 m^2/piglet, although no statistical difference was found. For the treatment group, the FCR of the group with large space allowance (0.5 m^2/piglet) was 6.0% and 3.2% higher than the other two groups, and a significant difference was detected from the group of 0.3 m^2/piglet ($P < 0.05$). There was no difference in average daily feed consumption of the piglets. Hence, it is concluded that the growth performance and feed conversion efficiency increased for the weaned piglets with the lower stocking density.

For the piglets in all surveyed compartments, the duration of lying, feeding and drinking behavior was not statistically differed under varied space allowances. On average, the piglets spent 18.5 h/day in lying, or 77% of the day, 1.95 h in feeding, and 0.18 h in drinking water. Considerable effect was found on aggression under different space allowances, showing that the smallest space allowance leaded to highest fighting behavior (Figure 1). For the pigs without any toys, the average fighting duration of piglets with 0.3 m^2/pig space allowance was almost twice higher than that of 0.5 m^2/pig ($P < 0.05$), and the aggression behavior also much more frequently happened. For the treatment group, a similarity was observed, namely the aggression increased as the decreasing of the space allowance. The results showed that smaller space allowance resulted in more aggressive behavior for the weaned piglets, during the first 48 h of mixing in particular. It was observed that it was more difficult for the pigs with lower social hierarchy to escape within smaller space allowance when aggression happened, which may lead to more losses in practical production.

Table 3. Production performance of the weaned piglets in different space allowances.

Group	Stocking Density, m^2/pig	ADG, kg/pig	ADFC, kg/pig	FCR, %
Control (A_1, B_1)	0.3	0.37±0.01a	0.71±0.07	52.1
	0.4	0.37±0.01a	0.70±0.07	52.9
	0.5	0.41±0.01b	0.75±0.08	54.7
Treatment (C_1, D_1)	0.3	0.35±0.00a	0.71±0.08	49.3
	0.4	0.38±0.01b	0.73±0.08	52.1
	0.5	0.41±0.01c	0.75±0.09	55.3

[a] Different letters indicate significant difference ($P < 0.05$) in the same column;
[b] ADG = Average Daily Gain, ADFC = Average Daily Feed Consumption, FCR = Feed Conversion Rate.

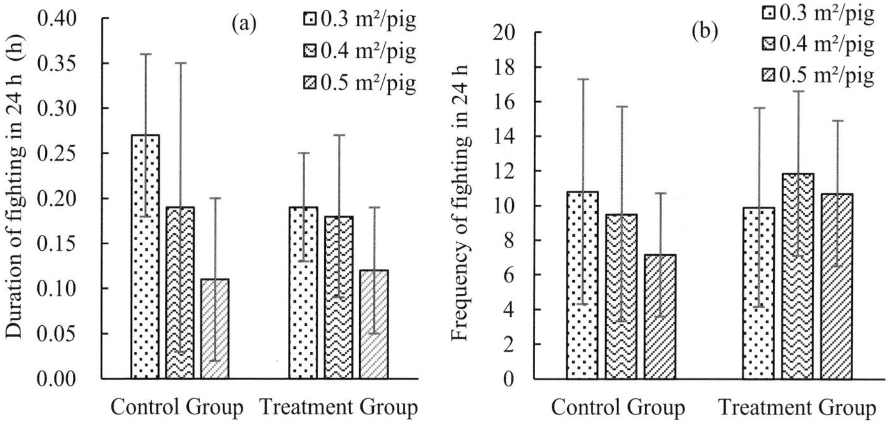

Figure 1. Influence of space allowance on duration (a) and frequency (b) of aggressive behaviors of the piglets with/without playing toys.

3.2.2. Influence of space allowance on the mixing behavior of the weaned piglets

Generally, the duration and frequency of most behavior of the weaned piglets with different stocking density during the first 48 h of mixing, including lying, feeding and drinking, did not greatly change except for the aggression, and larger space allowance was helpful in reducing the fighting behavior within the animals.

3.2.3. Influence of playing toys on the mixing behaviors under different space allowances

In our study, the behavior of the piglets to play toys was observed. The results showed that no significant influence of setting playing toys on lying behavior of the piglets with the same space allowance was found. The piglets in the treatment group spent less time in feeding, while there was no statistical difference. Significant difference on drinking and aggressive behaviors was found between the control and treatment groups. The drinking duration of the piglets in treatment was greatly reduced ($P < 0.05$). For the piglets with a stocking density of 0.3 m^2/pig, the duration of the fighting behavior in treatment group was significantly reduced by 0.08 h ($P < 0.05$). While for the stocking density of 0.5 m^2/pig and 0.4 m^2/pig, the aggressive behavior did not vary for the treatment and control groups. The frequency of lying, feeding, drinking and excreting behavior did not change neither, but the aggression did (Figure 2). Compared with the control, the frequency of aggressive behavior of the piglets with a 0.3 m^2/pig stocking density in the treatment was reduced ($P > 0.05$), while averagely increased by 2 times ($P < 0.05$), and 3 times per 24 h ($P < 0.05$) for the groups of 0.4 m^2/pig and 0.5 m^2/pig, respectively. In summary, the setting of playing toys inside the pens did impact the aggression of the piglets during the first 48 h of mixing, especially for group of 0.3 m^2/pig, but did not on lying, feeding and drinking.

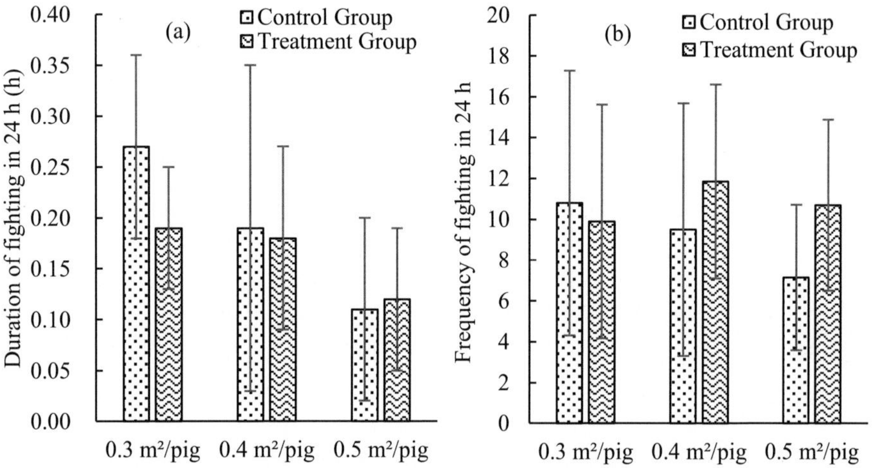

Figure 2. Duration (a) and frequency (b) of fighting behavior of the piglets housed in the pens with (treatment) and without (control) playing toys during the first 48 h of mixing.

3.2.4. Behaviors of the piglets with different space allowances in the first and second 24 h after mixing

Similarly to experiment 1, the lying, feeding, drinking and excreting behavior of the piglets did not remarkably change for the first and second 24 h after mixing, while significant effect on aggression was found. For both the control and treatment groups, average duration of lying behavior generally reduced for the second day of mixing, while the change was ignorable. Contrarily, the length of aggressive behavior of the piglets greatly reduced for all the groups ($P < 0.05$). In the control group, reductions were 74%, 75% and 72% for the piglets with the space allowances of 0.3 m^2/pig, 0.4 m^2/pig and 0.5 m^2/pig, and 60%, 50% and 67% in treatment group (Figure 3), respectively, which illustrated that the groups were stabilizing. Again, the frequency of lying, feeding, drinking and excreting did not obviously change for the second day, while on aggression it was significantly reduced ($P < 0.05$) by 59%, 55% and 50% for the groups of 0.3 m^2/pig, 0.4 m^2/pig and 0.5 m^2/pig, respectively.

This research studied the influence of different group sizes (10, 20, and 40 pigs/pen) and space allowances (0.3, 0.4, and 0.5 m^2/pig) on production performance and mixing behavior of weaned piglets via two 6-week experiments, and the effects of setting playing toys were also evaluated.

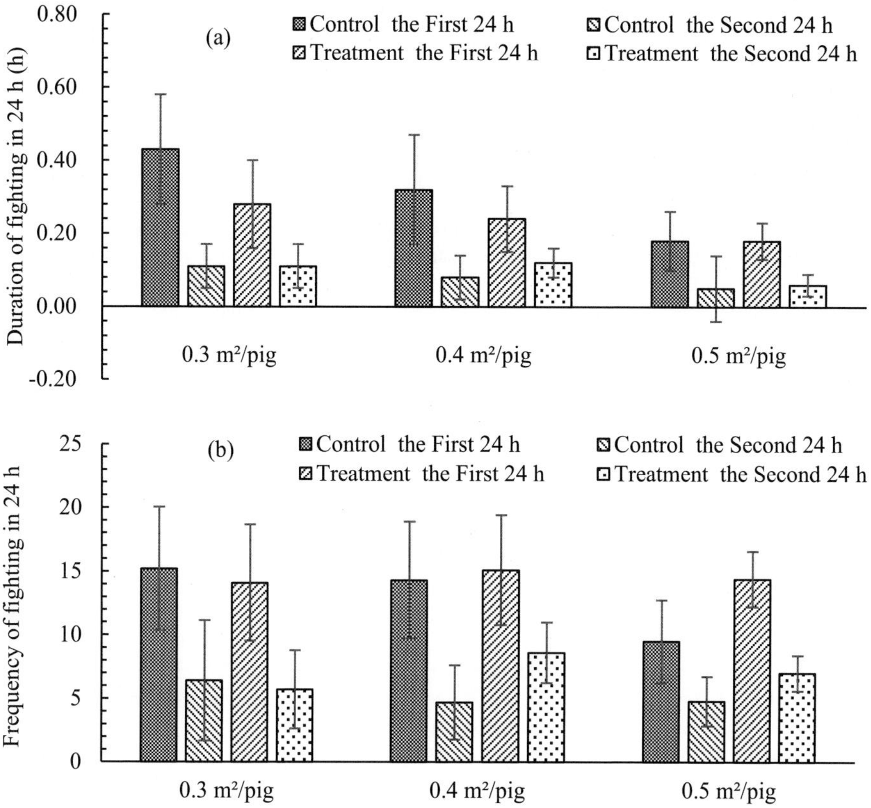

Figure 3. Changes of duration and frequency of fighting behavior of piglets for the first and second 24 h after mixing.

4. Conclusions

Production performance in the largest group size of 40 pigs/pen was better than the other two groups. Meanwhile, its aggressive behavior during the first 48 h of mixing was also significantly reduced, which differed from several previous studies. The setting of toys was helpful in reducing the fight behavior, especially for the piglets of smallest group size, while it did not impact other behavior during mixing. A great reduction of both duration and frequency of fighting behavior in the second 24 h after mixing ($P < 0.05$) was detected, indicated that the social hierarchy was under establishing and the group was becoming stabilized. Production performance of the piglets with the largest space allowance was significantly better ($P < 0.05$), and its fighting behavior was effectively reduced as well ($P < 0.05$). Again with the toys, a remarkable reduction on the aggression was observed, especially for the group with smallest space allowance ($P < 0.05$).

References

Andersen I.L., E. Nævdal, M. Bakken, 2004. Aggression and group size in domesticated pigs, Sus scrofa: "When the winner takes it all and the looser is standing small". Animal Behavior. 68, 965–975.

Birte L.N., B.L. Alistair, T. Colin, 1995. Effect of group size on feeding behavior, social behavior, and performance of growing pigs using single-space feeders. Livestock Production Science. 44, 73–85.

Bryant M.J., R. Ewbank, 1972. Some effects of stocking rate and group size upon agonistic behavior in groups of growing pigs. British Veterinary Journal. 128, 64–69.

Bryant M.J., R. Ewbank, 1974. Effects of stocking rate upon the performance, general activity and ingestive behavior of groups of growing pigs. British Veterinary Journal. 130, 139–149.

Dugatkin L.A., R.L. Earley, 2003. Group fusion: the impact of winner, loser, and bystander effects on hierarchy formation in large groups. Behavior Ecology. 14, 367–373.

Gehlback G.D., D.E. Becker, J.L. Cox, 1966. Effects of floor space allowance and number per group on performance of growing-finishing swine. Journal of Animal Science. 25, 386–391.

Forkman B., M.J. Haskell, 2004. The maintenance of stable dominance hierarchies and the pattern of aggression: support for the suppression hypothesis. Ethology. 110, 737–744.

Jensen A.H., D.E. Becker, R.S. Thatcher, 1962. Management of swine confined to slatted floors. Journal of Animal Science. 21(4), 1039.

Nielsen B.L., A.B. Lawrence, 1995. Effect of group size on feeding behavior, social behavior, and performance of growing pigs using single-space feeders. Livestock Production Science. 44, 73–85.

Patterson D.C., 1985. A note on the effect of individual penning on the performance of fattening pigs. Animal Production. 40, 185–188.

Randolph J.H., G.L. Cromwell, T.S. Stahly, 1981. Effects of group size and space allowance on performance and behavior of swine. Journal of Animal Science. 53, 922–927.

Simon P.T., W.H. Graham, A.E. Sandra, 2003. Assessment of sub-grouping behavior in pigs housed at different group sizes. Applied Animal Behavior Science. 83, 291–302.

Spicer H.M., F.X. Aherne, 1987. The effects of group size/stocking density on weanling pig performance and behavior. Applied Animal Behavior Science. 19, 89–98.

Schmolke S.A., Y.Z. Li, H.W. Gonyou, 2003. Effect of group size on performance of growing-finishing pigs. Journal of Animal Science. 81, 874–878.

Spoolder, H.A.M., S.A. Edwards, S. Corning, 1999. Effects of group size and feeder space allowance on welfare in finishing pigs. Animal Science. 69, 481–489.

Stephanie A.S., Y.Z. Li, W.G. Harold, 2003. Effects of group size on social behavior following regrouping of growing–finishing pigs. Applied Animal Behavior Science. 81, 874–878.

Turner S.P., S.A. Edwards, V.C. Bland, 1999. The influence of drinker allocation and group size on the drinking behavior, welfare and production of growing pigs. Animal Science. 68, 617–624.

Turner S.P., M. Dahlgren, D.S. Arey, 2002. Effect of social group size and initial live weight on feeder space requirements of growing pigs fed ad libitum. Animal Science. 75, 75–83.

Turner S.P., G.W. Horgan, S.A. Edwards, 2001. Effect of social group size on aggressive behavior between unacquainted domestic pigs. Applied Animal Behavior Science. 74, 203–215.

Hyun Y., M. Ellis, 2002. Effect of group size and feeder type on growth performance and feeding patterns in finishing pigs. Journal of Animal Science. 80, 568–574.

Assessment of Lighting Needs by Laying Hens via Preference Test

He Ma [a,b], Hongwei Xin [b,*], Yang Zhao [b], Baoming Li [a], Timothy A. Shepherd [b], Ignacio Alvarez [c]

[a] Key Laboratory of the Ministry of Agriculture for Agricultural Engineering in Structure and Environment, China Agricultural University, Beijing 100083, China

[b] Department of Agricultural and Biosystems Engineering, Iowa State University, Ames, IA 50011-3310, USA

[c] Department of Statistics, Iowa State University, Ames, IA 50011-3310, USA

* Corresponding author. Email: hxin@iastate.edu

Abstract

This study was conducted to investigate hen's needs for light intensity and circadian rhythm using a light tunnel with five identical compartments each at a different fluorescent light intensity of <1, 5, 15, 30 or 100 lux. The hens were able to move freely among the compartments. A group of four W-36 laying hens (23–30 weeks of age) was tested each time, and six groups or repetitions were conducted. Hen behaviors were continuously recorded, which yielded the data on time spent, feed intake, feeding time, and eggs laid at each light intensity and inter-compartment movement. The results show that the hens spent 6.4 h (45.4%) at 5 lux, 3.0 h (22.1%) at 15 lux, 3.1 h (22.2%) at 30 lux, and 1.5 h (10.3%) at 100 lux under light condition; and spent 10.0 h under dark (<1 lux). Daily feed intake was 87.3 g d^{-1} and distributed as 24.8 g hen^{-1} d^{-1} (28.4%) at <1 lux, 28.4 g hen^{-1} d^{-1} (32.5%) at 5 lux, 13.8 g hen^{-1} d^{-1} (15.8%) at 15 lux, 14.5 g hen^{-1} d^{1} (16.6%) at 30 lux, and 5.8 g hen^{-1} d^{-1} (6.7%) at 100 lux, respectively. Hen-day egg production rate was 96.0%; and most of the eggs were laid at <1 lux (61.9% of total) which was significantly different from other light intensities ($P < 0.05$). The overall photoperiod was 10 h darkness (<1 lux) and 14 h light or 14L: 10D, with a fairly constant hourly dark time of 25.0 ± 0.4 min throughout the day.

Keywords: Light intensity, laying hens, light preference, behavior, circadian rhythm

1. Introduction

Lighting and its properties (e.g., wavelength, intensity and duration) are a crucial factor affecting hen's growth, production, behavior, and welfare (Perry, 2003). Light intensity can affect egg size, feed intake and mortality. The recommended light intensity for commercial hen houses is 10–20 lux (Hy-Line, 2015). Some studies showed that lower light intensity could reduce the occurrence of cannibalism and feather pecking (Kjaer and Vestergaard, 1999). However, improper low light intensity (i.e., 1.1 lux) could cause some issues such as adrenal overweight (Siopes et al., 1984), body underweight (Hester et al., 1985), leg problems (Hester et al., 1985), and partial or complete blindness due to eye morphology change (Deep et al., 2010). Photoperiod is considered as one of the most critical environment factors affecting bird production. Lewis (2007) compared the performance of breeders reared in 11L: 13D *vs.* 16L: 8D, and found that birds under 11L: 13D had better feed conversion. Some intermittent lighting regimens have been reported to improve feed conversion (Ma, 2013) and reduce mortality by reducing duration of daytime lighting (Lewis et al., 1996). In commercial farms, light intensity and photoperiod are designed mainly to achieve high productivity, instead of considering actual light needs of the hens, which may cause some welfare issues (Prescott et al., 2003). Furthermore, desired light intensity and photoperiod from the hen's perspective are not fully understood and thus require investigation as concerns about animal welfare intensify.

Preference test is one of the best ways to assess the animal's biological or physiological demand. Some studies have been conducted on birds' light preference in several aspects. Davis et al. (1999) conducted a light intensity (6, 20, 60, and 200 lux) preference study using 2- and 6-week-old chicks and found that chicks preferred the brighter light at 2 weeks of age, but they spent more time under 6 lux at 6 weeks of age. Sherwin (1998) found that the light preference of male turkey was affected by the light intensity under which they were pre-acclimatized. Prescott and Wathes (2002) studied feeding preference under different light intensities (<1, 6, 20 or 200 lux) of ISA Brown hens and found that the birds chose to eat most of the time in the brightest condition (200 lux) and the least in the dimmest (<1 lux). No study was found on light intensity preference of W-36 hens, the most popular laying breed in USA.

The objective of this study was to investigate the preference of fluorescent light intensity by W-36 white laying hens by subjecting the birds to a range of light intensity (<1, 5, 15, 30 or 100 lux). An environmentally-controlled light tunnel, as described below, was constructed and used to address the objective.

2. Materials and Methods

2.1. System description

The systems for this study included a 5-interconnected compartment preference test light tunnel and a 2-interconnected compartment acclimation chamber. Both were located in an environmentally-controlled room at Iowa State University, Ames, Iowa, USA.

The light-proof preference test system, or light tunnel (LT) (Figure 1), was constructed with an angle iron frame. The LT was 366 cm long, 91 cm wide and 198 cm high. It was divided into five side-by-side compartments (identical in dimension, i.e., 61 cm L × 91 cm W × 198 cm H each) using plastic panels. The LT had white internal walls and black external walls. Each compartment had a perforated ceiling and a cage (61 cm L × 61 cm W × 91 cm H). A height-adjustable nipple drinker was installed at the back of the cage, and a feed trough was mounted to a load-cell weighing platform in front of the cage. Two fluorescent tube lights (GE, 15W) partially covered by aluminum foil were installed on top of the perforated ceiling to create different light intensities in each compartment. A LED rope light was fixed at 30 cm above the feed troughs along the entire LT. This arrangement was made so that all compartments or light intensity regimens had the same lighting intensity at the feed trough, except for the < 1 lux one (no feeder light). To eliminate the interference of the LED rope light with the main light intensity treatment, the LED rope light was covered to confine the light to the feeder area. The partition walls between two adjacent compartments had a black rubber-strip curtain pathway (36 cm H × 20 cm W each), allowing the hens to easily pass through. Access to the LT was done through the movable front plastic panel of each compartment. The five compartments shared a common hand-crank egg belt and a manure belt at the bottom of the LT, allowing eggs and manure moved to the end of the LT (Figure 1b) without opening the LT or disturbing the hens. A push-pull ventilation system was installed to provide 60 air changes per hour (ACH) to the LT.

A separate acclimatization chamber (216 cm L × 89 cm W × 159 cm H) was constructed and used to acclimatize the hens to the different light intensities under evaluation and to train the hens in passing the curtain pathway before they were transferred to the LT. The acclimatization system was also constructed using angle iron and black/white plastic panels. The system was divided into two identical compartments (74 cm L × 64 cm W × 46 cm H each) with a black rubber-strip curtain pathway (same as those used in the LT) in between. Each compartment had a feed trough and a nipple drinker. A plastic board was placed under the wire-mesh floor to catch manure. Eight fluorescent tube lights (GE, 15W) in pairs of two were installed overhead along the length direction of the chamber at the same height as those in the LT. Through partially covering the light tubes and operating proper light combination, four different light intensities of 5, 15, 30 and 100 lux were achieved inside the chamber.

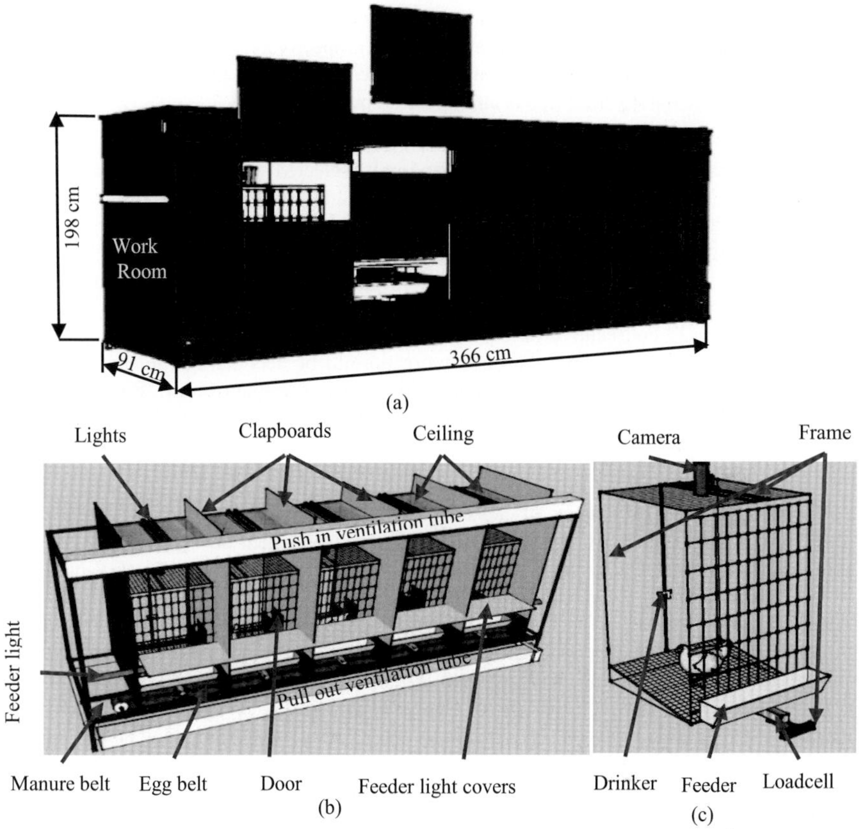

Figure 1. Schematic representation of the light tunnel (LT) for preference test: (a) outside view, (b) inside view, and (c) an individual compartment.

2.2. Experimental design

2.2.1. General information

Four batches of eight 23 week-old laying hens (Hy-Line W-36) procured from local commercial farms were used for the preference test. After brought back to the laboratory, the eight hens of each batch were kept for 8 days in the acclimatization chamber, where they were acclimatized and exposed to the five light intensities under evaluation. Following the acclimatization, four hens were randomly chosen and transferred to the preference test LT, where they remained for 27 (for trial 1) or 33 days (for trial 2–6). Prior to the first trial and between trials, both LT and training/acclimatization chamber were cleaned, disinfected and left empty for 3 days. The LT was maintained at 24.9 ± 0.5°C and 30 ± 5% relative humidity (RH) during the test periods. Feed and water were provided *ad-libitum*. Feeding and egg collection were performed at 8:30 AM every day, and manure was removed once a week.

2.2.2. Acclimatization

The curtain pathway was fully open on the first acclimatization day, then the curtain strips were gradually let down in the next four days (1/4 every day). From day 5 to day 8, the curtain was fully down. This arrangement trained the hens to freely pass through the curtain pathways in the LT during the preference test that followed. The hens were also exposed to multiple light intensities that were used in the LT. During the 8-day acclimatization period, the photoperiod within the day was 16L: 8D. The 16 lighting hours of each day were divided into four equal-length periods (4 h period^{-1}) of 5, 15, 30 and 100 lux, respectively. The order of light intensities was based on a 4 × 4 Latin Square arrangement, such that each time period received a given light intensity twice during the 8-day acclimatization period.

2.2.3. Preference test

After acclimatization, four birds were randomly selected and transferred to the LT for preference test. The five intensity levels of <1, 15, 30 and 100 lux were randomly assigned to the five compartments following a 5 × 5 Latin Square design, so that each compartment received a light intensity the same number of times per trial (balanced-randomized design). It should be mentioned that the <1 lux light intensity was always assigned to either of the two side compartments in trial 2. Nine or ten episodes were involved per trial. In trial 1, each episode lasted for 3 days (27 consecutive days of testing overall). For trials 2, 3, 4, 5 and 6, the first episode lasted for 6 days to give the birds more time to be acclimatized to the LT, and then lights were re-assigned every 3 days (33 consecutive days of testing overall). Between episodes, feed left in the trough in all five compartments was replaced and new feed added. Light intensity at the feeder level was 30 lux for all but the < 1 lux compartment (no feeder light).

2.3. Data collection

The compartment occupancy of hens was monitored using five infrared video cameras (GS831SM/B, Gadspot, Inc. Corp., Tainan City, Taiwan) mounted on the top of the respective compartments. The five cameras were connected to a video capture card (GV-600B-16-X, GeoVision Inc., Taipei, Taiwan) with Surveillance System software (Ver 8.5, GeoVision Inc., Taipei, Taiwan). Images of each compartment were continuously captured at 2-s intervals throughout the experiment period. The number of hens in each compartment was determined using image analysis in Matlab (R2013a, MathWorks, Inc., Torrance, CA, USA) and Excel 2013 (VBA program). To do so, original images were analyzed through cropping, binariyzation (with a black and white scale threshold of 0.75), boundary closing, hole filling, and spot removing to obtain the final image of hens in white and background in black. The total hen area, number of white regions, area of each region and number of compartments occupied by hens at a given moment were further analyzed in Excel to determine the number of hens in each image using an algorithm. The hen number data were analyzed in Excel to determine time spent and times of inter-compartment movement. Time spent (TS, h or min hen^{-1}) was calculated by summarizing hen number over a time period (hour or day), and inter-compartment movement (ICM, visits) were calculated by counting times of a hen entering the compartments. Visit duration (VD, min visits^{-1}) was calculated by time spent divided by inter-compartment movement

Load-cell sensors (RL1040-N5, Rice Lake Weighing Systems, Rice Lake, WI, USA) were used to monitor feeding activities and measure feed intake. Outputs of the load-cell sensors were continuously recorded at 1-s intervals through a LabVIEW program (version 7.1, National Instrument Corporation, Austin, TX, USA). An algorithm (in Matlab 2014a) was developed to calculate time at feeder (TAF, h hen^{-1} d^{-1}), feeding time (FT, h hen^{-1} d^{-1}), feeding rate (FR, g min^{-1} d^{-1}), and feeding time distribution (%). It should be noted that time at feeder was the overall time spent at the feeder (with or without feed consumption), while feeding time was the time spent on eating (with feed consumption). Eggs laid (EL, %) in each compartment were recorded daily.

Temperature and RH inside the LT were continuously measured during the experiment using thermocouples (± 0.5°C, Type-T, OMEGA Engineering, Inc., Stamford, CT, USA) and a RH sensor (HMT100, Vaisala, Inc., Woburn, MA, USA). Room temperature was also monitored using the same type of thermocouple.

2.4. Statistical analysis

The first four days of each trial and the first day of each remaining test episode were considered as acclimation periods. As such, only data on the last two days of each test episode were used for subsequent data analysis. Therefore, 16 days (trial 1) or 20 days (trials 2, 3, 4, 5 and 6) of data were summarized in the statistical analysis. The percentage of time spent (pts) was transformed using a log transformation [loge (pts / (1 - pts)] to stabilize the variance, allowing statistic model to be performed. All data were analyzed using PROC MIXED model in SAS 9.3 (SAS Institute Inc., NC, USA). The model included light intensity, compartment and their interaction as fixed effects. Trial and the three-way interaction among trials, day and compartment were considered as random effects. Finally a blocked-diagonal matrix for the error term was used, where blocks correspond to the interaction between trial and day. Effects were considered significant when P < 0.05. The statistical model has the form as Equation (1).

$$Y_{ijkd} = \mu + L_i + C_j + (LC)_{ij} + (CTD)_{jkd} + \varepsilon_{ijkd} \tag{1}$$

where, Y_{ijkd} represents the response variable on day d by trial k in compartment j with light intensity i, μ is the intercept, L_i is the light intensity (<1, 5, 15, 30, and 100 lux) effect, C_j is the compartment (1, 2, 3, 4 and 5) effect, $(LC)_{ij}$ is the interaction effect of light intensity and compartment T_k and $(CTD)_{jkd}$ are the random effects, and ε_{ijkd} is the error vector.

3. Results and Discussion

3.1. Time spent

The time spent (TS) in particular compartments was significantly affected by light intensity, but not by compartment or the interaction effect, indicating a strong preference of the laying hens for specific light intensities. Table 1 shows TS under different light intensities. Since the light intensity of <1 lux was considered as dark period, the total TS of hens in light period was 14 hours. During the light period, the hens spent most of their time staying in 5 lux (6.4 ± 0.5 h hen^{-1} d^{-1}, 45.3%) which was significantly higher (P < 0.05) than that in 15, 30 or 100 lux. Hens spent similar time in 15 lux (3.1 ± 0.4 h hen^{-1} d^{-1}, 22.2%) and 30 lux (3.0 ± 0.4 h hen^{-1} d^{-1}, 22.1%) (P = 0.988). Time spent in 100 lux (1.5 ± 0.2 h hen^{-1} d^{-1}, 6.1%,) was the lowest (P < 0.05). Therefore, the light intensity of 5 lux was preferred by the laying hens over 15 or 30 lux, and 100 lux was least preferred. This result seems consistent with the report by Davis et al. (1999) who observed that older poultry preferred dimmer light.

Animals generally behave to maximize their welfare and will preferentially choose the variant that is most likely to satisfy their requirements (Dawkins, 1990). During the light period, the W-36 hens spent most of the time in the lower light intensity of 5 lux, but they also spent part of the time in the higher light intensities of 15, 30 and 100 lux. This outcome suggests that the hens may be better served by a variable light intensity environment. It should be noted that the light set-up in the present study, where five levels of light intensity were provided continuously, is not comparable with the practice typical of commercial layer housing where birds are provided consecutive light (e.g., 16) or dark (8) hours per day. Therefore, further work is needed to verify the welfare effects of the light intensities or photoperiod patterns as preferred or displayed by the hens in the present study over an extended period.

3.2. Feed intake and feeding behavior

The feed intake (FI), time at feeder (TAF), and feeding time (FT) in particular compartments were significantly affected by the light intensity, but not by compartment or interactions. Feeding rate (FR) did not seem to be affected by light intensity (P = 0.057–0.998) or compartment (P = 0.5996). The distribution of FI in the five light intensities is shown in Table1. FI in <1 lux and 5 lux was significantly higher than that in 15, 30 or 100 lux (P < 0.05). Similar FI occurred in 15 lux and 30 lux (P = 0.901). Hens had the lowest FI in 100 lux (P < 0.05). Time at feeder (TAF) and feeding time (FT) followed similar trend to FI. The TAF averaged 4.4 ± 0.1 h and 3.4 ± 0.1 h of which were spent on actual feeding. This led to an average feeding rate (FR) of 0.43 ± 0.04 g min^{-1} d^{-1}.

Prescott and Wathes (2002) studied feeding preference under different light intensities (<1, 6, 20 or 200 lux) of ISA Brown hens and found that the birds chose to eat the most time in the brightest (200 lux) and least in the dimmest (<1 lux) light intensity. Their conclusion was that hens might be averse to eating in very dim light, presumably because the process of eating is normally guided visually. In the present study, feeding light (30 lux) was provided to all compartments except for the <1 lux regimen. The results showed that hens preferred to feed in the compartment in which they spent more time, which was somewhat contrary to the findings of Prescott and Wathes (2002). Moreover, the W-36 hen did not show aversion to feeding in <1 lux.

3.3. Egg production and eggs laid

Throughout the experiment period, 4 hens laid 3 or 4 eggs daily and the hen-day egg production averaged 96.0 ± 0.9%. Light intensity impacted the location of eggs laid (EL) in that the hens laid 61.9 ± 3.6% of total eggs in <1 lux, which was significantly higher than that in 5, 15, 30 or 100 lux ($P < 0.05$) (Table1). There were no significant differences among 5 lux (12.5 ± 2.3%), 15 lux (11.0 ± 2.3%), 30 lux (7.6 ± 1.6%) and 100 lux (7.0 ± 1.9%) (P = 0.184–0.795). Therefore, hens preferred to lay eggs in the <1 lux condition.

Millam (1987) found that hens preferred to nest in boxes at one end of a row compared to those in the middle. Appleby et al. (1984) found that some hens preferred brightly lit nest boxes (20 lux), while others preferred dimly lit nest boxes (5 lux), when given a free choice. When both situations were provided (compartments in a row and different light intensities) hens showed a strong motivation of laying eggs in <1 lux rather than end compartments (<1 lux was moved from compartment to compartment). It was concluded that hens preferred to lay eggs in dim light (<1 lux) over locations.

Table 1 Compartment use and feeding behavior of W-36 laying hens provided with free choice of light intensities (<1, 5, 15, 30 and 100 lux) (n = 116).

Items	Light intensity (lux)					Overall	SME	P-value
	<1	5	15	30	100			
TS (h hen^{-1} d^{-1})	10.0^{a}	6.4^{a}	3.1^{b}	3.0^{b}	1.5^{c}	24	0.4	<.0001
PTS per day (%)	41.5^{a}	26.5^{a}	13.0^{b}	12.9^{b}	6.1^{c}	100	1.7	<.0001
PTS in light period (%)	dark	45.3^{a}	22.2^{b}	22.1^{b}	10.4^{c}	100	1.5	<.0001
ICM (visits hen^{-1} d^{-1})	70^{ab}	85^{a}	70^{ab}	76^{ab}	64^{b}	365	5.9	<.0001
VD (min $visit^{-1}$)	8.5^{a}	4.5^{ab}	2.7^{b}	2.5^{b}	1.4^{b}	3.9	1.1	<.0001
FI (g hen^{-1} d^{-1})	24.8^{a}	28.4^{a}	13.8^{b}	14.5^{b}	5.8^{c}	87.3	1.7	<.0001
% of total FI (%)	28.4^{a}	32.5^{a}	15.8^{b}	16.6^{b}	6.7^{c}	100	1.9	<.0001
TAF (h hen^{-1} d^{-1})	1.2^{a}	1.4^{a}	0.7^{b}	0.8^{b}	0.3^{c}	4.4	0.1	<.0001
FT (h hen^{-1} d^{-1})	0.9^{a}	1.1^{a}	0.5^{b}	0.6^{b}	0.2^{c}	3.4	0.1	0.0245
FR (g min^{-1} d^{-1})	0.47	0.43	0.43	0.40	0.40	0.43	0.04	0.2522
EL (%)	61.9^{a}	12.5b	11.0^{b}	7.6^{b}	7.0^{b}	100	2.3	<.0001

Note: ^{a,b} Values within a row with different superscripts differ significantly at $P < 0.05$. SME = standard error of the mean. DTS = daily time spent. DPTS = daily percent of time spent. DICM = daily inter-compartment movement. VD = visit duration. DFI = daily feed intake. FI = feed intake. DTAF = daily time at feeder. DFT = daily feeding time. FR = feeding rate. EL = egg laid.

3.4. Inter-compartment movement (ICM) and visit duration (VD)

Light intensity, compartment and their interaction had effects on ICM. The ICM and VD under different light intensities are shown in Table 1. The only significant difference in ICM occurred between 5 lux (85 ± 6) and 100 lux (64 ± 6) ($P < 0.05$). The VD in <1 lux (8.5 ± 1.1 min $visit^{-1}$) was significantly greater than those of the other light intensities ($P < 0.05$). The ICM outcome confirmed that the hens experienced all the light intensities during the preference test. The VD can provide information on an animal's motivation to exit or enter a specific environment, hence reflecting the natural exploratory behavior of most animal species (Kristensen et al., 2000). For this study, when laying hens were first transferred to different light environments they were observed to explore different light intensities, and then chose the ones they preferred to stay in.

3.5. Distribution of dark and light

Hourly mean TS in each light intensity is shown in Figure 2. TS in the dark (<1 lux) and light (sum of 5, 15, 30 and 100 lux) was intermittent in each hour throughout the day. The hourly mean TS in the dark was 25.0 ± 0.4 min. Daily total mean TS in the dark (<1 lux) and light (sum of 5, 15, 30 and 100 lux) was 10.0 ± 0.7 h and 14.0 ± 0.7 h, respectively ($P < 0.001$).

Light intensity associated with the darkness definition has been reported to range from 0 to 4 lux (Coenen et al., 1988; Malleau et al., 2007). Light intensity level of <1 lux was considered as dark in the present experiment. Interestingly, the laying hens showed quite a consistent hourly rhythm in the usage of light and dark compartments, which yielded a daily overall light and dark distribution of 14L:10D. This light-dark pattern was considerably different from the photoperiod of 16L: 8D typically practiced in commercial operation in that the darkness was intermittently distributed in each hour of the day in the current study. The total dark period (10 h) was also longer than the typical 8 h typically used in commercial practices. Within the 10 h dark period (<1 lux), feeding activities and egg laying (Table 4) were also noticed, which might have contributed to the longer dark period. TS of laying hen at feeder in <1 lux was 1.2 h, but TS of egg laying and other behaviors (resting, drinking, walking, etc.) was unknown. The underlying reasons for this intermittent rhythm and longer dark period remain to be better understood.

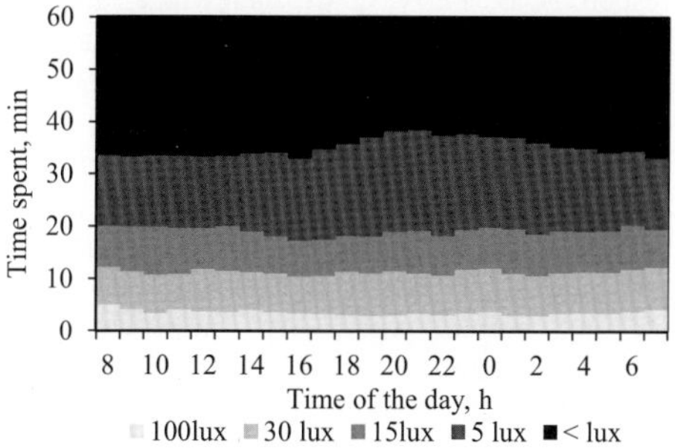

Figure 2. Distribution of time spent under different light intensity in each hour per day (n = 116).

4. Conclusions

The W-36 laying hen showed apparent preference when allowed to choose among a range of light intensities. Specifically, the following was observed. 1) The hen preferred to stay in dimer light (5 lux) during the light period, which made up 45.4% (6.4 h d^{-1}) of the total light-time budget. 2) The hen preferred to feed in < 1 lux (24.8 ± 2.4 g hen^{-1} d^{-1}, 28.4%) without feeding light and in 5 lux (28.4 ± 1.9 g hen^{-1} d^{-1}, 32.5%) with feeding light (30 lux), as compared with other light intensity levels. 3) The hens laid 61.9% of their eggs in < 1 lux compartment, reflecting their preference of laying egg in 'darkness' (< 1 lux).4) Within a day the hens spent a total of 10.0 ± 0.7 h (41.50%) in darkness and 14.0 ± 0.7 h (58.5%) in light. However, the time spent in darkness was intermittently and quite consistently distributed throughout the day (averaging 25.0 ± 0.4 min h^{-1}). Further study is needed to assess the light intensity preference of the laying hen as revealed in the present study with regards to impact on long-term welfare and production performance.

Acknowledgements

This study was funded by Iowa State University and China Agricultural Research System for laying hens (CARS-41). We thank the commercial hen farms for providing the birds and feed used in this study. The assistance by undergraduate students — Jace Klein, Haocheng Guo, Kyle Dresback in the system development and data processing was greatly appreciated. Author He Ma also thanks the China Scholarship Council (CSC) for supporting her 2-year research and study at Iowa State University.

References

Coenen, A. M. L., Wolters, E. M. T. J., Van Luijtelaar, E. L. J. M., and Blokhuis, H. J. (1988). Effects of intermittent lighting on sleep and activity in the domestic hen. Applied Animal Behaviour Science, 20(3–4), 309–318.

Davis, N. J., Prescott, N. B., Savory, C. J., and Wathes, C. M. (1999). Preferences of growing fowls for different light intensities in relation to age, strain and behaviour. Animal Welfare, 8(3), 193–203.

Dawkins, M. S. (1990). From an animal's point of view: motivation, fitness, and animal welfare. Behavioral and brain sciences, 13(01), 1–9.

Hester, P. Y., Sutton, A. L., Elkin, R. G., and Klingensmith, P. M. (1985). The effect of lighting, dietary amino-acids, and litter on the incidence of leg abnormalities and performance of turkey toms. Poultry Science, 64(11), 2062–2075.

Hy-Line 2015. Management Guide: W-36 Commercial Layers. In Hy-Line Company. http://www.hyline.com/UserDocs/Pages/36_COM_ENG.pdf. Accessed January 13, 2015.

Kjaer, J. B., and Vestergaard, K. S. (1999). Development of feather pecking in relation to light intensity. Applied Animal Behaviour Science, 62(2–3), 243–254.

Kristensen, H. H., Prescott, N. B., Perry, G. C., Ladewig, J., Ersboll, A. K., Overvad, K. C., and Wathes, C. M. (2007). The behaviour of broiler chickens in different light sources and illuminances. Applied Animal Behaviour Science, 103(1–2), 75–89.

Lewis, P. D., Gous, R. M., Ghebremariam, W. K., and Sharp, P. J. (2007). Broiler breeders do not respond positively to photoperiodic increments given during the laying period. British Poultry Science, 48(3), 245–252.

Ma, H., Li, B., Xin, H., Shi, Z., and Zhao, Y. (2013). Effect of intermittent lighting on production performance of laying-hen parent stocks. Paper presented at the 2013 ASABE Annual International Meeting, Kansas City, MO. http://lib.dr.iastate.edu/cgi/viewcontent.cgi?article=1325&context=abe_eng_conf.

Malleau, A. E., Duncan, I. J. H., Widowski, T. M., and Atkinson, J. L. (2007). The importance of rest in young domestic fowl. Applied Animal Behaviour Science, 106(1–3), 52–69.

Millam, J. R. (1987). Preference of turkey hens for nest-boxes of different levels of interior illumination. Applied Animal Behaviour Science, 18(3–4), 341–348.

Perry, G. C. (2003). *Lighting: Welfare of the laying hen.* Wallingford, UK: CABI. Vol. 27, pp. 299–311.

Prescott, N. B., and Wathes, C. M. (2002). Preference and motivation of laying hens to eat under different illuminances and the effect of illuminance on eating behaviour. British Poultry Science, 43(2), 190–195.

Prescott, N. B., Wathes, C. M., and Jarvis, J. R. (2003). Light, vision and the welfare of poultry. Animal Welfare, 12(2), 269–288.

Sherwin, C. M. (1998). Light intensity preferences of domestic male turkeys. Applied Animal Behaviour Science, 58(1–2), 121–130.

Effect of Group Size on Mating Behaviour of Layer Breeders and the Fertility of Eggs

Hongya Zheng, Baoming Li *, Weihao Cui, Zhengxiang Shi
Key Laboratory of Structure and Environment in Agricultural Engineering, Ministry of Agriculture,
China Agricultural University, Beijing 100083, China
* Corresponding author. Email: libm@cau.edu.cn

Abstract

This study was to investigate the effect of group size on mating behaviour and to identify the relationship between age variance and behaviour and fertility change. Beijing Jing Brown I layer breeders were housed in group size of 100 and 200 hens (100F and 200F) in perchery systems with a male to female ratio of 1:10. Observations were conducted respectively at 36 weeks, 40 weeks and 44 weeks of age. Frequencies of mating and wing flapping were significantly affected by group size. A remarkable higher frequency of mating was found in 100F groups in comparison to 200F ($P = 0.002$ for 36 weeks and $P = 0.031$ for 40 weeks). At 36 weeks, frequency of wing flapping in 100F was significantly higher ($P < 0.001$). Ages negatively impacted mating behaviours, as mating frequency declined with age in 200F groups ($F_{2,9} = 2.391$, $P = 0.147$). Mating and wing flapping were displayed the fewest in 40 weeks, significantly varied from 36 weeks ($P < 0.01$) in both group sizes. Fertility decreased with group size and was not affected by age variance. Average mating duration was alike between groups, and mating was better completed in 100F. For site selection, where the males stayed most were preferred, and for which we can tell that males in groups of current sizes were passive for mating. No exact relationship was identified between mating behaviour and fertility. More detailed observation on both males and females for a longer production period should be performed to confirm the conclusion about the effect of group size on mating behaviour and fertility.

Keywords: Layer breeders, mating, behaviour, group size, fertility, age

1. Introduction

Layer breeders are essential for commercial layers breeding. In China, layer stocks are mainly housed in conventional stair-step cages with artificial insemination (AI), as well as in a few types of floor systems. Colony cages are increasingly being used for layer breeder housing in recent years for less labor cost and higher stocking density.

Birds in barren cages are deprived of the ability to perform natural behaviours such as exploring, nesting, perching, dust bathing, or simply stretching their wings (Bareham, 1976; Baxter, 1994; Tanaka and Hurnlk, 1992). Colony cages are much more humane, usually with a colony size of 11 and 22 birds (Adams et al., 1978), 66 birds (Hughes, 1978) and 88 birds (Hughes and Holleman, 1976), with a sex ratio 1:10. However critics still remain. Such cages impair animal welfare because the space offered is insufficient to allow expression of the full behavioural repertoire, including foraging, dust bathing and extensive locomotion (FAWC, 2007). Increased activity or locomotion contributes to better bone quality (Newman and Leeson, 1998; Tactacan et al., 2009; Michel and Huonnic, 2003), and researches about effects of housing systems on physical condition of laying hens infer that aviary systems can offer some distinct advantages over conventional cages (Taylor and Hurnik, 1994; Colson et al., 2007). Moreover, although AI can improve fertility, the cost of implementing AI on a large scale is often cost prohibitive. De Jong and Guémené (2011) pointed out that AI involving regular handling of the birds may have negative implications for bird welfare, which may increase psychological stress both for the males and females.

When housing in group, group size is often concerned with welfare of birds either in cages or cage-free systems. The degree of aggression varies with group size (Al-Rawi and Craig, 1975;

Hughes and Wood-Gush, 1977) and is crucial for assessing welfare of hens (Appleby and Hughes, 1991). Pagel and Dawkins (1997) pointed out that hierarchy formation varied with group sizes. For having more than one male and female together in a cage, Adams et al. (1978) found that group size significantly affected average fertility in early laying period and having more than one male in a colony cage was advantageous to partially compensate for the ineffectiveness of penmates. Group size also impacts fearfulness and stress (Campo and Davila, 2002) and the degree of male competition (Bilcik and Estevez, 2005).

The aim of our study was to investigate the impacts of group size on mating behaviours of males and to determine behavioural traits and fertility associated with age variance in a perchery system. We hypothesized that mating activity was more frequent in large groups because dominance hierarchy is less obvious and higher mating frequency would be presented in large groups with a higher level of male competition. And fertility was predicted to increase with mating frequency but decrease with age. So fertility was expected to be higher in large groups.

2. Materials and Methods

2.1. Animals and housing

Birds were reared in conventional cage and transferred into the experimental perchery pen at the age of 13 weeks. Animals can feed and drink ad libitum. A total number of 600 Beijing Jing-Hong Brown parent laying breeders were housed in four groups, two flocks of 100 hens (100F) and the other two of 200 hens (200F). Cocks were housed in each group with a ratio of 1♂:10♀♀. Flocks were housed in one perchery unit or two with the floor stocking density of 18.3 birds m^{-2}. Each unit measuring 3 m width× 2 m length, incorporated an inverted trapezoidal frame of perches, four 0.7-m-wide mesh wire platforms, a 0.6-m-wide plastic mesh aisle with square node for inspection and routine activity of cocks, and two tiers of nest boxes with curtains along both sides of the pen (Figure 1). And at the bottom of the pen, there were manure belts, running once per day. Four 0.7-m-wide mesh wire platforms and the rest area of aisle floor were used for feeding, drinking and resting. Perches were sufficient to provide each bird more than 23 cm roosting space. Averagely, 4.6 hens shared one drinking nipple and each hen had 10 cm feed space and 10 cocks shared a feed trough of 2-m length.

2.2. Observation

At 36 weeks and then 40 weeks and 44 weeks of age, observations were conducted in a 4-day phase. From 14:00 to 16:00 and from 19:00 to 20:00, direct observations with instantaneous scans were conducted to record mating information, two days for each group. In addition the frequency of wing flapping had also been recorded during observation. Total number of mating behaviour was recorded in observation period and then transformed into percentage by dividing the overall number of cocks in the pen. When a mating behaviour took place, site selection, duration and behaviour completeness were recorded on a pre-prepared sheet. A mating behaviour was divided into four phases, containing courtship, mount, cloacal contact and ejaculation. Duration was from the beginning of courtship to ejaculation ending after the cock backing on the floor. Behaviour completeness was accordingly scored as 0, 1, 2, 3 and 4 ("0" means the cock doesn't finish any movement, "1" means only mounting is displayed, "2" means mounting and cock lowering his tail were finished, "3" means only ejaculation is not displayed, and "4" means all four phases are done). When only courting was displayed we did not score it as "1" because two birds did not get contact. Site selection was labeled as "A" (aisle), "PE" (platform edge), "PB" (platform bottom) and "P" (platform), detailed information was described in Figure 1.

2.3. Fertility examination

To evaluate the fertility of different competitive environment, all eggs except the broken ones laid on one day in each group were collected and were set for incubation at 37.5°C and 65% relative humidity once a testing week. On the 10th day after incubation we candled all the eggs and recorded the number of fertile eggs per pen, including blood eggs. Fertility was calculated as percentage.

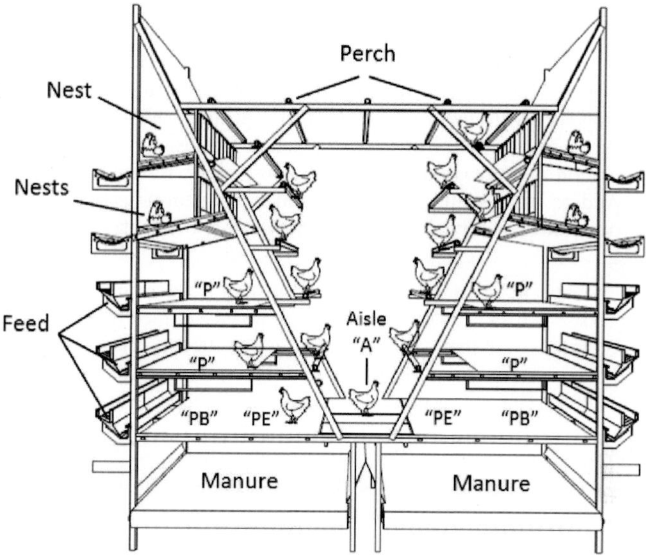

Figure 1. Structural drawing of the perchery system. "A", "PE", "PB" and "P" were separately short for aisle, area of platform edge, platform bottom and platform. Those contained all four possible sites for cocks.

2.4. Data processing and analyses

Mean frequency of mating and wing flapping per bird per hour for the total 6 hours of observation was calculated in three periods respectively. Site selection was presented with percentage for each group in different times. Durations were averaged based on the pooled data of group size in different weeks and percentages of different intervals were calculated. Repeated measures of General Linear Model was applied to determine if there was an effect of different group size, age, and the interaction, with age as the repeated measure, group size (100F and 200F) and the interaction of both as factors in the model. Before that, we use Shapiro-Wilk test and Levene to test normality and homogeneity of variances for data. After that, One-way ANOVA or Kruskal-Wallis tests were applied to investigate the difference of indicators at different ages for different group sizes. LSD post hoc test was used to detect the difference between weeks if applied. Data would be log-transformed before analysis in order to obtain a normal distribution and homogeneity of variances. All statistical analyses were performed using the SPSS 17.0 for windows and significance for all analyses was accepted at $P < 0.05$.

3. Results and Discussion

3.1. Effects on frequency of mating and wing flapping

Mean frequencies of mating and wing flapping were significantly affected by ages and group sizes respectively (all $P < 0.05$), but there was no relationship with the interaction of both factors (age×group size, $P > 0.05$).

3.1.1. Group size

Group sizes influenced both mating and wing flapping frequencies, but not in all three experimental weeks. At 36 and 40 weeks, a remarkable higher frequency of mating was found in 100F groups in comparison to 200F groups (Table 1), as indicated by Bilcik and Estevez (2005) that higher mating activity occurred in 1M groups other than 3M groups with more male competition. Only at 36 weeks, frequency of wing flapping in 100F was significantly higher than 200F groups and no distinct difference was detected for the rest two times (Table 1). Wing flapping was considered to be a probable indicator of courtship display movement (Wood-Gush, 1954; Duncan et al., 1990) and a sign of social dominance in the male group (Leonard and ZANETTE, 1998). On average, both mating and wing flapping activity were more active in 100F, which was contradicted with our previous hypothesis. Hughes et al. (1997) analyzed previous studies together with their own research to conclude that aggression was less frequent in large groups possibly because of infrequent resource defense. One possibility could be that individual competition in 200F groups is less obvious, which resulted in less competition for received hens. And as a consequence, less second mating towards the same hen and more missing mating occurred as observed which were not recorded systematically.

3.1.2. Age effect

Age negatively impacted mating behaviours, as mating frequency declined with age in 200F groups ($F_{2,9}$= 2.391, P = 0.147). While more mating was performed by males in 44 weeks than 40 weeks in 100F groups, and that was contrast to our prediction. During peak egg production, the effective and maximum duration of fertility of females were noted after 36 weeks (Gumułka and Kapkowska, 2005). That just accorded with our middle testing time in which less received females engaged in matings. For wing flapping, frequency was the lowest in 40 weeks, significantly differing from 36 weeks (P < 0.01) in both group sizes. Possibly the same situation as mentioned above. In terms of males, a decline in libido with age might account for the decrease of mating and wing flapping frequency (Duncan et al. 1990).

Table 1. Mean (±S.E.) frequencies of mating and wing flapping in different group sizes.

Frequency/male/hour	Week	Group size		
		100F	200F	P-value
Mating	36 weeks	1.22 ± 0.10^{Aa}	0.68 ± 0.04	0.002
	40 weeks	0.77 ± 0.05^{B}	0.58 ± 0.04	0.031
	44 weeks	0.81 ± 0.12^{b}	0.57 ± 0.04	0.094
	P-value	0.014	0.147	--
Wing flapping	36 weeks	2.29 ± 0.02^{A}	1.92 ± 0.01^{Aa}	<0.001
	40 weeks	0.98 ± 0.05^{Bb}	0.91 ± 0.10^{B}	0.547
	44 weeks	1.23 ± 0.09^{Ba}	1.16 ± 0.13^{b}	0.658
	P-value	<0.001	0.023	--

Different superscript of lower cases means significant difference between age groups (P < 0.05); Different superscript of capitals means significant difference between age groups (P < 0.01)

3.2. Effects on duration and site selection

3.2.1. Effects on duration

No significant differences of average mating duration were detected in different size groups ($F_{1,22}$ = 2.397, P = 0.136), and mean duration was maintained fairly consistent throughout the whole study (age×group size, P > 0.05; age, P > 0.05). Average mating duration in both group sizes were between 5 and 6 seconds once (Figure 2), while during observation, duration records fluctuated from 2 seconds to 15.4 seconds and even longer. More than half of durations were in the range of 2.2-6.6 seconds, which inferred a hurry mating which also could be inferred from the few courtships. The two experimental group sizes had no obvious impact on mating duration mainly because of a relative crowd density. Since a high involving density and adaption after transfer made females gather on platforms, so few would stay with males for much time. Of all observations, few mating consisted of courtship and directly began with males pecking on hens' head or rear approach.

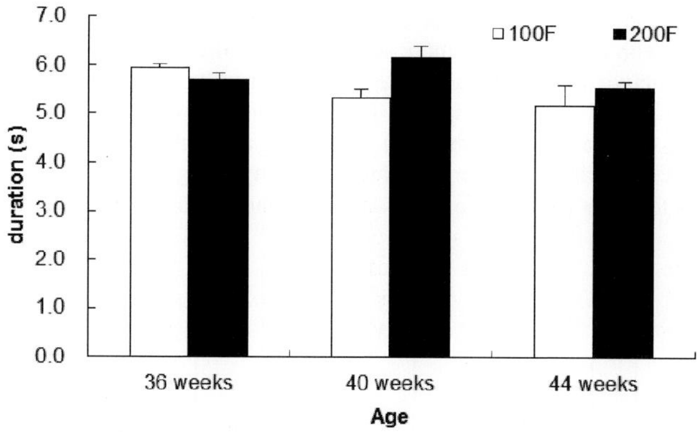

Figure 2. Mean (±S.E.) duration of matings of two group sizes at three periods.

3.2.2. Effects on site selection

Site selection for matings was not remarkably affected by group size (P > 0.05) in all observation periods (Figure 3). So did the age effect (P > 0.05). The majority of site selection were at "A" and "PE", mainly because the two sites were open-sided where males preferred to stay, so encountered hens showing reception were often mated with there. With birds aging, more mating behaviours in 200F groups were displayed at "A" and "PB", fewer at "PE" and "P". In 100F groups, it was different for "PB" site. Since adult males established their social dominance and would not change seriously with time (Berdoy et al., 1995), the most preferred "A" site accordingly would be occupied by high-ranking males in the group. Besides, females do actively respond to those superior males (Graves et al., 1985), an increasing number of hens would be attracted to the area where dominant males stay and accepted mating. Finally more mating occurred at the preferred site "A", and fewer in the other sites.

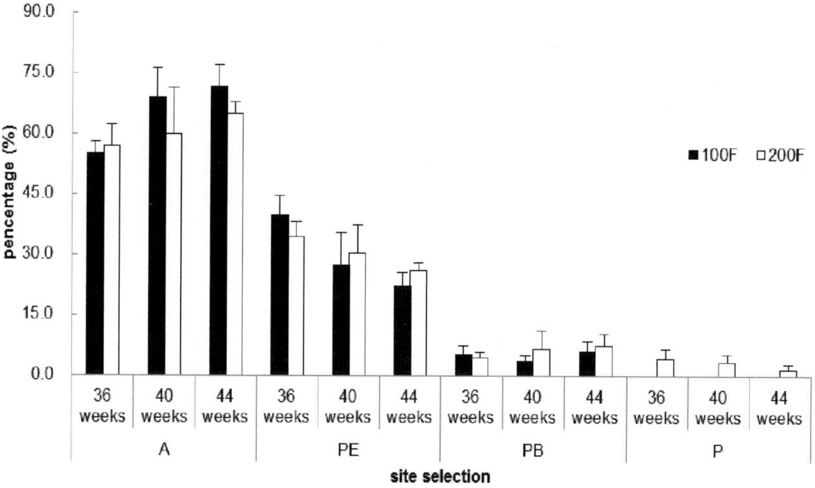

Figure 3. Site selection of groups in different observation periods, "A" (aisle), "PE" (platform edge), "PB" (platform bottom) and "P" (platform).

3.3. Effect on completeness

Results of completeness as described in observation session (2.2.) mainly consisted of score 3, which indicated that the majority of mating were not began with courtship, such as waltzing, wing flapping. As indicated by Kratzer and Craig (1980), flock density affected completed mating and males with more area had more activity. While, they found that small flocks showed more behavioural activity which was consistent with our results. No distinct variance was found between size groups and completed mating (score≥3) were lowest in 36 weeks during which mating frequency was the highest (Figure 4). There was a possibility that males in large flocks with less apparent hierarchy encountered greater stress on mating.

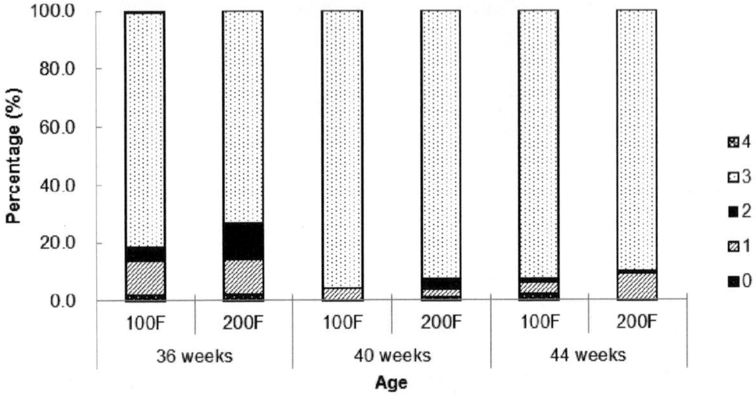

Figure 4. Percentage of completeness in groups of different times.

3.4. Effect on fertility

Fertility decreased with group size as inferred by Adams et al. (1978), who found that flock size negatively impacted on the fertility. And relationship between fertility and age could not be figured out because the whole experimental period was only one stage and was in the middle of the whole production cycle. When comparing data of 36 weeks against that of 44 weeks, we noticed that fertility dropped by 2% and 0.9% respectively for 100F and 200F (Figure 5). Such fertility variance between ages might due to the decreasing sperm penetration of hens (Bramwell et al., 1996; Gumułka and Kapkowska, 2005).

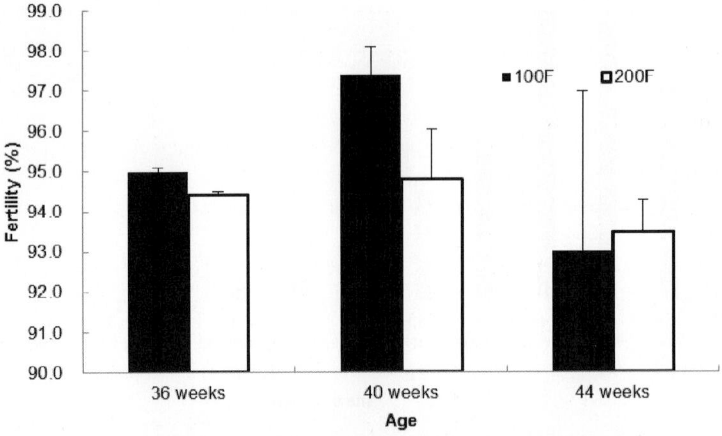

Figure 5. Fertility of groups in different times.

4. Summary

Frequencies of mating and wing flapping were significantly affected by group size and varied with ages. But there was no interaction of group size and age. Specifically, group size negatively impact mating and wing flapping revealed by our results. Fertility also decreased with group size and was not affected much by age during observation. Duration of each mating was quite changeable and no direct relationship with group size was detected. For site selection, home range of males was the place where most mating and wing flapping happened. And for which we can tell that males in groups of current sizes were passive for mating. No specific relationship was identified between mating behaviours and fertility. More detailed observation both about males and females in a longer production period should be performed to bring a more confirmed conclusion.

References

Adams, A.W., J.V. Craig, A.L. Bhagwat, 1978. Effects of flock size, age at housing, and mating experience on two strains of egg-type chickens in colony cages. Poultry Science, 57 (1), 48-53. http://dx.doi.org/10.3382/ps.0570048.

Al-Rawi, B., J.V. Craig, 1975. Agonistic behavior of caged chickens related to group size and area per bird. Applied Animal Ethology, 2 (1), 69-80. http://dx.doi.org/10.1016/0304-3762(75)90066-8.

Appleby, M.C., B.O. Hughes, 1991. Welfare of laying hens in cages and alternative systems: environmental, physical and behavioural aspects. World's Poultry Science Journal, 47 (2), 109-128. http://dx.doi.org/10.1079/WPS19910013.

Bareham, J.R., 1976. A comparison of the behaviour and production of laying hens in experimental and conventional battery cages. Applied Animal Ethology, 2 (4), 291-303. http://dx.doi.org/10.1016/0304-3762(76)90064-X.

Baxter, M.R., 1994. The welfare problems of laying hens in battery cages. Veterinary Record, 134 (24), 614-619. http://dx.doi.org/10.1136/vr.134.24.614.

Berdoy, M., P. Smith, D.W. Macdonald, 1995. Stability of social status in wild rats: age and the role of settled dominance. Behaviour, 132 (3), 193-212. http://dx.doi.org/10.1163/156853995X00694.

Bilcik, B., I. Estevez, 2005. Impact of male-male competition and morphological traits on mating strategies and reproductive success in broiler breeders. Applied Animal Behaviour Science, 92 (4), 307-323. http://dx.doi.org/10.1016/j.applanim.2004.11.007.

Bramwell, R.K., C.D. McDaniel, J.L. Wilson, B. Howarth, 1996. Age effect of male and female broiler breeders on sperm penetration of the perivitelline layer overlying the germinal disc. Poultry science, 75 (6), 755-762. http://dx.doi.org/10.3382/ps.0750755.

Campo, J.L., S.G. Davila, 2002. Influence of mating ratio and group size on indicators of fearfulness and stress of hens and cocks. Poultry Science, 81 (8), 1099-1103. http://dx.doi.org/10.1093/ps/81.8.1099.

Colson, S., C. Arnould, V. Michel, 2007. Motivation to dust-bathe of laying hens housed in cages and in aviaries. Animal, 1 (3), 433-437. http://dx.doi.org/10.1017/S1751731107705323.

De Jong, I.C., D. Guémené, 2011. Major welfare issues in broiler breeders. World's Poultry Science Journal, 67 (1), 73-82. http://dx.doi.org/10.1017/S0043933911000067.

FAWC, 2007. Opinion on Enriched Cages for Laying Hens. https://www.gov.uk/government/uploads/system/uploads/attachment_data/file/325142/FAWC_opinion_on_enriched_cages_for_laying_hens.pdf. Accessed June 3, 2015.

Graves, H.B., C.P. Hable, T.H. Jenkins, 1985. Sexual selection in Gallus: effects of morphology and dominance on female spatial behavior. Behavioural Processes, 11 (2), 189-197. http://dx.doi.org/10.1016/0376-6357(85)90060-9.

Gumułka, M., E. Kapkowska, 2005. Age effect of broiler breeders on fertility and sperm penetration of the perivitelline layer of the ovum. Animal Reproduction Science, 90 (1-2), 135-148. http://dx.doi.org/10.1016/j.anireprosci.2005.01.018.

Hughes, B.L., 1978. Efficiency of producing hatching eggs via artificial insemination and natural mating of broiler breeder pullets. Poultry Science, 57 (2), 534-537. http://dx.doi.org/10.3382/ps.0570534.

Hughes, B.L., K.A. Holleman, 1976. Efficiency of producing White Leghorn hatching eggs via artificial insemination and natural mating. Poultry Science, 55 (6), 2383-2388. http://dx.doi.org/10.3382/ps.0552383.

Hughes, B.O., N.L. Carmichael, A.W. Walker, P.N. Grigor, 1997. Low incidence of aggression in large flocks of laying hens. Applied Animal Behaviour Science, 54 (2), 215-234. http://dx.doi.org/10.1016/S0168-1591(96)01177-X.

Hughes, B.O., D.G.M. Wood-Gush, 1977. Agonistic behaviour in domestic hens: The influence of housing method and group size. Animal Behaviour, 25, 1056-1062. http://dx.doi.org/10.1016/0003-3472(77)90056-2.

Kratzer, D. D., and Craig, J. V. (1980). Mating behavior of cockerels: effects of social status, group size and group density. Applied Animal Ethology, 6 (1), 49-62. http://dx.doi.org/10.1016/0304-3762(80)90093-0.

Leonard, M.L., L. ZANETTE, 1998. Female mate choice and male behaviour in domestic fowl. Animal Behaviour, 56 (5), 1099-1105. http://dx.doi.org/10.1006/anbe.1998.0886.

Michel, V., D. Huonnic, 2003. A comparison of welfare, health and production performance of laying hens reared in cages or in aviaries. British Poultry Science, 44 (5), 775-776. http://dx.doi.org/10.1080/00071668.2003.10871381.

Newman, S., S. Leeson, 1998. Effect of housing birds in cages or an aviary system on bone characteristics. Poultry Science, 77 (10), 1492-1496. http://dx.doi.org/10.1093/ps/77.10.1492.

Pagel, M., M.S. Dawkins, 1997. Peck orders and group size in laying hens: `futures contracts' for non-aggression. Behavioural Processes, 40 (1), 13-25. http://dx.doi.org/10.1016/S0376-6357(96)00761-9.

Tactacan, G.B., W. Guenter, N.J. Lewis, J.C. Rodriguez-Lecompte, J.D. House, 2009. Performance and welfare of laying hens in conventional and enriched cages. Poultry Science, 88 (4), 698-707. http://dx.doi.org/10.3382/ps.2008-00369.

Tanaka, T., J.F. Hurnlk, 1992. Comparison of behaviour and performance of laying hens housed in battery cages and an aviary. Poultry Science, 71 (2), 235-243. http://dx.doi.org/10.3382/ps.0710235.

Taylor, A.A., J.F. Hurnik, 1994. The effect of long-term housing in an aviary and batty cages on the physical condition of laying hens - body-weight, feather condition, claw length, foot lesions, and tibia strength. Poultry Science, 73 (2), 268-273. http://dx.doi.org/10.3382/ps.0730268.

Wood-Gush, D.G.M., 1954. The courtship of the Brown Leghorn cock. The British Journal of Animal Behaviour, 2 (3), 95-102. http://dx.doi.org/10.1016/S0950-5601(54)80045-1.

Theme V: Animal Production Management and Welfare

Animal Welfare and Food Production in the 21st Century: Scientific and Social Responsibility Challenges

Candace Croney [a,*], Nicole Olynk Widmar [b], William M. Muir [c], Ji-Qin Ni [d]

[a] Department of Comparative Pathobiology, Purdue University, West Lafayette, IN 47906, USA

[b] Department of Agricultural Economics, Purdue University, IN 47906, USA

[c] Department of Animal Sciences, Purdue University, IN 47906, USA

[d] Department of Agricultural and Biological Engineering, Purdue University, IN 47906, USA

* Corresponding author. Email: ccroney@purdue.edu

Abstract

Global population growth creates the quandary of meeting increased demands for food, particularly in developing nations, while addressing ethical and sustainability concerns about food production that already challenge developed nations. The issue of farm animal welfare has rapidly become a hallmark of contemporary debates about animal agriculture. In the US, continuous confinement housing and behavioral restriction of animals have become primary areas of concern, although additional significant threats to animal well-being exist. These include inappropriate animal handling and other poor quality human-animal interactions on farms. Handling of non-ambulatory animals, and painful practices performed without analgesia, such as castration, tail docking and dehorning remain problematic. On-farm euthanasia methods and the timeliness of euthanasia decisions warrant attention, along with the distress, injury and mortality associated with animal loading and transport. Research on public perceptions of farm animal welfare in the US suggests that many citizens feel that agricultural animals are insufficiently protected. Concerns related to global trade and the pressure for multinational corporations to consistently meet social responsibility expectations make it imperative to address animal welfare even in nations where animal agriculture is still developing. It is therefore critical for the scientific and veterinary communities to be well versed in scientific advancements and challenges relative to farm animal welfare as well as the nature and reasons for public concerns. The latter is particularly important to facilitate improved communication, trust and perceived competence relative to current and emerging farm animal welfare issues.

Keywords: Livestock and poultry, ethics, sustainable production, consumer perceptions

1. Introduction

The anticipated expansion of the global population has led the United Nations to prescribe that world food production must dramatically increase to meet demands. Rising income levels in developing nations tend to result in dietary shifts that include more protein sources, a phenomenon that has been previously noted in South America and Asia particularly relative to fish production and consumption (Delgado et al., 2003). Because changes in consumption patterns, such as escalated demand for animal protein results in increased energy demands and impacts on natural resources, it has been suggested that "greener" methods of food production are needed to sustain the population (Freibauer, 2011). In developed nations, intensive production of livestock animals is already heavily scrutinized and criticized for its impacts on animals, environment and people; thus, its sustainability has been called into question. Broom (2010) notes that "no system or procedure is sustainable if a substantial proportion of people find aspects of it now, or of its consequences in the future, morally unacceptable." Consideration of "sustainable intensification" as a means of meeting food accessibility and quality demands requires consideration of many factors, including optimization of animal welfare, and deliberation about which food production systems might best fit the diverse and evolving definitions of sustainability. For example, in developed nations, where food security and accessibility are less problematic than in developing regions, consumers are increasingly concerned about food safety, animal welfare and environmental impacts of modern livestock production methods (Verbeke and Viaene, 1999; Baltzer, 2004; Botanaki et al., 2006). The relative affluence consumers enjoy in such countries, along with increased mobility and access to information have resulted both in greater demand for and apprehension about information related to many of our current food production practices. Because consumers use animal welfare as an indicator of food safety and quality (Harper and Makatouni, 2002), as well as a gauge of the social and ethical responsibility of the production source, it is unsurprising that farm animal welfare has become a focal issue in the US and other developed western nations, symbolizing the generally contentious nature of food production debates. Nonetheless, since animal welfare is an inherent, and in fact, critical component of sustainable agriculture, it is important to consider both the scientific and ethical challenges associated with meeting global food demands and corresponding human welfare issues, while attending to the quality of life experienced by animals raised for food.

2. Areas of Farm Animal Welfare Concern in the US: Scientific Challenges

2.1. Areas of concern and scientific challenges

The US livestock production faces a number of ongoing challenges relative to animal welfare, several of which have emerged as priority areas for research that have implications for agricultural sustainability (Table 1). For example, for at least the past decade, continuous confinement housing and ongoing behavioral restriction of animals have been primary areas of concern in the US with significant resources having been expended particularly to investigate alternatives to battery cage housing for laying hens and gestation stalls for sows. The use of sow stalls has particularly stymied the US swine industries, in part because scientific evidence exists that supports the use of stalls in regard to some aspects of sow well-being (e.g., individual monitoring of health and feed intake), even though infringements on other aspects of welfare (e.g., movement and social interactions) are evident (Gonyou, 2005). In addition, pressure from consumers, animal activists, various agri-food corporations and others have resulted in restrictions on the use of sow stalls, announcements to phase out their use entirely or to source pork only from producers who do not use them (Schulz and Tonsor, 2015; Tonsor et al., 2009). The US swine industry currently remains fragmented on the issue of sow housing, with segments of the industry adamantly defending the scientific basis and merits of sow stalls, particularly in light of reports of drastic reductions in the UK sow herds (Sheldon, 2013) and escalated imports of foreign pork products following enactment of national legislation banning the use of sow gestation stalls. In addition, many within the swine industries maintain that the primary driving factor in the push toward group housing is pressure from special interest groups opposed to animal production, resulting in further reluctance to move away from sow stalls.

Table 1. Scientific challenges of farm animal welfare in the US.

Areas of concern	Scientific challenge
Confinement animal housing	Animal behavioral restriction, safety, feeding
Sustainable animal production	Multiple dimensions of sustainable animal production systems; engineering innovation and economic feasibility
Mismatch between the animals and the environments	Direct and indirect genetic effects
Painful practices	Castration, tail docking, dehorning
Euthanasia	Timeliness and method
Indoor air quality and animal production	Economically feasible engineering innovation
Animal handling	Non-ambulatory animals and on-farm euthanasia

2.2. Role of genetics

As pressure mounts in the US to move toward group housing of livestock and poultry, the role of genetics in facilitating positive welfare outcomes for farm animals in various systems warrants renewed attention, given that at least some of the welfare problems animals experience are a result of a mismatch between the animals and the environments in which they are maintained, i.e., confined rearing or group housing (Muir et al., 2014). Such mismatches may manifest in a variety of ways, such as injurious feather pecking and cannibalism in laying hens, or tail or vulva biting in piglets and sows. However, to improve animal well-being, selection methods must consider both their direct and indirect genetic effects (Muir et al., 2013). A direct genetic effect is the influence of an animal's genes on its own performance, while indirect genetic effect is the effect of that animal's genes on another individual manifested through its phenotype, i.e., behavior (Muir, 2003; Muir and Cheng, 2004b). An important consequence of those effects is that the direct effect of genes can negatively impact others. Thus, one way to improve individual performance is increased competitive ability, which in turn decreases the performance and well-being of those interacting with that individual. Unfortunately, this is how most breeding programs are currently conducted. The way forward to improve both productivity and performance is therefore to change the unit and/or the context in which an individual is selected. Specifically, emphasis needs to shift from the individual's performance to that of the group (Muir and Cheng, 2004a; Muir, 1985).

Emphasizing group performance automatically includes both an individual's own performance and that of those with whom it interacts (Wade et al., 2010; Bijma et al., 2007; Ellen et al., 2007). The simplest method to achieve this goal is to house individuals in breeding programs by families rather than at random (Muir et al., 2013). Such grouping automatically includes both effects because the individuals are related and thus, their indirect genetic effects are reflected in their own phenotypes. In this context, one can still select the best performing individuals, but in the context of their own family. The entire group can also be selected as a unit, rapidly improving both performance and well-being, but at the cost of an increased rate of inbreeding (Muir et al., 2013). Thus, group selection can only be applied in very large breeding programs. Nevertheless, both group and individual selections in family groups has been highly successful in improving both performance and animal well-being in animals (e.g., Cheng et al., 2003; Craig and Muir, 1996; Hester et al., 1996; Muir and Cheng, 2013).

In short, evaluating the genetic basis for differences in expression of behavioral and physiological changes indicative of distress is likely to result in selection for healthier animals that are better adapted for contemporary production systems, and consequently, less likely to experience poor welfare conditions (Fraser et al., 2013).

2.3. Multiple dimensions of animal health and well-being

Development of sustainable animal production options requires attention to multiple dimensions that balance improved outcomes for animal health and well-being, environmental effects, impacts on worker health and safety, and effects on rural communities. As such, agricultural engineering approaches to optimize animal rearing environments are essential and ongoing. For example, the need to understand ammonia emissions in different types of laying hen housing systems as a function of factors such as litter quality, manure management, temperament, humidity, and air movement (Donham et al., 2010) have been well reviewed. However, production system designs that enhance the quality of animal environments must also include accommodations for animal social interactions and other species-typical behaviors, requiring a more integrated, multi-disciplinary approach than has previously been employed.

Recent research funded by the Coalition for Sustainable Egg Supply embodies the integrative approach to evaluating and improving agricultural animal production systems that is needed (Swanson, 2015). Multiple aspects of sustainability relative to different egg production systems were investigated, including animal welfare, implications for the environment and caretakers, as well as on the safety, quality and affordability of eggs (Swanson et al., 2015), illustrating the risks and benefits derived in each of these as a function of different laying hen housing environments. Similar multi-dimensional examinations of the various options for livestock management are needed.

2.4. Indoor animal living environment and engineering innovation

Research has revealed that indoor animal living environment, especially indoor air quality, does not only affect animal wellbeing but also animal safety, in intensive farm animal production systems. Major air pollutants in concern include ammonia, hydrogen sulfide, and organic dusts in animal agriculture (e.g., Collins and Algers, 1986; Gustafsson et al., 2013). Variations in indoor air quality could be correlated to several factors, among which were animal building structures and ventilation controls, manure handling and storage methods, animal densities, feed regimes, and farm management practices (Ni, 2015). However, engineering innovations that are favorable for animal welfare improvement can encounter various challenges. High-rise swine finishing and farrowing barns with composting manure pits greatly reduced in-building ammonia concentrations (Stowell et al., 2001; Dunn et al., 2003). However, their economic feasibility has hindered their applications.

On the other hand, an animal production technology may be contentious when viewed from different aspects of animal welfare. In the layer industry, although non-cage houses are considered more animal-welfare friendly, they generally have poorer in-house air quality (ammonia and dusts) than cage houses (Xin et al., 2011). Realizing the goal of developing diverse, sustainable livestock and poultry production systems clearly requires both innovation in bioengineering to optimize environmental conditions within animal housing areas and minimize external environmental impacts and better incorporation of animals' behavioral needs, so that welfare can be improved in intensive and extensive systems. Closer collaboration between agricultural engineers and animal welfare scientists, particularly, ethologists, is therefore needed.

2.5. Additional animal welfare challenges

It is important to note that while much work is needed to improve the housing of agricultural animals, a number of additional problems exist that are at least as significant in regard to potential infringement on animal well-being, although several of these have received comparatively less public attention (Table 1). These issues include inappropriate animal handling and other poor quality human-animal interactions on farms. Handling of non-ambulatory animals has improved at slaughter plants (Grandin, 2006), but continues to present a challenge on farms, particularly in areas, such as milking parlors, where the layout of the facility or location of the downed animal may complicate humane movement or euthanasia. Painful practices performed without analgesia, such as castration, tail docking and dehorning (Coetzee, 2011, 2013) also remain problematic. On-farm euthanasia methods and the timeliness of euthanasia decisions also warrant attention, along with the distress, injury and mortality that can occur during loading and transport of animals.

3. Public Perceptions of Animal Welfare: Social Responsibility Challenges

3.1. Ethical dimensions of food production

As the US public becomes more attuned to and concerned about the well-being of agricultural animals, understanding the basis for concerns and the sources of information that inform them becomes increasingly critical in regard to meeting public expectations. The ethical dimensions of food production have been noted to be of significant interest and importance to consumers in developed western nations (Croney and Anthony, 2014). Recent shifts in consumer attention to "ethically responsible" food or "ethical consumerism" as described by Singer and Mason (2006) are characterized by shoppers seeking out and selecting food products that appear to be in line with their values. For example, those focused on socially conscious consumerism appear to prioritize food purchases that seem to cause less societal and environmental harm than do conventional products. Typically, such consumers seek out labels, such as "organic" and "natural," believing them to be safer, more nutritious, and of higher quality (Harper and Makatouni, 2002); for foods of animal original, such labels may appear to connote products that are also more "animal-friendly."

3.2. US public perceptions of animal welfare

Given these developments, significant investments have been made by scientific, veterinary, non-governmental, producer and other agri-food industry groups to develop resource materials and assurance programs on farm animal welfare. However, a recent Purdue University study suggests that US residents do not necessarily look to these particular sources for information on animal welfare (McKendree et al., 2014). An online survey of 798 US households examined relationships between key household characteristics (demographics, geographic location and experiences), reported levels of concern about animal welfare, and sources of information people use to inform themselves on the topic. Because of the level of media attention dedicated to recent undercover videos of swine care practices on farms, specific questions pertaining to modern pork production were posed. Over half of those surveyed (56%) could not identify a specific source that they sought out for animal welfare information. Those who did have a source most commonly reported using information provided by the Humane Society of the United States (HSUS) and People for the Ethical Treatment of Animals (PETA). Respondents were more concerned about confinement housing of sows, identifying gestation and farrowing stalls as even more troubling than castration, teeth clipping or tail docking of piglets. Additionally, respondents reported acting on concerns about animal welfare, specifically pork production, with 14% reportedly decreasing their pork consumption in the past three years due to animal welfare concerns; for those reportedly decreasing consumption, it was by an average of 56% (McKendree et al., 2014).

These data collectively suggest that although US consumers are concerned about animal welfare, they either are unaware of, distrustful or disinterested in or unable to access much of the information produced and disseminated annually by those with expertise in animal welfare science and farm animal production. However, given the otherwise counter-intuitive finding that consumers were most concerned about swine housing conditions, a topic that has been prioritized by animal activist groups and successfully communicated through social and conventional media, it would appear that indeed, US consumers are being educated about animal welfare, but not by those credentialed on the topic who might be expected to be perceived as experts. The extent to which lack of awareness of these sources as well as distrust in scientists and their ethics may underlie the results of McKendree et al. (2014) require further investigation.

A study still underway at Purdue University is aimed at uncovering linkages between personal practices (such as, recycling, volunteerism, and donating to charities both domestically and abroad) and perceptions of animal welfare. This study seeks to discover potential relationships between personal ethical behaviors and perceptions of individual behavior and the beliefs imposed on food producers and retailers regarding appropriate ethical business practices and behavior.

3.3. Public attitudes in other countries

Surveys of public attitudes toward agricultural animal welfare conducted in the European Union and other western developed nations indicate comparable and often higher levels of concern relative to animal welfare. Similar surveys of the attitudes of Chinese consumers could provide useful insights into cultural similarities and differences between eastern and western consumers of animal products and help to identify potential challenges and opportunities that may be encountered by emerging Chinese commodity markets. Evidence that Chinese consumers may already resemble those in the US in some regards is presented by the findings of

Wang et al. (2014) who conducted a survey of 1489 Chinese consumers on preferences for pork traceability attributes such as traceability information, quality certification, appearance, and price. Participant responses indicated quality certification as highest priority, with "government certification," and traceability information that included farming, slaughter and processing reported as most preferred levels of quality certification. These results align with those of Ortega et al. (2014) and Widmar and Ortega (2014) who reported that US consumers had highest willingness to pay for government certification programs. Further, Chinese consumers' willingness to pay for these attributes was associated with factors such as age, income, and education level (Wang et al., 2014).

4. Animal Welfare in Developing Nations

4.1. An increasing importance

In the past few decades, animal welfare has begun to play an increasingly important role in discussions pertaining to international trade and policy (Bayvel, 2004). For example, following the establishment of the World Organization for Animal Health's (OIE) working group on animal welfare, recommendations on globally relevant agricultural animal welfare concerns have been incorporated into the OIE Strategic Plan. These include welfare standards for land, sea and air transport of animals, humane slaughter (including religious slaughter), and depopulation of food animals for disease control purposes.

4.2. Animal welfare and food production

Although in developing nations animal welfare may be viewed as a lower priority than immediate concerns about food production—namely safety, quality, security and accessibility, it is still an important consideration, particularly relative to long term implications for agricultural development and global trade (Rahman, 2004). For one thing, it is well established that sound animal health and welfare practices are the fundamental bases for the production of safe, affordable animal products. Further, concerns relative to global trade and the pressure for multi-national corporations to consistently meet high standards of animal welfare make it imperative to address animal welfare even in nations where animal agriculture is still developing.

Globally operating corporations involved in food production and/or distribution are likely to face mounting pressure to ensure that animal production standards in developing nations are scientifically and ethically in line with those established for developed nations. Despite the logistical and cultural differences that may be encountered in different parts of the world, companies that appear to operate with inconsistencies in apparent commitment to animal welfare and other aspects of corporate social responsibility are likely to face brand integrity challenges and potential economic losses (Croney and Anthony, 2014). Thus, at minimum, nations such as China that strive to grow their agricultural infrastructure and global competitiveness should consider incorporating appropriate animal welfare standards and assurance mechanisms into their plans for agricultural development and marketing (Rahman, 2004).

5. Discussion

5.1. Animal welfare and sustainable intensification

Achieving sustainable food systems requires considerable deliberation about how to meet the scientific challenges of feeding the growing global population while addressing the ethical dimensions of socially acceptable food production. Although it has been argued that intensive animal production is environmentally and socially irresponsible, increasing demand for high quality, consistent protein is likely to fuel efforts to achieve "sustainable intensification." This will require innovation that increases efficiency, reduces waste and mitigates or minimizes harm to animals, people and the environment (Freibauer, 2011). Animal welfare therefore must be viewed as an integral component of sustainable agriculture, rather than as a limiting factor in advancing the goals of global food security and sustainability.

5.2. Scientific research and public communication

A major lesson to be learned from the US and other developed western nations is the importance of not just engaging the scientific challenges of addressing animal welfare, but also of addressing the related ethical concerns and communicating public broadly on these subjects. While the importance of ethical considerations in livestock production are beginning to be acknowledged as a component of socially acceptable agriculture (Fraser, 1999; Harper and Makatouni, 2002; Broom, 2010; Croney and Anthony, 2014; Swanson et al., 2015), it is not uncommon for members of the US livestock and poultry industries to conflate "ethics" with "emotions." The resulting attempts to convey passion for farming may not only fail to properly address fundamental social concerns (e.g., the quality of life farm animals experience in intensive confinement systems) or articulate an ethical argument or framework for current practices, but may also serve to rationalize practices that are of concern to people, and in so doing actually alienate members of the public. Evaluation of the efficacy of different approaches to communicating about animal welfare is therefore needed.

5.3. Engaging target audiences

Surprisingly little has been published, however, on the role of public communications in addressing animal welfare concerns or the social acceptability of livestock and poultry production practices. However, the findings of McKendree et al., (2014) suggest that significant changes may be needed in the methods used by US industry, governmental and academic institutions to engage with the public on animal welfare. For example, depending on members of the public to visit industry or even academic sites to learn more about practices impacting animal care and welfare is unlikely to yield results given that few people reported any usage of these sources of information. Further, clarity of purpose in engaging members of the public is often lacking and requires consideration. For instance, a determination must be made as to whether the goal of engagement is simply to transfer information to the public, advocate for a particular position (e.g., support for meat eating and large scale production), assuage public concerns about animal welfare or create opportunities for dialogue. It is highly unlikely that a single message on animal welfare could achieve more than one of these goals. It is also imperative to consider the diversity of individuals and perspectives that are captured by the term "the public," as failure to do so may result in unproductive approaches targeted at a monolithic entity to which a single message or approach can be directed.

Not only must messages be tailored to the very different "publics" that exist, it is also important to consider whether the information provided actually targets the lay public. For example, Duffy et al. (2005) found that despite the belief of many groups that members of the consuming public are disconnected from and lack information about animal agriculture, consumers were generally not the primary audience for industry messages about food production. Rather, the intended targets were government entities or influential decision-makers in the food sector. Difficulties in effectively reaching consumers may be implicated, which includes the need to vastly simplify complex information and to transmit messages via television, other media and campaigns that can be cost-prohibitive and yield a poor return on investment for those members of the public who are not interested in learning about farming practices.

6. Conclusions

In conclusion, the challenges of addressing animal welfare are clearly not constrained to its scientific domains, although the latter presents an increasingly complex problem. A significant lesson in that regard is the problem of taking a reductionist approach to animal welfare, such as attempting to focus on single dimensions of problems, such as animal housing, without taking into consideration the inter-related issues such as feeding, managing behavior of animals in groups, as well as the environmental, human health, community and economic impacts of the housing systems in question. Properly addressing such issues is also dependent on incorporating relevant embedded ethical and social considerations and cooperatively engaging with diverse members of the public. Thus, it is critical for academics, veterinarians and all sectors of food animal production to be informed about current scientific advancements and challenges relative to agricultural animal welfare and to be considerate of changing public knowledge, concerns and values. Sustainability of livestock and poultry production in both developed and developing nations is ultimately dependent on sound scientific approaches to animal production combined with appropriate ethical considerations and two-way communication that engenders public trust and perceived competence on farm animal welfare.

References

Baltzer, K. 2004. Consumers' willingness to pay for food quality — the case of eggs. Acta Agriculturae Scandinavica Section C, Food Economics. 1, 78–90.

Bayvel, A.C.D. 2004. The OIE animal welfare strategic initiative — Progress, priorities and prognosis. In *Proceedings of Global conference on Animal Welfare: an OIE Initiative*. Vol. 14

Bijma, P., W.M. Muir, J.A.M. Van Arendonk. 2007. Multilevel selection 1: Quantitative genetics of inheritance and response to selection. Genetics. 175, 277–288.

Botonaki, A., K. Polymeros, E. Tsakiridou, K. Mattas. 2006. The role of food quality certification on consumers' food choices. British Food Journal. 108, 77–90.

Broom, D.M. 2010. Animal welfare: an aspect of care, sustainability, and food quality required by the public. Journal of Veterinary Medical Education. 37 (1), 83–88.

Cheng, H.W., P. Singleton, W.M. Muir. 2003. Social stress in laying hens: Differential effect of stress on plasma dopamine concentrations and adrenal function in genetically selected chickens. Poultry Science. 82, 192–198.

Coetzee, J.F. 2011. A review of pain assessment techniques and pharmacological approaches to pain relief after bovine castration: Practical implications for cattle production within the United States. Applied Animal Behaviour Science. 135 (3), 192–213.

Coetzee, J.F. 2013. A review of analgesic compounds used in food animals in the United States. Veterinary Clinics of North America: Food Animal Practice. 29 (1), 11–28.

Collins, M.B. Algers. 1986. Effects of stable dust on farm-animals - a review. Veterinary Research Communications. 10 (6), 415-428.

Craig, J.V., W.M. Muir. 1996. Group selection for adaptation to multiple-hen cages: Beak-related mortality, feathering, and body weight responses. Poultry Science. 75, 294–302.

Croney, C., R. Anthony. 2014. Food Animal Production, Ethics, and Quality Assurance. In Encyclopedia of Food and Agricultural Ethics (Eds). Paul B. Thompson and David M. Kaplan. Springer Reference. 1–10.

Delgado, C.L., N. Wada, M.W. Rosegrant, S. Meijer, M. Ahmed. 2003. Fish to 2020: supply and demand in changing global markets. International Food Policy Research Institute, Washington and the World Fish Center, Penang, Malaysia.

Donham, K., R. Aherin, D. Baker, G. Hetzel, 2010. Safety in swine production systems. In: Pork Industry Handbook. PIH 16-01-02. Purdue Extension, West Lafayette, Indiana. 7 p.

Duffy, R., Fearne, A., Healing, V. 2005. Reconnection in the UK food chain: bridging the communication gap between food producers and consumers. British Food Journal. 107 (1), 17-33.

Dunn, J.L., J.-Q. Ni, B.E. Hill, M.L. Henry, A.J. Heber, T.-T. Lim, 2003. Odor nuisance potential and pig performance of a conventional and a high rise™ swine farrowing barns. In: *Proceedings of Air Pollution from Agricultural Operations III*. Research Triangle Park, North Carolina, USA, 12–15 October. Ed., Keener, H. 303–310.

Ellen, E.D., W.M. Muir, F. Teuscher, P. Bijma. 2007. Genetic improvement of traits affected by interactions among individuals: Sib selection schemes. Genetics. 176, 489–499.

Fraser, D. 1999. Animal ethics and animal welfare science: bridging the two cultures. Applied Animal Behaviour Science. 65 (3), 171-189.

Fraser, D., I.J, Duncan, S.A. Edwards, T, Grandin, N.G. Gregory, V. Guyonnet, Paul H. Hemsworth, S. M. Huertas, J.M. Huzzey, D.J. Mellor, J.A. Mench, M. Špinka, H.R. Whay. 2013. General principles for the welfare of animals in production systems: the underlying science and its application. The Veterinary Journal. 198 (1), 19–27.

Freibauer, A. 2011. Sustainable Food Consumption and Production in a Resource-Constrained World. 3rd Standing Committee on Agricultural Research (SCAR) Foresight Exercise. European Commission, Brussels, Belgium. http://ec.europa.eu/research/agriculture /conference/pdf/feg3-report-web-version.pdf. Accessed Sept. 7, 2015.

Gonyou, H.W. 2005. Experiences with alternative methods of sow housing. Journal of the American Veterinary Medical Association. 226(8), 1336–1340.

Grandin, T. 2006. Progress and challenges in animal handling and slaughter in the US. Applied Animal Behaviour Science. 100 (1), 129–139.

Gustafsson, G., S. Nimmermark, K.H. Jeppsson. 2013. Control of emissions from livestock buildings and the impact of aerial environment on health, welfare and performance of animals - a review. In *Livestock Housing: Modern Management to Ensure Optimal Health and Welfare of Farm Animals*. Wageningen, The Netherlands: Wageningen Academic Publishers. 261-280.

Harper, G.C., A. Makatouni. 2002. Consumer perception of organic food production and farm animal welfare. British Food Journal. 104, 287–299.

Hester, P.Y., W.M. Muir, J.V. Craig, J. L. Albright. 1996. Group selection for adaptation to multiple-hen cages: Production traits during heat and cold exposures. Poultry Science. 75, 1308–1314.

McKendree, M.G.S., C.C. Croney, N.J.O Widmar. 2014. Effects of demographic factors and information sources on United States consumer perceptions of animal welfare. Journal of Animal Science. 92 (7), 3161–3173.

Muir, W.M. 1985. Relative efficiency of selection for performance of birds housed in colony cages based on production in single bird cages. Poultry Science, 64, 2239–2247.

Muir, W.M. 2003. Indirect selection for improvement of Animal Well-Being. In W. M. Muir, and S. Aggrey (ed.) Poultry Genetics Breeding and Biotechnology. CABI Press, Cambridge MA USA. 247–256.

Muir, W.M., H. Cheng. 2004a. Breeding for Productivity and Welfare. p. 123–138. Welfare of the Laying Hen, No 27. Poultry Science Symposium Series. CABI Press, Cambridge MA.

Muir, W.M., H.W. Cheng. 2004b. Breeding for productivity and welfare of laying hens. Welfare of the Laying Hen. 27, 123–138.

Muir, W.M., H.W. Cheng. 2013. Genetics and the Behaviour of Chickens. In Temple Grandin (ed.) Genetics and the Behaviour of Domestic Animals. Academic Press.

Muir W.M., P. Bijma, A. Schinckel. 2013. Multilevel selection with kin and non-kin groups, experimental results with Japanese quail (Coturnix japonica). Evolution 67, 1598–1606.

Muir, W.M., H.W. Cheng, C. Croney. 2014. Methods to address poultry robustness and welfare issues through breeding and associated ethical considerations. Frontiers in Genetics. 5, Article 407.

Ni, J.-Q., 2015. Research and demonstration to improve air quality for the U.S. animal feeding operations in the 21st century — A critical review. Environmental Pollution. 200, 105–119.

Ortega, D. L., H.H. Wang, N.J.O Widmar. 2014. Welfare and market impacts of food safety measures in China: Results from urban consumers' valuation of product attributes. Journal of Integrative Agriculture, 13 (6), 1404–1411.

Rahman, S.A. 2004. Animal welfare: a developing country perspective. Proceedings of Global Conference on Animal Welfare: An OIE Initiative. Vol. 14.

Schulz, L. L., G.T. Tonsor. 2015. The US gestation stall debate. Choices, 30(1), 1-7.

Sheldon, M. 2013. Competitiveness in pork production — A view from Europe. In *Banff Pork Seminar Proceedings*, Banff, Alberta, Canada, 15–17 January. Department of Agricultural, Food and Nutritional Science, University of Alberta. 51–55.

Singer, P. and J. Mason, 2006. The Ethics of What We Eat: Why our Food Choices Matter. US: Rodale Inc.-Holtzbrinck Publishers.

Stowell, R.R., S.F. Inglis, H. Keener, D. Elwell, 2001. Ammonia emissions and hydrogen sulfide levels during handling of manure-laden drying bed material in a High-Rise™ hog facility. In *ASAE Annual International Meeting*. St. Joseph, Mich.: ASAE, Sacramento, California.

Swanson, J.C., J.A., Mench, D. Karcher, D. 2015. The coalition for sustainable egg supply project: An introduction. Poultry Science. 94 (3), 473–474.

Tonsor, G.T., C. Wolf, N. Olynk. 2009. "Consumer Voting and Demand Behavior Regarding Swine Gestation Crates." Food Policy. 34, 492–498.

Verbeke, W., J. Viaene. 1999. Consumer attitude to beef quality labels and associations with beef quality labels. Journal of International Food and Agribusiness. 10 (3), 45–65.

Wade, M.J., P. Bijma, E.D. Ellen, W. Muir. 2010. Group selection and social evolution in domesticated animals. Evolutionary Applications. 3, 453–465.

Wang, S., L. Wu, D. Zhu, H. Wang, H, L. Xu. 2014. Chinese consumers' preferences and willingness to pay for traceable food attributes: The case of pork. In *2014 Annual Meeting, Agricultural and Applied Economics Association*. Minneapolis, Minnesota, July 27–29. No. 165639.

Widmar, N.J.O., D.L. Ortega. 2014. Comparing Consumer Preferences for Livestock Production Process Attributes Across Products, Species, and Modeling Methods. Journal of Agricultural and Applied Economics. 46 (3), 375–391.

Xin, H., R.S. Gates, A.R. Green, F.M. Mitloehner, P.A. Moore, C.M. Wathes, 2011. Environmental impacts and sustainability of egg production systems. Poultry Science. 90 (1), 263–277.

Improving Farm Animal Productivity and Welfare, by Increasing Skills and Knowledge of Stock People

Hans Spoolder *, Marko Ruis

Department of Animal Welfare, Wageningen UR Livestock Research, Wageningen, The Netherlands

* Corresponding author. Email: hans.spoolder@wur.nl

Abstract

Animal welfare and production results are closely linked. There is a lot of variation in both, even if the circumstances under which the animals are kept appear to be similar. This means that the farmer himself plays an important role in maintaining a good level of welfare, and a good level of production. Taking care of animals is a job for which skills and training are required. Some of the most important attributes are a sound technical understanding of what animals need, a keen eye for the signals animals send regarding their health and welfare and the personal characteristics to act on information. Stock people should be supported to do their job well, e.g., through training on technical issues, but also by training on how to learn about the 'signals' that animals send to us. A correct assessment of the welfare status, e.g., through the Welfare Quality® protocol, will also facilitate the issues which need to be addressed through good farm management. Finally, the attitude of the stock person to his or her work is of crucial importance to animal wellbeing. Training which addresses attitudes is available, and has also proven to be effective in increasing farm productivity.

Keywords: Animal welfare, animal performance, stock person training, animal signals, Welfare Quality®

1. Introduction

A large variation exists in technical production results among farms — even if the circumstances under which the animals are kept appear to be similar (e.g., Peet-Schwering et al, 2009; Fraser, 2014). This indicates that the farmer himself plays an important role: on many farms there is room for improvement in skills and knowledge related to husbandry and management. In this paper we will address the role of the farmer or stockperson, and suggest that improvements in productivity will go hand in hand with improvements in farm animal welfare.

2. What is Animal Welfare?

2.1. Definitions of welfare

The concept of 'animal welfare' has been around for several decades now, but its definition still remains notoriously vague. It has, for example, been defined as "living in reasonable harmony with its environment, from a physiological as well as an ethological point of view" (Lorz, 1973). Broom (1986) describes animal welfare as "its state as regards its attempts to cope with its environment". Duncan (1996) and Fraser and Duncan (1998) suggest that mental states (such as pleasure, pain, fear and frustration) should be included. Nevertheless, despite the many definitions, "the quality of life as the animal perceives it" (Bracke, 1999) seems to be a central and shared concept.

The problem with all of these definitions is that they do not make it easier to understand what it is and how it can be assessed, or as professor John Webster puts it: "they tend to have the self-referential flavour of 'a rose is a rose'" (Webster, 1994). His suggestion of a set of 5 'freedoms' based on earlier work by the Brambell Committee in 1965 offers criteria that every animal friendly system should meet (Farm Animal Welfare Council, 1993). These freedoms, which consider the biological functioning as well as the mental state aspects defining animal welfare, have recently been interpreted by the World Organization for Animal Health (OIE) as "an animal is in a good state of welfare if (as indicated by scientific evidence) it is healthy, comfortable, well nourished, safe, able to express innate behaviour and is not suffering from unpleasant states such as pain, fear and distress" (OIE, 2010). They provide the basis for a scientific assessment of welfare from the animal's point of view, but in themselves are still insufficient to be used as a tool to measure it.

Parallel to these developments, the animal welfare concept has evolved from an initial, almost exclusive consideration of the animal, towards a multidimensional concept. The definition and assessment of animal welfare involves not only animal welfare science, but also economic and social science aspects (Whay, 2007), as well as moral and ethical considerations (e.g., Duncan and Fraser, 1997; Aparicio and Vargas, 2006; Cauldfield and Cambridge, 2008). The public debate on what is appropriate animal welfare has led to questions on what animals need as a minimum requirement (which can vary from e.g., 'free of suffering' to 'having a decent a life of quality') and society accordingly consider to be ethically right. Moreover, some management procedures we routinely practice in animal husbandry meet with societal opposition as they are considered not acceptable at all. An example is the killing of male day-old chicks from layer breeds just after hatching (which are not profitable as regards the production of meat). The killing itself may not cause suffering to the animals, when it is done properly. However, for many people this is not in line with an ethically sound and responsible treatment, respecting the animals' 'intrinsic value'. Although increasing societal awareness regarding the production processes in the livestock industry may thus result in new and different pressures on the industry, there are also increasing opportunities for farmers able and willing to adjust their way of livestock keeping. The rise in the number of niche products specifically addressing societal concerns is a clear example of that.

3. How Does Welfare Relate to Productivity?

3.1. A happy animal performs well

It is often said that a happy animal performs well. A nutritious diet, a healthy living environment and the absence of stress will promote growth and fertility in all livestock species. In addition to positive effects of welfare on animal performance, factors that affect welfare negatively may also have negative consequences for the quality of the product. For example, stress during transport to the slaughter house may reduce meat quality. The basis for this relationship lies in the way animals cope with stress. Spoolder and Waiblinger (2009) describe the sequence of events in relation to the handling of livestock. In short, when confronted with a stressful stimulus animals (and humans!) will react with a so called 'fight-or-flight' response (Cannon, 1914). This immediate reaction involves autonomic and neuroendocrine elements which mobilise the body's reserves so it can respond adequately to the challenge. It may include elevated levels of adrenaline, increased heart rate, blood pressure and body temperature. If the stressor persists the physiological responses move into a second phase, which is still relatively short term but predominantly corticosteroid mediated. The increased levels of corticosteroids allow extra energy (glucose) to be generated from, for example, muscle proteins. If the stressor is still not removed and cannot be avoided, a long term or chronic stress response ensues. This response also involves corticosteroids and has a negative effect on the animal's physiology. Chronic stress may result in reduced reproductive performance (e.g., Hemsworth et al., 1986), reduced growth (through poorer efficiency: e.g., Hemsworth and Coleman (1998)) and impaired immunity (e.g., Clarke et al. 1992).

3.2. Is a well performing animal always happy?

If a happy animal performs well, is the reverse also true? There has been much criticism on this suggestion, supported by several examples which seem to indicate that high production does not necessarily agree with high welfare. For example, broiler chickens generally growth fast (which is good for farm productivity), but at such a speed that it is claimed their bones and leg muscles do not keep up and are not strong enough to support the rapidly increasing weight (which is bad for their welfare). High productivity in dairy cows can lead to metabolic disorders. Increasing the litter size of pigs will increase the number of pigs weaned per sow per year, but also the number of weak piglets that die from either starvation or overlaying by the sow.

There are no easy answers to these dilemmas, and we need to continue to work on alternatives that will provide good performance as well as good welfare. These alternatives do not always have to be of a technical nature. The stockperson him or herself can make a lot of difference in how animals deal with the pressures of production.

4. Attributes a Professional Stock Person Needs

Not every person is a good stock person by nature. Taking care of animals is a job for which skills and training are required (Fraser, 2014). Some of the most important attributes are discussed below.

4.1. A sound technical understanding of what animals need

Sound animal welfare starts with a good understanding of the animal and its needs. Species specific knowledge about physical, nutritional and hygiene requirements is essential to be able to provide the right circumstances for animals to thrive. Physical requirements include a proper living environment, divided in functional areas for different activities, e.g., resting, feeding, exploring and defecating. A healthy living environment also includes maintaining a good hygiene, the right temperature and the right amount and quality of light, Diets should be well-balanced for the specific nutritional requirements of animals. The exact requirements will change over time and are related to e.g., age, breed, gender and season.

4.2. A keen eye for the signals animals send regarding their health and welfare status

However, 'technical knowledge' alone is not sufficient. As husbandry circumstances vary considerably, not only between farms but also over time, it is important that animal caretakers have the skills to observe and record signals requiring corrective action. This means they need to know animals and understand their normal behaviors, as well as recognize changes in their appearance, posture and activities.

A good example is the use of a cow's 'flight zone' when moving animals. The flight zone is the point at which a grazing animal "no longer can tolerate the approach of a person or other animals and moves away" (Grandin 2008, p. 33). The size of the flight zone differs between individual animals, depending on breed, age and experience. Animal handlers use the flight zone to move cattle in the right direction. However, penetrating an animal's flight zone to deeply may cause it to escape over the side of the race (Grandin, 1987), or turn back on the handler if it does not see another way out (Hutson, 1993).

Another example is related to aggressive behaviour in breeding sows. Agonistic interactions between sows cause skin lesions, leg injuries and potentially have a negative effect on fertility. Aggression occurs predominantly because of competition for access to a limited resource, or to establish a social relationship between unfamiliar animals (c.f. Spoolder et al. (2009)). Although the interactions themselves are not always observed during day-to-day inspection rounds, the resulting skin lesions often are. They can differ substantially in terms of location and severity, and a good stock person will know which kind of lesions is related to which of the two main causes of aggression. This will allow him to apply the right management strategy to solve the problem.

A third example is related to health of broiler chickens. The quality of broiler's foot pads is an important indicator of litter quality. The drier and looser the litter, the less problems there are with foot pad health. Foot pad lesions are painful and not only bad for the welfare of the birds: broilers with affected feet will also eat and grow less, and have poorer feed conversion rates (De Jong et al., 2014). Understanding broiler behaviour will help the care taker to initiate action when problems occur.

4.3. The personal characteristics to act on information

Possessing the right knowledge and skills is essential, but applying them is equally important. The response by stockpeople to welfare issues which require attention is of course partly determined by the situational variables (physical possibilities and constraints) on the farm (e.g., Spoolder and Waiblinger, 2009). In the present paper we will not address these variables, as they are important, but rather specific to each situation and farm. However, the human behavioural responses are also determined to a large extent by his or her personality and attitude (Ajzen 1988; Schiefele 1990).

Personality can be defined as an individual's unique, relatively consistent pattern of thoughts, feelings, and behaviors. Humane personality is relatively stable over time (Costa and McCrae 1986), and can be related to animal productivity (e.g., Seabrook 1995; Waiblinger et al. 2002). Seabrook (1972) identified a personality profile of a 'confident introvert' to be the 'ideal' stockperson, in terms of achieving the highest milk yield in dairy cow herds.

Attitudes are learnt and changeable by new information or experiences (Azjen 1988). They are often seen to be the most important causal factor of a person's behaviour towards social objects (Schiefele 1990). The influence of stockpersons' attitudes on their interactions with the animals and on subsequent behaviour and productivity of the animals, has been shown in pigs (e.g., Hemsworth et al. 1994; Coleman et al. 2000) and cattle (e.g., Lensink et al. 2000; Waiblinger et al. 2002). In their review, Spoolder and Waiblinger (1997) suggest that the stockpersons' attitude towards animals is not only limited to the direct interactions, but also involves indirect aspects such as attention to detail, readiness to solve problems, decisions in management and housing.

There are other characteristics of the stockperson which may also have effects on his/her behaviour. Coleman et al. (1998) suggests that empathy with animals may influence the development of beliefs about pigs and the way they are handled. According to Lenssink *et al.* (2000) gender also seems to have an effect on attitudes and behaviour to animals, with women showing more positive attitudes and interactions than men.

5. Supporting and Encouraging Stock People

5.1. Sound technical knowledge to the farming industry

In last decades a lot of knowledge based on scientific research has become available. However, much of this knowledge does not reach farmers and stock people. The 'translation' from scientific results to the daily farm practice, including the dissemination methods, therefore needs special attention. Science must meet the social-psychological dimensions of the farmer in the role of entrepreneur, livestock keeper and stockman in order to make the knowledge transfer work (Pompe and Ruis, 2015). Farmers may have reservations about the success of implementation of knowledge: it may be too remote and not easy to implement. Or it is not clear how welfare and business interests match and whether potential losses may incur. An important incentive for supplying more animal welfare is accompanied with little cost or even a financial benefit (Lawrence and Stott, 2010).

Dissemination methods are more successful when there is an exchange of knowledge between science and practice, bridging a gap between both worlds. Where scientists tend to focus on valid and reliable knowledge and information, practice aims more at changes, innovation and usability. Combining both leads to better understanding and smarter interaction, and enhances successful valorisation of welfare knowledge (Pompe and Ruis, 2015). Participation matches the more interactive perspective of entrepreneurial process towards implementation and innovation. For science, interaction with farmers and stock keepers will prospect the continuity in data valorisation and new data gathering.

5.2. Means to identify the welfare status

To measure = to know. The collection of data on a regular basis will allow management decisions to be taken on the basis of facts, rather than assumptions. Implicit to our emphasis on stockperson skills in the previous paragraphs, is that he or she bases their judgement of the welfare situation not on data regarding the circumstances under which the animals are kept, but on so called 'Animal Based Measures' (ABMs). The European Food Safety Authority (EFSA, 2012) defines an animal based measure as "a response of an animal or an effect on an animal used to assess its welfare. It can be taken directly on the animal or indirectly and includes the use of animal records. It can result from a specific event, e.g., an injury, or be the cumulative outcome of many days, weeks or months, e.g., body condition". EFSA concludes in her statement on ABM's that they are the most appropriate indicators of animal welfare, and suggest that a carefully selected combination of animal-based measures can be used to assess the welfare of a target population in a valid and robust way (EFSA, 2012). Furthermore, the use of ABMs for welfare assessment purposes has the advantage of being applicable in different situations, not being restricted to a given structure or facility (Messori et al, *in press*).

5.2.1. Animal 'signals'

Proper day-to-day management of animals starts with recognising the animal based measures or 'signals' in practice. Irrespective of the species, this means being alert when in the barn or house, watching and listening to the animals, and paying attention to their behaviour as a group as well as their individual behaviour. Do they vocalise normally? How is their gait? Are the animals resting together or far apart? Observing this will provide information about the animals' health, nutritional status, social situation, etc. There are many references in scientific literature supporting this approach and relating animal behaviour to production efficiency. To aid livestock keepers on a practical level, a series of books on 'animal signals' should be mentioned. They have been written for caretakers of poultry (Bestman et al., 2009), cows (Hulsen, 2005), pigs (Hulsen et al., 2005) and horses (Steenbergen and Hulsen, 2013), and include points of departure allowing for more animal-oriented procedures. In addition to the books, training sessions are offered to practice picking up signals and to use them to monitor and improve the care, nutrition and housing of farm animals.

5.2.2. The Welfare Quality® assessment protocols

Comprehensive welfare assessment protocols based on animal based measures were pioneered by the EU funded project Welfare Quality®. This project ran between 2004 and 2010, and was executed by 44 universities and research institutes from thirteen European and four Latin American countries. The two main goals of the project were to develop a standardized method to objectively assess animal welfare on farms, and to reliably transform the assessment results into information for consumers on livestock products (not classified, acceptable, enhanced, excellent). Figure 1 presents the relationships between these different aspects of welfare assessment.

At the basis of the assessment are the Five Freedoms, which were operationalized in the following four principles: good feeding, good housing, good health and appropriate behavior. These four principles were further divided into 12 criteria. Behind each criterion are specific and measurable indicators or 'measures'. In total there are about 30 measures for each species. Most are ABMs, but there are also some measures related to environmental parameters. Protocols were developed for poultry, pigs and cattle (Welfare Quality®, 2009a, b and c).

The Welfare Quality® protocols have been used extensively in research projects, and their approach has been applied to other species as well, even including dolphins (Clegg et al., 2015). There are currently a large number of projects, mainly in Europe, which develop the Welfare Quality® protocols into format that are applied by the industry.

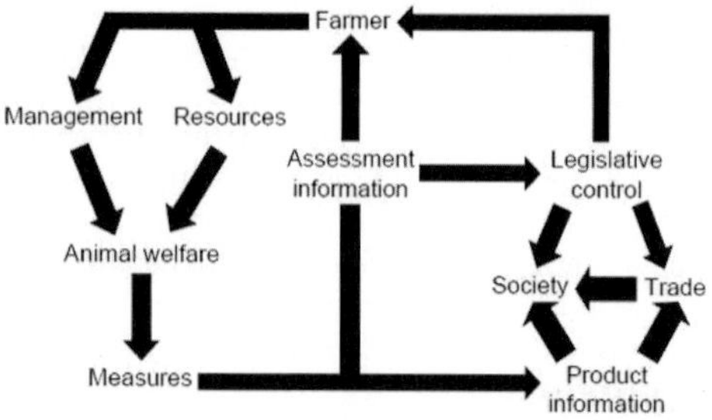

Figure 1. Conceptual model showing the relationship between welfare assessment, the management decisions to be taken by the farmer, and information regarding animal welfare on the product (Blokhuis et al., 2010).

5.3. Targeting attitudes

Knowing what needs doing, and being able to identify a problem are two crucial attributes of a professional stock person. The third critical component is the right attitude towards using this knowledge.

It is a challenge to alter well-established attitudes and beliefs of farmers and stock people with regard to certain farm routines, and targeting denial and offence. In this case it is important to emphasize the credibility of knowledge. Preferably the information only consists of facts, and is not any derived from 'ivory tower' laboratory research, but also from research carried out in commercial farms ('real life' knowledge). Training programs based on cognitive-behavioural intervention techniques have proven to be effective in changing stock people's attitudes and behaviour towards farm animals An example is specific training on improving attitude and behaviour of stock people towards their animals (human-animal interactions), In Australia, and the United States, and also in Europe, specific cognitive-behavioural intervention programmes have produced substantial improvements in the attitude and behaviour of stock people and a marked reduction in the level of fear of humans by farm animals (e.g., Boivin and Ruis, 2011, Coleman et al., 2000, Ruis et al., 2010). These intervention programmes, ProHand® (Australia, US; Coleman et al., 2000) and Quality Handling© (Europe; Boivin and Ruis, 2011) make stock people aware of the importance of their attitudes and behaviour towards farm animals and to change them, if necessary, as well as to maintain these changes. This should help to improve the human-animal relationship and minimise the stress of animals during handling by improving the quality of handling. In the end, this should positively affect animal welfare, the ease of handling, animal productivity, and job satisfaction (Hemsworth and Coleman, 2010).

6. Conclusions

Only animal caretakers with a sound understanding of the behaviour of the animals, the skill to observe and identify undesirable changes and the technical knowledge to implement remedial actions will be able to provide and maintain a high level of animal welfare and productivity.

References

Ajzen, I., 1988. Attitudes, Personality and Behaviour, The Dorsey Press, Chicago.

Aparicio M.A., J.D. Vargas, 2006. Considerations on ethics and animal welfare in extensive pig production: breeding and fattening Iberian pigs. Livestock Science. 103, 237-242.

Bestman, M.W.P., M.A.W. Ruis, K. van Middelkoop, J. Heijmans. 2009. Poultry Signals - A practical guide for poultry-oriented management. Roodbont, Zutphen.

Blokhuis H.J., I. Veissier, M. Miele, R.B. Jones, 2010. The Welfare Quality project and beyond: safeguarding farm animal well-being. Acta Agriculturae Scandinavica A, Animal Science 6, 129-140.

Boivin, X., M.A.W. Ruis, 2011. "Quality Handling" a training program to reduce fear and stress in farm animals. In: 4th Boehringer Ingelheim Expert Forum on farm animal well-being: May 27th 2011, Seville (Spain), 20-27.

Bracke, M.B.M., 2001. Modelling of animal welfare: the development of a decision support system to assess the welfare status of pregnant sows. PhD Thesis, Wageningen University. 150 pp.

Broom, D.M., 1986. Indicators of poor welfare. British Veterinary Journal. 142, 524–526.

Cannon, W.B., 1914. The emergency function of the adrenal medulla in pain and the major emotions. American Journal of Physiology. 3, 356-372.

Cauldfield, M.P., H. Cambridge, 2008. The questionable value of some science-based 'welfare' assessments in intensive animal farming: sow stalls as an illustrative example. Australian Veterinary Journal. 86, 446-448.

Clark, I.J., P.H. Hemsworth, J.L. Barnett, A.J. Tilbrook, 1992. Stress and reproduction in farm animals. In *Stress and Reproduction*. Sereno Symposium Publications, Vol. 86. Raven Press, New York. Eds., K.E. Sheppard, J.H. Boublik and J.W. Funder. 239-251.

Clegg, I.L.K., J.L. Borger-Turner, H.C. Eskelinen, 2015. C-Well: The development of a welfare assessment index for captive bottlenose dolphins (Tursiops truncatus). Animal Welfare. 24, 267-282. doi: 10.7120/09627286.24.3.267

Coleman, G.J., P.H. Hemsworth, M. Hay, 1998. Predicting stockperson behaviour towards pigs from attitudinal and job-related variables and empathy. Applied Animal Behaviour Science. 58, 63-75.

Coleman, G.J., P.H. Hemsworth, M. Hay, M. Cox, 2000. Modifying stockperson attitudes and behaviour towards pigs at a large commercial farm. Applied Animal Behaviour Science. 66, 11-20.

Coleman, G.J., P.H. Hemsworth, M. Hay, M. Cox., 2000. Modifying stockperson attitudes and behavior towards pigs at a large commercial farm. Applied Animal Behaviour Science. 66, 11–20.

Costa, P.T., R.R. McCrae, 1986. Personality stability and its implications for clinical psychology. Clinical Psychology Reviews. 6, 407-423.

De Jong, I.C., H. Gunnink, J. van Harn, 2014. Wet litter not only induces footpad dermatitis but also reduces overall welfare, technical performance, and carcass yield in broiler chickens. Journal of Applied Poultry Research. 23 (1), 51-58.

Duncan, I.J.H., 1996. Animal welfare as defined in terms of feelings. Acta Agriculturae Scandinavica A. 27, 29–35.

Duncan, I. J .H., D. Fraser, 1997. Understanding animal welfare. In *Animal Welfare*. CAB International, Wallingford. Eds., M.C. Appleby and B. O. Hughes.

EFSA Panel on Animal Health and Welfare, 2012. Statement on the use of animal-based measures to assess the welfare of animals. EFSA Journal. 10 (6), 2767. doi:10.2903/j.efsa.2012.2767.

Farm Animal Welfare Council, 1993. Second Report on the Priorities for Research and Development in Farm Animal Welfare. MAFF, Tolworth.

Fraser D., I.J.H. Duncan, 1998. 'Pleasures', 'pains' and animal welfare: toward a natural history of affect. Animal Welfare. 7, 383–396.

Fraser, D., 2014. Could animal production become a profession? Livestock Science. 169, 155–162.

Grandin, T., 2008. *Humane livestock handling*. Adams, MA: Storey Publishing.

Grandin, T., 1987. Animal handling. In: *Farm animal behavior.* Veterinary clinics of north America: food animal practice 3. Eds., E.O. Price. 323-338.

Hemsworth, P.H., G.J. Coleman, 1998. Human-Livestock Interactions: The Stockperson and the Productivity of Intensively Farmed Animals. CAB International, Wallingford,.

Hemsworth, P.H., G.J. Coleman, 2010. Human-Livestock Interactions: The Stockperson and the Productivity and Welfare of Farmed Animals. 2nd Edition CAB International, Oxon UK.

Hemsworth, P.H., J.L. Barnett, C. Hansen, 1986. The influence of handling by humans on the behaviour, reproduction and corticosteroids of male and female pigs. Applied Animal Behaviour Science. 15, 303-314.

Hemsworth, P.H., G.J. Coleman, J.L. Barnett, 1994. Improving the attitude and behaviour of stockpersons towards pigs and the consequences on the behaviour and reproductive performance of commercial pigs. Applied Animal Behaviour Science. 39, 349-362.

Hulsen, J., 2005. Cow Signals : a practical guide for dairy farm management. Roodbont, Zutphen.

Hulsen, J., K. Scheepens, T. van Schie, 2005. *Pig Signals – Practical guide for pig farmers.* Roodbont, Zutphen.

Hutson G.D., 1993. Behavioural principles of sheep handling. In *Livestock handling and transport.* CAB international Wallingford UK. Eds., T. Grandin. 127-146.

Lawrence, A.B., A.W. Stott, 2010. Profiting from animal welfare: An animal based perspective. Journal of the Royal Agricultural Society of England, 170.

Lensink, J., A. Boissy, I. Veissier, 2000. The relationship between farmers' attitude and behaviour towards calves, and productivity of veal units. Annales de Zootechnie. 49, 313-327.

Lorz A., 1973. *Tierschutzgesetz – Kommentar von A. Lorz.* (Animal Welfare Law – Comments from A. Lorz). Verlag Beck, München.

Messori, S., E. Sossidou, M. Buonanno, B. Mounaix, S. Barnard, V. Vousdouka, P. Dalla Villa, K. de Roest, H. Spoolder, 2015. A pilot study to develop an assessment tool for sheep welfare after long journey transport. *Accepted for publication in Animal Welfare.*

OIE, 2010. Terrestrial Animal Health Code. http://web.oie.int/eng/normes/mcode /en_chapitre_1.7.1.pdf

Pompe, V., M. Ruis, 2015. Participative monitoring of the welfare of veal calves. Proceedings of the 66th annual meeting of the European Association for Animal Production (EAAP), Warsaw, Poland. September 2015.

Ruis, M.A.W., G.J. Coleman, S. Waiblinger, I. Windschnurer, X. Boivin, 2010. A multimedia-based cognitive-behavioural intervention program improves attitudes and handling behaviours of stockpeople in livestock farming. Proceedings of the British Society of Animal Science (BSAS), Annual Conference, Belfast, UK, 12-14 April. http://edepot.wur.nl/173182

Schiefele, U., 1990. *Einstellung, Selbstkonsistenz und Verhalten* (Setting, self-consistency and behaviour). Verlag für Psychologie, Göttingen.

Seabrook, M.F., 1972. A study to determine the influence of the herdsman's personality on milk yield. Journal of Agricultural Labour Science. 1, 45-59.

Seabrook, M.F., 1995. Behavioural interaction between pigs and humans. Jounal of the British Pig Veterinary Society. 34, 31-40.

Spoolder, H.A.M, M.J. Geudeke, C.M.C. van der Peet-Schwering, N.M. Soede, 2009. Group housing of sows in early pregnancy: a review of success and risk factors. Livestock Science. 125: 1-14.

Spoolder, H.A.M., S. Waiblinger, 2009. Pigs and Humans. In *The Welfare of pigs.* Springer Science + Business Media B.V., Netherlands. Eds., J.N. Marchant-Forde. 211-236.

Steenbergen, M., J. Hulsen, 2013. *Horse Signals – Look, think, act.* Roodbont, Zutphen.

Van der Peet-Schwering, C.M.C., A.I.J. Hoofs, N.M. Soede, H.A.M. Spoolder, P.H. Vereijken, 2009. *Groepshuisvesting van zeugen tijdens de vroege dracht* (Group housing of sows during early gestation). Lelystad, Wageningen UR Livestock Research, Report 283. P.p. 91

Waiblinger, S., C. Menke, G. Coleman, 2002. The relationship between attitudes, personal characteristics and behaviour of stockpeople and subsequent behaviour and production of dairy cows. Applied Animal Behaviour Science. 79, 195-219.

Webster J, 1994. *Animal Welfare – a Cool Eye towards Eden*. Blackwell Science, Oxford.

Welfare Quality®, 2009a. *Welfare Quality® assessment protocol for poultry (broilers, laying hens)*. Welfare Quality® consortium, Lelystad, Netherlands.

Welfare Quality®, 2009b. Welfare Quality® assessment protocol for pigs (sows and piglets, growing and finishing pigs). Welfare Quality® consortium, Lelystad, Netherlands.

Welfare Quality®, 2009c. *Welfare Quality® assessment protocol for cattle*. Welfare Quality® consortium, Lelystad, Netherlands.

Whay, H.R. 2007. The journey to animal welfare improvement. Animal Welfare. 16, 117-122.

Alternative Gestation Housing for Sows: Reproductive Performance, Welfare and Economic Efficiency

Laurie Connor

Department of Animal Science, Faculty of Agricultural and Food Sciences,

University of Manitoba, Winnipeg, Manitoba, Canada

Email: Laurie.Connor@umanitoba.ca

Abstract

Increasingly, the global expectation is that gestating sows will be kept in group-housing alternatives to gestation stalls. Unlike stall housing, there are many options available for group-housing, each with strengths and limitations. It has become apparent from research and experience that success with group housing is dependent on key features and management of any given system that can provide adequate individualized feeding and minimal distress. Therefore key concepts for group housing focus on feed delivery systems, quantity and quality of space, strategies for mixing sows into groups, the ability to attend to individual sows and stockmanship. In concert, these factors will impact reproductive performance, sow welfare and economic efficiency. Although conflicting results can be found in the literature, overall it can be concluded that compared to stall housing, reproductive performance is not reduced in well-managed group-housing systems. The assessment of animal welfare is confounded by different, but overlapping concepts of what constitutes good welfare. However, it is clear that sow well-being is dependent on pen-environments and management strategies that minimize aggression, pain and long-term distress. Particular attention is directed towards flooring quality and lameness as impacting welfare as well as economic efficiency. Well-trained, compassionate stockpeople are a fundamental component of ensuring animal well-being and system success. The operational factors involved in economic efficiency of sow group-housing are fundamentally the same as with stall housing, though initial capital investment will differ from system to system.

Keywords: Sow group-housing, loose housing, welfare in group-housing, sow productivity

1. Introduction

The global direction to housing sows in groups as an alternative to stall confinement has largely been driven by legislative and consumer assumption that group housing is better for the animals. While there is scientific and experiential evidence to both support and dispute this assumption, the onus is on the pork industry to provide assurance of good animal welfare. Conceivably, that assurance will come with being able to provide an environment for sows free of chronic or frequent distress, where their full biological potential can be realized; and from producers' standpoint, essential production targets for economic sustainability are met. However, there is no 'ideal' or 'standard' group housing system to recommend that will make this assurance easier. Rather, the diversity of group housing design options and associated management systems available has made it difficult to conduct system-based comparisons of value to producers. The emphasis, therefore, needs to be on the components of the system as they singly or in combination influence functionality, animal welfare and economic efficiency.

This paper will provide an overview of our current understanding of key group-housing concepts and components that impact sow productivity and well-being as well as economics

2. General Concepts

A major challenge in determining the most suitable group-housing system is that, unlike standard gestation stall systems, there are numerous options. Considering only the general categories of feeder-type, flooring type, time of mixing sows into groups and whether the groups are managed as static or dynamic, results in a large number of potential combinations or options to choose from. Each of these categories would be further broken down when making final design decisions and the process can seem daunting. But, the final choices will impact functionality of the system, sow welfare and productivity.

Within recent years several publications and tools have become available focused on assisting producers with their decision-making (e.g., National Pork Board, 2013; Connor et al, 2014; Manitoba Pork: Tools for Group Housing). Key concepts and system features fundamental to the success of any group housing system are sow-centric. Including the critical points identified by Johntson and Li (2013) these include: quantity and quality of space; feeding strategies to control sow body condition; ability to attend to or segregate individual sows; timing and method of mixing sows into groups - whether dynamic or static groups; and stockperson skills. This latter point, stockperson skills, cannot be overemphasised in the success of any animal housing system, but particularly for group-housing. General concepts of feed delivery systems will be discussed here and the remaining critical features and concepts will be addressed in subsequent sections.

2.1. Feeding strategies and system

The method of feeding is usually central to the design of a sow group-housing system. Within each of the two general categories of competitive and non-competitive feeding systems there are several options, each with strengths and limitations (Levis and Connor, 2013; Connor et al, 2014). Since pregnant sow feed intake is usually restricted in order to maintain appropriate body condition, feed availability can be a source of high competition and the choice of feeding system affects feeding-related aggression (Spoolder et al., 2009)

Non-competitive feeding systems are designed to allow individual sows to consume their daily allotment, protected from other sows. The most common designs incorporate a gated stall which encloses the sow during feeding. These include various designs of electronic sow feeder (ESF) systems, closable feeding stalls, and walk-in-lock-in stalls where there is sufficient room for the sow to lie down and rest, as well. In the context of minimizing agonistic behaviour and assuring welfare associated with adequate feed intake, an appropriately designed non-competitive feeding system is recommended. There are various ESFs and feed-stall systems available, but it is beyond the scope of this paper to discuss pros and cons of each system and controlled studies comparing systems are lacking.

Competitive feeding systems include common floor feeding as well as partial or full stalls without closed back gates. Such systems are pre-disposed to more dominant sows displacing subordinate animals to access feed resulting in inter-sow aggression and variable body condition (Edwards, 1992). While competitive feeding systems can be managed successfully (Gonyou, 2005), they tend to require much higher levels of management associated with feeding than do the non-competitive systems and may be more appropriate for small-group operations.

Therefore, the choice of feeding system, accessibility for each sow and management of feed delivery can influence aggressive behaviour and welfare, body condition variation within the group, productivity and economic outcomes.

3. Reproductive Performance

The transition to group housing from stalls is often fraught with concerns that reproductive performance will be compromised when sows are housed in groups throughout gestation. However, meta-analysis of scientific data in a literature review published over 10 years ago (McGlone et al., 2004a) concluded that reproductive performance in well-managed group housing systems was not compromised compared to stall systems. More recently, McGlone (2013) reviewed research published in the intervening years and came to essentially the same conclusion. In several countries of the EU, where modern sow-group housing has long been practiced, average reproductive performance of sows exceeds 28 piglets per sow per year (Boyle et al., 2012).

Comparisons of the different group-housing systems, relative to reproductive performance, are few and not very conclusive. Variations in reproductive performance across commercial farms in the Netherlands lead Spoolder et al. (2009) to conduct a detailed literature review assessing success and risk factors for sow-group housing. They concluded that factors within each system, such as feed intake, floor quality, space allowed per animal and management factors resulting in chronic stress, have bigger impacts on sows' performance than the housing system per se. Time of mixing sows and lameness issues relative to reproductive performance will be discussed here; stockperson skills and attitude can also influence reproduction (Hemsworth and Coleman, 2011), but will be discussed under the welfare section.

3.1. Grouping sows

The effect of stress on reproduction will depend on the timing, duration and severity of the stress response (Munsterhjelm, et al, 2008), at times making causal determination difficult. Management post-mating, and feeding system seem to be more critical to conception failure and rebreeding than housing system per se (Kemp and Soede, 2012; Spoolder et al., 2009). Both post-mating management and feed delivery represent opportunities for high levels of aggression and stress. Fighting at introduction of unfamiliar sows into groups can be very intense while they form their social hierarchy (Arey and Edwards, 1998; Velarde, 2007)). However, not all animals in the group engage in active fighting. Avoidance and escape behaviours tend to be the more common activity of choice when the pen environment is set-up to accommodate retreat (Gonyou and Lang, 2013). Therefore, all animals in the group may not experience the same level of distress at the time of mixing, hence variations in biological response. Results comparing early gestation mixing of sows to grouping after the first month of pregnancy are inconclusive (Verdon et al., 2015). In practice, commercial operations mixing sows within four days of insemination (e.g., Boyle et al., 2012; Connor, 2012) or also later in gestation (Buhr, 2010) can show very good reproductive performance; suggesting the move to group housing when managed properly is unlikely to cause any long term reduction in performance as many producers feared. This was corroborated by a survey of producers conducted by Buhr (2010) wherein time of mixing sows into groups varied, but reproductive outcomes were similar. Even when pen-housed sows demonstrated relatively higher lesion scores than sows in stalls, there was little impact on sow productivity (Salak-Johnson et al, 2007). Similarly, studies of factors associated with static and dynamic sow-group management found no impact on farrowing rate and litter characteristics (Anil et al., 2006; Strawford et al., 2008; Li and Gonyou, 2013) even when more skin injuries were noted (Anil et al., 2006; Li and Gonyou, 2013).

Based on the scientific literature, Spoolder et al. (2009) concluded that it is not easy to form conclusions on the best time to mix sows into groups for optimal reproductive performance. However, it is generally accepted that embryo survival is particularly vulnerable to maternal stress (Varley and Stedman, 1994) and that stress should be minimal during the 3 week period associated with embryo expansion and implantation (about days 8–30 after insemination). This has led to management practices of mixing sows either within a few days of insemination or around days 28–35 of gestation when confirmed pregnant. Although mixing at day 35 of pregnancy may result in lower frequency of aggressive behaviour (Stevens et al., 2015) the public pressure to eliminate stall housing altogether means that more investigations are needed to determine best strategies for minimizing distress for sows grouped immediately at weaning or within a week of insemination.

3.2. Lameness and reproduction

Feet and leg soundness are crucial, particularly in group housing, since they dictate the ease with which an animal can access resources and avoid aggressive interactions. In recent years, sow lameness has been of increasing concern in terms of reproductive performance, welfare and economics. One survey of commercial pig farms found that 48% of pregnant sows and gilts and 39% of replacement gilts were affected by lameness (Quinn et al., 2013). But, are lameness and reproductive performance linked? Next to reproductive performance leg and feet soundness have been identified as the second major reason for culling sows (Stalder, 2012). Lameness of sufficient severity to result in culling is likely accompanied by pain, difficulty accessing feed and water in a pen situation and compromised mobility, in general. It is conceivable that other reasons for culling such as low body condition score and reproductive failure are actually associated with lameness and/or chronic pain. A survey of slaughter sows in the USA. (Knauer et al, 2007) found that 84% had one or more hoof lesions. While not all hoof lesions cause lameness, Pluym et al. (2013) noted that lesions of the white line, heel and sole, in particular, were directly associated with lowered reproductive performance including increased number of stillborns and crushed piglets.

Lameness is often associated with pain to varying degrees (reviewed by Heinonen et al., 2013). Apart from affecting sows' locomotor abilities, the physiological consequences to pain include inflammation of the affected area and elevated immune responses which can also contribute to decreased reproductive performance and longevity.

Therefore, lameness either directly or indirectly decreases sow herd productivity and economic efficiency. It has a negative impact on sow longevity, directly increases the number of sows culled for non-productive reasons, decreases the number of pigs produced per sow per year and increases cost of production. Lame sows can be expensive sows. Even before culling, costs may be associated with productivity decline, stockperson attention, treatment and possibly on-farm euthanasia. Culled animals must be replaced with gilts which will likely produce smaller litters, initially.

Therefore, lameness is not only a welfare issue, but also a productivity and economic issue. Causes of lameness are multifactorial, but in terms of loose-housed sows predisposing factors of inappropriate flooring, inter-sow aggression and quality of space strongly influence the incidence and severity (Spoolder et al., 2009; Bench et al., 2013b). Objective early detection methods (Conte et al, 2014) and prevention through attention to appropriate housing environment and management, as well as animal selection will reap benefits for welfare assurance and productivity of the whole system.

4. Welfare

A major driver for the global trend to housing sows in group pens as opposed to stalls is the perception that group-housing will improve welfare. The World Organization for Animal Health (OIE, 2015) defines animal welfare as "how an animal is coping with the conditions in which it lives. An animal is in a good state of welfare if … it is healthy, comfortable, well nourished, safe, able to express innate behaviour, and if it is not suffering from unpleasant states such as pain, fear and distress". Yet, for group-housing of sows, the goal of achieving more natural behaviours of freedom to move around and to socialize with conspecifics is countered by the necessary farming practice of mixing together unfamiliar animals in a confined space. The high levels of aggression that can occur at sow-group formation (Velarde, 2007) can compromise welfare when accompanied by fear, injury, pain and distress, even if of short duration. Further, the diversity of designs and management of group housing systems makes generalized conclusions about welfare in alternative housing systems near impossible.

Well-being or good welfare is the outcome of a complex of interactions involving not only physical and social environments but also how the animal perceives these environmental variables; the animal's perception will be influenced by its genetics, prior experience, stage of reproduction and interactions with stockpeople. There are clear metrics to determine if a barn design meets expected criteria for dimensions/space and organization of materials. However, when dealing with assessments of animal comfort and welfare within the barn, scientifically and in practice, there are different but overlapping concepts of what constitutes animal welfare that are used. These include: the more traditional biological functioning measured by health and productivity; affective state of the animal (pain, distress, discomfort) which present more subjective assessment; and the natural living concept considering ability to express species specific behaviour (Fraser, 2008). Depending on the approach or assessment metric different conclusions regarding welfare can result. For example, the aggression between unfamiliar sows at group formation (Velarde, 2007) will likely never be eliminated completely. Because this fighting causes injuries, pain, stress and fear, at least for some animals in the group, it can be seen to adversely affect sow welfare (Verdon et al., 2015). However, this short-term/acute distress for the most part seems to be "acceptable" with considerable effort directed toward strategies to minimize the negative impact (see reviews by Spoolder et al., 2009; Bench et al., 2013a,b; Verdon et al., 2015) in order to establish stable hierarchies and gain the longer term benefits of the loose-housing environment.

Recent literature reviews have focused on factors associated with animal environments that promote success in terms of health and reproductive performance, as well as physical and physiological indicators of pain, injury and distress (Spoolder et al., 2009; Bench et al., 2013a,b; Verdon et al., 2015). Lameness in sows, a major welfare issue (Heinonen et al., 2013) was discussed previously in this paper related to reproduction. This section will consider space characteristics, mixing strategies and stockpeople.

4.1. Quantity and Quality Space

While economic considerations may dictate providing the minimum space needed, from an animal welfare and productivity standpoint minimum may not be the best investment. An "ideal" space allowance per sow has not been clearly identified since the quantity of space required is affected by many factors such as configuration of the space, type of flooring, feeding system, sow-group parity, and group size and dynamics. However, each animal must have ready access to key resources - feed and water, dunging area, appropriate lying space as well as sufficient space to avoid or engage in social interactions, i.e., to form a stable hierarchy. Crowding makes it difficult for sows to access key resources, decreases their ability for effective avoidance behaviour, promotes aggressive interactions, is associated with more injuries (Bench et al., 2013a) and elevated levels of cortisol (Hemsworth et al., 2013) - all indicators of compromised welfare. Insufficient space for grouped sows compromises lower ranking individuals in particular, forcing them into less desirable areas of the pen, such as alleyways and dunging areas (Weng et al., 1998). With increasing space allowance, sows also engage in more exploratory behaviour (Gonyou and Rioja-Lang, 2013; Bench et al., 2013a) which is a positive welfare indicator.

Currently, the European Union (EU) directive (Council directive 2008/120/EC) requires a minimum floor space allowance of 1.67 and 2.25 m^2 per animal for gilts and sows, respectively. In North America current recommendations range from 1.4 to 1.7 m^2 for gilts and 1.8 to 2.2 m^2 for sows (Gonyou et al., 2013; National Farm Animal Care Council, 2014) depending on floor type. A recent review by Verdon et al., (2015) noted that while research demonstrates 1.4 m^2 per sow is inadequate, there is a paucity of information defining space effects in the range of 1.8 to 2.2 m^2. Since effects are also a function of quality as well as quantity of space, recommendations on space allowance should be seen as guidelines, only. Determination of actual space needed per animal is best determined in practice by observing animal behaviour and outcomes in a particular group system.

Pen space quality is a function of the pen layout or configuration, including components of the pen environment and flooring. High quality pen space allows for avoidance and escape behaviours, ease of movement and access to defined feeding, drinking, dunging and resting areas, as well as ease of lying down and getting up from resting. Strategically placed visual barriers such as low partition space dividers in large-group pens can create refuge from aggressors and facilitate sub-group resting areas (Gonyou and Rioja-Lang, 2013; Verdon et al., 2015). Pen layouts should allow natural activities while avoiding bottlenecks or cul-de-sacs (Thibault, 2004). Areas where sows may need to pass each other to reach resources, such as walkways, should be wide enough that access for subordinate sows cannot be blocked by dominant animals; this is defined as 3 m by EU standards (Council directive 2008/120/EC) but a minimum of 2.1 m has been recommended in North America (Gonyou et al., 2013), even though the average sow is about 1.8 m long (McGlone et al., 2004b).

Floor type and characteristics, as well as the availability of material to manipulate, impact space quality and animal well-being. Lying and walking comfort as well as the incidence of lameness and hoof injuries are associated with flooring (Barnett et al., 2001). The hard surface of concrete, its abrasiveness and slipperiness, along with slat and gap widths in slatted floor areas affect leg soundness (Kilbride et al., 2009) and sow welfare. Slatted floors are associated with increased incidence of lameness in sows, particularly after regrouping-induced fighting (Anil et al, 2005). Sows kept on partially slatted flooring reportedly have lower lameness scores than those on total slats (Vermeij et al, 2009).

Well-defined slat and gap widths most appropriate for sows are not readily available in the scientific literature. Although the EU directs minimum slat widths of 80 mm with gaps not exceeding 20 mm (Council directive 2008/120/EC) there seems a paucity of scientific data to support or refute these numbers. Most experiential information in Canada favours a slat width of 127 mm and gap width of 19 mm (Peet, 2011) for sow flooring. The functional challenge is to have slat widths of sufficient width and surface characteristics for sows to walk comfortably and gap widths that will not catch toes or claws but will allow manure passage.

Pen amendments such as bedding or rubber mats can improve sow comfort (Rioja-Lang et al., 2013) decrease the incidence of lameness (Seddon et al., 2013; Connor, 2013) and increase longevity (Fynn, 2010).

4.2. Mixing sows into groups

The most intense injury-inducing aggression at the time of mixing unfamiliar sows into groups is usually greatest during the first day, although it can take several days or more for the hierarchy to become stable. All during this time, some sows at least, can be experiencing varying degrees of distress apart from obvious physical injury. One strategy to lower the negative impacts is initially grouping in a dedicated mixing pen which has suitable solid non-slippery flooring, large space allowance per animal and space dividers to facilitate avoidance and escape behaviours (Gonyou and Rioja-Lang, 2013). After a few days when social ranking is established, the group is relocated to a home pen, with less but adequate space, for the remainder of gestation.

Another area of concern is whether to keep sows in static or dynamic groups. There are many factors involved in making this decision and some evidence that aggression is more of a problem in dynamic groups (Arey and Edwards, 1998) wherein breeding cohorts are introduced into the larger group at intervals and farrowing groups removed, thus disrupting the social ranking each time. However, from the literature there seems to be no consensus on whether static or dynamic groups per se are best in terms of reproductive performance or welfare (Verdon et al., 2015). Mixing strategies and sow group dynamics/organization are key component of successful sow group management (Gonyou and Lang, 2013; Johnston and Li, 2013; Verdon et al., 2015).

4.3. The human factor

Regardless of the housing system, the impact of stockpeople on sow reproductive performance and well-being has long been recognized (Hemsworth et al., 1986). Negative interactions leading to increased levels of fear can compromise welfare and productivity (Hemsworth, 2003; Hemsworth and Coleman 2011). Training stockpeople to interact with pigs in a positive manner during handling will reduce fear stress responses and help ensure animal well-being (as reviewed by Spoolder et al., 2009 and Verdon et al., 2015). Hemsworth and Coleman (2011) concluded that the attitude of stockpeople towards surveillance and attention to welfare and production issues may be of greater consequence in group-housing than in stall housing systems. Recognition of the importance of the human element by appropriate selection and training of stockpeople will greatly enhance the ability to assure animal welfare. Interestingly, anecdotal accounts demonstrate that stockpeople involved with conversions from stall to group-housing of sows often find rewards such as improved ease of handling sows and a better appreciation of the animals and their individual differences.

5. Economics of Group-Housing

While comparisons on fixed costs associated with new construction of group-housing systems or conversion from stalls are available (Meyer, 2012), these will vary regionally. In terms of economic efficiency, factors influencing the operating costs of group housing will be similar to those in stall housing and be largely related to the biological efficiency of the sows, feed utilization, medical treatment and labour. Values associated with these factors will vary depending on the feeding system used, the quantity and quality of space provided, herd health status, the stockpeople and the general well-being of the animals. For example, as noted previously, issues resulting in lameness or compromised welfare result in lowered performance and decreased longevity in the herd which lowers economic returns. Feeding system choice can impact operating costs if it does not allow for strict regulation of individual sow feed intake, While competitive feeding systems (e.g., floor feeding, open stalls) may require lower initial investment, the competitive nature of these systems can result in variation in sow body weight, which will compromise overall performance and/or utilization of more feed either from the producer providing more to compensate for the body condition disparities or due to wastage.

In considering the economic effect of transitioning from stalls to group housing in the U.S.A. Buhr (2010) and Meyer (2012) determined a significant cost for the actual conversions, but Buhr (2010) concluded from a producer-survey that the adoption of pen gestation housing had not changed the labour or productivity. Increased labour costs and lowered sow productivity have been major industry concerns with transitioning to sow group housing.

6. Conclusions

High levels of sow reproductive performance, welfare and economic efficiency can be achieved using group-housing as an alternative to the gestation stall. The diversity of feeding system and pen design options make choosing the "best' system impossible. Rather, it is clear that producer selection of a system must be based on attention to key concepts that will allow each animal ready access to adequate feed, a pen environment and management strategies that minimize aggression and chronic distress, encourages the rapid establishment of a social hierarchy soon after sows are mixed, and provides skilled stockpeople attentive to signs of animal health and welfare. Particular attention must be given to the feeding system, sow mixing strategies and timing, space quantity and quality and the selection and training of stockpeople. However, relative to the focus of this paper, there remains an urgent need for scientifically supported recommendations on flooring characteristics as they impact feet and leg soundness, lameness mitigation strategies, optimum space allowance for different activities, and strategies for early mixing of sows that minimize distress.

References

Anil, L. S.S. Anil, J. Deen, S.K. Baidoo, R.D. Walker. 2006. Effect of group size and structure on the welfare and performance of pregnant sows in pens with electronic sow feeders. Can. J. Vet. Res. 70: 128-136.

Anil, L., S.S. Anil, J. Deen, S.K. Baidoo, J.E. Wheaton. 2005. Evaluation of well-being, productivity and longevity of pregnant sows housed in groups in pens with an electronic sow feeder or separately in gestation stalls. Am. J. Vet. Res. 66(9):1630-1638.

Arey, D.S. and S.A. Edwards. 1998. Factors affecting aggression in sows after mixing and the consequences for welfare and production. Livest. Prod. Sci. 56:61-78.

Barnett, J.L., P.H. Hemsworth, G.M. Cronin, E.C. Jongman, G.D. Hutson. 2001. A review of the welfare issues for sows and piglets in relation to housing. Aust. J. of Agric. Res. 52: 1-28. http://doi:10.1071/AR00057.

Bench, C.J., F.C. Rioja-Lang, S.M. Hayne, H.W. Gonyou. 2013a. Group housing with individual feeding –I: How feeding regime, resource allocation, and genetic factors affect sow welfare. Livest. Sci. 152:208-217

Bench, C.J., F.C. Rioja-Lang, S.M. Hayne, H.W. Gonyou. 2013b. Group housing with individual feeding –II: How space allowance, group size and composition, and flooring affect sow welfare. Livest. Sci. 152:218-227

Boyle, L. C. Carroll, G. McCutcheon, S. Clarke, M. McKeon, P. Lawlor, T. Ryan, R. Ryan, T. Fitzgerald, A. Quinn, J. Calderon Diaz, D. Lemos Teixeira. 2012. *Towards 2013: Updates, implication and options for group housing pregnant sows*. Pig Development Department, Teagasc. Agriculture and Food Development Authority, Fermoy, Co. Cork. 77p. http://www.teagasc.ie/publications/2012/1291/Towards_January2013_Updates_implicationsand_optionsfor_grouphousing_pregnantsows.pdf

Buhr, B.L. 2010. Economic impact of transitioning from gestation stalls to group housing in the U.S. pork industry. Staff Seminar Series. Department of Applied Economics, University of Minnesota. Staff Paper P10-4, 79p. http://ageconsearch.umn.edu/bitstream/61604/2/P10-04.pdf

Connor, M.L. 2012. Evaluation of a Commercial Group Sow Housing Alternative.

Final Report to Manitoba Rural Adaptation Council Inc. http://umanitoba.ca/faculties/afs/ncle/pdf/FR_Evaluation_commercial_group_sow_housing_alternative.pdf

Connor, L. 2013. Sow Housing: risk factors and assessment techniques for lameness, productivity and longevity in group and individually housed gestating sows. Final Report to Swine Innovation Pork on Project CSRDC 1004.

Connor, L., J. Goodridge and M. Fynn. 2014. Options for successful group-housing of sows. Manitoba Pork. http://manitobapork.com/wp-content/uploads/2014/03/Edge_MP_Options_Group_Housing_Booklet_cp4.pdf

Council Directive 2008/120/EC of 18 December 2008 laying down minimum standards for the protection of pigs. http://eur-lex.europa.eu/legal-content/EN/TXT/?qid=1430308928912&uri=CELEX:32008L0120. Accessed Sept. 3, 2015

Conte, S., R. Bergeron, H. Gonyou, J. Brown, F.C. Rioja-Lang, M.L. Connor, N. Devillers. 2014. Measure and characterization of lameness in gestating sows using force plate, kinematic, and accelerometer methods. J. Anim. Sci. 92(12):5693-5703. http://doi: 10.2527/jas.2014-7865.

Edwards, S.A. 1992. Scientific perspectives on loose housing systems for dry sows. Pig Vet. J. 28:40-51.

Einarsson, S., Y. Brandt, N. Lundeheim, A. Madej. 2008. Stress and its influence on reproduction in pigs: a review. Acta Vet. Scand. 50:48. http://doi:10.1186/1751-0147-50-48

Elmore Pittman, M.R., J.P. Garner, A.K. Johnson, R.D. Kirkden, B.T. Richert, E.A. Pajor. 2011. Getting around social status: Motivation and enrichment use of dominant and subordinate sows in a group setting. Appl. Anim. Behav. Sci. 133:154-163.

Fraser, D., 2008. Understanding animal welfare. Acta Veterinaria Scandinav ica, 50 (Suppl. 1), S1. http://doi.org/10.1186/1751-0147-50-S1-S1

Fynn, M. 2010. Comparing two systems of sow group-housing: animal welfare and economics. MSc Thesis. University of Manitoba. http://mspace.lib.umanitoba.ca/handle/1993/4219?mode=full

Gonyou, H.W. 2005. Experiences with alternative methods of sow housing. J. Am. Vet. Med. Assoc. 226:1336-1339. http://doi:10.2460/javma.2005.226.1336.

Gonyou, H. and F. Rioja-Lang. 2013. Management to control aggression in group housing. National Hog Farmer, Oct 15, 2013. http://nationalhogfarmer.com/animal-well-being/management-control-aggression-group-housing.

Dealing with Airborne Transmission of Animal Diseases –A Review

Qiang Zhang

Department of Biosystems Engineering, University of Manitoba, Winnipeg, Manitoba R3T 5V6, Canada
Email: Qiang.Zhang@UManitoba.ca

Abstract

This paper presents a review of the research advances in airborne transmission of animal diseases. Airborne transmission is an important route of animal disease transmission. In contrast to transmission by direct and indirect contact, airborne transmission is less understood and much more difficult to deal with. Most biosecurity measures for preventing animal disease spread have been focused on direct and indirect contact, with little consideration of airborne transmission. Airborne transmission of animal diseases involves several complicated biological and physical processes, including pathogen shedding and aerosolization, atmospheric dispersion of pathogen-laden aerosols, decay of aerosolized pathogens in the atmosphere, and initiation of infection. Our understanding of airborne transmission of diseases is still limited. Some technologies, such as filtration and air ionization, for reducing and preventing airborne transmission of animal diseases are promising.

Keywords: Infectious animal disease, aerosol, dispersion, filtration, air ionization

1. Introduction

Animal diseases have numerous economic and social impacts and are becoming an emerging global threat. An example is avian influenza which has caused huge economic losses to the poultry industry globally and a significant threat to public health. There are three main routes of animal disease transmission: direct contact between animals; indirect contact through contaminated objects; and airborne transmission. Most biosecurity measures for preventing animal disease spread have been focused on direct and indirect contact, with little consideration of airborne transmission (Desrosiers, 2011; Seo and Lee, 2014). Many epidemiological studies and experimental data have shown that airborne transmission of disease pathogens plays an important role in their epidemiology (Desrosiers, 2011). Airborne transmission of animal diseases is closely related to emission and dispersion of bioaerosols generated in animal operations. In animal operations, bioaerosols are produced from animals, feed, bedding, and feces and they consist of a complex mixture of biological, microbial, and inorganic particles. When emitted to the atmosphere from an infected animal facility, these bioaerosols act as a carrier for pathogens that travel in the air and may infect the nearby naïve facilities. In contrast to transmission by contact, airborne disease transmission is less understood and much more difficult to control. A simple but critical research question is: how far can the pathogens associated with bioaerosols travel in the atmosphere and still be viable to cause infection? The objective of this paper is to review the research advances in airborne transmission of animal diseases, including the mechanisms by which pathogens are aerosolized and dispersed, initiation of infection by aerosol inhalation, airborne transmission models for forecasting disease spread, and technologies for reducing and preventing airborne transmission of animal disease. PRRS (porcine reproductive and respiratory syndrome) is used as an example in discussion.

2. Mechanisms of Airborne Disease Transmission

2.1. Processes of airborne transmission

Airborne disease transmission involves several critical physical and biological processes, including pathogen shedding, aerosolization and emission; aerosol dispersion and pathogen decay in the atmosphere; and infection of naïve animals (Figure 1). Diseased animals shed pathogens which are carried by exhaled aerosols and/or dust particles, and these aerosolized pathogens are emitted to the atmosphere. Once in the atmosphere, the pathogen-laden aerosols may be transported by air currents (winds) to naïve facilities, causing disease outbreaks. The distance of disease spread by airborne transmission is closely related to both physical movement (dispersion) of aerosols and biological decay (survival) of pathogens. Individual aerosols containing pathogens will face three possible outcomes during dispersion (Figure 1): (1) the aerosol settles out of the air before it reaches a naïve facility; (2) pathogens die during dispersion; and (3) the aerosol reaches a naïve facility with viable pathogens in it. Only Outcome 3 can potentially result in disease transmission. For Outcome 3 to occur, the following conditions must be satisfied: $\mathbf{t_A} \geq \mathbf{t_T}$ and $\mathbf{t_V} \geq \mathbf{t_T}$, where $\mathbf{t_A}$ is the aerosol suspension time in the atmosphere; $\mathbf{t_T}$ the aerosol travel time from the infected to the naïve facility; and $\mathbf{t_V}$ the pathogen survival time. The aerosol suspension time $\mathbf{t_A}$ is dictated by the aerodynamic size of aerosol and air movement. The aerosol travel time $\mathbf{t_T}$ depends on aerosol dispersion in the atmosphere. The pathogen survival time $\mathbf{t_V}$ may be affected by many factors, including the type of pathogen, aerosol properties, and environmental conditions (humidity, temperature, solar radiation, etc.). Once aerosols carrying viable pathogens reach the location of a naïve facility (Outcome 3), the possibility for them to induce infection depends on: (1) whether aerosols can enter the naïve facility (building); and (2) the critical amount of aerosols inhaled by animals to cause infection.

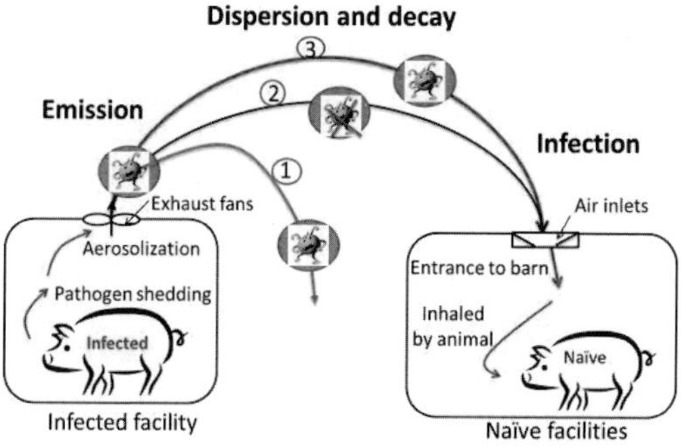

Figure 1. An airborne animal disease transmission model.

2.2. Pathogen aerosolization and emission

The pathogens shed by diseased animals may stay in the air as a single particle, but in most cases, attach to aerosols (aerosolized) (WUR, 2015). In other words, diseased animals shed pathogens which are aerosolized and emitted to the atmosphere through air movement, such as ventilation. Pathogen shedding and aerosolization are extremely complicated, involving the pathogen growth kinetics within animals and the physical processes of aerosolization.

2.2.1. Pathogen growth kinetics

Pathogen growth kinetics within animals is complicated and variable. It may never be possible to describe the growth of a microorganism with a single set of constants (Jannasch and Egli, 1993). Even for the common human disease pathogens, such as influenza viruses, information concerning the growth kinetics during an infection is limited (Baccam et al., 2006). For viral pathogens, virus concentrations in an infected animal typically increase exponentially during the growth period and then decay exponentially as the immune system destroys the virus (Baccam et al., 2006). Taking PRRSV (PRRS virus) as an example, the genome load in the blood is detectable about three days post-inoculation and then increases rapidly until day 14, after which it decreases gradually and persists for several weeks (Charpin et al., 2012). Besides pathogen species and animal hosts, the environmental conditions may also affect the viral growth kinetics, which is not well understood. Lowen et al. (2006) observed a pronounced difference in the viral growth kinetics between animals (guinea pigs) kept at 5°C and 30°C. Specifically, the animals kept at 5°C exhibited a substantial lag time in the exponential decay.

Lowen et al. (2007) hypothesized that inhalation of colder air somehow favored growth of virus in the mucosa of animals.

2.2.2. Aerosolization

Diseased animals shed pathogens in different routes, including expiratory droplets/aerosols, saliva, nasal secretions, urine, semen, and feces. Depending on the shedding route, pathogens may be aerosolized in liquid or solid (dust) aerosols. The main source of liquid aerosols for airborne disease transmission is expiratory aerosols generated by sneezing and coughing, which has been a research focus of human airborne diseases, such as influenza and tuberculosis. Pathogens present in or on the mucosae of the upper respiratory tract are expelled through moving air caused by coughing or sneezing, or even through normal exhalation, producing small droplets that contain pathogens (Blachere et al., 2009). The expelled aerosols laden with pathogens will either quickly settle due to gravity or suspend in the air (airborne), depending on their sizes (aerodynamic diameters or terminal velocities). Suspended aerosols are the cause of airborne transmission, while aerosols settled on surfaces may be the cause of indirect contact transmission. It is generally agreed that aerosols with an aerodynamic diameter less than 5 μm are airborne (Xie et al., 2007). Although many researchers have studied the droplet concentrations and sizes by cough, sneeze, speech and breath, the results were inconsistent (Han et al., 2014). For example, Knight (1973) reported that a sneeze in humans produced approximately 2×10^6 particles, of which more than 75% being smaller than 2 μm. But Han et al. (2014) measured much large aerosols in sneezes by using a laser particle size analyser - they reported that the geometric mean of the droplets was 360.1 mm for unimodal distribution and 74.4 mm for bimodal distribution with geometric standard deviations of 1.5 and 1.7, respectively. The inconsistences in aerosol sizes reported in the literature were probably attributed to the influence of the measurement methods, the experimental devices and the evaporation effects. In particular, evaporation plays a significant role in formation of airborne aerosols. Wells (1934) argued that droplets derived from respiratory fluid with diameters less than 100 μm would rapidly evaporate to become "droplet nuclei" before they settle due to gravity. These droplet nuclei can remain suspended in air for an indefinite time. The droplet nuclei size depends on the initial droplet size, the amount of nonvolatile content, and the humidity of air, as well as the pathogen size in case the pathogen is relatively large (such as bacteria) (Yang et al., 2012).

While liquid (wet) expiratory aerosols have been studied extensively for transmission of human diseases, solid (dry) particles (dust) are considered to be critical in animal disease transmission. Solid particles are produced by animals (skin and feathers), feed, bedding, and feces. Pathogens in solid particles may come directly from feces, or indirectly from other shedding routes, such as saliva, nasal secretions, urine, and large expiratory droplets that settle on dusty surfaces. It is not clear which form of aerosols, liquid or solid, plays more important role in airborne transmission of animal diseases. Some questions are: what pathogens are more likely be aerosolized in liquid or solid aerosols? How long can pathogens survive in different forms of aerosols? Taking PRRS virus as an example, is it aerosolized in expiratory aerosols or in solid particles? Cho et al. (2006) used a mask system to collect exhaled air from PRRS infected pigs and the virus was recovered in the exhaled air samples. This meant that PRRSV was aerosolized in the wet expiratory aerosols. However, in a similar study by Hermann et al. (2008), pigs were inoculated by seven pathogens (or strains) and all pathogens were detected in swabs of the upper respiratory tract, but only two were detected in expired air individually sampled from acutely infected pigs. They concluded that either that the acutely infected pigs did not shed viruses in the expiratory aerosols or that the quantity of virus exhaled was below the detection threshold of sampling or measurement systems, or both.

Zhao et al. (2014) pointed out that although wet aerosolization mimics the fate of microorganisms expelled from animal respiratory tracts, the large droplets settle on surfaces and the small ones shrink into dry nuclei; both processes occurs in very short time frames. Therefore, the microorganisms in wet aerosols may only be transported over short distances and induce infections in limited areas. Microorganisms released from dry sources may be more closely associated with long distance transmission (Zhao et al., 2014).

2.2.3. Pathogen shedding and emission rate

A typical emission model used in air pollution control describes the quantity of a pollutant emitted per unit time (e.g., g/s). However, the pathogen emission involves not only the quantity of pathogens shed by animals but also their infectivity. The emission rate for virus pathogens is commonly expressed as $TCID_{50}$ per unit time. $TCID_{50}$ (the median Tissue Culture Infective Dose) is defined as the amount of a pathogenic agent that will produce pathological change in 50% of cell cultures inoculated, i.e., it represents both the quantity and infectivity of pathogens.

Pathogen shedding is affected by many factors, including the host animal, pathogen species/isolates, coinfections, and others (Chao et al., 2006). Shedding is affected even by the method of inoculating animals for determining pathogen shedding in experimental studies. An experimental study conducted by Gloster et al. (2008) showed that the amount of aerosolised FMD virus emitted was inversely correlated to the administered dose of virus. Specifically, pigs infected with a lower dose of FMD virus inoculum emitted more airborne virus than those infected with a higher dose. They explained that this inverse correlation between the virus emission and the administered dose was possibly related to the extent and speed with which the animals' innate immune mechanisms were triggered. Therefore, it is extremely difficult to theoretically quantify pathogen emission. Empirical models are commonly used to estimate pathogen production and emission. For example, using the experimental data reported by Lowen et al. (2006), Halloran et al. (2012) proposed the following model to estimate the shedding of influenza virus by individual animals (guinea pigs):

$$C_{\exp} = \begin{cases} C_0 10^{k_g t} & 0 < t \le t_{peak} \\ C_{\max}\left[10^{-k_d(t-t_{peak})}\right] & t > t_{peak} \end{cases} \qquad (1)$$

where C_{exp} is the pathogen concentration in the expiratory fluid (droplet), PFU mL^{-1} (PFU: plaque-forming unit) (or $TCID_{50}$ mL^{-1}); C_0 is the initial pathogen concentration in the expiratory fluid, PFU mL^{-1}; C_{max} is the maximum observed concentration in the expiratory fluid, PFU mL^{-1}; t is the time, s; t_{peak} is the time at which the maximum concentration occurs, s; and k_g and k_d are the rate constants for the growth and decay periods, respectively, s^{-1}.

In essence, this model (equation 1) shows that the virus shedding follows the same pattern as the viral growth kinetics, i.e., increasing exponentially during the growth phase ($0<t<t_{peak}$), and then decreasing exponentially. Once the pathogen concentration in the expiratory fluid is known, the emission rate can be determined as (Nicas et al., 2005):

$$G = C_{\exp} E V \qquad (2)$$

where G is the pathogen emission rate, PFU h^{-1} (or $TCID_{50}$ h^{-1}); E is the expulsion event rate, h^{-1}; and V is the volume of the airborne aerosols per expulsion event, mL.

2.3. Aerosol dispersion and pathogen decay in atmosphere

Pathogen-laden aerosols move with air currents (winds) in the atmosphere. The process of movement is commonly termed as dispersion, which is the combined effect of diffusion due to turbulent eddy motion and advection due to the wind within a layer of air near the Earth's surface (Stockie, 2011). Dispersion dictates how far the aerosolized pathogens travel. The dispersion process strongly depends on meteorological conditions, including wind intensity, wind direction, solar radiation, and atmospheric stability, as well as topography. Specifically, pathogen-laden aerosols move in a plume in the direction of the wind, and spread (disperse) in both the vertical and cross-wind directions. The plume shape and the aerosol concentration at a given time and a downwind location are dictated by meteorological conditions, such as atmospheric stability and wind velocity (Casal et al., 1995). Generally speaking, the highest concentrations near the ground level occur in evening and night when the atmosphere is usually stable, while greater turbulence exists in the atmosphere during the day, resulting in lower concentrations (Casal et al., 1995). Dee et al. (2010) observed that winds of low velocity in conjunction with the presence of periodic gusts were significantly associated with high-risk days of PRRS transmission by aerosol. They summarized that significant environmental conditions present on PRRSV-positive air days (when virus was detected in air samples) included cool temperatures, higher relative humidity and pressure, slow moving winds and low sunlight levels.

Viability of aerosolized pathogens during dispersion depends on both the nature of the pathogen and the environmental conditions, as well as the form of aerosols (wet vs. dry aerosols). The air humidity has been identified as one of the most significant factors affecting the survival of pathogens in the atmosphere. For example, foot and mouth disease caused by a picornavirus is more prevalent after periods of wet weather and transmission distance can be many miles (Knight 1980). Knight (1980) explained that lipid enveloped viruses may be inactivated by surface effects when exposed to high relative humidity, while lipid-free viruses appear to be damaged in aerosols by removal of structural water. Viruses with structural lipids are generally stable at low relative humidity, whereas those without lipids are more stable at high relative humidity.

Temperature is another important environmental factor for the survival of pathogens in the air during dispersion. In studying the seasonality of human influenza transmission, a general hypothesis is that the lower temperature coupled with low humidity in wintertime enhances virus transmission (Halloran et al., 2012). Hermann et al. (2007) conducted experiments to study the effect of humidity and temperature on the survival of PRRSV and observed that that aerosolized PRRSV was more stable at lower temperatures and temperature had a greater effect on the half-life of PRRSV than did relative humidity. Based on the experimental data, they developed an empirical equation for predicting the half-life of PRRSV as a function of temperature and relative humidity:

$$T_{1/2} = 339.037e^{-(0.0839T+0.00754RH)} \tag{3}$$

where $T_{1/2}$ is the half-life of PRRSV, min.; T is the air temperature, °C; and RH is the relative humidity, %.

Besides humidity and temperature, biological decay of airborne microorganisms may be affected by several other environmental factors, including radiation (UV, γ-ray, X-ray), air ions, and gases in the atmosphere (O_2, O_3, CO, SO_2 and NO_x) (Zhao et al, 2014). For example, Dee et al. (2010) identified low sunlight level as a significant factor in promoting airborne transmission of PRRSV. Tong and Lighthart (1997) found the bacterium populations in the atmosphere decreased with radiation intensity in a log-linear fashion.

2.4. Initiation of infection

Once aerosols carrying viable pathogens reach the location of a naïve facility (Outcome 3, Figure 1), the possibility for them to induce infections depends on: (1) whether aerosols can enter the naïve facility (building); and (2) the critical amount of aerosols inhaled by animals to cause infection. Much research has been conducted on the critical infectious doses, in particular for human diseases, while limited information is available on the entrance of pathogen-laden aerosols into buildings. The entrance of aerosols into a naïve facility is a complicated physical process, dictated by such factors as local winds; building size, shape and orientation; and ventilation (mechanical vs. natural ventilation, ventilation rate, and air inlet configuration).

Infectious dose 50 (the median infectious dose) (ID_{50}), defined as the dose that will infect 50% of an experimental group of hosts within a specified time period, is commonly used to quantify the amount of pathogens required to cause an infection in the host, whereas the minimal infectious dose (MID) is also used but to a lesser extent. It should be mentioned that the infectious dose varies not only from (animal and pathogen) species to species but also with the transmission route. For example, a study conducted by Hermann et al. (2005) to determine the infectious dose of PRRSV (isolate VR-2332) showed that the ID_{50} was $10^{5.3}$ $TCID_{50}$ for oral exposure and $10^{4.0}$ $TCID_{50}$ for intranasal exposure, whereas the ID_{50} was only $10^{2.2}$ $TCID_{50}$ for the intramuscular exposure. Small aerosols carrying pathogens reach the lower respiratory tract upon inhalation, and therefore infectious dose by aerosol inhalation is much less (100-fold for influenza viruses) than that by inoculation with intranasal drops. For example, Guan et al. (2013) reported that the ID_{50} for aerosolized virus was about 2 $\log_{10}$ and 5 $\log_{10}$ lower than by nasal or oral inoculation, respectively, for a low pathogenic avian influenza virus (LPAI H9N2).

A dose-response curve is useful for describing the probability for infection to be produced when an animal is exposed to a specific dose of an airborne pathogen (Cutler et al., 2011). Exposure to a specific dose may either produce infection or not, and the probability for a randomly chosen animal to become infected for a specific exposure dose may be calculated by such models as logit (log-logistic) and probit (log-normal) models which are considered plausible models for PRRS (Hermann et al., 2005; 2009). Taking the probit model as an example, the probability of infection by PRRSV is calculated as follows (Hermann et al., 2009):

$$P = \left(2\pi\right)^{-1/2} \int_{-\infty}^{\alpha+\beta x} e^{-u^2/2} du \qquad -\infty < x < \infty \tag{4}$$

where P is the probability that an individual chosen at random will become infected for a given dose; x is the exposure dose, $TCID_{50}$; and α and β are two empirical constants, and u is an intermediate variable.

The exposure dose can be calculated from the pathogen concentration in the air, the animal breath rate, the tidal volume, and the time duration of exposure.

3. Measures for Dealing with Airborne Disease Transmission

Similar to dealing with any disease outbreaks, dealing with an outbreak of airborne disease requires several emergency measures, including rapid detection and diagnosis, restriction of the movements of persons and animals, and avoiding the spread of the disease by other means (Casal et al., 1995). This review focuses on some important measures specific to the airborne transmission.

3.1. Forecasting disease spread by airborne transmission

It should be noted that airborne disease can be spread through different routes of direct and indirect contact, such as people and animal movement. This review focuses on forecasting spread of airborne diseases by aerosol transmission. This forecasting can be used to assist in decision making to direct surveillance efforts both geographically and temporally (Sørensen et al., 2000). "Geographically, the extent of the region exposed to an infectious 'plume' can be quickly defined and the farms at risk under the plume identified. Temporally, an estimate can be made of the periods during which infection was most likely, the resulting period of incubation and when the earliest clinical signs can be expected. Based on this information, the timing and frequency of clinical inspections of livestock on farms at risk can be optimized so that control and eradication measures are implemented without delay." (Sørensen et al., 2000). Dispersion modelling (a mathematical description of aerosol transport in the atmosphere) has been used to predict spread of airborne diseases for both humans and animals (e.g., Sørensen et al., 2000; Gloster et al., 2010; Halloran et al., 2012; Seo et al., 2014). Dispersion models for human disease transmission focus mostly on short ranges (i.e., within a building and between adjacent buildings) (e.g., Halloran et al., 2012), while the interest in animal disease transmission is generally long ranges (between farms) (e.g., Sørensen et al., 2000). Since the early 1980s, several dispersion models have been developed for FMD (foot-and-mouth disease) (Gloster et al., 1981; Sørensen et al., 2000; Gloster et al., 2010; Seo et al., 2014). These models are mostly based on Gaussian or Lagrangian dispersion theories, and a few based on CFD (computational fluid dynamics) for short distances (several km) (Seo and Lee, 2014). Gloster et al. (2010) conducted a study to compare six models that are used in UK, Denmark, New Zealand, Australia, Canada, and USA for predicting the spread of FMD. They concluded that generally all six models could be used to assess windborne spread of FMD in terms of providing advice to making control and eradication decisions in the event of an FMD outbreak. However, discrepancies existed between the models, and they suggested that virus decay and local scale topography should be considered.

3.2. Air cleaning – filtration

A number of studies have evaluated the effectiveness of air filtration in reducing and preventing airborne transmission of animal diseases, such as PRRS and M hyo (Mycoplasma hyopneumoniae) (Dee et al., 2005; 2010; Batista et al., 2009). A filtration system is typically installed at the air inlet to stop pathogen-laden aerosols from entering the facility. However, a filtration system may also be installed to filter the exit air from an infected facility to prevent the spread of pathogens, as a measure of biocontainment. A typical filtration system consists of a low cost protective pre-filter for dust removal, and several stages (layers) of high efficiency filters. Depending on their mechanisms of filtration, filters are categorized as mechanical filters and antimicrobial filters, both types having been shown to be effective in reducing and preventing airborne transmission of animal diseases (Dee et al., 2010). Mechanical filters capture airborne particles when they come into contact with the filter media and adhere to the media fibres. Antimicrobial filters also rely on fibres to first mechanically capture pathogens-laden aerosols and then use antimicrobial agents that are integrated into the media fibres to inactivate the pathogens (Pouliot et al., 2011).

The first adopters of air filtration systems in pig facilities were the insemination centres in Brittany, France, where HEPA (high efficiency particulate air) filters were installed in positively ventilated buildings (Ricard et al., 2013). In North America, the first HEPA filtration system was installed in two artificial insemination centres in 2003 in Quebec, Canada. Since then HEPA filters are used in almost all major artificial insemination centres in Quebec, and in some farrow-to-finish facilities (Ricard et al. 2013). A key performance parameter for filters is the efficiency. Several filter efficiency rating standards are used in the industry, including the MERV (minimum efficiency reporting value) established by the American Society of Heating, Refrigerating and Air Conditioning Engineers (ASHRAE) and the European Standards (EN779 and 2012 and EN1822-1). The "standard" filter efficiency for use in the pig facilities is MERV 16, but recently MERV 14 filters have been used to reduce the cost (Devries, 2012).

Dee et al. (2005) conducted experiments to evaluate the ability of an air filtration system in reducing airborne transmission of PRRSV. They found that the filtration system consisted of a pre-filter, a first-stage filter of EU8 rating, and a second-stage filter of EU13 rating was highly effective at reducing aerosol transmission of PRRSV. Dee et al. (2010) compared three filtration systems: i) a two-stage mechanical system consisted of a fiberglass pre-filters of MERV 4, and stage 2 pleat-in-pleat V-bank fiberglass filters rated at MERV 16 (EU9); ii) a similar two-stage system with stage 2 filters rated at MERV 14 (EU 8); and iii) an antimicrobial filter consisting of 10 layers of polypropylene fabric impregnated with a mixture of virucidal and bactericidal compounds. Their test results showed that all three filtration systems performed equally in their ability to prevent aerosol transmission of both PRRSV and M hyo from infected to susceptible populations. Dee et al. (2012) reported the results of a long-term sustainability study of air filtration as a means to reduce PRRS transmission. They analysed the effect of air filtration on the occurrence of PRRSV infections in 38 herds from September 2008 to January 2012. Their results indicated that new PRRSV infections in filtered breeding herds were significantly lower than in non-filtered control herds, the odds for a new PRRSV infection in breeding herds before filtration was about eight (8) times higher than that after filtration was initiated, and the median time to new PRRSV infections in filtered breeding herds was significantly longer than that in non-filtered herds (30 months vs. 11 months).

While it has been shown that filtration is effective in reducing and preventing airborne transmission of animal diseases, challenges still exists in using this technology, including the pressure drop across the filters; filter clogging; and high capital and operational costs. When air passes the filters, excessive pressure drop across the filters could interfere with building ventilation, such as reduced fan performance and airflow rate, and increased infiltration, which in turn would reduce the efficiency of the filtration system because air entering the building through infiltration bypasses the filters. As aerosol particles cumulate on the filter medium, the filter is gradually clogged. Clogging may drastically change the filter performance and lifespan (Ricard et al., 2013). When estimating the cost of a filtration system, the total cost should be considered, which includes the purchase and installation, energy for operation, maintenance and disposal (Ricard et al., 2013).

4. Conclusions

1. Airborne transmission of animal diseases involves several biological and physical processes, including pathogen shedding and aerosolization, atmospheric dispersion of pathogen-laden aerosols, decay of aerosolized pathogens in the atmosphere, and initiation of infection. Our understanding of airborne transmission of diseases is still limited.

2. Understanding and dealing with airborne transmission of animal diseases require knowledge from several disciplines, including biological sciences (e.g., microbiology, and animal and veterinary sciences) and engineering (e.g., fluid mechanics).

3. Filtration is a promising technology for reducing and preventing airborne transmission of animal diseases.

References

Batista, L., L. Urizar, and F. Pouliot, 2009. Evaluation of the effectiveness of Noveko's antimimicrobial commercial filter after 16 months of installation in a commercial swine unit. Final report. Québec, QC, Canada: CDPQ, 6p.

Casal, J., E. Planas-Cuchi, J.M. Moresob, and J. Casalb. 1995. Forecasting virus atmospherical dispersion-studies with foot-and-mouth disease. Journal of Hazardous Materials. 43:229–244.

Charpin, C., S. Mahé, A. Keranflec, C. Belloc, R. Cariolet, M-F. Le Potier, and N. Rose. 2012. Infectiousness of pigs infected by the Porcine Reproductive and Respiratory Syndrome virus (PRRSV) is time-dependent. Veterinary Research 43:69. doi:10.1186/1297-9716-43-69.

Cho, J.G., S.A. Dee, J. Deen, C. Trincado, E. Fano, Y. Jiang, K. Faaberg, M.P. Murtaugh, A. Guedes, and J.E. Collins. 2006. The impact of animal age, bacterial coinfection, and isolate pathogenicity on the shedding of porcine reproductive and respiratory syndrome virus in aerosols from experimentally infected pigs. Canadian Journal of Veterinary Research. 70:297–301.

Cutler, T.D., W.G. Chong, S.J. Hoffb, A. Kittawornrata, and J.J. Zimmermana. 2011. Median infectious dose (ID50) of porcine reproductive and respiratory syndrome virus isolate MN-184 via aerosol exposure. Veterinary Microbiology. 151 (3–4): 229–237.

Dee, S., J.P Cano, G. Spronk, D. Reicks, P. Ruen, A. Pitkin, and D. Polson. 2012. Evaluation of the long-term effect of air filtration on the occurrence of new PRRS infections in large breeding herds in swine-dense regions. Viruses. 4:654–662; doi:10.3390/v4050654

Dee, S., S. Otake, and J. Deen. 2010. Use of a production region model to assess the efficacy of various air filtration systems for preventing airborne transmission of porcine reproductive and respiratory syndrome virus and Mycoplasma hyopneumoniae: Results from a 2-year study. Virus Research 154:177–184.

Dee, S.A., L. Batista, J. Deen, and C. Pijoan. 2005. Evaluation of an air-filtration system for preventing aerosol transmission of porcine reproductive and respiratory syndrome virus. The Canadian Journal of Veterinary Research. 69:293–298.

Desrosiers, R. 2011. Transmission of swine pathogens: different means, different needs. Animal Health Research Reviews. 12(1):1–13, doi: 10.1017/S1466252310000204.

Devries, S. 2012. Air filtration for swine operations. 31st Annual Centralia Swine Research Update: I-25–I-26.

Gloster, J., A. Jones, A. Redington, L. Burgin, J.H. Sørensen, R. Turner, M. Dillon, P. Hullinger, M. Simpson, P. Astrup, G. Garner, P. Stewart, R. D'Amours, R. Sellers, and D. Paton. 2010. Airborne spread of foot-and-mouth disease – model intercomparison. The Veterinary Journal. 183(3):278–286.

Gloster, J., J. Blackall, R.F. Sellers, and A.I. Donaldson. 1981. Forecasting the airborne spread of foot-and-mouth disease. Veterinary Record. 108:370–374.

Guan, J., Q. Fu, M. Chan, and J. L. Spencer. 2013. Aerosol transmission of an avian influenza h9n2 virus with a tropism for the respiratory tract of chickens. Avian Diseases 57:645–649.

Halloran, S.K., A.S. Wexler, and W.D. Ristenpart. 2012. A comprehensive breath plume model for disease transmission via expiratory aerosols. PLoS ONE 7(5):e37088. doi:10.1371/journal.pone.0037088.

Han, Z.Y., W.G. Weng, and Q.Y. Huang. 2014. Characterizations of particle size distribution of the droplets exhaled by sneeze. Journal of Royal Society Interface 10:20130560. http://dx.doi.org/10.1098/rsif.2013.0560.

Hermann, J. R., S.L. Brockmeier, K.L. Yoon, and J.J. Zimmerman. 2008. Detection of respiratory pathogens in air samples from acutely infected pigs. Canadian Journal of Veterinary Research 72:367–370.

Hermann, J.R., S.J. Hoff, C. Munoz-Zanzi, K.J. Yoon, M. Roof, A. Burkhardt, and J.J. Zimmerman. 2007. Effect of temperature and relative humidity on the stability of infectious porcine reproductive and respiratory syndrome virus in aerosols. Veterinary Research. 38:81–93. DOI: 10.1051 / vetres:2006044

Hermann, J., C.A. Munoz-Zanzi, and J.J. Zimmerman. 2009. A method to provide improved dose-response estimates for airborne pathogens in animals: An example using porcine reproductive and respiratory syndrome virus. Veterinary Microbiology. 133:297–302.

Hermann, J.R., C.A. Munoz-Zanzi, M.B. Roof, K. Burkhart, and J.J. Zimmerman. 2005. Probability of porcine reproductive and respiratory syndrome (PRRS) virus infection as a function of exposure route and dose. Veterinary Microbiology. 110:7–16.

Jannasch H.W. and T. Egli. 1993. Microbial growth kinetics: a historical perspective. Antonie van Leeuwenhoek 63: 213–224, 1993.

Knight, V. 1973. Airborne transmission and pulmonary deposition of respiratory viruses. In Viral and Mycoplasmal Infections of the Respiratory Tract. V. Knight, Ed.: 1–9. Lea & Febiger. Philadelphia.

Knight, V. 1980. Viruses as agents of airborne contagion. Annals of the New York Academy of Sciences. 353:147–56.

Lowen, A.C., S. Mubareka, J. Steel, and P. Palese. 2007. Influenza virus transmission is dependent on relative humidity and temperature. PLOS Pathogens. 3: e151.doi:10.1371/journal.ppat.0030151.

Lowen A.C., S. Mubareka, T.M. Tumpey, A. Garcia-Sastre, P. Palese. 2006. The guinea pig as a transmission model for human influenza viruses. Proceedings of the National Academy of Sciences. 103: 9988–9992.

Nicas M., W.W. Nazaroff, A. Hubbard. 2005. Toward understanding the risk of secondary airborne infection: emission of respirable pathogens. Journal of Occupational and Environmental Hygiene. 2(3):143–54.

Pouliot, F. M-A. Ricard, and V. Dufour. 2011. Canadian Swine Buildings and Inlet Air Filtration Systems. Technical Guide, Canadian Swine Health Board, Ottawa, Canada

Ricard, M-A, E. G. Tremblay, V. Dufour, and F. Pouliot, 2013. Air Filtration in Swine Barns: Overview of the Current Situation. Centre de Développement du Porc du Quebec Inc., Quebec, Quebec, Canada

Seo, I-H. and I-B. Lee. 2014. Web-based forecasting system of airborne livestock virus spread simulated by OpenFOAM CFD. Proceedings International Conference of Agricultural Engineering, Zurich, 06-10.07.2014.

Seo, I-H., I-B Lee, O-K. Moon, N-S. Jung, H-J. Lee, S-W. Hong, K-S. Kwon, and J.P. Bitog. 2014. Prediction of the spread of highly pathogenic avian influenza using a multifactor network: Part 1 Development and application of computational fluid dynamics simulations of airborne dispersion. Biosystems Engineering 121:160–176.

Sørensen, J.H., D.K.J. Mackay, C.Ø. Jensen, and A.I. Donaldson. 2000. An integrated model to predict the atmospheric spread of foot-and-mouth disease virus. Epidemiology and Infection. 124:577–590.

Stockie, J.M. 2011. The mathematics of atmospheric dispersion modeling. SIAM Review. 53(2):349–372.

Tong, Y. and B. Lighthart. 1997. Solar radiation has a lethal effect on natural populations of culturable outdoor atmospheric bacteria. Atmospheric Environment. 31(6):897-900.

Wells, W.F. 1934. On air-borne infection. Study II. Droplets and droplet nuclei. American Journal of Hygiene. 20:611–618.

WUR. 2014. Airborne transmission of pathogens and the relation to dust in livestock production. Project Description. Wageningen UR, https://www.wageningenur.nl/en/show/Airborne-transmission-of-pathogens-and-the-relation-to-dust-in-livestock-production.htm. Accessed July 28, 2015.

Xie, X., Y. Li, A.T.Y. Chwang, P.L. Ho, and W. H. Seto. 2007. How far droplets can move in indoor environments – revisiting the Wells evaporation–falling curve. Indoor Air. 17: 211–225.

Yang, Y., G.N. Sze-To, and C.Y.H. Chao. 2012. Estimation of the aerodynamic sizes of single bacterium-laden expiratory aerosols using stochastic modeling with experimental validation, Aerosol Science and Technology. 46:1, 1–12, DOI: 10.1080/02786826.2011.604108

Zhao, Y., A.J.A. Aarnink, M.C.M. de Jong, and P.W.G. Groot Koerkamp. 2014. Airborne microorganisms from livestock production systems and their relation to dust. Critical Reviews in Environmental Science and Technology. 44 (10):1071–1128.

Modeling Disinfection of Vehicle Tires Inoculated with Enteric Pathogens on Animal Farms Using Response Surface Methodology

Yitian Zang [a], Xingshuo Li [a], Baoming Li [a], Changhua Lu [b], Wei Cao [a,*]

[a] Key Laboratory of Agricultural Engineering in Structure and Environment, Ministry of Agriculture, China Agricultural University, P.O. Box 67, Beijing 100083, China

[b] Key Laboratory of Animal Diseases Diagnostic and Immunology, Ministry of Agriculture, National Center for Engineering Research of Veterinary Bio-products, Nanjing, China

* Corresponding author. E-mail: caowei@cau.edu.cn

Abstract

To reduce the risk of enteric pathogens transmission and drug residues in animal farms, the disinfection effectiveness of slightly acidic electrolyzed water (SAEW, pH 5.85–6.53) for inactivating *Escherichia coli* and *Salmonella enteritidis* mixture on the surface of vehicle tires was evaluated. The coupled effects of the tap water washing time (2 to 4 min), SAEW treatment time (3–7 min) and available chlorine concentrations (ACCs) of 80 to 140 mg/l on the reductions of *E. coli* and *S. enteritidis* mixture on tires were investigated using a central composite design of the response surface (RS) methodology. The established RS model had a goodness of fitting quantified by the parameters of R^2 (0.984), lack of fit test ($P > 0.05$). The maximum reduction of 1.38 $\log_{10}$ CFU/cm^2 (92.9%) for *E. coli* and *S. enteritidis* mixture was obtained by the vehicle tire treated with 4 min tap water washing for followed by 5 min SAEW treatment for at an ACC of 140 mg/l. The established RS model shows the potential of SAEW in disinfection of bacterial cells on tires.

Keywords: Slightly acidic electrolyzed water, disinfection, vehicle tires, response surface methodology, animal farms

1. Introduction

As the increasing demand for efficient and sustainable poultry production and the global demand for reduced antimicrobial drug consumption in poultry industry, the importance of poultry disease prevention has increased significantly (Chauvin et al., 2005). Disinfection is a generally agreed concept to prevent the introduction of both endemic and epidemic infections (Böhm 1998). The application of disinfection practices in poultry production to people and vehicles is equally important, because it can reduce disease transmission risk as the movement of people and vehicles between and within farms (Capua and Marangon 2006; Barrington et al., 2006).

The process of transport has long been considered an important risk for pathogens entry into farms. The entering farm vehicles, which could go through a long road before entry into the poultry farm with tires directly contacting with the road, can be easily contaminated with shed chicken manure and pathogenic microorganisms, such as S. *enteritidis*, E. *coli*, etc. through poor biosecurity measures and personal hygiene. *Salmonella enteritidis* and *Escherichia coli* are pathogens that occur naturally in the poultry farms and can be frequently found in farms (Rosanowski et al., 2012). Moreover, the disinfection of tires would be a more difficult task in the case of the manure soiling interference presence (Davies et al., 2001; Guard-Petter., 2001). When the contaminated tires enter into the farm, they can directly contact with the road, which is used by all farm personnel and vehicles, and augment additional opportunities for bird and human infection. Therefore, it is extremely important that the tires with manure soiling being sanitized before entering into poultry farms to reduce the risk of disease transmission into farms.

Slightly acidic electrolyzed water (SAEW) with a pH value of 5.0–6.5, produced by electrolysis of a dilute hydrochloric acid in a chamber without a membrane, is one of the emerging environment friendly antimicrobial disinfectants (Hricova et al., 2008; Huang et al.,

2008). It has been proven to be safe and cheap and exhibits high bactericidal and fungicidal efficacy (Abadias et al., 2008; Cao et al., 2009). Moreover, SAEW has the advantage of possessing broad spectrum antimicrobial activity, reduced corrosion of surfaces and minimization of the potential for damage to human health (Abadias et al., 2008).

Currently, some studies on disinfection efficacy of SAEW in medicine, agriculture, and food industry were reported (Cao et al., 2009) and the published literature involved in using SAEW to decontaminate pathogenic microorganisms in poultry production, including air and surface sterilization in layer houses (Hao et al., 2013), inactivation of *Salmonella enteritidis* contaminated shell eggs (Cao et al., 2009). However, little was reported regarding the efficacy of SAEW treatment for the tires of the entering vehicle under manure soiling interference circumstance.

The purpose of this work was to evaluate effectiveness of SAEW to inactivate pathogenic microorganisms on tires under the chicken manure interference circumstance. For this purpose, response surface methodology (RSM) was used to describe the inactivation of *E. coli* and *S. enteritidis* mixture by SAEW on tires as a function of the tap water cleaning time, SAEW treatment time and available chlorine concentrations (ACCs).

2. Materials and Methods

2.1. Bacterial cultures

The strains of *E. coli* (ATCC 25922) and *S. enteritidis* (CVCC 2184) used were obtained from the Department of National Center for Medical Culture Collections (CMCC, Beijing, China) and the China Veterinary Culture Collection (CVCC, Beijing, China), respectively. Stock cultures of each pathogen were transferred into tryptic soy broth (TSB, Beijing Land Bridge Technology Company Ltd., Beijing, China) and incubated for 24 h at 35°C. Following the incubation, a 10-ml of each culture was pooled into a sterile centrifuge tube and centrifugated at 3000 × g and 4°C for 10 min. The supernatant was decanted, and the pellets were resuspended in 10 ml of sterilized 0.85% NaCl solution, washed three times and resuspended in 10 ml of the same solution to obtain a final cell concentration of about 10^9 CFU/ml. The bacterial population in each cocktail culture was confirmed by plating 0.1 ml portions of appropriately diluted culture on tryptic soy agar (TSA, Beijing Land Bridge Technology Company Ltd., Beijing, China) plates and incubating the plates at 35°C for 24 h.

2.2. Preparation of slightly acidic electrolyzed water

Slightly acidic electrolyzed water was produced using a non-membrane generator (Zhouji Biosafety Technology Co., Ltd., Beijing, China) to electrolyzing NaCl (10 g/l) solution. The SAEW with a pH of 6.15–6.53, an oxidation-reduction potential (ORP) of 974–989 mV, and different ACCs of 80–140 mg/l was produced by the generator. The physicochemical properties of SAEW were measured before use. The pH and ORP values were measured using a dual scale pH/ORP metre (CON60, Trans-Wiggens, Singapore) with a pH electrode (PE02; range 0.00–14.00) and an ORP electrode (ORP06; range from -999 to +999 mV). The ACC was determined using a digital chlorine test system (RC-2Z, Kasahara Chemical Instruments Co., Saitama, Japan) that had a detection range of 0–320 mg/l. The solutions were placed into an atmospheric pressure manual sprayer (2000 ml, Taizhou Huangyan Li-mao Plastic Factory, Zhejiang, China) for sample treatment.

2.3. Inoculation of tires

Liquid manure of 6% total solids was prepared with an addition of 30 g chicken manure (obtained from pheasantry with no bedding) in 0.5 l sterile distilled water and then inactivated by an autoclave (YXQ-LS-18SI, Shanghai Boxun Industrial Co., Ltd., Shanghai, China). Five ml of the 6% sterilized liquid manure was shocked and then combined with equal portions of the prepared culture cocktails to obtain the final populations of contaminated culture about 10^8–10^9

CFU/ml and to achieve 3% concentration (Elassaad et al., 1993) of liquid chicken manure soiling interference in disinfection.

Tires (750-16, Qingdao Hongxinyu Rubber Co., Ltd., Qingdao, China) were obtained from an agricultural three-wheel car (ATX3000, Foton Motor Co., Ltd., Beijing, China) in Beijing. The tires were washed with tap water to remove soil and then trimmed to approximately 5 × 5 cm^2 in size and packed in a polyethylene bag for the experiment. Before inoculation, the tires were inactivated by the autoclave and then air-dried under a biosafety hood (DH-920, Beijing East Union Hall Instrument Manufacturing Co., Ltd., Beijing, China) at the room temperature for 50 min. Each tire was inoculated by spreading 0.1 ml of the prepared contaminated culture inoculum with a pipette tip onto the front side surface. Subsequently, all inoculated tires were air-dried under the biosafety hood for 30 min at room temperature to allow the bacterial attachment. The final concentration of *E. coli* and *S. enteritidis* mixture inoculated on the tires was 5 $\log_{10}$ CFU/cm^2 on average. Samples for each treatment were prepared and sampled at least in duplicate.

2.4. Treatment of tire samples with SAEW and microbiological determination

Inoculated tires were first sprayed with tap water using the atmospheric pressure manual sprayer to clean and then sprayed with SAEW by another atmospheric pressure manual sprayer to disinfection under different conditions. After the treatment, moisten sterile swabs with neutralizing agent (0.1% $Na_2S_2O_3$) were used to collect the surface samples. The sterilized cotton swabs, which had been wiped back and forth for 20 times on the tire surfaces, were immediately transferred into 5 ml neutralizing agent (0.1% $Na_2S_2O_3$) tubes for microbiological analyses. The tubes were shaken on a platform shaker at 1800 rpm (MIR-S100, Sanyo Electric Biomedical Co., Ltd., Osaka, Japan). Surviving bacteria were determined by serial dilutions in sterile 0.1% peptone water and then plated in duplicate (0.1 ml) on tryptic soy agar plates. The plates were incubated at 37°C for 24 h to counting of colonies. Moreover, un-inoculated tires yielded no colonies on the agar. Two trials with three replicates per trial in each treatment conditions were done.

Inactivation of *E. coli* and *S. enteritidis* mixture on the tire by the coupled effects of cleaning times, treatment times and ACC was studied using response surface methodology (RSM). Twenty trials of experiments were designed by central composite design (CCD) to determine the cleaning times in range of 2–4 min, treatment times of 3–7 min and ACC of 80–140 mg/l for SAEW treatments.

2.5. Statistical analysis

All treatments were replicated three times and results were reported as means. The response value was expressed as the log reductions between final load after the treatments and the initial inoculate per tire. An analysis of variance and an estimation of response surface were performed using the software Design Expert (Version 8.0.5, Stat-Ease Inc., Minneapolis, MN, USA).

The statistical significance and adequacy of each coefficient value were calculated and tested using the Fisher F-test and Lack of fit test.

2.6. Model validation

In order to validate the adequacy of the inactivation model, additional random eight conditions were carried out under the range of the experimental domain. The line of correlation ($y = x$) was made to evaluate the performance of RS model between the actual experimental values, which were obtained from the eight conditions and the predicted values calculated from the RS model.

3. Results and Discussion

3.1. Establishment and evaluation of response surface model

The response result measured in terms of log reductions under different combined conditions is shown in Table 1. Response surface analysis demonstrated log reductions ranged from 0.25 to 1.41 $\log_{10}$ CFU/cm^2. The response model was established and represented as follows:

$$R = 1.06 + 0.22x_1 + 0.13x_2 + 0.23x_3 + 0.097x_1x_2$$
$$0.034x_1x_3 + 0.046x_2x_3 - 0.14x_1^2 - 0.085x_2^2 - 0.057x_3^2 \quad (1)$$

where R is the response value ($\log_{10}$ CFU/cm^2); x_1 is the treatment time, x_2 and x_3 are the cleaning time and ACC, respectively.

Table 1. Observed and predicted reduction of *E. coli and S. enteritidis* mixture on the surface of tires by RS model according to the central composite design arrangement.

Trial no.	x_1 Treatment time (min)	x_2 Cleaning time (min)	x_3 ACC* (mg/l)	Observed Value** (log CFU/cm^2)	Predicted value (log CFU/cm^2)
1	5.0	3.0	110	1.06 (0.08)	1.06
2	3.0	2.0	140	0.73 (0.13)	0.75
3	1.6	3.0	110	0.30 (0.02)	0.30
4	7.0	4.0	140	1.41 (0.04)	1.46
5	5.0	3.0	60	0.52 (0.06)	0.51
6	5.0	3.0	110	1.03 (0.07)	1.06
7	5.0	1.3	110	0.59 (0.08)	0.60
8	5.0	3.0	110	1.10 (0.04)	1.06
9	5.0	4.7	110	1.07 (0.15)	1.03
10	5.0	3.0	160	1.30 (0.24)	1.28
11	5.0	3.0	110	0.94 (0.03)	1.06
12	3.0	4.0	80	0.25 (0.04)	0.28
13	5.0	3.0	110	1.13 (0.05)	1.06
14	7.0	4.0	80	0.99 (0.20)	0.98
15	5.0	3.0	110	1.10 (0.08)	1.06
16	3.0	4.0	140	0.92 (0.04)	0.90
17	7.0	2.0	80	0.60 (0.02)	0.62
18	7.0	2.0	140	0.94 (0.11)	0.92
19	3.0	2.0	80	0.35 (0.01)	0.31
20	8.4	3.0	110	1.06 (0.03)	1.04

*ACC: available chlorine concentration.

**Mean value (standard deviation)

The analysis of the variance of the quadratic model indicated the goodness of fit of the regression equation. The correlation coefficient (R^2) and adjusted R^2 were 0.984 and 0.969, respectively, indicating that the model fitted the experimental data, a 98.4% of the total variation can be explained by the established model and there was a good agreement between the observed and predicted values. The F-value of 68.52 and probability value ($P < 0.0001$) indicated that the treatments were highly significant and the lack of fit test ($P > 0.05$) was not significant. The linear coefficients (x_1, x_2, and x_3), the quadratic term coefficients (x_1^2, x_2^2 and x_3^2) and the cross coefficients (x_1x_2, x_2x_3) were significant ($P < 0.05$). The other term coefficient (x_1x_3) was not significant.

3.2. Validation of the model

The experimental values had a significant agreement with the predicted values at R^2 of 0.97 and a statistical significance level of $P < 0.0001$ (Table 2). Therefore, the model was proven to be applicable for predicting the disinfection of *E. coli* and *S. enteritidis* mixture on the tires for SAEW treatment parameters of cleaning times in the range of 2–4 min, treatment times of 3–7 min and ACCs of 80–140 mg/l.

Table 2. Observed and predicted reduction values of E. *coli* and S. *enteritidis* mixture on the surface of tires by RSM under the additional random eight conditions.

x_1 Treatment time (min)	x_2 Cleaning time (min)	x_3 ACC* (mg/l)	Observed Value** ($\log_{10}$ CFU/cm^2)	Predicted value ($\log_{10}$ CFU/cm^2)
3.0	3.7	80	0.29 (0.02)	0.33
3.0	4.0	90	0.41 (0.06)	0.43
3.0	2.0	105	0.59 (0.17)	0.55
3.5	3.7	140	0.87 (0.15)	1.04
4.4	3.9	140	0.99 (0.25)	1.23
4.1	3.6	140	1.31 (0.10)	1.17
3.0	3.6	119	0.63 (0.02)	0.77
3.0	3.9	127	0.69 (0.13)	0.82

*ACC: available chlorine concentration.

**Mean value (standard deviation)

3.3. Response surfaces

The effect of cleaning time, treatment time and ACC for SAEW treatment on the inactivation of *E. coli* and *S. enteritidis* mixture is shown in Figures 1–3 by response surfaces. In each Figure, one factor was maintained constant at its central point. Figure 1 shows the effect of treatment time and cleaning time on the inactivation of E. *coli* and S. *enteritidis* mixture while keeping the ACC at its central point (110 mg/l). It was observed that the log reduction related to increase in treatment time and cleaning time, and the log reductions changed significantly ($P <$ 0.05) with the interactions between cleaning time and treatment time. The interaction of treatment time and ACC at a constant of clean time (3 min) is shown in Figure 2. Figure 2 shows that the log reduction increases directly with increase in treatment time and ACC, but the log reductions did not significantly change ($P > 0.05$) with the interactions between treatment time and ACC. The reduction as a function of clean time and ACC at a fixed treatment time of 5 min was given in Figure 3, which shows that as clean time and ACC increases, the reduction in population increases.

The more reduction of pathogens and the significant interactions between cleaning time and treatment time were likely due to the manure soiling, which was a strongly limiting factor for disinfection of SAEW, and if it was removed more by the increasing cleaning time, a more effective disinfection along with the increasing treatment time would be obtained. This is similar with the previous results (Toshihiro et al., 2000; Ozer et al., 2006).

The log reductions changed significantly ($P < 0.05$) with the interactions between ACC and cleaning time was seen in Figure 3. Moreover, it also can be explained by the presence of manure soiling. Several authors have reported that the organic soiling could change the formation of combined available chlorines (Böhm 1998; Oomori et al., 2000). The combined available chlorines had much lower bactericidal activity than the free form at an identical chlorine concentration. Therefore, the increasing cleaning time which could remove more manure soiling decreased the changing of the combined available chlorines formation, and then the disinfection effective of the ACC would be reinforced.

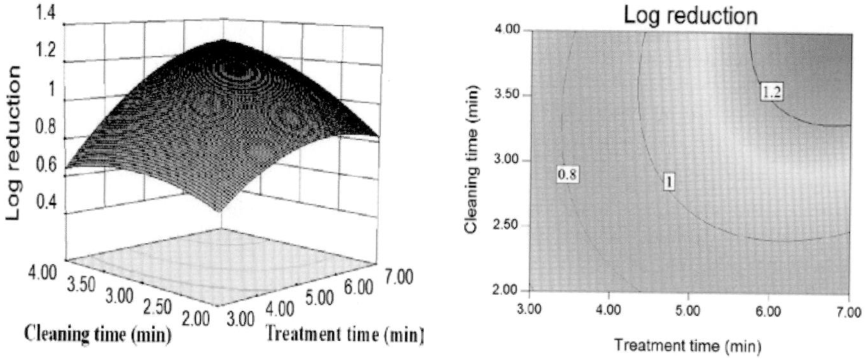

Figure 1. Response surface plots describing the effects of treatment time (x_1) and cleaning time (x_2) of SAEW at ACC 110 mg/L on inactivation of bacteria.

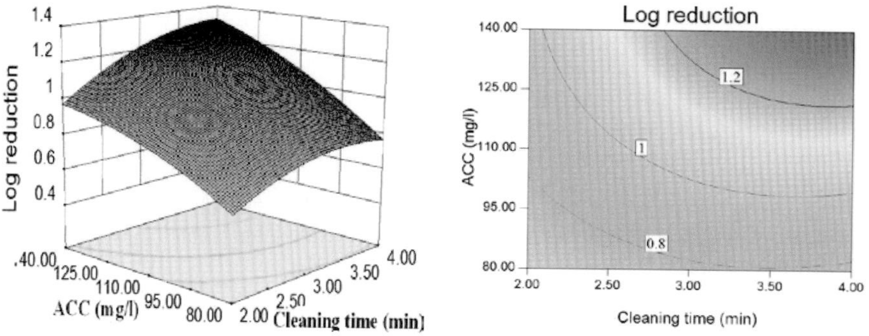

Figure 2. Response surface plots describing the effects of treatment time (x_1) and ACC (x_3) of SAEW at cleaning time 3 min on inactivation of bacteria.

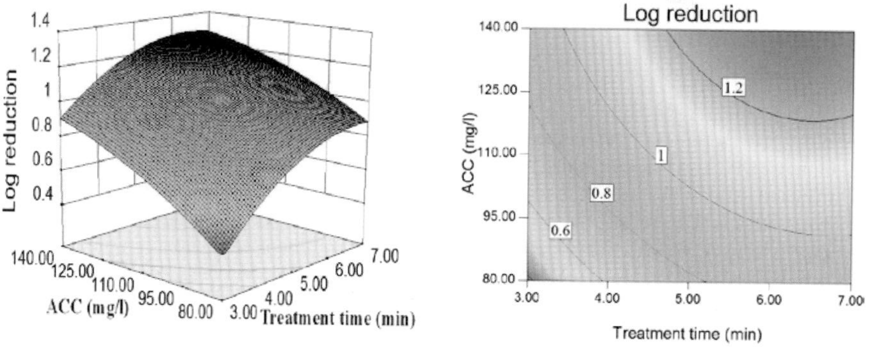

Figure 3. Response surface plots describing the effects of cleaning time (x_2) and ACC (x_3) of SAEW at treatment time 5 min on inactivation of bacteria.

4. Conclusions

In conclusion, the established RS model had a good statistical performance and was suitable to predict the population of *E. coli* and *S. enteritidis* mixture as a function of cleaning time, treatment time and ACC. A maximum log reduction of 1.38 $\log_{10}$ CFU/cm^2 (92.9%) for *E. coli* and *S. enteritidis* mixture was obtained after the treatment with tap water for 4 min and then SAEW treatment for 5 min at an ACC of 140 mg/l. Moreover, the result shows the potential of SAEW in disinfection of bacterial cells on tires and in promoting the implementation of disinfection measures to control and reduce the transmission risk of the disease.

Acknowledgements

This work was supported by the Earmarked Fund for Modern Agro-industry Technology Research System (CARS-41), the National Natural Science Foundation of China (grant number: 21106179), and Animal Disease Prevention and Food Safety Key Laboratory of Sichuan Province.

References

Abadias, M., J. Usall, M. Oliveira, I. Alegre, I. Vinas, 2008. Efficacy of neutral electrolyzed water (NEW) for reducing microbial contamination on minimally processed vegetables. Int J Food Microbiol, 123,151– 158.

Barrington, G.M., A.J. Allen, S.M. Parish, A. Tibary, 2006. Biosecurity and biocontainment in alpaca operations. Small Rumin Res, 61, 217–225.

Böhm, R. 1998. Disinfection and hygiene in the veterinary field and disinfection of animal houses and transport vehicles. Int Biodeterior Biodegradation, 41, 217–224.

Cao, W., Z.W. Zhu, Z.X. Shi, C.Y. Wang, B.M. Li, 2009. Efficiency of slightly acidic electrolyzed water for inactivation of Salmonella enteritidis and its contaminated shell eggs. Int J Food Microbiol, 130, 88–93.

Capua, I., S. Marangon, 2006. Control of avian influenza in poultry. Emerg Infect Dis,12 (9),1319–132.

Chauvin, C., I. Bouvarel, P.A. Beloeil, J.P. Orand, D. Guillemot, P. Sanders, 2005. A pharmaco-epidemiological analysis of factors associated with antimicrobial consumption level in turkey broiler flocks. Vet Res, 36 (2), 199–211.

Davies, R., M. Breslin, 2001. Environmental contamination and detection of Salmonella enterica serovar Enteritidis in laying flocks. Vet Rec, 149, 699–704.

Elassaad, F.E, L.E. Stewart, L.E. Carr, S.W. Joseph, E.T. Mallinson, G.E Berney, 1993. Decontamination of poultry transport cages. St Joseph Mich: ASAE. 90-3010.

Hao, X.X, B.M. Li, Q. Zhang, B.Z. Lin, L.P. Ge, C.Y. Wang, W. Cao, 2013. Disinfection effectiveness of slightly acidical electrolyzed water in swine barns. J Appl Microbiol, 115, 703–710.

Hricova, D., R. Stephan, C. Zweifel, 2008. Electrolyzed water and its application in the food industry. J Food Prot, 71, 1934 –1947

Huang, Y.R, Y.C. Hung, S.Y. Hsu, Y.W. Huang, D.F. Hwang, 2008. Application of electrolyzed water in the food industry. Food Control, 19, 329– 345.

Oomori, T, T. Oka, T. Inuta, Y. Arata, 2000. The efficiency of disinfection of acidic electrolyzed water in the presence of organic materials. Anal Sci, 16, 365–369.

Rosanowski, S.M., C.W. Rogers, N. Cogger, J. Benschop, M.A. Stevenson, 2012. The implementation of biosecurity practices and visitor protocols on non-commercial horse properties in New Zealand. Prev Vet Med, 107, 85–94.

Toshihiro, O., O. Takumi, I. Tooru, A. Yoji, 2000. The efficiency of disinfection of acidic electrolyzed water in the presence of organic materials. Anal Sci, 16, 365–369.

Wang, J.J., Z.H. Zhang, J.B. Li, T. Lin, Y.J. Pan, Y. Zhao, 2014. Modeling Vibrio parahaemolyticus inactivation by acidic electrolyzed water on cooked shrimp using response surface methodology. Food Control, 36, 273–27.

Experiences with Precision Livestock Farming in European Farms

Daniel Berckmans
Animal and Human Health Engineering (previous M3-BIORES)
Department of Bio Systems, KU Leuven, Heverlee-Leuven, B-3001 Belgium
Email: Daniel.berckmans@biw.kuleuven.be

Abstract

Precision Livestock Farming (PLF) is the fully automated continuous monitoring of (individual) animals by using modern information and communications technology (sensors, cameras, microphones, etc.) as part of the management system. PLF assists livestock producers through automated, continuous monitoring and improvement of the status of the animals. A number of PLF tools have been developed at laboratory levels as prototypes and as commercial products that are now available for farmers. The overall objective of a running EU-PLF project is to support farmers using PLF solutions in commercial farms and to experience/analyse how these products can improve value for the farmers in practice. The objective of this paper is to share some results of this project. Twenty farms (5 broiler, 10 fattening pig, and 5 milking cow farms) spread over Europe have been selected and PLF equipment has been installed. In these farms, sensor data, images and sound data are stored and PLF algorithms run on these data. During farm visits manual scores are rated by using the Welfare Quality Protocols to score these farms on animal welfare. The PLF systems calculated "PLF scores" based upon fully automated continuous monitoring. It is analysed whether the PLF scores do correspond to the Welfare Quality scores that were rated in parallel. Farmers have been using the technology while visits and meetings have been organised to understand their response to the use of these systems on the commercial farms. Results of these events are reported and discussed.

Keywords: Automated monitoring, information and communications technology, commercial farms, animal welfare, welfare quality protocols

1. Introduction

The size of livestock farms has been increasing rapidly in recent times and will further increase in future. The increasing income for many people in Brazil, Russia, India and China and corresponding changing diets will increase the demand for animal products in the years to come. Accepted forecasts so far expect the worldwide demand for animal products to increase with 40 to 70% by 2050 (FAO, 2013; FAO 2015). This results in larger numbers of animals worldwide and at the same time a decreasing number of farmers. The logic consequences are with larger herds, the farmer has to take care of higher numbers of animals while he cannot spend more time. The main problem is that the farmer has very small or no time left to care for each individual animal.

At the same time the larger herds raise more concerns about animal health and animal welfare. Today there are more questions about the risk of disease transfer from livestock to humans and this makes animal health a high priority. Moreover, there are significant data showing that the use of antibiotics is far too high (Aarestrup, 2012). More than 50% of all antibiotics used worldwide are used in the livestock sector. In Europe over 25,000 people are dying every year due to infections that used to be treated with antibiotics (Verstringe, 2015).

To implement more sustainable livestock production systems another serious problem is yet far from being solved: the environmental impact of current livestock systems is too high. For example, it is estimated that more than 92% of the ammonia in the environment is due to animal production (Asman and Janssen, 1987). Today in naturally ventilated animal housing there is no accurate way to measure the ventilation rate through the building and consequently no accuracy in measuring gas emissions. An error of plus and minus 400 $m^3\ h^{-1}$ is just not good enough to measure ventilation rate through naturally ventilated livestock houses. A significant portion of the livestock houses worldwide are ventilated naturally.

The European Union (EU) has invested large sums of money in the Welfare Quality® project, which aimed to develop a methodology to score animal welfare on farms (www.welfarequality.net). The EU seeks to implement this in practice (EFSA, 2012) by means of new directives. However implementing more directives onto the farmers will increase again the cost for livestock farming. The European farmers are already subject to many regulations and laws and related cost must be gained from better productivity. The cost/benefit of a yearly visit to score animal welfare must be questioned when it comes to getting this paid by someone. When farmers have to manage more animals in the same available working time they need to have high productivity to cover their costs. Consequently farmers get more problems to make their living out of their livestock business.

It has been over 20 years now the technology of monitoring animals with ICT technology is under development in laboratory conditions (Vanderstuyft et al., 1991). The European Committee for Precision Livestock Farming has just organised the 7th ECPLF2015 conference. Over 1000 scientific papers have been published in the proceedings of these two annual conferences. Most of that research was done in laboratories or in farms under laboratory conditions. Instead of developing more technology it is now time to bring this Precision Livestock Farming (PLF) approach into real farms and start testing and implementing in commercial farms.

The European Commission has started the European EU-PLF project on value creation through Precision Livestock Farming. This project allowed installing PLF technology in 20 farms spread all over Europe to make farmers using and experiencing the technology.

The objective of this paper is to share experiences from several discussions in the past years and from this European project with PLF technology in commercial pig and broiler farms. For farms of fattening pigs and broilers this is the first time that PLF technology is tested in commercial farms over Europe. The focus is discussing some very important issues and questions rather than more technical results.

2. Materials and Methods

2.1. PLF installations in commercial farms

Precision livestock farming has the objective to create a management system based on continuous automatic real-time monitoring and control of production/reproduction, animal health and welfare, and the environmental impact of livestock production. Precision livestock farming is based on the assumption that continuous direct monitoring or observation of animals will enable farmers to detect and control the health and welfare status of their animals at any given time.

Technological development and progress have advanced to such an extent that accurate, powerful and affordable tools are now available to implement this technology in commercial farms. The available technology includes cameras, microphones, sensors (such as 3D accelerometers [including gyroscopes], temperature sensors, skin conductivity sensors and glucose sensors), wireless communication tools, Internet connections and cloud storage. Modern technology makes it possible to place cameras, microphones and sensors sufficiently close that they can replace the farmers' eyes and ears in monitoring individual animals.

The aim of PLF is to combine all the available hardware with intelligent software in order to extract information from a wide range of data. Precision livestock farming can offer a management tool that enables a farmer to monitor animals automatically and to create added value by helping to secure improved health, welfare, yields and environmental impact.

In 10 fattening pig farms and 5 broiler farms the eYenamic (Fancom BV Netherlands) camera system eYenamic), Figures 1 and 2, and the Soundtalks (Soundtalks NV, Belgium) sound analysis system was installed. The farms were spread over The farms are spread over different counties and climate conditions (Spain, Italy, the Netherlands, France, Belgium, Sweden, Hungary, Denmark, Ireland, Germany, Israel) in different types of livestock houses with different ways of managing the animals. This set up in commercial livestock houses allows collecting data from 60 productions cycles during the project and this or each of the considered species.

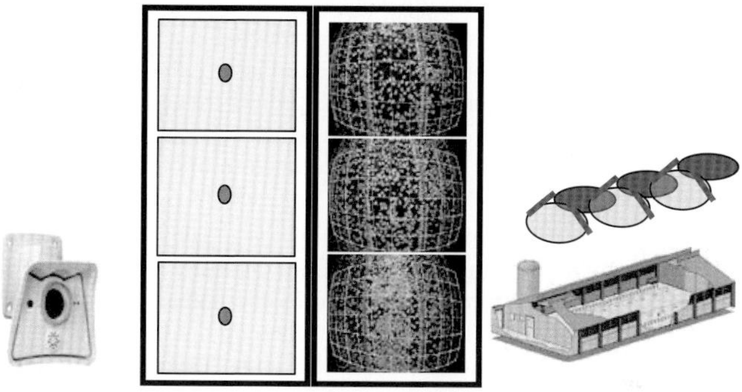

Figure 1. The eYenamic camera system for monitoring broiler behaviour.

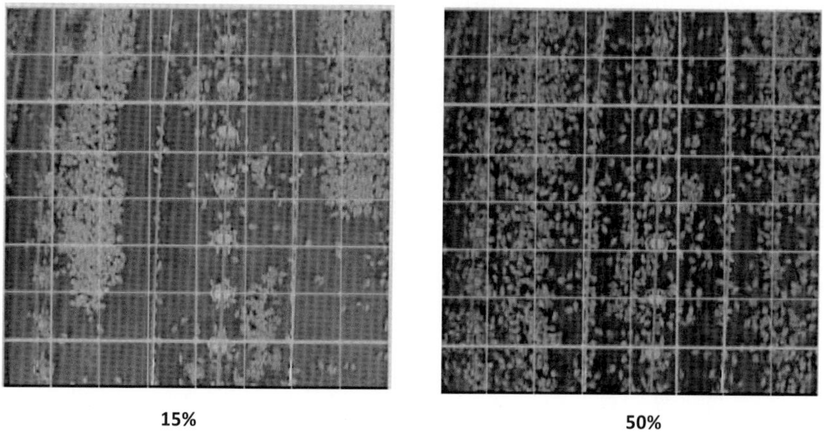

Figure 2. Images used by real time eYenamic to calculate animal activity, zone occupation index and animal distribution.

The advantage of image and sound technology is that the animals are not touched and no possible influence on the animal is induced since there is no contact. This is what we will focus on in this paper.

It has been described in literature that sound analysis of animals offers high potential to monitor different aspects of the status of livestock (Van Hirtum and Berckmans, 2002; Guarino et al., 2008; Vandermeulen et al., 2013). Algorithms to distinguish between pig coughs and other sounds in the environment have already been developed and validated in laboratory environment. Initially, these algorithms were able to distinguish between sounds, then between pig coughs and other sounds and then between pathological and non-pathological coughs. To bring the technology to real farms the focus on simpler target variables is required and validation must be carried out in real livestock houses.

Literature shows that image analysis on livestock can disclose a lot of (real time) information of individual and interacting animals (Van Der Stuyft, 1991; Kashiha et al., 2013 and 2014). In the actual settings the eYenamic camera system is calculating 2 feature variables (Berckmans, 2013): activity of the birds and distribution of the birds.

3. Results and Discussion

3.1. Agreement of PLF technology with reference measurements

Validation of the sound algorithms in the field, conducted in a piggery in Lombardy (Italy), has shown that the algorithms developed were able to classify the cough correctly in 86% of cases (Guarino et al., 2004). In the running EU-PLF project trained assessors were visiting the pig houses and counted the number of coughs. In livestock houses in the Netherlands, Hungary, France and Spain a total number 447 coughs was counted in a total of 4 livestock houses in 6 different measuring periods spread over all seasons. For each dataset this number of coughs was compared with the number of coughs resulting from the PLF technology.

It is shown that the number of coughs resulting from the PLF technology by Soundtalks is in good agreement with the number of coughs counted by the human assessor (Berckmans et al., 2015). Moreover these authors show that the PLF technology is faster in detecting respiratory problems than the farmer which is normal since the sound technology is monitoring 20,000 samples continuously, analysing and giving early warnings continuously. The PLF technology can detect problems up to 10 days before the farmer has noticed them while some problems were never noticed by the farmer.

It is shown that the output of real time image analysis on pigs correlates well with the human observations of behavioural activities (Ott et al., 2014). The results from the real time image analysis show that the eYenamic system can detect 95.3% of all problems that were experienced and noted by the farmer (See Figure 3) (Kashiha et al., 2013).

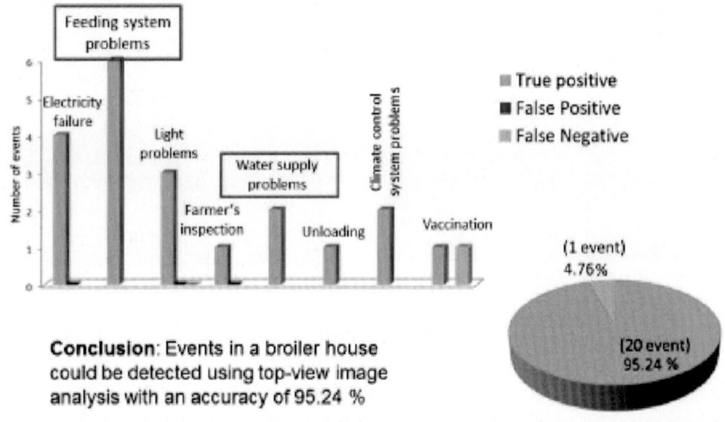

Figure 3. Results from the eYenamic algorithm to detect problems during a broiler fattening period of over 42 days.

In general it is seen now that not only in laboratory conditions but also in commercial livestock houses both the PLF image and sound technology can deliver results of target variables that show good agreement with reference measurements of assessors scores.

Another interesting finding is that again both the real time sound analysis and the image analyses allow giving early warnings to the farmer. For sound this can go up to 10 days before the event or problem is noticed by the farmer. This is not a surprise since the PLF technology allows monitoring the animals 24 hours a day and 7 days a week. In this way the continuous monitoring by cameras, microphones and/or sensors delivers a basis for immediate actions and improvement on the (individual) animals when they need attention or help (Berckmans, 2006). This is very different from the time a farmer can spend to monitor his animals. Let's assume that a farmer would be able to spend 4 hours a day (!) in monitoring his animals. A broiler farmer with 120,000 birds could then spend 0.12 second per animal per day. A pig farmer with 5,000 pigs could spend 2.9 seconds per pig and a cow farmer with 1,000 cows could spend 14 seconds per animal per day!

From several workshops with farmers we could hear that these PLF systems in practice have several ways to create value. A first one is the time saving issue resulting from the continuous automated monitoring. The PLF systems are replacing the eyes and ears of the farmer. As a result the farmer can focus now on solving problems and not too much on controlling his animals. A second advantage was described as the impact on social life since now the farmer can participate more in the activity of the community since he is not continuously needed at his farm: the technology will give an early warning or alarm when needed. A third point is the social recognition for the farmer. He wants to show the consumers and the public how he is treating his animals. No need to enter his farm with hidden cameras since all information is available and used continuously for management decisions.

4. Conclusions

A lot of PLF technology has been developed for different species over the past years. The potential to use ICT technology, sensors, cameras and microphones is steadily increasing. Most PLF technology however was developed and tested in laboratory conditions. Instead of developing more technology it is time to test the accuracy and results of this technology in commercial farms. From several publications and results in the running EU-PLF project we can conclude that both the image analysis and sound analysis allow generating results that agree with reference measurements and scores of assessors in commercial fattening pig farms and broiler farms. An important advantage of the PLF technology is the continuous automated monitoring 24 h/day and 7 days/week. It was experienced that this allows giving early warnings up to 10 days before events for the sound technology.

The running EU PLF project will deliver a Blueprint for stakeholders, with a focus on farmers, on how to implement the PLF systems in farms and which problems can be expected. This PLF-Blueprint is like a manual for farmers and companies who want to bring this technology to the farm. The format will be an e-course on a website that is free and available for everybody. Moreover the e-course will has a second level for researchers and other stakeholders interested in the PLF technology.

The PLF and livestock sectors are a matter that deserves the interest of many more stakeholders than only farmers. Farmers organisations, veterinarians, technology providers, feeder companies, breeding companies, slaughter houses, retailers, consumers, the general public, regional and national governments, politicians, the press, schools from first to highest degree and last but not least researchers should contribute to the debate on how the livestock sector should evolve.

At this moment one of the most important issues to be considered is the workable business model for the PLF technology. This includes the question on who owns the data and who owns the information from the data. Since value is created by combining different types of expertise from different stakeholders, the business model should reflect the contribution for values at different levels. This means that the technology, including the requited platform for operational implementation, of the PLF technology in the food chain will be a new income for farmers rather than a cost since it is all done based upon their data. Farmers and farmers' organizations must be well aware of this!

Acknowledgements

The author gratefully acknowledges all his partners in the project and the European Community for financial participation in Collaborative Project EU-PLF KBBE.2012.1.1-02-311825 under the Seventh Framework Programme.

References

Aarestrup, F.m 2012. Sustainable farming: Get pigs off antibiotics. Nature. 486 (7404): 465–66. doi:10.1038/486465a.

Asman W.H. and A. Janssen, 1987. A long-range transport model for ammonia and ammonium for Europe. Atmospheric Environment. 21 (10), 2099–2119.

Berckmans D., 2013. Basic principles of PLF: Gold standard, labelling and field data. In *Proceedings of the 6th European Conference on Precision Livestock Farming, ECPLF2013*, Leuven, 21-29.

Berckmans D., 2006. Automatic on-line monitoring of animals by precision livestock farming. In *Livestock Production and Society*, edited by R. Geers and F. Madec, 51–54. Wageningen Academic Publishers.

Berckmans D., 2015. Animal sound …talks! Real-time sound analysis for health monitoring in livestock. In *Proceedings of the 2015 International Symposium on Animal Environment and Welfare*, Chongqing, China October 23-26, 2015, 11 pages.

EU-PLF, 2015: European Collaborative Project EU-PLF KBBE.2012.1.1-02-311825 under the Seventh Framework Programme.

European Food Safety Authority (EFSA), 2012. Technical meeting on animal welfare: workshop on the use of animal-based measures for dairy cows, pigs and broilers, 4–5 July, Parma (Italy). EFSA, Parma.

FAO. 2013. *FAO Statistical Yearbook* 2013: World Food and Agriculture. Rome.

Guarino, M., P. Jans, A. Costa, J.-M. Aerts, and D. Berckmans, 2008. Field test of algorithm for automatic cough detection in pig houses. Computers and Electronics in Agriculture. 62 (1): 22–28. doi:10.1016/j.compag.2007.08.016.

Kashiha A., C. Bahr, S. Ott, C. Moons, T. Niewold, F. Tuyttens, and D. Berckmans, 2014. Automatic monitoring of pig locomotion using image analysis. Livestock Science, 159, 141-148.

FAO. 2015 Mottet A., 2015. Keynote speech at the 7'h European Conference on Precision Livestock Farming, Milan, September 15-18, 2015. FAO 2015.

Ott S., C. Moons, M. Kashiha, C. Bahr, F. Tuyttens, D. Berckmans, and T. Niewold, 2014. Automatic video analysis of pig activity at pen level highly correlates to human observations of behavioural activities. Journal of Livestocok Science. 160, 132-137.

Vandermeulen, J-W-D., D. Berckmans, V. Exadaktylos, C. Bahr, and D. Berckmans. 2013. The pig cough monitor: From research topic to commercial product. In *Proceedings of Precision Livestock Farming '13*, 717–23. Leuven. https://lirias.kuleuven.be/handle/123456789/418316.

VanderStuyft E., C.P. Schofield, j.M. Randall, P. Wambacq, and V. Goedseels, 1991. Development and application of computer vision systems for use in livestock production. Computer Electronic Agriculture. 6, 243-265.

Van Hirtum A., and D. Berckmans, 2002. Assessing the sound of cough towards vocality. Medical Engineering and Physics. 24(7-8):535-540.

Verstringe S., 2015. Keynote speech at the 7th European Conference on Precision Livestock Farming, Milan, September 15-18, 2015. FAO 2015.

Theme VI:

Animal Production Systems, Equipment, and Techniques

Coalition for a Sustainable Egg Production Project — A Holistic Approach to Address Egg Production Systems

Darrin M. Karcher

Department of Animal Science, Michigan State University, East Lansing, MI 48824, USA

Email: dkarcher@msu.edu

Abstract

Food production systems are continuing to be scrutinized demanding an increased level of accountability related to sustainable production. The definition of sustainability is broad and can be defined differently for each individual. At the same time, efforts are being lead to regulate or eliminate specific types of animal housing resulting in a dictation of how to produce the food source. However, there has been limited research attention to explore the sustainability of the food system, specifically egg production. Therefore, the Coalition for a Sustainable Egg Supply was formed with the intent to investigate the sustainability of three different housing systems: conventional cage, enriched colony cage, and cage-free aviary. Five different sustainable areas: animal health and well-being, food safety and quality, environment, worker health and safety, and food affordability, would be assessed at a commercial level on the same flocks at the same time. Michigan State University and University of California, Davis scientists lead the research with additional scientists from Iowa State University and the USDA-ARS Egg Safety and Quality Research Unit. The first research flock was placed in April 2011 with the second of the two-flock study housed in July 2012. A comprehensive final report was released in March 2015. The Coalition's work resulted in science-based data that will provide guidance to the future of egg production and purchasing decisions within the U.S.

Keywords: Laying hens, environment, well-being, food safety, housing system

1. Introduction

The Coalition for a Sustainable Egg Supply (CSES) was comprised of a multi-stakeholder group encompassing egg suppliers, food manufacturers, research institutions, non-governmental organizations, restaurant, foodservice, and retail food companies who collaborated on a commercial-scale study of housing alternatives for U.S. egg-laying hens. In order to sustain the U.S. egg supply, the coalition members believed a balanced and objective evaluation of egg production needed to be investigated between alternative housing systems. Three different egg-laying systems (conventional cage, enriched colony cage, and cage-free aviary) were evaluated from five different sustainability areas: behavior and well-being, environmental, food safety, worker health and safety, and food affordability. While some research within respective areas suggests benefits to one type of housing system, the commercial-scale research evaluated all areas simultaneously allowing for trade-offs within an area to be identified. Additionally, Swanson et al. (2015) discussed the background on the formation of the CSES project keeping in mind the biggest caveat: The data collected is a snapshot in time relative to the commercial sites familiarity with these new types of housing systems.

2. Materials and Methods

Several publications report in detail the various components of the project. The methods employed for data collection were consistent across the two different flocks. Each sustainability area will be outlined with specific references to the publication for more information.

2.1. Housing and Management

Karcher et al. (2015) and Zhao et al. (2015a) discussed the housing and management employed in the CSES. Three different hen housing systems were evaluated: conventional cage (CC), enriched colony cage (EC), and a cage-free aviary (AV). Beyond the CC, the other two housing systems had enrichments available to the hens. The EC had nest area, perches, and scratch pad while the AV had nest area, perches, and litter (floor) access. Two different commercial flocks were monitored with the initial flock placed in April 2011 and the second flock housed in July 2012 with hens being depopulated at 78 weeks of age. The Lohmann LSL Classic White, a white Leghorn laying hen, was used in all three housing systems in the two different flocks.

The pullets were appropriately reared in an aviary rearing system for the AV and pullet conventional cage rearing for the CC and EC. The AV and EC commercial houses were constructed for the project with the CC was a preexisting house on the laying hen site, which was located in the Midwest, USA. Therefore, the CC had a population of 200,000 hens while the AV and EC had 50,000 hens each. The birds per housing unit were: CC had six hens per unit at 516 cm^2, EC had 60 hens per housing unit at 748 cm^2, and AV had hens at 852 (outer row) and 1,704 (inner row) per colony unit with 929 cm^2 per bird.

The hens nutritional requirements were formulated by the commercial company's nutritionist with hens fed to maximize the production efficiency. The feeding schedule consisted of two feedings a day for the CC and EC and five feedings for the AV. Additionally, the scratch auger delivered feed to the scratch pad in the EC system with each feeding. The manure was removed every 3–4 days via belts with the exception of the AV system, which had the litter that was removed at the end of the lay cycle.

2.2. Animal health and well-being

The sustainability area can be broken down into hen physical condition and health, resource use within the housing system, and physiological stress. The physical condition of 100 hens from each housing system was evaluated using the Welfare Quality Assessment protocol for poultry (2009). The hens were evaluated for 14 different traits, e.g., keel abnormalities, plumage cover, foot health, at peak, middle, and end of lay (Blatchford et. al, 2015). At the beginning and end of lay, 120 hens per housing unit were sacrificed and keel, tibiae, and humeri were extracted for bone quality assessment. Additionally, blood was collected from the hens prior to euthanasia to assess different bone markers as an indicator of overall skeletal health (Regmi et al., 2015). Throughout the production cycle, daily mortality was necropsied twice per week from each housing system with a quarter of the daily mortality from the CC to account for the different house populations. Necropsy findings were classified as pullet placement and housing type, behavioral issues, skeletal, egg production, and infection.

The resource use was assessed at peak, middle, and end of lay in the AV and EC using video and in-person observations. The nest area use, scratch pad, perch use, and litter use were evaluated. Additionally, hen flight and landing successes in the AV were quantified. The methodological approaches can be found in Campbell et al., (2015a,b).

The physiological stress was assessed through heterophil to lymphocyte (H:L) ratios at pullet placement, peak, middle, and end of lay. Adrenal glands were taken at placement and end of lay to assess long-term stress. Additionally, when blood was collected for H:L ratio assessment, hematocrits were evaluated for bird dehydration.

2.3. Food safety and quality

The egg quality samples were collected monthly and egg weight, shell dynamic stiffness (shell acoustical characteristics), static compression shell strength, Haugh unit/albumen height, yolk index, vitelline membrane strength and deformation (elasticity), and whole egg total solids were assessed (Karcher et al., 2015). Once a quarter eggs were washed and stored at 4°C for 12 weeks with egg quality assessed at 28, 42, and 84 days of storage (Jones et al., 2014).

There were 20 sampling points were identified as replicates in each of the housing systems. During each of four sample times during the second flock environmental swabs and egg shell pools (3–6 shells each) were collected from each of the replicates. The indicator populations of total aerobic organisms and coliforms were enumerated for up to 10 swabs/shell pools for each sample type by housing system combination each collection period. The prevalence (presence or absence) of *Salmonella* spp. and *Campylobacter* spp. was determined for every swab/shell pool collected. Manure scraper swabs were only assessed for pathogen prevalence and were denoted per housing system proper and not assigned to unique sampling replicates.

Twenty laying hens per housing system were sampled monthly over the course of the flock cycle. Serum was collected, processed, and analyzed for immunoglobulin G while crop lavage samples were collected and analyzed for immunoglobulin A.

2.4. Environment

Continuous environmental monitoring of all systems and their manure storage facility was carried out over both flock cycles using two state-of-the-art monitoring systems (a photoacoustic multi-gas analyzer and Tapered Element Oscillating Microbalances equipped with particulate matter, or PM, head). The areas monitored were indoor thermal environment (temperature and relative humidity – RH), ventilation rate, air quality (NH_3, CO_2, CH_4, N_2O, PM_{10}, $PM_{2.5}$). In addition, feed samples, eggs, and laying hens were taken periodically to determine the nutrient composition. The collected data and nutrient composition was used for emissions modeling and nutrient mass balance calculations. Moreover, properties of the hen manure removed from the houses and the storage were monitored. For detailed description of the monitoring systems refer to Zhao et al. (2014; 2015a,b) and Shepherd et al. (2015).

2.5. Worker health and safety

Workers' respiratory health and exposure to air pollutants in the three housing systems was assessed. Each worker wore a backpack holding exposure monitors during his or her work shift. The exposure monitors sampled ammonia, particles of all sizes that can be taken in through the nose (referred to as inhalable), and smaller particles (referred to as PM2.5), which can travel deep into the lungs. Workers were assigned to work an equal number of days in a random pattern in each of the three barns for 13 to 15 days in each of three seasons: summer, winter and spring. A total of 124 sampling day air samples were collected. Workers' lung function and breathing symptoms before and after work were monitored on the same days that personal air sampling was conducted. From these measurements, changes across a shift in workers' respiratory symptoms and pulmonary function were recorded. Further details can be found in Mitchell et al. (2015) and Arteaga et al. (2015). The ergonomic challenges were evaluated by classifying tasks into three categories indicating the level of risk to the body position. Three main stressors (force, repetition and posture) were also assessed.

2.6. Food affordability

The egg cost of production data represent the first on-farm cost comparisons across three housing systems on a single farm, at the same location and employing the same accounting definitions and consistent cost measurement for each housing system. Farm managers provided specific cost of production data measured weekly, biweekly and monthly during commercial operations over the two flock cycles. The economic data provided were used to compare feed costs, labor costs, pullet costs, calculated energy costs, capital costs and miscellaneous operating costs and the sum of all available costs across the three housing systems (Matthews and Sumner, 2015).

3. Results and Discussion

The information in results and discussion are collected from the Coalition for Sustainable Egg Supply Final Research Results Report (2015). Additionally, within each section references to published data are included.

3.1. Laying hen performance

Production performance is both an indicator of overall flock health and a major component of economic and environmental sustainability. Daily feed usage, water consumption, hen-day egg production, and mortality data were collected during each 28-day period of each flock cycle; a summary is provided in Table 1.

Table 1. Production summary for two commercial flocks housed in different housing environments.

Production Parameter	Conventional Cage (CC)	Cage-free Aviary (AV)	Enriched Colony (EC)	Lohmann Management Guide Ref.
Egg per hen housed (Flock 1)	352	340	363	354.2
Egg per hen housed (Flock 2)	371	345	382	354.2
Average Hen-day Prod. (%, Flock 1)	87.3	86.6	90.5	86.8
Average Hen-day Prod. (%, Flock 2)	90.0	88.0	94.3	86.8
Water use, L/100 hen-day (Flock 1)	1.54	1.27	1.36	-
Water use, L/100 hen-day (Flock 2)	1.53	1.29	1.33	-
Water/Feed, kg/kg (Flock 1)	2.06	1.64	1.73	-
Water/Feed, kg/kg (Flock 2)	2.05	1.74	1.76	-
FC, kg/dozen eggs (Flock 1)	1.44	1.49	1.42	-
FC, kg/dozen eggs (Flock 2)	1.40	1.44	1.38	-
FC, kg feed/kg egg (Flock 1)	2.02	2.12	1.99	2.0–2.1
FC, kg feed/kg egg (Flock 2)	1.96	2.04	1.94	2.0–2.1
78-wk body weight, kg (Flock 1)	1.56	1.53	1.55	1.71–1.86
78-wk body weight, kg (Flock 2)	1.67	1.60	1.59	1.71–1.86

The hen-day percentage production was above the calculated 86% management guide average from 19–78 weeks for hens in all three housing systems in both flocks. Hen-day production at 78 weeks of age was higher than the Lohmann LSL White management guide (2013) target of 77% for both flocks. Flock 2 overall performed better than Flock 1, showing an increase in eggs per hen housed, average hen-day production, and better feed conversion. The Lohmann LSL management guide (2013) reports that cumulative mortality should be around 6% for the flock; mortality in both the CC and EC was slightly lower than this in both flocks, but mortality in the AV was double that figure. Mortality in both the AV and CC were about the same in Flock 2 as they were in Flock 1. In the EC, however, mortality decreased from Flock 1 to Flock 2, resulting in EC mortality being similar to that of CC (Karcher et al., 2015).

3.2. Animal health and well-being

Cumulative hen mortality in the EC and CC was slightly lower than the 6% Lohmann LSL management reference, but double that percentage in the AV. Major mortality causes in all systems were hypocalcemia and egg yolk peritonitis. More AV hens died from being caught in the structure, vent cannibalized or excessively pecked. The EC and AV systems offered hens more behavioral freedom than CC with the nesting area and perches generally well used (Blatchford et al., 2015; Campbell et al., 2015a,b). In the AV the litter was used for dust bathing, but the EC scratch pad was not well used for dust bathing or foraging and accumulated manure. Nest use by AV hens was variable, with a proportion of eggs laid in the enclosure or on the litter.

Bone health/strength measures indicate EC hens had more keel abnormalities than CC hens, particularly during late lay. The AV reared pullets had more keel bone damage at placement than those reared in CC, and keel breaks were more prevalent in the AV hens during lay. Pullets in the AV rearing system had better bone quality at placement in their tibiae and femurs than pullets reared in the CC rearing system; this good bone quality was maintained throughout the lay cycle. Bone quality in CC and EC was not as good, although it improved somewhat in EC during the lay cycle (Regmi et al., 2015).

Measures of stress overall did not indicate acute or chronic stress. The EC hens had slightly less feather loss than CC hens, while the AV hens had the best feathering. Feather cleanliness of EC and CC was similar, but AV hens had slightly dirtier feathers. The EC hens had shorter claws and fewer foot problems (e.g., hyperkeratosis) than in CC, and no severe foot problems (e.g., bumblefoot). Incidence of foot problems in AV was lowest, but those problems were more severe. Air temperatures in AV and EC were similar to CC, and the hens were never observed panting. Indoor air quality (PM and ammonia) for EC was similar to CC but worse in AV. However, there were no signs of hen health problems associated with poor air quality in any housing system. Feed and water consumption by hens and body weights were similar across systems (Blatchford et al., 2015; Karcher et al., 2015).

3.3. Food safety and quality

Housing system type did not influence the rate of egg quality decline through 12 weeks of extended storage, and current U.S. egg quality standards/grades are adequate to describe eggs for all three of the housing systems (Jones et al., 2014). It is not uncommon for poultry to shed Salmonella spp. or other coliforms thus the prevalence (presence or absence) of Salmonella spp. and Campylobacter spp. were determined for every swab/egg shell pool collected from each system. Jones et al. (2015) reported that hens in all housing systems were shedding Salmonella spp. at a similar rate; the prevalence of Salmonella spp. on egg shells was very low and did not differ between housing systems. The AV had higher levels of environmental Campylobacter spp. recovery (drag swab). Salmonella spp. was detected at similar levels of prevalence in the EC and CC production environments however AV were more positive. The manure scraper had low levels of Campylobacter spp. recovery in all systems, but AV drag swabs and EC scratch pad swabs had high levels of Campylobacter spp. recovery. AV floor shells had the greatest levels of total aerobes and coliforms. Aerobic organisms were also elevated on AV nest box and system shells. Previous studies indicate total aerobe levels are greater on eggs produced in high dust environments. Eggs laid on litter (in AV only) have greater shell microbial levels than eggs laid on system wires or in nest boxes. Coliforms are indicators of fecal contamination which is linked to many human pathogens. In the EC system wire egg shell coliform levels were detected at levels similar to CC. The coliform level in AV nest box egg shells was similar to the EC. The coliform levels were low for all shell samples, excluding the AV floor shells, which had the highest levels of total coliforms.

In each flock, 20 laying hens per housing system were sampled monthly over the course of the flock cycle. Serum and crop lavage samples were collected and analyzed allowing for the measurement of antibodies to *Salmonella Enteritidis* given in vaccines and to any Salmonellas that are present in the housing environment. While there were seasonal differences evident, no differences between housing systems were detected.

3.4. Environment

Zhao et al. (2014; 2015b) and Shepherd et al. (2015) reported the environmental measures across the different housing systems. Briefly, ammonia and PM concentrations were significantly higher in the AV house than in the EC or CC house. The PM concentrations were roughly 8–10 times higher in the AV than either the CC or EC. The PM emissions from the EC and CC house remained low and similar year round, whereas the AV house had 6–7 times more PM emissions than the other two types of housing. The higher AV PM levels and emissions were caused by hens' behavioral activities on the litter floor. Poor indoor air quality may lead to eye and respiratory tract irritation in workers and hens. Farm-level ammonia emissions were lowest for the EC system, approximately half that of CC or AV, due to its lower stocking density and drier manure. Ammonia emissions from manure storage accounted for two-thirds of farm-level emissions.

Greenhouse gas (GHG) emissions, including methane, carbon dioxide, and nitrogen dioxide, were low for all systems due to relatively dry manure. Manure removed from the EC house was drier and had a slightly higher nitrogen content than that removed from the CC or AV house. In the AV house, 77% of manure was deposited on the belts and the rest on the litter floor when hens had free access to the litter floor. Manure on the AV litter floor had to be removed separately, either mechanically or manually onto the manure belt. With respect to natural resource use, the EC house had similar energy use and feed efficiency to the CC house. The AV house may need supplemental heat during cold days, and when coupled with lower AV feed efficiency, creates a larger carbon footprint than EC or CC, as feed supply accounts for approximately 80% of total carbon footprint in the egg-supply chain. In addition, more natural resources are needed per bird space in the construction of AV houses.

3.5. Worker health and safety

Airborne PM can make its way into workers' airways, with smaller particles being deposited deep into the lungs. Mitchell et al. (2015) and Arteaga et al. (2015) reported the worker respiratory health and exposures as observed in the different housing environments. Essentially in the EC and CC houses, workers were exposed to significantly lower concentrations of airborne particles than when working in the AV house. Inhalable particle and PM 2.5 concentrations were higher in AV house due to the litter on the floor. The overall daily mean indoor ammonia concentration was well below the recommended limit of 25 ppm for the CC (4.0 ppm), EC (2.8 ppm) and AV (6.7 ppm). Ammonia concentrations only exceeded 25 ppm in the winter of Flock 1 in each house but for less than 10% of the work shift. In the AV there was worker exposure to significantly higher concentrations of endotoxin than in CC or EC. High use of mask/respirator by workers, and similar concentrations of exposures in both CC and EC, was associated with similar cross-shift lung health outcomes. Average mask use was higher by workers in the AV protecting them from higher exposures and greater respiratory consequences.

Ergonomic stressors assessed included force, repetition, and posture. Loading and unloading of cages in EC and CC systems required extreme body positions and significant twisting. Gathering floor eggs in AV required extreme body positions for extended periods and exposure to respiratory hazards. With respect to access, EC and CC systems posed significant hazards normally and at population/depopulation. The EC workers stepped on the cage front instead of ladders to reach hens and worked from unapproved platforms and railings. There were no access issues in the AV. During unloading, the cage modules were placed in the aisles blocking them in the event a rapid evacuation was needed, and AV workers placed themselves inside the wire enclosures and locked the doors behind them, reducing the ability for a rapid evacuation.

3.6. Food affordability

Matthews and Sumner (2015) described the assumptions for calculations and the differences between housing systems that may impact the cost of production. Feed for hens comprised the largest share of operating costs for each of the housing systems. Feed consumption per dozen eggs was similar across the systems, increasing somewhat over the life of the flock. Feed cost per dozen eggs produced in the AV was higher because production per hen placed in that system declined more over the life of the flock. The cost per dozen eggs for pullets placed in the AV were substantially higher than the other systems, due to higher rearing costs, higher hen mortality and lower production per hen in that system. The EC had higher weekly labor costs (per dozen eggs) than did the CC, though costs did not rise over the life of the flock as they did with the AV. An EC with more hens per house might be more efficient and reduce labor costs per dozen eggs produced. The labor costs per dozen eggs produced were highest in the AV, primarily due to greater labor costs for egg collection. Higher hen mortality and other hen health issues were also contributing factors. The EC had total capital costs per dozen eggs that were 107% higher than CC, largely the result of higher construction costs and fewer hens housed in comparison to CC. The AV had total capital costs per dozen eggs that were 179% higher than

CC, largely the result of higher construction costs and fewer hens housed in comparison to CC. The EC had total operating costs per dozen eggs that were 4% higher than CC. Coupled with higher capital costs, EC had total costs per dozen eggs produced that were 13% higher than CC. The AV had total operating costs per dozen eggs that were 23% higher than the CC. Coupled with higher capital costs, the AV had total costs per dozen eggs produced that were 36% higher than the CC.

4. Conclusions

The CSES project was a unified effort of various stakeholders to invest in research evaluating the trade-offs that exist in different laying hen housing systems. The sum total of the results does not give a clear indication of which housing system is the best. This is due to two main constraints: (1) The research is a snapshot in time of the production systems attempting to control as many variables as conceivable in a commercial facility and (2) Individual and business values are different with respect to sustainability. Therefore, the research will provide information for each food system investor to decide what type of housing system fits their definition of sustainability. Overall, this research was a stepping-stone into a new realm of thinking related to conducting research at a commercial scale.

Acknowledgements

This research would not have been possible without the numerous individuals involved: Drs. Joy Mench, Nicholas Kenyon, Frank Mitloehner, Marc Schenker, Daniel Sumner, Cassandra Tucker, Ruihong Zhang, Janice Swanson, R.M. “Mick” Fulton, Roger Haut, Michael Orth, Janice Siegford, Juan Pedro Steibel, Hongwei Xin, Deana Jones, and Zaid Abdo.

References

Arteaga, V., D. Mitchell, T. Armitage, D. Tancredi, M. Schenker and F. Mitloehner. 2015. Cage versus noncage laying-hen housings: respiratory exposures. Journal of Agromedicine. 20 (3), 245–255. http://dx.doi.org/10.1080/1059924X.2015.1044681

Blatchford, R. A., R. M. Fulton, and J. A. Mench. 2015. The utilization of the Welfare Quality® assessment for determining laying hen condition across three housing systems. Poultry Science. http://dx.doi.org/10.3382/ps/pev227

Campbell D. L. M., M. M. Makagon, J. C. Swanson, J. M. Siegford. 2015a. Laying hen movement in a commercial aviary: Enclosure to floor and back again. *Poult Sci.* July 2015:pev186. doi:10.3382/ps/pev186.

Campbell D. L. M., M. M. Makagon, J. C. Swanson, J. M. Siegford. 2015b. Litter use by laying hens in a commercial aviary: dust bathing and piling. *Poultry Science*. September 2015:pev186doi:10.3382/ps/pev183.

Coalition for Sustainable Egg Supply. 2015. Final research results publication. The Center for Food Integrity, Gladstone, Missouri. March.

Jones, D. R., D. M. Karcher, and Z. Abdo. 2014. Effect of a commercial housing system on egg quality during extended storage. Poultry Science. 93, 1282–1288.

Jones, D. R., N. A. Cox, J. Guard, P. J. Fedorka-Cray, R. J. Buhr, R. K. Gast, Z. Abdo, L. L. Rigsby, J. R. Plumblee, D. M. Karcher, C. I. Robison, R. A. Blatchford, and M. M. Makagon. 2015. Microbial impact of three commercial laying hen housing systems. Poultry Science. 94, 544–551.

Karcher, D. M., D.R. Jones, Z. Abdo, Y. Zhao, T.A. Shepherd, and H. Xin. 2015. Impact of commercial housing systems and nutrient and energy intake on laying hen performance and egg quality parameters. Poultry Science. 94, 485–501.

Lohmann Tierzucht. Lohmann LSL Classic Layers management Guide. http://www.ltz.de/de-wAssets/docs/management-guides/en/LTZ-Management-Guide-LOHMANN-LSL-Classic-ENGLISH-2013.pdf. Accessed March 4, 2015.

Matthews, W.A. and D.A. Sumner. 2015. Effects of housing system on the costs of commercial egg production. 2015. Poultry Science. 94, 52–57.

Mitchell, D., V. Arteaga, T. Armitage, F. Mitloehner, D. Tancredi, N. Kenyon and M. Schenker. 2015. Cage versus noncage laying-hen housings: worker respiratory health. Journal of Agromedicine. 20 (3), 256–264. http://dx.doi.org/10.1080/1059924X.2015.1042177

Regmi,P., T.S. Deland, J.P. Steibel, C.I. Robison, R.C. Haut, M.W. Orth and D.M. Karcher. 2015. Effect of rearing environment on bone growth in pullets. Poultry Science. 94, 502–511.

Shepherd, T., Y. Zhao, H. Li, J. P. Stinn, M. D. Hayes, and H. Xin. 2015. Environmental assessment of three egg production systems – Part II: Ammonia, greenhouse gas, and particulate matter emissions. Poultry Science 94, 534–543.

Swanson, J.C., J.A. Mench and D.M. Karcher. 2015. Editor's Choice: The Coalition for Sustainable Egg Supply project: An Introduction. Poultry Science. 94, 473–474.

Welfare Quality. Welfare Quality® assessment protocol for poultry (broilers, laying hens). Welfare Quality® Consortium, Lalystad, Netherlands. 2009.

Zhao, Y., H. Xin, T. A. Shepherd, and H. Li. 2014. Concentrations of ammonia, greenhouse gases and particulate matters in conventional cage, aviary, and enriched colony laying-hen houses. Proc. 2014 ASABE Annual International Meeting, Montreal, Canada.

Zhao, Y., T. A. Shepherd, J. Swanson, J. A. Mench, D. M. Karcher, and H. Xin. 2015a. Comparative evaluation of three egg production systems: Housing characteristics and management practices. Poultry Science. 94, 475–484.

Zhao, Y., T. Shepherd, H. Li, and H. Xin. 2015b. Environmental assessment of three egg production systems – Part I: Monitoring system and indoor air quality. Poultry Science. 94, 518–533.

An Automated Tracking and Monitoring System for Laying-Hen Behavioral Research in an Enriched Colony System

Hongwei Xin [a,*], Yang Zhao [a], Wilco Verhoijsen [a,b], Lihua Li [a,c]

[a] Department of Agricultural and Biosystems Engineering, Iowa State University, Ames, Iowa 50011, USA

[b] Farm Technology Group, Wageningen University, Wageningen, The Netherlands

[c] College of Mechanical and Electrical Engineering, Agricultural University of Hebei, Baoding, Hebei, China

* Corresponding author. Email: hxin@iastate.edu

Abstract

Alternative housing systems for egg production are emerging and increasingly adopted by egg producers in Europe and North America. Enriched colony and aviary housing systems are among those of consideration and adoption. Along with shift to the new housing systems comes the need for specification of the new design parameters that will accommodate the behavioral needs of hens and at the same time maximize the resource utilization and production efficiencies. However, information is seriously lacking regarding the behavioral and production responses of hens to resources allocation in the housing systems, such as feeder space, nest area, and perches. To enable the much-needed research for establishing these system design parameters, a new laboratory equipped with automated tracking and monitoring capabilities for an enriched colony system is being developed at Iowa State University. The research facility features continuous monitoring of water and feed use, perch use, nestbox use, time spent at and frequency of visit to feeders by individual hens in a group-housed setting, and the hens feeding simultaneously. The automated instrumentation system employs ultra-high-frequency radio frequency identification (UHF RFID) sensing, load-cell weighing, video imaging, special engineering assembly, and the associated operation control and data acquisition systems. The first experiment underway that is using the facility is to determine the behavioral responses of laying hens to feeder space – an important design parameter that has received much attention among both producers and animal welfare certification organizations. This paper describes the design and development of the research facility with emphasis on the automated behavioral monitoring system.

Keywords: Alternative hen-housing, egg production, animal welfare, high-frequency RFID

1. Introduction

Concerns and movements toward improving welfare/well-being of production animals have led to mandatory or voluntary shift, to various degrees, from conventional cage to alternative housing systems in Europe and North America. Examples include shift from individual stalls to group-pen housing for gestating sows, banning of conventional cages by the European Union as of January 1, 2012, and implementation of California Proposition 2 as January 11, 2015. For egg production, the systems that are emerging and increasingly adopted in Europe and the United States include enriched colony house (ECH) and cage-free multi-level aviary (AV) systems. In a recent multi-institution/disciplinary study, the two alternative housing systems along with the conventional cage (CC) system were holistically evaluated with regards to animal health/well-being, environmental impact, egg quality and safety, economic efficiency, and worker's ergonomics (Swanson et al., 2015; Karcher et al., 2015; Zhao et al., 2015a,b, Shepherd et al., 2015). The results reveal that each housing system has its advantages and limitations regarding the five aspects of sustainability (http://www2.sustainableeggcoalition.org/). In the meantime, a number of issues associated with the new housing systems have surfaced that must be addressed through scientific research (P.Y. Hester, Chair of United Egg Producers Scientific Advisory Committee on Animal Welfare, Personal Communication, 2014). These issues are primarily surrounding the allocation of resources in the alternative housing systems that are needed in the establishment of industry's animal welfare guidelines. For instance, with the introduction of the enrichment elements and a much larger group size, what is the adequate feeder space for the hens? Do the hens feed at the same time? Similarly, what is the proper amount of nestbox area for the group? The same applies to the scratch pads area. The amount of resources can significantly affect the design of the housing system which in turn will have implications on food safety, ease of management, production efficiency, and ultimately sustainability of the operation. For instance, installing additional feeders inside the hen colonies to meet the feeder space "guideline" may cause management complications (difficult to inspect) and potential feed contamination (defecation in the feed). To answer these critical questions, research facilities that allow for quantification of behavioral and performance responses of the animals to these design and management factors are imperative.

The objective of this work was to develop a research facility equipped with an automated tracking and monitoring system for laying-hen behavioral and production response studies in an enriched colony housing, whereby design and management factors can be modified and the responses of the group-housed hens characterized. This paper describes design and development of the research facility.

2. Materials and Methods

2.1. The hen room and the enriched colony modules

The hen room has dimensions of 11.35 m × 6.55 m × 4.28 m (W × D ×H), which can fit four 2-tier Big Dutchman (Holland, Michigan, USA) enriched colony modules, 3.73 m L × 1.91 m W each (Figures 1, 2 and 3). Two variable fans each with a maximum airflow rate of 1500 m^3/hr placed in the back wall create a negative static pressure for the room. With preconditioning of the ventilation air, this airflow capacity was designed to accommodate a maximum of 960 hens in the room.

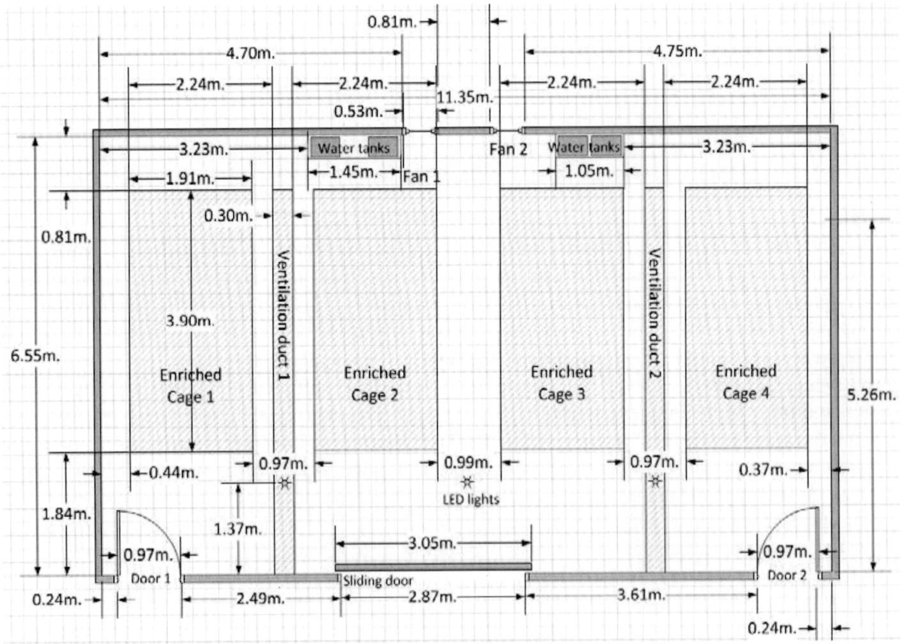

Figure 1. Top view of the hen room.

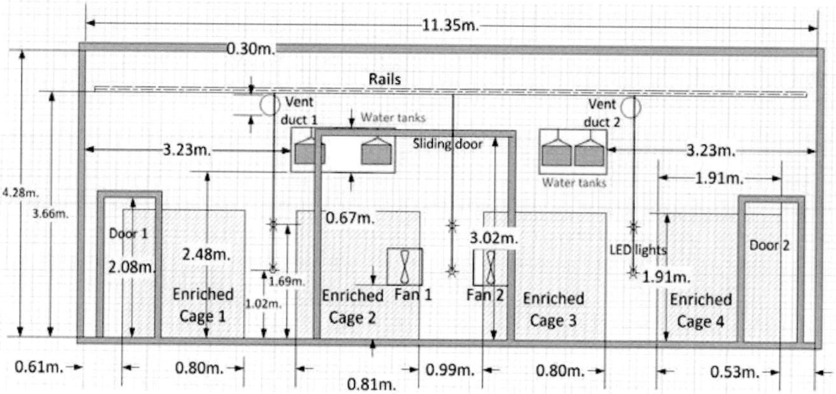

Figure 2. Front view of the hen room.

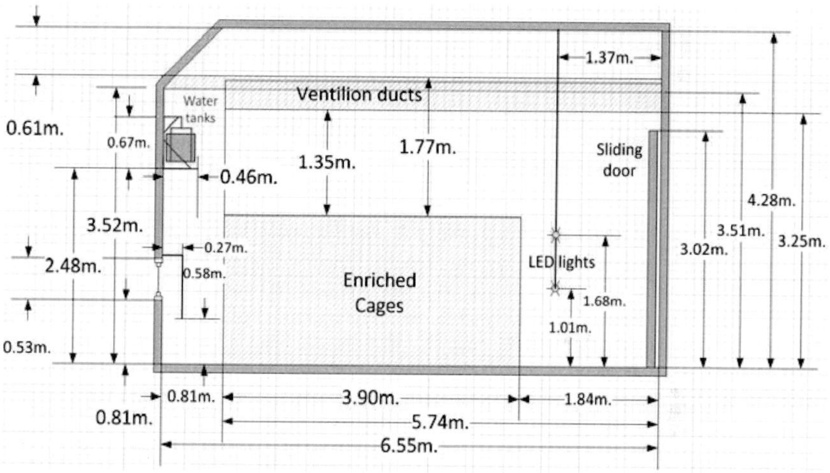

Figure 3. Side view of the hen room.

Two perforated ventilation ducts, positioned between the first and the second colony modules and between the third and fourth colony modules, were designed to provide a uniform air distribution (Figure 4). The inlets of the ventilation ducts were connected to the control room, where air could be preconditioned (heated or cooled) before entering the hen room. Programmable LED lights and the controller (Once Innovation, Plymouth, Minnesota, USA) were used to provide the lighting.

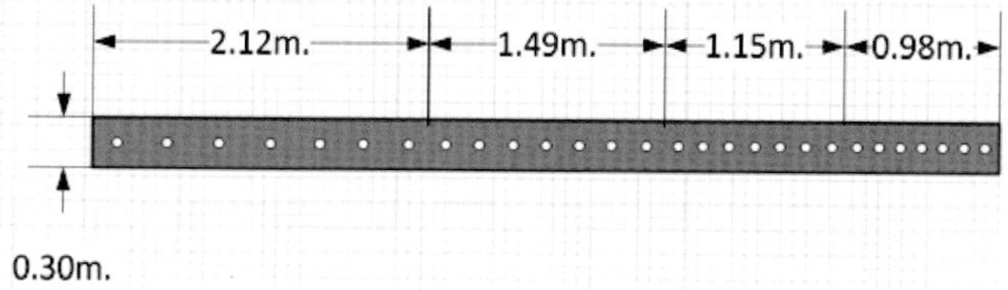

Figure 4. Schematic drawing of the design of the ventilation air ducts.

A plane-view schematic representation of the two-tier enriched colony modules (provided by Big Dutchman) is shown in Figure 5. Each tier has multiple perches and a feed trough on each side. Beneath the feed trough is an egg collection belt. Below each tier is a manure belt used to collect the manure which will be removed every 3-4 days. Each tier normally houses 60 laying hens but may hold up to 120 at increased stocking densities.

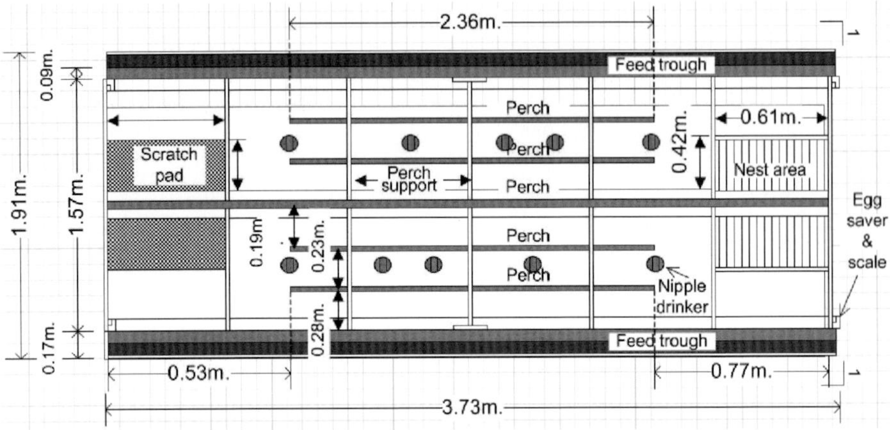

Figure 5. Top view of the enriched colony module.

2.2. Measurement and data acquisition systems

2.2.1. Environmental conditions

Air temperatures at two locations in the hen room, two locations in the pre-conditioning room, and one outside location are measured with copper-constantan (T type) thermocouples (0.1°C resolution) that are connected a compact FieldPoint module (National Instruments, Austin, Texas, USA). Relative humidity (RH) of the hen room, collocated with the temperature measurements, is measured with RH sensor (Vaisala model HMT100, Vaisala Inc., Woburn, Massachusetts, USA). Carbon dioxide (CO_2) levels at the two locations of the hen room are measured with Vaisala GMT222 CO_2 sensors. The temperature, RH and CO_2 sensors are checked and calibrated, as deemed necessary, prior to and during data collection, following standard operation protocols of quality control and quality assurance. A LabVIEW program is built and used to measure and record the temperature, RH and CO_2 concentration data that are retrieved to the hosting computer.

2.2.2. Drinking behaviors and water consumption

In commercial poultry houses, water consumption is measured with flow meters that may have pulse output as well as digital readings. The measurement is either for the entire house (e.g., floor-raised birds) or part of the house (i.e., one cage rows), where a large number of birds are involved. However, such a water metering system is inadequate for accurately measuring water use by a small number of birds, as is the case here (60 to 120 laying hens per tier). To accommodate this need, a system was designed that consists of a suspension S-shaped load-cell, a frame, a water storage tank, and an automatic water refill. These load cells (type LC101-100, Intertechnology Inc., Toronto, Ontario, Canada) each has a measurement range of 0 to 45 kg with an accuracy of 0.0075 mV which is equal to 11 g, and are powered by a 12 VDC supply. Calibrations were performed to relate the load-cell mV output to the load (g) for each load-cell weighing system.

The drinking water system of each enriched colony tier is connected to a 45 liter water storage tank. Since gravity is used for the water delivery, the tank needs to be suspended at an adequate elevation to provide the proper pressure and thus water flow to the nipple drinkers. The tank is suspended on a frame mounted against the room wall (Figure 6). Between the tank and the frame is the load-cell used to measure the weight of the tank, visible in the right picture of Figure 6 (the back side). The water inlet tube is inserted inside the tank about half way, and a cap prevents dust from falling into the water tank.

The water tank will be refilled once a day during the dark period when there is no drinking activity. The water supply to the tank is controlled by a normally-closed solenoid valve (left picture of Figure 6) that is opened once energized by a 24 VDC supply. A LabVIEW program is used to control this process.

Figure 6. Water-weighing system. Picture to the left is the overall picture, whereas picture to the right is a close-up of the load cell and the water inlet connection.

2.2.3. Feeding behaviors and feed use

Feeding behaviors as well as daily feed intake are important parameters that will affect the design and management of a housing system. To quantify these welfare and production traits, a tracking and measurement system was designed that identifies and determines the number of hens at the feeder at a given moment, frequency of hen's visit to the feeders, and average daily feed intake (Figure 7). These features were accomplished by using load-cell sensors mounted underneath the feeders and an ultra-high-frequency radio frequency identification (UHF RFID) system. Specifically, three 1040-I-7kg Tedea load-cells (Nationwide Scale LLC, Prospect Park, New Jersey, USA) were used, each with a 7 kg capacity and 2 mV/V sensitivity. The outputs of the individual load-cells were added to obtain the total weight of the feeder.

Figure 7. Feed intake measurement system. A plastic feed trough is connected to three load-cells placed under the egg collection rack and supporting the feeder through the vertical bolts.

The RFID system (TransTech Systems, Aurora, Oregon, USA) consisted of four elements – antennas, tags, readers, and a database. Figures 8 and 9 provide both photographical and schematic illustrations of the system elements. Each of the 60 hens in the colony will wear a miniature RFID tag (PT-103, tie-wrap tag passive Gen 2 UHF, TransTech Systems) on its neck.

The antennas (SlimLine 8060, 65cm long × 8.6 cm wide × 0.8 cm thick each) for the feeders were placed at the bottom inside the tough and covered with plastic plates. Because the original metal feed trough interferes with the RFID measurement, it was replaced with a plastic feed trough. Each feed trough had six antennas aligned in series covering the length of the feeder. The antennas were situated such that they will detect the hens at the feeder when their heads are inside the trough. All 12 feeder antennas (6 per feeder × 2 feeders) were connected to three 4-channel readers (ThingMagic Mercury M6) that were further connected to the hosting computer via Ethernet connection (Figure 9).

Figure 8. The RFID antennas placed inside the feed trough (left), covered with plastic plate (middle), and the reader (right) used to monitor feeding behaviors of individual birds.

2.2.4. *Nestbox use and laying behavior*

Visit to the nestbox by the individual birds and the number of egg laid inside the nestbox are monitored using the RFID and load-cell weighing systems. A square-shaped antenna (A1030, 30cm × 30cm × 0.65cm thick) was placed underneath the pad of each nextbox. The antennas were situated such that they will detect the hen when her body is inside the nextbox. Both nextbox antennas were connected to the same type of 4-channle ThingMagic Mercury M6 reader as described above that is also connected to the hosting computer via Ethernet connection.

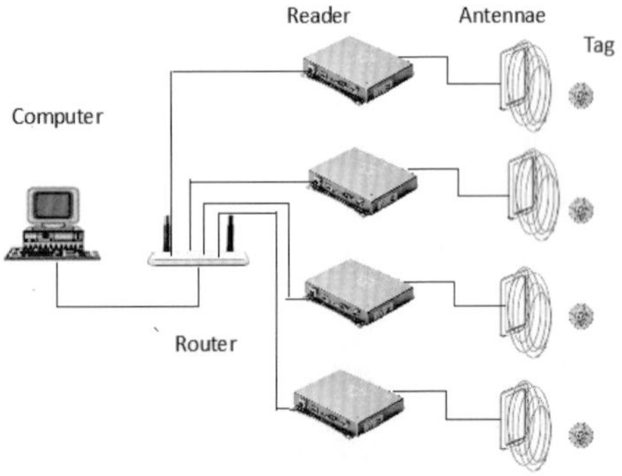

Figure 9. Schematic representation of the RFID system for tracking hen feeding behaviors.

2.2.5. *Perching and scratching behaviors*

A video recording system (IP-PRO 16 Camera 1080P kit, Backstreet Surveillance, Cupertino, California, USA) is used to evaluate the perching and scratching behaviors of the hens and to provide supplemental information on feeding and nesting behaviors. Two video cameras were mounted to the ceiling to cover the top tier area.

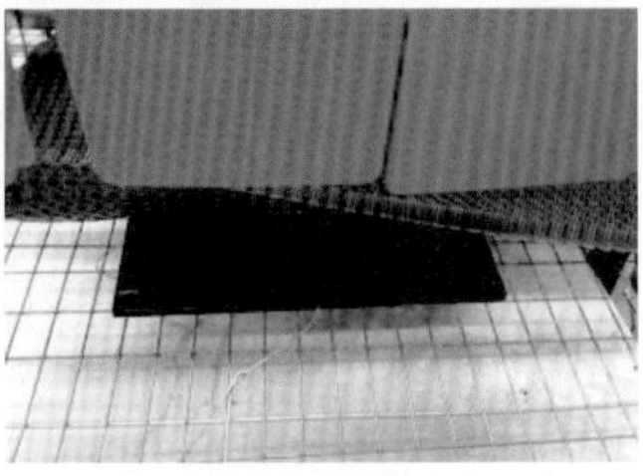

Figure 10. RFID antenna in nestbox.

2.3. Data acquisition system

Output signals from the thermocouples, RH, CO_2, and load-cell sensors are connected to the respective Compact Fieldpoint modules (cFP-TC-120, cFP-AI-111, National Instruments, Austin, Texas, USA). A digital output module (cFP-DO-401) is used to control the solenoid valves for refilling the water tank. The Compact Fieldpoint (cFP-2020) is interfaced with a hosting computer via an Ethernet connection using a LabVIEW program.

RFID involves detecting and identifying a tagged object through the data it transmits. In this research facility, each bird has an RFID tag affixed to its neck. When the bird's head is inside the feed trough or when the bird is inside the nextbox, its identity is detected and the duration of activity (feeding or nextbox use) is recorded. This tracking is done through the coupling of the antennas between the tag and the readers. The readers are connected to a host computer that has the necessary intelligence to further process the tag data through RJ45 (10/100 Base-T Ethernet). The Tag/Transponder Protocols was EPCglobal Gen2 (ISO 18000-6C) with Digital Rights Management (DRM). The data acquisition program was written using C# (C Sharp) based on Application Programming Interface (API), stored as text files. Data analysis and processing was realized using the SQL server and EXCEL VBA programs.

Two cameras (IP Pro 3 Megapixel Bullet, DSS-BFR3MP, Backstreet Surveillance) were used to monitor the hen behaviors. The two cameras were wired to an 8-port power-over-ethernet (POE-108, Backstreet Surveillance) injector. The video files were stored in 8 terabyte storage (4 drives) of an NVR system (DSS-NVR5816, Backstreet Surveillance). For processing, each frame of the video files will be extracted and the hens' behavior of interest can be analyzed using the code developed in Matlab (R2015b, MathWorks, Natick, Massachusetts, USA).

3. Results and Discussion

3.1. The feeder load-cell weighing system

The *in-situ* load-cell weighing system was examined for weight outputs from summation of weight from three individual calibration equations versus weight from an overall calibration equation. It was found that use of individual calibration equations produces higher quality data than use of the overall calibration equation. Specifically, use of the individual outputs was more immune to the changing distribution of the feed load in the feed trough.

3.2. The RFID tracking system

A preliminary test was conducted to test the effectiveness of attaching the miniature RFID tag around the hen's neck and its communication with the antenna placed inside feeder trough. The test involved six hens (with tag ID of 1, 2, 13, 16, 20, and 26, respectively) housed in a litter-floor pen. The feed trough was attached to the outside of the wire-mesh-walled pen. To verify the identification precision of the RFID system, video images were taken and used. Over a 3-day testing period, results from random instantaneous sampling of data revealed a 100% agreement between the RFID system and the video system in detecting the hens feeding events. Figure 11 illustrates the individual feeding events of the six hens. In this particular case, the cumulative time of feeding or at the feeder for the six hens during the light period was, respectively, 4.4 h, 3.6 h, 2.7 h, 6.1 h, 1.5 h, and 3.5 h (mean of 3.6 h). The preliminary data showed considerable variability among the individual birds (coefficient of variation of 43%). Characterization of the inherent biological variability among the individuals is one of the justifications and advantages of such a unique individual-tracking system for group settings. The performance of the tracking system remains to be fully tested in the experimental ECH setting.

3.3. First application of the monitoring/tracking system

The inaugural experiment with this research facility is to assess the impact of feeder space on feeding behaviors and feed intake of laying hens in an enriched colony housing. The hens will be housed at a constant stocking density while the feeder space is varied. Through this study we hope to answer the following questions: (1) Do hens in the enriched colony feed at the same time? If not, what is the maximum number or percentage of hens feeding at the same time? (2) What is the adequate feeder space to accommodate the feeding behavioral needs and daily feed intake of the hens? (3) What is the individual variability in feeding behaviors for the group-housed hens? What is the nestbox use of the hens and percentage of eggs laid in the nest boxes?

How is the perch resource utilized by these bird? Results of the experiment will be available at a later date.

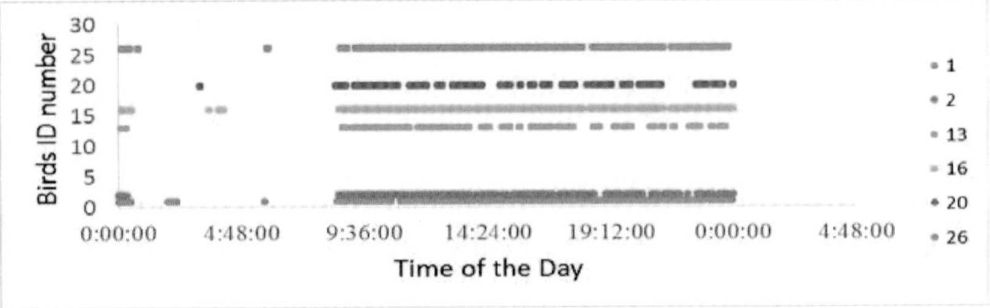

Figure 11. Feeding events of six birds (ID 1, 2, 13, 16, 20, 26) (light hour: 09:00–00:00 h).

Acknowledgements

The enriched colony modules used in this research facility were donated by Big Dutchman Company and are greatly appreciated. Financial support for purchase of the electronic and mechanical devices and supplies for the behavioral monitoring/tracking system is provided in part by the Iowa Egg Council Endowed Professorship fund awarded to Hongwei Xin. Assistance by undergraduate research assistants – John Remus I, Evan Anderson, and Kyle Dresback, in the system setup is also acknowledged.

References

Karcher, D.M., D.R. Jones, Z. Abdo, Y. Zhao, T.A. Shepherd, and H. Xin. 2015. Impact of commercial housing system and nutrition and energy intake on laying hen performance and egg quality parameters. Poultry Science 94(3):485-501.

Shepherd, T.A., Y. Zhao, H. Li, J.P. Stinn, M.D. Hayes, and H. Xin, 2015. Environmental assessment of three laying-hen housing systems– Part II: ammonia, greenhouse gas, and particulate matter emissions. Poultry Science 94(3):534-543.

Swanson, J.C., J.A. Mench, and D. Karcher, 2015. The Coalition for Sustainable Egg Supply project: An Introduction. Poultry Science 94(3):473-474.

Zhao, Y., T. A. Shepherd, J. Swanson, J. A. Mench, D.M. Karcher, and H. Xin, 2015. Comparative evaluation of three laying-hen housing systems: description of the production systems and management practices. Poultry Science 94(3):475-484.

Zhao, Y., T.A. Shepherd, T.A., H. Li, J.P. Stinn, M.D. Hayes, and H. Xin. 2015. Environmental assessment of three laying-hen housing systems–Part I: monitoring system and indoor air quality. Poultry Science 94(3):518-533.

Effect of Weakly Alkaline Electrolyzed Water on Production Performance, Egg Quality and Biochemical Parameters of Layers

Jiafa Zhang, Baoming Li *, Li Ni, Liu Yang

Key Laboratory of Agricultural Engineering in Structures and Environment, Ministry of Agriculture, College of water Resources and Civil Engineering, China Agricultural University, Beijing, 100083, China

* Corresponding author. Email: libm@cau.edu.cn

Abstract

With the development of animal husbandry, it seems more and more important to search for effective measures to improve animal health. The objective of this study was to evaluate the effect of weakly alkaline electrolyzed water on production performance, egg quality and biochemical parameters of laying hens in comparison with that of tap water. One hundred and forty-four numbers of Jinghong's 50-week-old commercial layers of were divided into two groups and each treatment had 3 duplicates with 24 layers. Layers in control group drank tap water, while those in experimental group drank weakly alkaline electrolyzed water with pH value between 9.3 and 9.5. Results showed that drinking weakly alkaline electrolyzed water could significantly improve the feed conversion rate (FCR) and Haugh unit ($P < 0.05$), and, at the same time, result in a downward trend in feed-egg ratio ($P < 0.1$). However, no significant differences were observed in terms of the production performance, such as feed intake, drinking amount, body weight and egg quality including the weight and color of the yolk, as well as the color, strength and thickness of the egg shell. The amounts of urea nitrogen (UREA), cholesterin (CHO), triglyceride (TG) and alanine transaminase (ALT) showed a major decrease in the experiment group when compared with the control group ($P < 0.05$), while the level of aspartate aminotransferase (AST) and serum creatinine concentration ($P > 0.05$) remained almost the same. In conclusion, drinking weakly alkaline electrolyzed water had little effect on production performance and eggs quality of layers, but it can obviously increase the biochemical parameters such as ALT, UREA, CHO and TG of layers.

Keywords: Feed conversion rate, Haugh unit, antioxidation, functional drinking water, healthy growth

1. Introduction

The safety of potable water is crucial to the health of animals in the livestock industry. In the past few decades, decreased quality of tap water due to worldwide pollution has become a major social concern which gives rise to the study of drinking electrolyzed water. Japanese scholars started studying drinking electrolyzed water in 1931. It was applied in agriculture for the first time in 1954 (Shirahata, 2002). Because alkaline electrolyzed water is rich in hydrogen, Japanese scholars define alkaline electrolyzed water with pH values between 8 and 10 as a functional drinking water. It is believed that water rich in hydrogen can remove reactive oxygen species (ROS), which is beneficial to animal health (Shirahata et al., 2007).

Weakly alkaline electrolyzed water is produced near a cathode. Studies find that weakly alkaline electrolyzed water contains hydrogen (atoms and molecules), mineral nanoparticles, and mineral nanoparticle hydrides, which are all active agents for scavenging ROS and reducing oxidative damage (Shirahata et al., 2012). Clinical data suggest that weakly alkaline electrolyzed water helps treat diseases related to oxygen stress (Hayashi and Kawamura, 2002).

Weakly alkaline electrolyzed water with different pH values has different effects on antioxidant capacity. Constantly drinking weakly alkaline electrolyzed water makes cats' have higher antioxidant capacity than those drinking tap water (Hu, 2012). Li et al. (2011) reported that water rich in hydrogen could sustain the level of antioxidant enzymes and glutathione in mice treating with radiation exposure, and reduce producing lipid. Weakly alkaline electrolyzed water, as a kind of reduced drinking water, can scavenge intracellular ROC in a hamster pancreatic p cell line HIT-T15 cells, significantly improve the secretion of insulin, and suppress the oxidative damage caused by alloxan-induced type 1 diabetes in model mice (Li et al., 2002, 2011, 2012). However, there are still no reports regarding the efficacy of drinking weakly alkaline electrolyzed water on the production performance of laying hens. Therefore, the objective of this study is to evaluate the effect of weakly alkaline electrolyzed water on the production performance, egg quality and biochemical parameters of laying hens in comparison of that of tap water.

2. Materials and Methods

2.1. Materials and Grouping

The weakly alkaline electrolyzed water was generated by using an electrolyzed water generator (Ruiande Environmental Protection Equipment Co. Ltd., Beijing, China). One hundred and forty-four 50-week-old "Jinghong" commercial layers, were purchased from Huadu Yukou Poultry Industry Co. Ltd. (Beijing). All layers were divided into two groups randomly, namely the control group and the experimental group. Each group had three repetition and twenty-four hens for each repetition. Layers in control group drank tap water, while those in experimental group drank weakly alkaline electrolyzed water. The formal experiment was preceded by a week of preliminary feeding.

2.2. Indicators and Methods

Six indicators including feed intake, drinking amount, body weight, egg production rate, feed conversion rate, feed-egg ratio, of production performance were measured. Specifically, feed intake and drinking amount were measured on a weekly basis while egg production rate and egg weight were measured every day. At the end of the sixth week, twelve layers were fed in individual cages. There was a seven-day preliminary feeding before the three-day formal experiment to measure the feed conversion rate.

On the last day of the sixth week, thirty fresh eggs were chosen randomly from each group, and examined for the egg quality in the ministry of agriculture poultry quality supervision, inspection and testing center (Beijing). The egg yolk weight was weighed by T-500 electronic scale. Egg shell index was measure by egg form coefficient measurement instrument (FHK, Japan). A CM-2600d portable spectral color measurement meter (Konica Minolta, Tokyo, Japan) was used to test the color of egg shell. Eggshell strength analyzer (Robotmation, Tokyo, Japan) was used to measured egg shell strength. Egg weight, egg yolk color and Haugh unit were all measured with EMT-5200 Egg quality analyzer (Robotmation, Tokyo, Japan). A micrometer caliper was used to measure the thickness of egg shell.

Two milliliter blood sample was collected from the vein of nine random layers from each group. The sample was analyzed by colorimetric method (Hitachi 7160 automatic biochemical analyzer, Japan).

2.3. Statistical Analyses

The data were analyzed by one-way analysis of variance (ANOVA) using SPSS 20.0 (International Business Machines Corporation, New York, America). Independent samples T-test was used to separate means at a level of significance of $P < 0.05$. The trend levels are presented at a level of $P = 0.1$.

3. Results and Discussion

3.1. Effect of weakly alkaline electrolyzed water on production performance of layers

Table 1 showed that drinking weakly alkaline electrolyzed water tend to reduce feed-egg ratio of laying hens compared with the tap water group ($P < 0.1$). Weakly alkaline electrolyzed water reduced drinking amount and feed intake, but improved egg laying rate of laying hens. However, no significant differences were observed between weakly alkaline electrolyzed water and tap water in theses aspects. ($P > 0.05$). The feed conversion rate of layers in weakly alkaline electrolyzed water group was significantly higher than that in the tap water group ($P < 0.05$). Similar to the previous report, adding antioxidant into feed could significantly reduced feed-egg ratio and improve egg production ratio (Zhang et al., 2012).

Table 1. Effect of alkaline electrolyzed water on production performance of layers.

Production performance	Unit	Control group	Experimental group
Drinking amount	ml per day per layer	227 ± 9^{a}	226 ± 16^{a}
Feed intake	g per day per layer	120 ± 4^{a}	119 ± 3^{a}
Egg production rate	%	83.3 ± 3.5^{a}	86.6 ± 3.1^{a}
Body weight	g	2068.3 ± 200.8^{a}	2088.5 ± 193.6^{a}
Feed-egg ratio	kg of feed per kg of gain	2.34 ± 0.95^{A}	2.18 ± 0.49^{B}
Feed conversion rate	%	62.5 ± 1.1^{b}	64.85 ± 0.6^{a}

Note: Means within the same row with different lowercase letters superscripts are significantly different ($P < 0.05$). Values with different capital letter superscripts mean having a trend ($P < 0.1$).

3.2. Effect of weakly alkaline electrolyzed water on egg quality of layers

As shown in table 2, Haugh Unit in weakly alkaline electrolyzed water group was significantly higher than that in the tap water group ($P < 0.05$). Weakly alkaline electrolyzed water improved egg yolk weight, egg shell color, egg yolk color, eggshell thickness and eggshell strength, but the influence was not significantly different from that of tap water.

Haugh unit could indicate whether the egg was fresh. Previous studies showed that the higher haugh unit was, the higher the consistency of protein was in the egg, which indicated a better egg quality (Zhang, 2004). Zhou (2011) found increasing haugh unit could increase the antioxidant capacity of laying hens. Wang et al. (2011) added 100 mg kg^{-1} and 150 mg kg^{-1} WONGs polysaccharide, a kind of antioxidants, into feed, and observed that the haugh unit increased 13.81% and 13.60% respectively. Wu et al. (2009) studied natural vitamin E as an antioxidant additive in feed and found it would increase the level of haugh unit significantly. All these results showed that adding antioxidants could increase the antioxidant capacity of laying hens as well as influence the haugh unit of eggs indirectly. This was in line with the result of effect of weakly alkaline electrolyzed water on haugh unit.

Table 2. Effect of weakly alkaline electrolyzed water on egg quality of layers

Egg quality	Unit	Control group	Experimental group
Egg yolk weight	g	16.6 ± 1.0^{a}	16.8 ± 1.1^{a}
Egg shell index		1.34 ± 0.02^{a}	1.33 ± 0.03^{a}
Egg shell color		57.85 ± 2.00^{a}	58.93 ± 2.13^{a}
Egg shell strength	$kg \cdot cm^{-2}$	3.060 ± 0.534^{a}	3.138 ± 0.545^{a}
Egg yolk color		8.7 ± 0.6^{a}	8.8 ± 0.4^{a}
Haugh unit		86.2 ± 2.8^{b}	88.2 ± 2.8^{a}
Thickness of egg shell	mm	0.345 ± 0.017^{a}	0.346 ± 0.017^{a}

Note: Means within the same row with different lowercase letters superscripts are significantly different ($P < 0.05$). Values with different capital letter superscripts mean having a trend ($P < 0.1$).

3.3. Effect of weakly alkaline electrolyzed water on biochemical parameters of layers

Table 3 showed that drinking weakly alkaline electrolyzed water instead of tap water could significantly reduce alanine aminotransferase (ALT), urea nitrogen (UREA), total cholesterol (CHO) and triglyceride (TG) ($P < 0.05$). Drinking weakly alkaline electrolyzed water could also reduce aspertate aminotransferase (AST) and creatinine (CRE), but no significant difference was found ($P > 0.05$).

Blood is an important part of animals, which plays a key role in the metabolism of the body material and waste transportation. The composition of blood can reflect the change of the body's metabolism indirectly (Zhao et al., 1997). In this experiment, layers drinking weakly alkaline electrolyzed water experienced a dramatic decrease of the amount of ALT in blood. At the same time, the ratio of cereal third transaminase and aspertate aminotransferase also decreased significantly, indicating that drinking weakly alkaline electrolyzed water could substantially reduce the risk of liver cell damage. Tsai, et al. (2009) proved that weakly alkaline electrolyzed water had the function of preventing liver damage through the carbon tetrachloride induced liver damage test, which was consistent with the results of this study.

Liver, as the major organ of triglyceride metabolism, can be damaged by the production of oxygen radical. Triglyceride can accelerate the oxidation of low density lipoprotein (LDL) and injure the endothelial cells, which all increase the risk of atherosclerotic (Xiong et al., 2008). The result was similar with Zhang (2014), which showed that water rich in hydrogen had the function of selectively neutralizing reactive oxygen and could effectively prevent atherosclerosis.

Table 3. Effect of weakly alkaline electrolyzed water on biochemical parameter of layers

Serum metabolites	Unit	Control group	Experimental group
ALT	$U\ L^{-1}$	8.237 ± 2.964^{a}	6.235 ± 2.177^{b}
AST	$U\ L^{-1}$	188.720 ± 34.065^{a}	179.936 ± 23.409^{a}
CRE	$\mu mol\ L^{-1}$	36.427 ± 4.209^{a}	34.627 ± 2.943^{a}
UREA	$mmol\ L^{-1}$	0.810 ± 0.129^{a}	0.578 ± 0.082^{b}
CHO	$mmol\ L^{-1}$	7.511 ± 1.087^{a}	5.596 ± 0.790^{b}
TG	$mmol\ L^{-1}$	17.327 ± 0.933^{a}	15.780 ± 1.356^{b}

Note: Means within the same row with different lowercase letters superscripts are significantly different ($P < 0.05$). Values with different capital letter superscripts mean having a trend ($P < 0.1$).

The UREA is the final product from metabolism of protein and amino acid in vivo, and it is synthesized by ornithine cycle. The content is negatively related to the utilization rate of amino acid or dietary protein level (Liu et al., 2014). It was consistent with the increase of the feed conversion rate. Modern medicine has proved that as the cholesterol level in blood exceeds the normal level, animal will be more susceptible to cardiovascular diseases (Xia, 2011.). Gadek et al. (2006, 2009) let four hundred and eleven diabetes mellitus patients drink 2L weakly alkaline electrolytic water every day and six days later, it was observed that their blood indexes had no significant change. However, as the intake of weakly alkaline electrolytic water increased, the level of CHO and CRE dropped significantly. The longer the patients drank weakly alkaline electrolytic water, the more of them would experience a symptomatic improvement. In this study, the serum cholesterol of laying hens drinking weakly alkaline electrolytic water was significantly reduced, which was consistent with the result of Gadek et al. (2009). Drinking weakly alkaline electrolytic water could also reduce the serum creatinine level of layers but it was not significantly different from the effect of tap water.

4. Conclusions

Weakly alkaline electrolyzed water is functional water because of its higher activation energy and smaller water cluster structure. Long-term consumption of it can significantly improve feed conversion rate, therefore improving feed utilization rate and reducing feed costs. Due to the antioxidant function of the weakly alkaline electrolyzed water, it can significantly increase the haugh unit of eggs while reducing urea nitrogen, cholesterin, triglyceride and alanine transaminase. The experiment shows that long-term consumption of weakly alkaline electrolyzed water can reduce the feed cost, increase the freshness of the eggs and contribute to the healthy growth of layers.

References

Gadek, Z., Y. Li, S. Shirahata, 2006. Influence of natural reduced water on relevant tests parameters and reactive oxygen species concentration in blood of 320 diabetes patients in the prospective observation procedure. In *Animal Cell Technology: Basic & Applied Aspects*. Springer Netherlands. 377-385.

Gadek, Z., T Hamasaki, S Shirahata, 2009. "Nordenau Phenomenon"–application of natural reduced water to therapy. In *Animal Cell Technology: Basic & Applied Aspects*. Springer Netherlands. 265-271.

Hayashi, H., M Kawamura, 2002. Clinical applications of electrolyzed-reduced water. In *Animal Cell Technology: Basic & Applied Aspects*. Springer Netherlands. 31-36.

Hu, J.X., 2012. Notoginseng alcohol extract and alkaline water electrolysis oxidation action research. MS Thesis. The Fourth Military Medical University, Xian, China.

Liu, J., G.Y. Fang, F.R. Liao, D.C. Ye, Z.Y. Dong, 2014. The effect of low protein amino acid balanced diet on growth performance, and serum protein apparent digestibility of fattening pig under the environment of high temperature. Journal of Fujian Agriculture. 29(12), 1159-1163.

Li, Y., T. Nishimura, K. Teruya, T. Maki, T. Komatsu, T. Hamasaki, T. Kashiwagi, S. Kabayama, S. Shim, Y. Katakura, K. Osada, T. Kawahara, K. Otsubo, S. Morisawa, Y. Ishii, Z. Gadek, S. Shirahata, 2002. Protective mechanism of reduced water against alloxan-induced pancreatic β-cell damage: Scavenging effect against reactive oxygen species. Cytotechnology. 40(1-3), 139-149.

Li, Y., T. Hamasaki, N. Nakamichi, T. Kashiwagi, T. Komatsu, J. Ye, K. Teruya, M. Abe, H. Yan, T. Kinjo, S. Kabayama, M. Kawamura, S. Shirahata, 2011. Suppressive effects of electrolyzed reduced water on alloxan-induced apoptosis and type 1 diabetes mellitus. Cytotechnology. 63(2), 119-131.

Li, Y., T. Hamasaki, K. Teruya, N. Nakamichi, Z. Gadek, T. Kashiwagi, H. Yan, T. Kinjo, T. Komatsu, Y. Ishii, S. Shirahata,2012. Suppressive effects of natural reduced waters on alloxan-induced apoptosis and type 1 diabetes mellitus. Cytotechnology. 64(3), 281-297.

Qian, L.R., 2011. The radiation protection effect of rich H_2 solution for cells and hematopoietic system. MS Thesis, The Second Military Medical University, Shanghai, China.

Shirahata, S., 2002. Reduced water for prevention of diseases. In *Animal Cell Technology: Basic & Applied Aspects*. Springer Netherlands. 25-30.

Shirahata, S., Y. Li, T. Hamasaki, Z. Gadek, K. Teruya, S. Kabayama, K. Otsubo, S. Morisawa, Y. Ishii, Y. Katakura, 2007. Redox regulation by reduced waters as active hydrogen donors and intracellular ROS scavengers for prevention of type 2 diabetes. In *Cell Technology for Cell Products*. Springer Netherlands. 99-101.

Shirahata, S., T. Hamasak, K. Teruya, 2012. Advanced research on the health benefit of reduced water. Trends in Food Science & Technology. 23(2), 124-131.

Tsai, C.F., Y.W. Hsu, W.K. Chen, W.H. Chang, C. C. Yen, Y.C. Ho, F.J. Lu, 2009. Hepatoprotective effect of electrolyzed reduced water against carbon tetrachloride-induced liver damage in mice. Food and Chemical Toxicology. 47(8), 2031-2036.

Wang, C.J., H.F. Wang, H. Chen, L.R. Huang, Q.X. Yang, Y. Gao, 2011. The effect of astragalus polysaccharides on antioxidant performance and egg quality of laying hens. Journal of Animal Nutrition. 23(2), 280-284.

Wu, D., Y.M. Zhou, T. Wang, 2009. The effect of Natural and synthetic vitamin E on production performance antioxidant and egg quality of laying hens. Jiangsu Agricultural Science. 6, 276-278.

Xia, S.Y., 2011. The effect of alfalfa meal fodder adding cellulose enzyme on production performance, egg quality and nutrient digestibility of laying hens. MS Thesis, Henan Agricultural University, Zhengzhou, China.

Xiong, M., Y.Y. Bi, D.L. Zhang, J. Song, H.L. Yang. Y. Xu. J.P. Ouyang, 2008. The effect of ferulic acid sodium on metabolic of cholesterol and triglyceride metabolism in the process of atherosclerosis caused by hyperlipidemia. Journal of Chinese Pathophysiology. 24(9), 1670-1675.

Zhang, C.L., 2014. The research on preparation of rat model of atherosclerosis and effect of rich hydrogen water on atherosis. Hebei Medical University, Shijiazhuang, China.

Zhang, J.L., 2004. The research on egg quality of Xinyang layers. Northwest Agriculture and Forest University, Xian, China.

Zhang, Z.Y., J.M. Li, H.T. Liu, W. Li, J. Zhang, 2012. Thermal stress compound products to the layer chicken production performance and the influence of serum antioxidant index. Forage and Livestock: New Feed. 12, 12-14

Zhao, G.X., Z.S. Zhang, Y.D. Wang, Y.Q. Li, H.Z. Zhu, Z.L. Li, 1997. The effect of low protein diet adding amino acids on production performance and blood biochemical indexes of feeding rabbit. Forage and Livestock. 2, 9-11.

Zhou, J.C., 2011. The effect of oxidative stress on egg quality. Feed Industry. 32(18), 45-46.

Alternative Approaches for Bedding Materials Used in the Housing of Dairy Calves and Cows

Peter D. Krawczel *, Christa A. Kurman, Heather D. Ingle, Randi A. Black, Nicole L. Eberhart

Department of Animal Science, the University of Tennessee, Knoxville, Tennessee 37996, USA

* Corresponding author. Email: krawczel@utk.edu

Abstract

The objective of the first study was to determine the effect of bedding on the behavior and performance of Holstein and Jersey calves housed using individual hutches bedded with gravel, rubber mats, or sand. The second study evaluated the behavioral response of late lactation Holstein cows to two types of sand bedding: (1) new, clean sand or (2) sand reclaimed from the sand settling lane of the farm's manure management system. In study 1, 23 Holstein calves and 38 Jersey calves were blocked by birth date and randomly assigned to 1 of 3 bedding treatments (gravel, rubber mat, or sand). Data were collected for 6.5 to 10 weeks following birth, depending on when calves were weaned. In study 2, Cows (n = 64) were divided into 8 groups of 8 cows with treatments imposed using a replicated crossover design with 7-d periods. Data loggers were used to measure daily lying time (h d^{-1}), number of lying bouts (n d^{-1}). In both studies, data were analyzed using mixed models in SAS. In study 1, there was a significant effect of breed for all response variables. Mean lying time, BW gain, or grain intake were not affected by treatment, week, or treatment × week interactions for Holstein and Jersey calves. In study 2, there were minor differences in overall lying times, less than 25 min d^{-1}, with no other differences in lying behaviors. This suggests any of these beddings may be used without compromising the welfare of pre-weaned dairy calves or lactating cows.

Keywords: Dairy cow, dairy calf, bedding, housing, welfare, behavior

1. Introduction

Animal welfare can be defined by 3 parameters: natural living and behavior, biological function, and affective state (Fraser, 2008). If any one of these areas is deficient the welfare of the animal can be considered to be lacking (Fraser, 2008). Deficient animal welfare can cause decreased animal production indicated by factors such as decreased feed intake or increased incidence of disease caused by stress which can lower weight gain (Neindre, 1993). Animal welfare wad particularly important when housing animals in intensive production systems, which differ from their natural environment (Dellmeier et al., 1985). Around 75% of dairy calves in the United States were housed individually in commercial hutches during the pre-weaning period to reduce disease transmission (USDA, 2010). The housing situation for lactating cows was slightly more complex. In small herds (less than 100 lactating cows), the majority of farms (63%) used tie-stall housing. However, in medium (100 to 499 lactating cow) or large herds (more than 500 lactating cows), the majority of farms use freestall housing (68 and 73%, respectively; USDA, 2010). The ideal bedding surface for dairy cattle was non-slippery, well drained, soft, and easy to keep clean (Webster, 1984a). Bedding was crucial to maintaining animal comfort in intensively managed dairy calves indicated by the fact that provision of bedding increased essential behaviors such as lying and increased indicators of production such as weight gain (Camiloti et al., 2012; Anderson et al., 2006; Fregonesi et al., 2007).

Dairy calves spent up to 70-80% of their daily time budget lying down (Chua et al., 2002) and lactating cows spend 50 to 60% of their day resting (Krawczel et al., 2013). Lying behavior was an important indicator of animal comfort and welfare in adult dairy cattle and calves indicated by the fact that lying behavior took up a large portion of the time budget of adult cattle and calves as well as the fact that lying behavior can increase production parameters (Ito et al., 2009; Chua et al., 2002; Haley et al., 2000; Mogensen et al., 1997). Adequate lying time (12 – 14 h d^{-1}) increased milk production and reduced stress in adult cattle and may have increased weight gain in dairy heifers (Haley et al., 2000; Mogensen et al., 1997). One aspect of dairy animal management that might increase lying time in cattle was providing a soft, dry bedding surface. When given a choice between sawdust with various levels of moisture and bare concrete, calves chose to lie down on the driest surface available and always avoided bare concrete because concrete provided a hard surface (Camiloti et al., 2012). Collectively these studies suggested that calves preferred bedding that provided a soft, dry surface.

It is possible that bedding surface may affect biological function of calves. Biological function goals for dairy calves include weight gain and increased grain consumption. In order to be bred as early as possible (around 15 months of age), calves should be managed to maximize weight gain (Davis Rincker et al., 2011). Maximizing growth in pre-weaned calves also affected future milk production (Van Amburgh, 2008). Since providing bedding may increase lying behavior, which in turn can influence weight gain, the addition of bedding may allow for improved weight gain in dairy calves. When weaned calves were given access to either a bare floor or a bedded floor during winter in North Dakota, calves with access to bedding had greater weight gain during the coldest months of the year (Anderson et al., 2005). In contrast calves housed on several types of organic and inorganic surfaces (gravel, sand, rice hulls, straw, and wood shavings) in a study taking place from August-October in Arkansas showed no difference in weight gain (Panivivat et al., 2004). Calves housed on a variety of inorganic surfaces (slatted floors, rubber mats, and concrete) did not differ in weight gain or feed efficiency (the timeline of this study was not specified; Yanar et al., 2010).

While results have been somewhat mixed, previous research on bedding suggests that both calves and cows prefer a soft, dry surface. To address this issue, we have recently completed two research projects evaluating various bedding materials, which are reviewed in this paper. The objective of the first study was to determine the effect of bedding on the behavior and performance of Holstein and Jersey calves housed using individual hutches bedded with gravel, rubber mats, or sand. The second study evaluated the behavioral response of late lactation Holstein cows to two types of sand bedding: (1) new, clean sand or (2) sand reclaimed from the sand settling lane of the farm's manure management system.

2. Materials and Methods

2.1. Experiment 1

Twenty-three Holstein and 38 Jersey calves born from September 2012 to March 2013 were enrolled on this study. All calves received 4 L of colostrum within 12 h of birth and were housed in commercial plastic calf hutches (Calf-Tel, Hampel Animal Care, Germantown, WI) with an outside wire enclosure and racks to hold buckets for starter grain and water. Calf starter utilized at both farms contained 18% protein and 2% crude fat (tag values, Tennessee Farmers Co-op, LaVergne, TN). All calves were fed 4 L waste milk once daily and water was provided ad libitum. Calves were blocked by birth date and randomly assigned to 1 of 3 hutch bedding treatments: gravel (typically 2.5 cm in diameter), rubber mat (1.3 cm thick) over a gravel base, or river sand (5 cm deep) over a gravel base (gravel used for base was typically 2.5 cm in diameter). Calf hutches measured 2.1 m × 1.2 m inside and had an attached outside area measuring 1.1 m × 1.7 m surrounded by a 1.3 m high wire fence. All animal procedures were approved by the University of Tennessee IACUC committee.

Holstein calves (gravel n = 8, mat n = 7, sand n = 8) were housed at the Middle Tennessee Research and Education center (MTREC) in Spring Hill, TN. Calf hutches at MTREC were placed on a gravel pad measuring approximately 12 m × 54 m. Holstein calves were offered 0.21 ± 0.02 kg calf starter initially and this was increased in 0.2 kg increments as needed to ensure ad libitum access. Holstein calves were weaned at 6.5 ± 0.2 wk of age. Jersey calves (gravel n = 14, mat n = 11, sand n = 13) were housed at the Dairy Research and Education Center (DREC) in Lewisburg, TN. The DREC calf hutches were placed on a gravel pad measuring approximately 21 m × 61 m. At 6 wk of age, milk for Jersey calves was reduced to 2 L once d^{-1}. All Jersey calves were offered 0.2 kg calf starter initially and this was increased in 0.2 kg increments as needed to ensure ad libitum access. All Jersey calves were weaned at 10 wk of age.

Mean lying time (min d^{-1}) and lying bouts (n d^{-1}) of all calves were determined using the HOBO pendant G accelerometer data logger (Onset Corporation, Bourne, MA; Bonk et al., 2013). Loggers were attached to the lateral side of calves' hind legs above the pastern joint with vet wrap. Calves were fitted with a logger after birth during the move to individual hutches. Loggers were removed and replaced each week on alternating legs to prevent skin irritation. Loggers recorded posture at 1-min intervals for 6 d wk^{-1} throughout the study. Lying time and number of lying bouts were calculated as described by (Bonk et al., 2013).

On the day of birth, all calves were weighed and time of birth and time of first colostrum were recorded. Calving ease was also recorded on a scale of 1-5 with 1 being no problems and 5 being extremely difficult birth (Cappel et al., 1998). All calves were removed from the hutch each week to be weighed until weaning to determine body weight gain using a mobile scale (Tru-Test, Tru-Test Limited, Auckland, NZ). Calf starter grain intake from all calves was determined for 3 d each week by daily grain offered (Monday, Tuesday, and Wednesday) and refused (Tuesday, Wednesday, Thursday). Calf health was evaluated by farm staff at each location according to the University of Wisconsin calf health scoring chart guidelines (Lago et al., 2006). Type and duration of any treatment administered to calves was recorded.

The experimental unit was the individual calf. The effects of bedding type (gravel, mat, or sand) and breed (Holstein or Jersey) were determined using a mixed model ANOVA with repeated measures in SAS (SAS institute, Cary, NC). There was an effect of breed ($P < 0.001$) for all variables so data from the two breeds were analyzed separately. Cortisol data were log transformed, log transformed means and backtransformed means are presented. All values are reported as mean ± SE.

2.2. Experiment 2

This study was conducted at the University of Tennessee's Little River Animal and Environmental Unit (Walland, TN) during August-September 2014 and January-February 2015. Late-lactation Holstein dairy cows (n = 64) were split evenly between seasonal experiments. For each season, cows were assigned to 1 of 4 groups with each group consisting of 8 cows and balanced by DIM (Summer: 268.1 ± 11.9 d; Winter: 265.5 ± 6.0 d) and parity (Summer: 2.0 ± 0.2). Cows were milked 2 × daily starting at 0700 and 1730 h. Cows were housed in a 4-row freestall barn providing access to one stall per cow (8 useable stalls per pen). Available stalls were split evenly between feed bunk alley and back alley. Feed bunk headlocks were blocked off to allow 8 useable headlocks per pen. Cows were fed fresh TMR 2× daily (0700 h and 1530 h), and feed was pushed up 2 × daily to maintain access. Water was available ad libitum.

Bedding treatment consisted of the provision of either (1) clean sand (not previously used for bedding of cows) and (2) recycled sand (sand reclaimed from the dairy's flushing system). Recycled sand was stored in an uncovered area, while clean sand was stored under a covering. Bedding was added daily to maintain level with the curb. Stalls were raked 2 × daily during each milking. Cows were placed on clean or recycled sand for one week, then a crossover occurred and treatments switched within groups. Due to an injury one cow was removed during the second week of the study and data from that week was excluded from analysis.

IceTag data loggers (IceRobotics Ltd., Edinburgh, Scotland) were attached to the rear fetlock while in the milking parlor two days prior to start of the experiment to allow for habituation. Data loggers remained on for the duration of the study. Data loggers collected daily lying times, lying bout frequency, lying bout length, and total steps. Each measurement was averaged for a weekly assessment.

Udder hygiene scores were scored on days 0, 3, and 7 of each week in the experimental pen. At the beginning and between treatments, udders were cleaned with a brush. Each cow was visually scored from the rear on a scale of 0 to 3. A score of 0 was splashes of manure covered < 50% of the area, score of 1 was fresh splashes of manure covered > 50% of the area, score of 2 was dried and fresh manure covered > 50% of the area and a score of 3 was the entire area was covered with dried caked manure (adapted from Cook, 2011).

Statistical analyses were performed using SAS 9.4 (Cary, NC). A mixed model of analysis was used to determine significant effects of sample day, treatment, and treatment sequence of events. The GLIMMIX procedure was used to determine mean separation. Due to the numerous amounts of zeros, a frequency table (PROC Frequency) was used to report percentages of each variable. For bacterial counts, to correct for skewed data a log transformation was used, while back-transformed least square means are reported. All values are reported as mean ± SE.

3. Results

3.1. Experiment 1

Two Holstein calves housed at MTREC were removed for medical reasons (1 euthanized after going down, 1 died due to bloat), 4 Jersey calves housed at DREC were removed for medical reasons (4 died due to scours) and 1 Jersey calf was removed for non-medical reasons (stolen). There was an average low temperature of 0 ± 0.5 ° C and an average high temperature of 16 ± 0.6° C at MTREC (Holsteins) and an average low temperature of 0 ± 0.4°C and an average high temperature of 15 ± 0.4°C at DREC (Jerseys) over the 30 wk study period.

Holstein calves spent more time lying (979.9 ± 13.5 min d^{-1}) than Jersey calves (917.5 ± 10.8 min d^{-1} lying; $P < 0.001$). Mean lying time for Holstein calves (968.9 ± 35.5 min d^{-1}) was not affected by trt ($P = 0.73$), wk ($P = 0.44$), or trt × wk interactions ($P = 0.89$). Mean lying time for Jersey calves (916.2 ± 16.0 min d^{-1}) was also not affected by trt ($P = 0.95$) or wk ($P = 0.31$). However, there was a trend for an effect of trt × wk ($P = 0.08$).

Holstein calves had a higher number of lying bouts d^{-1} (13.1 ± 0.4 bouts d^{-1}) than Jerseys (10.3 ± 0.3 bouts d^{-1}; $P < 0.001$). Mean lying bouts for Holstein calves (figure 2a) was not affected by trt ($P = 0.11$), wk ($P = 0.14$) or trt × wk ($P = 0.47$). Mean lying bouts for Jersey calves housed in hutches with rubber mats (11.6 ± 0.6 bouts d^{-1}) were higher ($P = 0.01$) than for calves housed on gravel (10.0 ± 0.6 bouts d^{-1}) or sand (9.2 ± 0.6 bouts d^{-1}). There was a tendency for an effect of wk ($P = 0.08$) and no effect of trt × wk ($P = 0.17$).

Holstein calves gained an average of 3.8 ± 0.9 kg wk^{-1} during wk 1 which increased to a maximum of 6.0 ± 1.7 kg wk-1 during wk 5 before decreasing to 2.7 ± 1.5 kg wk^{-1} during wk 7. Mean body weight gain (2.2 ± 0.2 kg wk^{-1}) of Jersey calves was not affected by trt ($P = 0.18$) but there was an effect of wk ($P < 0.001$) and trt × wk ($P = 0.04$). Jersey calves gained 1.1 ± 0.3 kg wk^{-1} during wk 1, which increased to a maximum of 3.6 ± 0.3 during wk 6.

3.2. Experiment 2

Lying time, lying bout length, number of lying bouts, and total steps taken were not affected by treatment in summer, but there was a treatment effect for total steps taken during winter (Table 1). Cows on clean sand had a higher number of total steps taken. There was a sequence effect for number of lying bouts in the summer ($P = 0.04$, clean-to-recycled: 9.971 ± 0.553; recycled-to-clean: 11.724 ± 0.574). There were no other sequences of events effects (summer – lying time: $P = 0.60$, lying bout length: $P = 0.12$, total steps: $P = 0.85$; winter-lying time: $P = 0.30$, number of lying bouts: $P = 0.90$, lying bout length: $P = 0.19$, total steps taken: $P = 0.40$).

Hygiene scores differed in the summer segment ($P = 0.03$, clean: 0.05 ± 0.03; recycled: 0.13 ± 0.03). Only 5% of cows were scored something other than a zero while on clean sand, while 12% of cows on recycled sand scored something other than a zero during the summer. There was no treatment effect in the winter segment ($P = 0.40$), but there was a sample day effect where scores steadily increased until d 7 of the first week then dropped for the remainder of the second week. All means reported are small as most scores cows were scored a zero for the duration of the study.

4. Discussion

In the literature, calves are recorded lying for as much as 1020-1140 min d^{-1} (17 - 19 h d^{-1}) or as little as 364-735 min d^{-1} (6–12 h d^{-1}; Chua et al., 2002; Yanar et al., 2010; Stefanowska et al., 2002; Sutherland et al., 2013; Camiloti et al., 2012). It was evident from these values that calves exhibited a wide range of lying times, indicating that further study was needed to determine the relationship between calf comfort and lying time for dairy calves as well as the effect of management on comfort and lying time. Several factors in previous studies influenced calf lying times. Bedding material type influenced lying time as calves on organic surfaces spent as much as 776 minutes more lying down when compared to calves on inorganic surfaces (Stefanowska et al., 2002). Calves in the current study were all housed on inorganic materials, which may explain why they had lower lying times (917–980 min d^{-1} or 15–16 h d^{-1}) than calves housed on organic surfaces such as straw (lying times of 1020-1140 min d^{-1} or 17–19 h d^{-1}) in previous studies. Pre-weaned calves housed on a similar bedding material (river stones) spent around 1080 min d^{-1} (18 h) lying down, however, these calves were housed in groups, a factor which has previously been shown to increase lying time (Sutherland et al., 2013). A possible reason that calves spend more time lying on organic bedding surfaces is that these surfaces are often soft and absorbent. Housing type also influenced lying time with calves in group housing spent more time lying down than calves in individual housing, this effect was believed to be due to increased space allowance for calves in group housing (Tripon et al., 2012). Although previous research indicates that calves lie down more on a soft, dry surface, calves in the current study did not have increased lying times on sand or mats which provided a softer surface than gravel (Camiloti et al., 2012). However, previous research indicated a soft surface might not be necessary for lying behavior. Calves housed on river stones or sawdust did not differ in lying time at 1 wk of age or 6 wk of age (Sutherland et al., 2013). Our results were similar to those obtained when calves housed on either concrete or rubber mats did not differ in lying time with calves spending around 1013 min d^{-1} or 17 h d^{-1} lying down, which was close to the 917-980 min d^{-1} or 15-16 h d^{-1} of time spent lying observed in our calves (Hänninen et al., 2005). Calves on the current study may not have increased lying time on rubber mats or sand when compared to gravel because gravel provided good drainage and a dry surface. Calves in the current study were lighter weight than previously studied adult cattle and may not have required a soft surface for comfort (Drissler et al., 2005; Hänninen et al., 2005). Although lying time may not differ on different surfaces, a difference in lying behavior can still be observed. Bulls housed on concrete slats, rubber slats, and rubber mats had similar lying times, however, bulls on concrete slats had significantly more interruptions in lying behavior than those housed on rubber flooring (Graunke et al., 2011). One way to study lying behavior outside of lying time is by measuring lying bouts.

Holstein calves may have had a greater number of lying bouts than Jersey calves because they were significantly larger and heavier and therefore may have felt more constrained by calf hutch hardware causing increased changes in position. A similar effect was observed in adult dairy cattle housed on pasture or in stalls with cows in stalls showing a greater number of interruptions in lying behavior (Krohn and Munksgaard, 1993). Holstein calves had a slightly higher number of lying bouts d^{-1} than the 10–12 bouts d^{-1} observed in previous studies (Dellmeier et al., 1985). Jersey calves fell more closely in the range of 10–12 lying bouts d^{-1} regardless of treatment. It is possible that calves on the current study housed on mats had a higher number of lying bouts because the mats did not allow for drainage and became dirty; this effect was observed with calves housed on mats being much dirtier and wetter than those housed on gravel or sand. Interestingly calves had more variation in lying bouts during the first 3 wk of the study, this may have been due to calves adapting to their environment as the study progressed. Wet bedding was demonstrated to uncomfortable for calves and a dirty coat can cause discomfort and even wounds on the skin (Camiloti et al., 2012; Graunke et al., 2011). There is a limited amount of research, which suggests that cattle may transition from lying to standing more in environments which were aversive (Krohn and Munksgaard, 1993). It is possible that this greater surface area caused Jersey calves to be more sensitive to heat loss by conduction through the wet flooring than Holstein calves.

Holstein calves had a greater body weight and gained more weight than Jersey calves because of their larger frame size, a similar effect was observed in Holstein and Jersey calves born to pasture raised cows (Dhakal et al., 2013). These data indicated that calves gained weight and continued to grow throughout the study period regardless of bedding treatment. These data are consistent with previous research which determined that calves housed on river stones or sawdust did not differ in weight gain (Sutherland et al., 2013). This indicated that animals were healthy and were utilizing provided nutrients from milk and calf starter while undergoing rumen development (Berends et al., 2012b). Research suggested that Holstein calves' growth rate should not exceed 0.7 kg d^{-1} and Jersey calves' growth rate should not exceed 0.4 kg d^{-1} in order to reduce the risk of fat deposition in the mammary gland which can caused decreased milk production (Johnsson, 1988). Holstein calves met this recommendation as they gained appoximately 0.6 kg d^{-1}, Jersey calves also met the recommendation as they gained approximately 0.3 kg d^{-1}. It was important to note that this was a maximum recommendation and neither calf breed exceeded it, which indicated that they grew at an appropriate rate to reduce deposition of fat in the mammary.

5. Conclusions

There were no biologically significant differences in behavior or performance among bedding treatments. It is likely that calves had adequate lying time because lying time is correlated with weight gain in pre-weaned calves and calves on this study gained weight at the recommended level as the study progressed regardless of treatment. Grain intake is associated with calf growth and calves on all treatments also consumed an increasing amount of grain as they approached weaning. Use of recycled sand did not affect the lying behavior of dairy cows. There was also little effect on udder hygiene. These results as a whole suggest that bedding alone may not negatively affect welfare of pre-weaned Holstein and Jersey calves or lactating Holstein cows.

Acknowledgements

The authors are grateful for the assistance of the staff at the Little River Animal and Environment Unit for their assistance. This research was supported by State and Hatch Funds allocated to the College.

References

Anderson, V.L., R.J. Wiederholt, and J.P. Schoonmaker, 2006. Effects of bedding feedlot cattle during the winter on performance, carcass quality, and nutrients in manure. NDSU Carrington Research Extension Center Feedlot Research Report. Volume 29. Oct 10, 2006.

Berends, H., C.G. Van Reenen, N. Stockhofe-Zurwieden, and W.J.J. Gerrits. 2012b, Effects of early rumen development and solid feed composition on growth performance and abomasal health in veal calves. J. Dairy Sci. 95 (6), 3190–3199.

Bonk, S., O. Burfeind, V.S. Suthar, and W. Heuwieser, 2013. Technical note: Evaluation of data loggers for measuring lying behavior in dairy calves. J. Dairy Sci. 96 (5), 3265–3271. doi:10.3168/jds.2012-6003.

Camiloti, T.V., J.A. Fregonesi, M.A.G. von Keyserlingk, and D.M. Weary, 2012. Effects of bedding quality on the lying behavior of dairy calves. J. Dairy Sci. 95 (6), 3380–3383.

Cappel, T., A. Bueno, and E. Clemens, 1998. Calving difficulty and calf response to stress. Neb. Beef Cattle Rep. 1 (1), 17–19.

Chua, B., E. Coenen, J. van Delen, and D.M. Weary, 2002. Effects of Pair Versus Individual Housing on the Behavior and Performance of Dairy Calves. J. Dairy Sci. 85 (2), 360–364.

Davis Rincker, L.E., M.J. VandeHaar, C.A. Wolf, J.S. Liesman, L.T. Chapin, and M.S. Weber Nielsen, 2011. Effect of intensified feeding of heifer calves on growth, pubertal age, calving age, milk yield, and economics. J. Dairy Sci. 94 (7), 3554–3567. doi:10.3168/jds.2010-3923.

Dellmeier, G.R., T.H. Friend, and E.E. Gbur, 1985. Comparison of Four Methods of Calf Confinement. II. Behavior. J. Anim. Sci. 60 (5), 1102–1109.

Dhakal, K., C. Maltecca, J.P. Cassady, G. Baloche, C.M. Williams, and S.P. Washburn, 2013. Calf birth weight, gestation length, calving ease, and neonatal calf mortality in Holstein, Jersey, and crossbred cows in a pasture system. J. Dairy Sci. 96 (1), 690–698. doi:10.3168/jds.2012-5817.

Drissler, M., M. Gaworski, C.B. Tucker, and D.M. Weary, 2005. Freestall maintenance: Effects on lying behavior of dairy cattle. J. Dairy Sci. 88 (7), 2381–2387.

Fraser, D., 2008. Understanding animal welfare. Acta Vet. Scand, 50 (1), S1. doi:10.1186/1751-0147-50-S1-S1.

Haley, D.B., J. Rushen, and A.M. de Passillé, 2000. Behavioural indicators of cow comfort: activity and resting behaviour of dairy cows in two types of housing. Can. J. Anim. Sci. 80 (2), 257–263. doi:10.4141/A99-084.

Hänninen, L., A.M. de Passillé, and J. Rushen, 2005. The effect of flooring type and social grouping on the rest and growth of dairy calves. Appl. Anim. Behav. Sci. 91 (3-4)193–204. doi:10.1016/j.applanim.2004.10.003.

Graunke, K.L., E. Telezhenko, A. Hessle, C. Bergsten, and J.M. Loberg, 2011. Does rubber flooring improve welfare and production in growing bulls in fully slatted floor pens? Anim. Welf. 20 (2), 173–183.

Ito, K., D.M. Weary, and M.A.G. von Keyserlingk, 2009. Lying behavior: assessing within- and between-herd variation in free-stall-housed dairy cows. J. Dairy Sci. 92 (9), 4412–4420. doi:10.3168/jds.2009-2235.

Johnsson, I.D., 1988. Nutrition and lactation in the dairy cow. Butterworths, London. 171-192.

Krohn, C.C., and L. Munksgaard, 1993. Behaviour of dairy cows kept in extensive (loose housing/pasture) or intensive (tie stall) environments II. Lying and lying-down behaviour. Appl. Anim. Behav. Sci. 37 (1), 1–16.

Lago, A., S.M. McGuirk, T.B. Bennett, N.B. Cook, and K.V. Nordlund, 2006. Calf Respiratory Disease and Pen Microenvironments in Naturally Ventilated Calf Barns in Winter. J. Dairy Sci. 89 (10), 4014–4025. doi:10.3168/jds.S0022-0302(06)72445-6.

Mogensen, L., C.C. Krohn, J.T. Sorensen, J. Hindhede, and L.H. Nielsen, 1997. Association between resting behaviour and live weight gain in dairy heifers housed in pens with different space allowance and floor type. Appl. Anim. Behav. Sci. 55 (1-2), 11–19.

Neindre, P.L, 1993. Evaluating housing systems for veal calves. J. Anim. Sci. 71 (5), 1345–1354.

Panivivat, R., E.B. Kegley, J.A. Pennington, D.W. Kellog, and S.L. Krumpelman, 2004. Growth Performance and Health of Dairy Calves Bedded with Different Types of Materials. J. Dairy Sci. 87 (11), 3736–3745.

Stefanowska, J., D. Swierstra, A.C. Smits, J.V.V.D. Berg, and J.H.M. Metz, 2002. Reaction of Calves to Two Flooring Materials Offered Simultaneously in One Pen. Acta Agric. Scand. Sect. - Anim. Sci. 52 (2), 57–64. doi:10.1080/09064700212076.

Suárez, B.J., C.G. Van Reenen, N. Stockhofe, J. Dijkstra, and W.J.J. Gerrits, 2007. Effect of Roughage Source and Roughage to Concentrate Ratio on Animal Performance and Rumen Development in Veal Calves. J. Dairy Sci. 90:2390–2403. doi:10.3168/jds.2006-524.

Sutherland, M.A., M. Stewart, and K.E. Schütz, 2013. Effects of two substrate types on the behaviour, cleanliness and thermoregulation of dairy calves. Appl. Anim. Behav. Sci. 147 (1-2), 19–27. doi:10.1016/j.applanim.2013.04.018.

Tripon, I., L.T. Cziszter, M. Bura, S. Acatincai, D. Gavojdian, and S. Erina, 2012. The effect of space allowance on drinking and resting behaviour in six months of age calves. J. Food Agric. Environ. 10 (2), 1356–1358.

USDA, 2010. Facility Characteristics and Cow Comfort on U.S. Dairy Operations, 2007. USDA–APHIS–VS, CEAH. Fort Collins, CO.

Van Amburgh, M., 2008. Early Management and Long-term productivity of Dairy Calves. Tempe, AZ.

Webster, J., 1984a. Calf Husbandry, Health, and Welfare. Collins, London, UK.

Yanar, M., T.Z. Kartal, R. Aydin, R. Kocyigit, and A. Diler. 2010. Effect of Floor type on Holstein Calves. J. Anim. Plant Sci. 20 (3), 175–179.

Using Broiler Sound Frequency to Model Weight

Ilaria Fontana [a], Emanuela Tullo [a,*], Martijn Hemeryck [b], Marcella Guarino [a]

[a] Department of Health, Animal Science and Food Safety (VESPA), University of Milan, Via Celoria 10, 20133 Milan, Italy

[b] SoundTalks, Kapeldreef 60, 3001 Heverlee, Belgium

* Corresponding author. Email: emanuela.tullo@unimi.it

Abstract

Chicken weight provides information about growth and feed conversion in order to identify deviations from the expected homogeneous growth trend of the birds. Precision Livestock Farming (PLF) can support the farmer through the use of sensors, cameras and microphones. Previous studies showed a significant correlation ($P < 0.001$) between the frequency of vocalisation and the age and weight of the broiler. In this study, recordings were made in an automated, non-invasive way through the entire life of the birds, to evaluate the frequency variation of the sounds emitted during production cycles. In total, sound data collected during 8 production cycles (in an intensive broiler farm – 30,000 birds reared per round) were analysed. Sound data were manually and automatically compared with the weight of the birds automatically measured. Sound analysis was performed based on the amplitude and frequency of the sound signal in audio files recorded at farm level. The aim of this study was to sample automatically broiler vocalisations under normal farm conditions, to identify and model the relation between animal sounds and growth trend, and develop a tool to automatically detect the growth level of the animals based on the frequency of the vocalisation. The model used to predict the weight as a function of the Peak Frequency (PF) confirmed that the animal weight could be predicted by the frequency analysis of the sounds emitted at farm level although a more accurate editing of the audio file is necessary.

Keywords: Broiler, vocalisation, growth trend, frequency analysis, Precision Livestock Farming

1. Introduction

The world population is expected to grow throughout the rest of the century and will reach 11 billion by the year 2100 (Mountford and Rapoport, 2015), as a consequence, the world's food scenario is rapidly changing. Poultry is one of the lowest cost sources of animal protein in the world, and the demand for poultry meat is growing every year globally (Tullo et al., 2013). Livestock/crop production is becoming increasingly industrialised worldwide (Tullo et al., 2013), and nowadays the global production of broiler chickens is estimated to be of the order of 61 billion animal slaughtered per year according to FAOSTAT 2015 updated to 2013 production (FAOSTAT http://faostat3.fao.org/download/Q/QL/E). The progresses in farming technologies led to have broilers with a higher growth rate than before due to the controlled environment and the limited physical exercise (Rauw et al., 1998; Rizzi et al., 2013).

Chicken weight provides information about the growth and the feed conversion efficiency of the flock to identify deviations from the expected homogeneous growth trend of the birds (Mollah et al., 2010), also showing details about the health and welfare status of the animals. Furthermore, in close relation to chicken broiler production, growth trend is an important indicator to evaluate the body weight of the animals (Rizzi et al., 2013). Previous study showed that the Peak Frequency (PF) of the sounds emitted by the animals is inversely proportional to the bird age and the bird weight (Fontana et al., 2014).

The sounds emitted by the animals, linked to their age and weight, might be used as an early warning method/system to evaluate the health and welfare status of the animals at farm level. Therefore, animal vocalisation could be useful for assessing the animal's condition (Vandermeulen et al., 2015). Automated animal monitoring (with images or sounds) could potentially be used to support farmers in achieving farm production, and animal health and welfare goals (Costa et al., 2013), and has potential for wide application in animal husbandry (Halachmi et al., 2002; Ismayilova et al., 2013). The non-invasive nature of the audio and video equipment allows its use in long term monitoring of animals, without disturbing them (Aydin et al., 2013). Moreover, Precision Livestock Farming (PLF) may also be used to aid the management of some complex biological production processes, to measure the growth rate and to monitor the animal activity (Halachmi et al., 2002; Ismayilova et al., 2013; Tullo et al., 2013).

The aim of this study was to record and analyse both manually and automatically broiler vocalisations under normal farm conditions, to identify and model the relation between animal sounds and growth trend, for further development of a tool being able to automatically detect the growth of the animals based on the frequency of the vocalisation emitted by the birds at different ages.

2. Materials and Methods

2.1. Manual collection and analysis of sound data

Sound recordings were made at regular intervals, for the entire life of the birds, in two commercial broiler farms, one located in the UK (following the RTFA-ACP standard) and the other one located in the Netherlands. Sound recordings were made during three production cycles, two collected in the UK farm and one collected in the Dutch farm, following an automated, non-invasive and non-intrusive method in both farms.

At day one of each production cycle, 30,000 chicks were placed into the farms. In the UK farm, a professional handheld solid state recorder (Marantz PMD 661 MK II) was used, connected to a directional microphone (Sennheiser K6 / ME66" with a frequency response of 40-20,000 Hz, ± 2.5 dB). Depending on the height/age of birds, the microphone was placed at a height comprised between 0.4 m and 0.8 m, in order to keep the same distance among animals and microphones during the entire data-collecting procedure. Sound recordings were made for one continuous hour during each experimental session from day 1 to day 38 of the birds' life. Sound recordings in the Dutch farm were performed with the commercially available system (Pig Cough Monitor-SoundTalks) described in section 2.2. Broiler weights were collected automatically with a "step-on scale", placed on the floor of the houses, and were continuously stored in an online server in order to follow the growth changes in the birds.

The entire data collection from the UK farm consisted in 16 days of sound recordings for round 1, 15 days of sound recordings for round 2 for a total of 55 h 20 min of recordings. The entire data collection in the Dutch farm consisted in 42 days of recordings (24/7). In total 1008 h of recordings were collected during the round in the Dutch farm. Only 18 h of sound recordings, collected in days 1, 9, 16, 23, 30 and 37 were used for the sound analysis, so to perform a consistent comparison between the sounds collected in both farms. The weights collected in conjunction with the days chosen for the sound analysis were included in the statistical analysis. Sound analysis was performed based on the amplitude and frequency of the sound signal in audio files recorded at farm level. Sound recordings were manually analysed and labelled using sound analysis software: Adobe® Audition™ CS6. Every hour long recorded digital file was cut into shorter files of 10 minutes each in order to simplify the sound analysis. Sound labelling involved the manual extraction and classification of sounds coming from the flock on the basis of the amplitude and frequency of the signal (Tullo et al., 2013). The labelling procedure was done combining the acoustic analysis and the visual observation of the spectrogram of the whole audio file.

The Fast Fourier Transform (FFT) was used to perform the frequency analyses using a Hamming window with a FFT dimension of 256 sampled points. The frequency range was band pass filtered between 1000 Hz to 13,000 Hz. The lower frequency limit was set at 1000 Hz to remove the low frequency background noise and the upper limit was set at 13,000 Hz to cut off the high frequency noise. This choice is appropriate, as broilers are sensitive to a frequency range between 60 and 11,950 Hz (Appleby et al., 1992; Tefera, 2012).

Audio datasets from the two farms were analysed following two different methods. The labelling in the UK farm focused on the extraction of the peak frequency (PF = representing the frequency of maximum power) from individual vocalisation, while the extraction of PF from the Dutch farm was performed on the entire files (Fontana et al., 2014). The audio files, 600 individual vocalisations (50 sounds per day), chosen randomly and selected from 12 days of recordings were manually labelled and analysed. For each vocalisation sound, the PF was manually extracted and averaged for the week of age of the birds. Regarding the files recorded in the Dutch farm, the frequencies of maximum power (PF) were extracted considering the entire audio files, focusing the attention on the general sounds included between the lower (1000 Hz) and upper cut-off frequencies (13,000 Hz).

Data obtained with the manual labelling were used to predict the variation in the weight according to the changes in PF of the vocalisation of the birds (expressed in Hz) using the statistical tool PROC REG (SAS 9.3) with the following model: weight = PF.

2.2. Automated collection and analysis of sound data

The sound acquisition was performed with the SoundTalks Pig Cough Monitor. This system consists of condenser microphones (type Behringer C4) and a sound card (type ESI Maya 44). The microphones are phantom-powered and connected according to a balanced principle to allow recording over long cables with limited susceptibility to noise. The recordings are performed at 16 bit integer precision and with sampling frequency of 22.5 kHz (standard WAV file format). The sound card is placed in an embedded board (x64 architecture), running a GNU/Linux operating system. The embedded board has a passive cooling (fanless) and installed in a protective sealed enclosure. Each microphone is enclosed in a flexible plastic cover to shield it from the harsh environment whilst still allowing a proper frequency response in the range of interest. An extra acoustic shielding the size of a mortar tub is placed over the microphone to only capture the near-field sound sources, i.e., the animals directly beneath the recording microphone. The system runs a suite of diagnostics software to log the sound recording quality, its temperature and its processing load. A remote monitoring of the system is possible over the internet via a wired or wireless connection. The microphone is typically suspended from the roof at a height adjustable with a winch. For this setup, the microphone was suspended at 1 m from the ground level. The recordings are continuous, with the audio grouped in recordings of five minutes. All raw recordings were stored online on external hard drives for subsequent post processing.

A single peak frequency was determined for each 5-minute length raw audio recording. For each recording, a power spectral density (PSD) was calculated using Welch's method 2. The PSD was calculated with a frequency resolution of 256 bins and an overlap of 50%. The window used was the Hanning window. The spectral content was transformed to a single sided representation, resulting in a linear frequency axis of 129 bins, ranging from DC to 11.025 kHz. The peak frequency was determined as the argument of the maximum spectral content obtained. For the determination of this maximum, only frequency bins over 1000 Hz was used, effectively filtering out any lower frequency noise to remove the ventilation and feeder lines noises. Furthermore, the peak frequencies extracted automatically were manually edited in order to avoid the influence of the outliers. The frequencies chosen as a threshold were 1,100 Hz and 3700 Hz according to a previous study (Fontana et al., 2014). In this way, the sounds automatically collected during the dark period (background noises, without vocalisations) were eliminated from the data set.

Automated collected sounds were then associated to the daily weights of broiler chickens, which were also automatically collected. Five rounds, from April 2014 to September 2014, of weights and frequencies data were used in the analysis. In total 29,700 audio files were used for the automated frequency analysis. Observed and expected weights were compared with the TTEST procedure (SAS 9.3), firstly on the general trend and successively week by week.

3. Results and Discussion

Table 1 shows the data found and used in the manual collection and analysis procedure of sound data collected during three rounds in the two farms. For each round, the age of the birds, the PF (± St. Dev.) and the average weight (± St. Dev.) of the broilers are reported. The mean PF values of the UK farm are referred to average values obtained focusing the attention on individual vocalisations recorded at the same days of the production cycle during the two different rounds. Meanwhile, the PF values of the sounds collected in the Dutch farm were acquired considering the entire sound file recorded.

Table 1. Data found and used in the manual collection and analysis procedure of sound data collected during three rounds in the two farms. Age of the birds, PF (± St. Dev.) and average weight (± St. Dev.) of the broilers collected in the three rounds in the two farms.

Farm	Week	Round	Age	Mean PF (Hz) (± s.d.)	Mean weight (g) (± s.d.)
K	1	1	1	3545 (± 365)	44.56 (± 1.5)
	2	1	8	3059 (± 94)	198.64 (± 10.1)
	3	1	15	2618 (± 360)	550.3 (±21.7)
	4	1	22	2329 (± 605)	1039.5 (±68.6)
	5	1	29	1943 (± 569)	1529 (±120.5)
	6	1	36	1506 (± 434)	2104.28 (±208.5)
Dutch	1	1	1	3388 (± 258)	55.6 (± 2.2)
	2	1	9	2972 (± 0)	260.7 (± 11.36)
	3	1	16	2225 (± 130)	689.38 (± 16.09)
	4	1	23	2067 (± 47)	1103.26 (± 11.41)
	5	1	30	2024 (± 0)	1570.86 (± 11.86)
	6	1	37	1077 (± 0)	2022.11 (± 38.05)
Dutch	1	2	1	3621 (± 402)	40.72 (± 4.9)
	2	2	9	2953 (± 353)	231.42 (± 12.1)
	3	2	16	2474 (± 384)	608.66 (±26.7)
	4	2	23	1955 (± 520)	1092.84 (± 74.4)
	5	2	30	1902 (± 585)	1731.6 (± 130.3)
	6	2	37	1475 (± 493)	2275.44 (± 247.0)

Figure 1 shows the regression analysis performed to predict the weight of the broilers as a function of the PF emitted. The regression model is significant ($F = 90.90$, $P < 0.001$), indicating that the model accounts for a significant portion of variation in the data. The R^2 indicates that the model accounts for 85% of the variation in weight. The confidence interval of the observed values (between l95% and u95% in Figure 1) shows a 95% probability that the true linear regression of the population lay within the confidence interval of the regression line calculated from the sample data. The regression coefficients predicting the weight as a function of the PF, was used to predict the broiler expected weight on sound data collected and analysed automatically. The regression coefficients used were:

$$\text{Weight} = 2887.8 - 0.8861 \times pf \quad (R^2 = 0.85) \tag{1}$$

where, PF was the Peak Frequency of the sounds (Hz) emitted by the birds.

Automated weights for each round considered in the validation of the regression line are reported in Table 2. The datasets of round 3, 4 and 5 are not complete due to technical problem occurred during data collection, principally due to the loss of internet connection. Sound data were collected in real time and stored through internet connection in order to easily access the recordings; for this reason, the lack of internet connection caused a considerable loss of data.

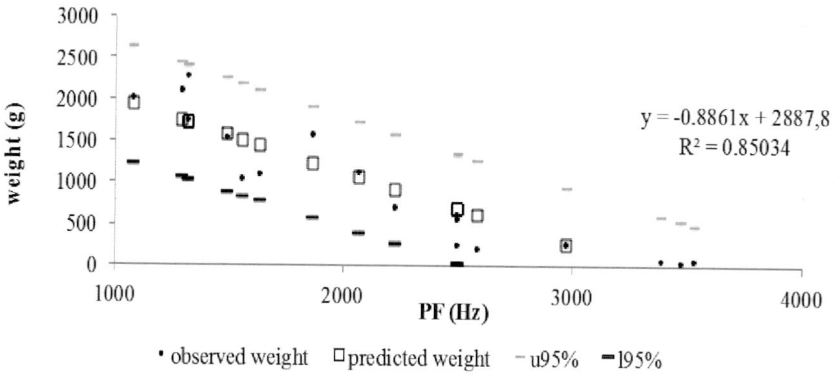

Figure 1. Linear regression to predict the weight of the broilers as a function of the PF emitted. The u95% and l95% show the 95% confidence interval of the observed values.

Table 2. Mean and standard deviation of the automated measured weights at the beginning of each week of the cycle production (rounds) used for the validation.

Round	Weeks	Mean (g)	STD (g)
1	1	73.52	14.49
	2	280.58	34.36
	3	589.82	54.36
	4	1000.76	92.98
	5	1507.69	105.05
	6	1894.33	66.54
2	1	75.11	15
	2	296.99	48.54
	3	667.58	54.92
	4	1148.38	60.57
	5	1500.69	71.26
	6	2064.19	89.13
3	2	344.63	37.89
	3	773.9	351.07
	4	1141.72	58.84
	5	1615.53	77.33
	6	2511.03	363.3
4	4	1108.19	56.36
	5	1569.7	35.86
	6	2039.76	102.7
5	1	66.41	13.31
	2	277.75	44.51
	3	682.25	68.43

The trend in PF reported in Figure 2 shows clearly how the frequency of the vocalisation changes according to the broiler age. At the beginning of the round, the peak frequencies resulted on average 3,136±63 Hz, while around day 40 the average peak frequencies resulted 1,230±70 Hz (Fontana et al., 2014). In this case, round 3, 4 and 5 were not complete due to internet connection problems on farm during the data collection.

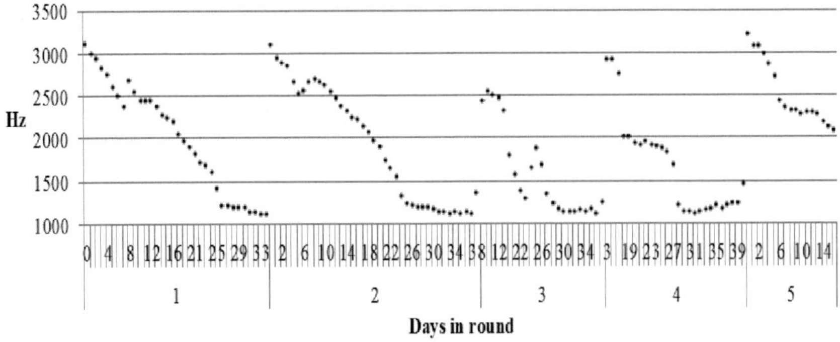

Figure 2. Peak frequencies trends during the five considered rounds. Data in rounds 3, 4 and 5 were not complete for the whole rounds.

Expected weights for each week of age were obtained applying the equation (1) to PF data extracted automatically from sounds file collected in the broiler house. Observed and expected values are reported in Table 3.

Table 3. Observed and expected weights of broiler chickens obtained applying the equation weight = 2887.8 - 0.8861 × pf

Week	Expected weight	Observed weight
1	30.55	71.68
2	704.46	285.11
3	826.71	604.14
4	1215.90	1079.41
5	1736.62	1494.87
6	1868.04	2001.91

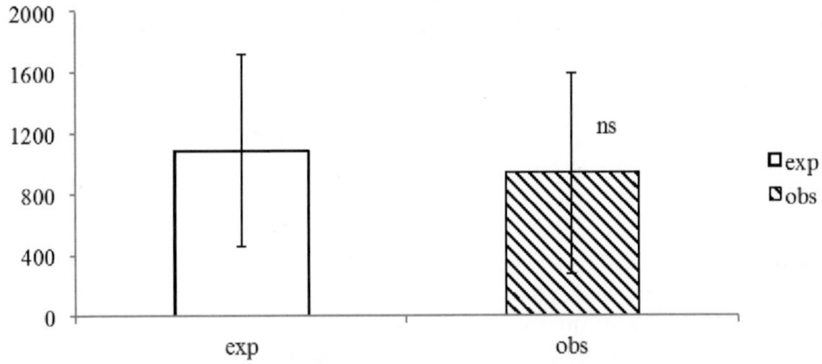

***: P < 0.001; **: P < 0.01; *: P < 0.05; ns: P > 0.05

Figure 3. Levels of significance of the differences between general observed and predicted weight.

The TTEST procedure was performed to evaluate the general difference between observed and expected values. The results (Figure 3) show no significant difference between expected and observed values. However, when the TTEST procedure was performed on expected and observed weights week by week, the results showed different results (Figure 4). Indeed, considering the expected weights for each week of age as a separate value, the statistical difference varied considerably.

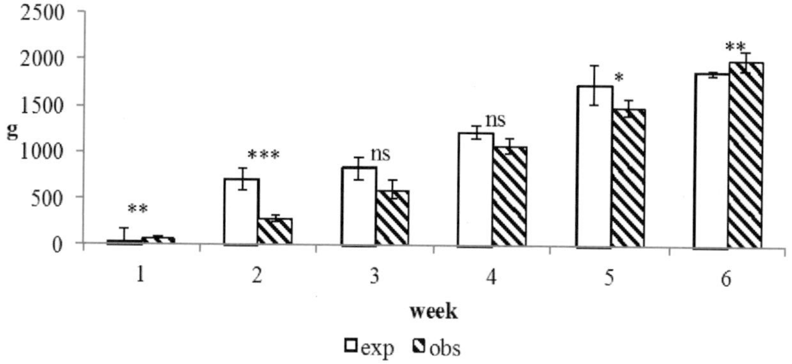

***: P < 0.001; **: P < 0.01; *: P < 0.05; ns: P > 0.05

Figure 4. Levels of significance of the differences between observed and predicted weight performed week by week.

As reported in Figure 4, only expected weights for the week 3 and 4 resulted are not significantly different from the observed ones. The remaining predicted values differed significantly from the observed values. This discrepancy between observed and predicted values could be ascribed to different reasons. First of all, the lack of completeness of considered data, both for weight and peak frequencies, that could have affected negatively the results. Then, the original data was used to estimate the regression coefficients to predict the weight as a function of the PF. Indeed, PF values were obtained from single vocalisations manually chosen and extracted from the totality of the audio files, based on the clarity of sound and spectrum. In this study, the peak frequencies were automatically extracted from 5 minutes recorded audio file, applying a filter to exclude frequencies below 1,000 Hz and mildly edited to remove outliers relative to the dark periods. This editing procedure was used to test whether the automated recognition of PF of the vocalisation could be precise enough in predicting the weight of the broiler chickens, but a more accurate extraction procedure is needed in order to reduce the difference between predicted and observed .

4. Conclusions

The results of the present study confirmed how the PF of the sounds emitted by the broilers changes according to the age and the weight of the birds. However, the automated frequency analysis of the audio files should be improved.

The results indicate that the PF of the sounds emitted by the animals is inversely proportional to the age and the weight of the broilers. The model implemented to predict the weight as a function of the PF by frequency analysis of the sounds emitted at farm level was proven to be reasonably accurate although a more accurate editing of the audio file is recommended. The simplification of the labelling procedure of audio files through the use of the automated PF extraction procedure, even though imprecise, could be the basis for the creation of an accurate weight prediction algorithm based on sounds emitted by the broilers. However, up to now, the audio files have to be checked manually before performing the PF analysis automatically, in order to have reliable results.

Acknowledgements

This project was funded by the European project no: 311825 EU-PLF (Animal and farm-centric approach to precision livestock farming in Europe), co-financed by the European Commission.

References

Appleby, M.C., B.O. Hughes, H.A. Elson, 1992. *Poultry production systems: behaviour, management and welfare*. Wallingford, Oxon, UK: C.A.B. International.

Aydin, A., C. Bahr, D. Berckmans, 2013. A relational study of gait score with resting behaviours of broiler chickens. In: *Precision Livestock Farming 2013 - Papers Presented at the 6th European Conference on Precision Livestock Farming, ECPLF 2013*. 597–606.

Costa, A., G. Ismayilova, F. Borgonovo, T. Leroy, D. Berckmans, M. Guarino, 2013. The use of image analysis as a new approach to assess behaviour classification in a pig barn. Acta Veterinaria Brno. 82 (1), 25–30.

Fontana, I., E. Tullo, A. Butterworth, M. Guarino, 2014. Broiler vocalisation analysis used to predict growth. In: *Proceedings of Measuring Behavior 2014*. Wageningen, The Netherlands., August 27-29. Wageningen, The Netherlands. Ed., A.J. Spink, L.W.S.L., M. Woloszynowska-Fraser & L.P.J.J. Noldus. .

Halachmi, I., J.H.M. Metz, A. Van't Land, S. Halachmi, J.P.C. Kleijnen, 2002. Case study: Optimal facility allocation in a robotic milking barn. Transactions of the American Society of Agricultural Engineers. 45 (5), 1539–1546.

Ismayilova, G., A. Costa, I. Fontana, D. Berckmans, M. Guarino, 2013. Labelling the behaviour of piglets and activity monitoring from video as a tool of assessing interest in different environmental enrichments. Annals of Animal Science. 13 (3), 611. http://dx.doi.org/10.2478/aoas-2013-0037.

Mollah, M.B.R., M.A. Hasan, M.A. Salam, M.A. Ali, 2010. Digital image analysis to estimate the live weight of broiler. Computers and Electronics in Agriculture. 72 (1), 48–52. http://dx.doi.org/http://dx.doi.org/10.1016/j.compag.2010.02.002.

Mountford, A., H. Rapoport, 2015. Migration Policy, African Population Growth and Global Inequality. The World Economy.

Rauw, W.M., E. Kanis, E.N. Noordhuizen-Stassen, F.J. Grommers, 1998. Undesirable side effects of selection for high production efficiency in farm animals: a review. Livestock Production Science. 56 (1), 15–33. http://dx.doi.org/Doi 10.1016/S0301-6226(98)00147-X.

Rizzi, C., B. Contiero, M. Cassandro, 2013. Growth patterns of Italian local chicken populations. Poultry Science. 92 (8), 2226–2235.

Tefera, M., 2012. Acoustic signals in domestic chicken (Gallus gallus): a tool for teaching veterinary ethology and implication for language learning. Ethiopian Veterinary Journal. 16 (2), 77–84.

Tullo, E., I. Fontana, M. Guarino, 2013. Precision livestock farming: An overview of image and sound labelling. In: Precision Livestock Farming 2013 - Papers Presented at the 6th European Conference on Precision Livestock Farming, ECPLF 2013. 30–38.

Vandermeulen, J., C. Bahr, E. Tullo, I. Fontana, S. Ott, M. Kashiha, M. Guarino, C.P.H. Moons, F.A.M. Tuyttens, T.A. Niewold, D. Berckmans, 2015. Discerning pig screams in production environments. PLoS ONE. 10 (4), e0123111. http://dx.doi.org/10.1371/journal.pone.0123111.

Technical Resources for Swine Production Systems in the U.S.

Teng-Teeh Lim *, Joseph M. Zulovich

Food Systems and Bioengineering, University of Missouri, Columbia, MO 65211-5200, USA

* Corresponding author. Email: limt@missouri.edu

Abstract

The swine production industry in the US has gone through a lot of changes in the last few decades. In order to stay competitive, the swine operations and their managers are experts in using improved breeding stock to grow pigs faster and produce leaner pork with less feed and pollution. Climate-controlled buildings, computerized information systems, and excellent disease control programs are just some of the tools needed to stay on top of business. There have been useful resources that are helpful to provide consideration of either starting or expanding a swine enterprise, determining type of operation that best meet one's management and production goals. This paper discusses important topics to be covered when designing or improving an operation, and highlights some available publications, websites, and models. The resources provide critical information and insight that might be required as part of a business plan for producers considering different types and sizes of pork production systems. For example, the Missouri Swine Enterprise Manual provides suggestions and recommendations to enhance pork production, including: breeding, genetics, economics, health, nutrition, meat quality, structures, ventilation, and environmental nutrient management. There are many examples of different types and sizes of sample operations, each was presented as a SWOT analysis (Strength, Weakness, Opportunity and Threat). It is important to make the information available to stakeholders and producers who can benefit from them, to avoid duplicating the effort, and more importantly, to keep improving the production systems for a better environment and better profitability.

Keywords: Business plan, sample operations, SWOT analysis, recommendations, pork production

1. Introduction

Livestock and poultry production in the Unites States has moved away from the small barnyard type of operations. Many of the modern operations are utilizing highly specialized buildings and engineering technologies for mass production, and are often called "Concentrated Animal Feeding Operations" (CAFOs). Prior to the late 1990s, hundreds of thousands of small hog farms were the main streams of pork production, but nowadays a large portion of the pork production comes from a small number of CAFOs. According to a U.S. Department of Agriculture (USDA) report, operations with 2,000 or more hogs accounted for 87 percent of the inventory in the U.S. (USDA, 2013).

In order to stay competitive, the swine operations use improved breeding stock to grow pigs faster and produce leaner pork with less feed and pollution. For example, the producers use to acquire breeding stock from many different independent breeding stock producers. The genetic makeup of hogs varied widely, and the many different management and feeding systems resulted in large variations in the nation. However, the concentrated and intensified operations means fewer and specific companies are offering selected breeding stocks with excellent genetics for pigs to grow faster and more lean meat, within a well-maintained environment.

Climate-controlled buildings, computerized information systems, and excellent disease control programs are just some of the tools needed to stay on top of business. There also have been more challenges associated with hog production. Rising costs, lower profit margin, environmental concerns, disease, food safety issues, and social controversies have made hog production more challenging. The producers need to be responsive to the new challenges, thus staying informed of the latest news, research findings, technology development, and utilizing the right information and tools to make the best decisions are critical to their operations.

It took the U.S. pork production decades to evolve to its current stage, a lot of lessons have been learned in the process. There have been useful resources that are helpful to provide consideration for either starting or expanding a swine enterprise or determining the type of operation that best meet one's management and goals. The experience and resources can be meaningful and useful to other countries, especially those who are experiencing similar production growth and system evolution. It is important to make the information available to stakeholders and producers, to avoid duplicating the effort, and more importantly, to keep improving the production systems for a better environment. The objectives of this paper are to highlight and discuss some of the available U.S. publications, websites, and models that provide additional insight and considerations that might be required as part of a business plan for producers considering different types and sizes of pork production systems.

2. Materials and Methods

When producers are considering starting, changing or expanding hog production in their portfolio of farm enterprises, many decisions will have to be made in advance. It is critical to make the best decisions for starting/operating an operation that best fit the overall and long-term goals. Often the new or expansion operation is limited by the capitalization requirements, and the mix of capital, management, labor, and type of operation can determine the vulnerability of your operation to the market variation. Fortunately there have been experiences shared by previous entrepreneurs that are documented. This information can be obtained through publications, websites, and models.

This section lists the available resources that range from assessing one's strength and interests, to types of operation, finance and business arrangement, facility layout, construction and operation, nutrition and feeding, herd health, disease control, meat quality, environmental, manure nutrient management, regulatory topics, and labor. These U.S. resources are often provided and organized by University Extension Teams (such as the University of Missouri Swine Resource Guide, http://swine.missouri.edu/), USDA Extension Service (http://www.extension.org/swine), the commodity groups (Pork Checkoff factsheets, https://www.pork.org/fact-sheets-brochures/), and certain land grant universities (http://www.extension.org/).

The subsections discuss a topic of importance and provide relevant resources (factsheets, website, decision making tools, computer programs, etc.). There are also comprehensive industrial handbook or manual type of materials that are written as educational documents which provide insight and considerations that might be required as part of a business plan for producers. The Missouri Swine Enterprise Manual (http://extension.missouri.edu/p/MM102) contains different pork production systems, an interactive cash flow model with base-line economic considerations for 14 different operation scenarios, and many other practical topics that are critical to the success of a swine operation (Swine_Focus_Team, 2003).

2.1. Swine operation options

Resources for determining the type and size of an operation is a very important part of the planning stage by providing considerations that best meet one's management and goals. The Missouri Swine Enterprise Manual lists 14 types and sizes of sample operations to start the entire educational series. The sizes for the different operations are nominal sizes, and are related to the approximate inventory of the different operation, Table 1.

Under each type of common commercial operation, explanations were given that highlights the important information, such as the range of pig weights, if pigs need to be procured, labor requirement assumptions, and certain tasks that are not included in the labor assumptions. An analysis of Strength, Weakness, Opportunity and Threat (SWOT) is also included for each type of operation that is presented. The analysis provides a systematic explanation of the different types of operation. An example of the SWOT is provided in Table 2, for the 4000-hd wean-to-finish operation. The University of Missouri Swine Resource Guide, (http://swine.missouri.edu/econ/) offers a few economic impact factsheets for some of these operations. The factsheets, in a summary form, report estimations for construction and equipment costs, gross incomes, labor, taxes, insurance, and even manure values and land based needed for the manure application.

Table 1. The 14 types of swine operation models provided in the Missouri Swine Enterprise Manual.

Operation Type	Herd Size	Special Note
Farrow-to-Wean	1,200-Sows	
Farrow-to-Wean	2400-Sows	
Farrow-to-Finish	150-Sow	Monthly production
Farrow-to-Finish	300-Sow	Biweekly production
Farrow-to-Finish	600-Sow	Weekly production
Farrow-to-Finish	1200-Sow	Weekly production
Feeder-to-Finish	2000-Head	
Feeder-to-Finish	4000-Head	
Feeder-to-Finish	8000-Head	
Wean-to-Finish	2000-Head	
Wean-to-Finish	4000-Head	
Wean-to-Finish	8000-Head	
Nursery	2300-Head	
Nursery	4600-Head	

Table 2. Summary of a "Strength, Weakness, Opportunity and Threat analysis" for a 4000-hd wean-to-finish operation.

Analysis	Description
Strength	- Does not require full time employment - Can complement other farm enterprises or an off-farm job - Capital investment is comparable to a 300 sow farrow-to-finish operation - Produces a volume that marketing contracts can be obtained - Labor efficiencies are improved when compared to the 2,000 head facility
Weakness	- Is a CAFO and requires a regulatory operating permit - Capital investment is larger which increases financial risk
Opportunity	- Can be a contract basis to minimize risk and allow potential expansion, which also reduces time needed for feed acquisition, pig procurement, and marketing decisions, and expertise to assist with health and production issues is provided - Labor to clean facilities and move pigs is reduced - Commingling of pig sources may be allowed because of the high antibody presence in early-weaned pigs
Threat	- Neighbor relations and community acceptance become more complicated as unit size increases - Contract production can be subject to cancellation or not be renewed

2.2. Business arrangement and finance consideration

After analyzing the different type and size of operation, information is needed to identify the size, phase of production and business arrangement that fit one's available resources and production goals. The Missouri Swine Enterprise Manual also provides detailed numbers of initial investment, annual revenue and expenses, annual net income and net income ratio for each of the 14 pork production models. There is also discussion of different business arrangements and advantages of each, while two particular types, the sow coops and contract production were discussed in detail.

Estimates of annual revenue and expenses are important to consider, to make sure there is sufficient capital for building, and have enough to populate and feed the animals. The initial investment can be obtained through normal lending channels such as banks or other credit services. These lending institutions will often require a detailed balance sheet and cash flow projection. The manual includes some of the more important ratios and their normal range of values, and includes a spreadsheet sample that is useful in calculating and analyzing financial ratios.

Using a model to simulate the effects of certain events or decisions to an operation is a great way to better understand the risks and decisions associated with a new operation. The Missouri Swine Enterprise Manual provides an educational software package that models different swine enterprise types and allows changing the nutrition, facilities, prices, financing and time on a particular swine production system. The software also simulates a disease outbreak, by changing assumptions for average daily gain, feed efficiency, death loss and/or breeding and farrowing efficiencies.

2.3. Facility layout, design, construction, and operation

There are many factsheets and publications that specify the requirements, recommendations, and details regarding the swine operation facility. The Missouri Swine Enterprise Manual dedicates an entire chapter that discuss the considerations. The common goals are to provide and maintain the thermal and indoor environmental conditions needed for good pig performance throughout the year. An effective swine production building must first be properly sited to minimize any off site impacts; while considering site limitations and management preferences.

2.3.1. Site selection and tools

When designing a new operation, one must consider topography, local weather, utilities location and source; vehicle traffic patterns; manure management; biosecurity; and other factors such as flood plan and sensitive watershed area, can affect the physical location of facilities. Most of the considerations will be site specific and are generally unique to each farm. A Missouri Extension Factsheet, "Selecting a Site for Livestock and Poultry Operations, EQ378" provides concise considerations for selecting a site for an animal feeding operation (Pfost and Fulhage, 2009).

There are open-access online tools that one can use to identify a piece of property on interactive map (similar to maps seen in internet browser software), to gather a series of information. The Missouri Animal Feeding Operation (AFO) Site Assessment is such a tool that provides information including the exact property location, any community and biosecurity concern (listing sensitive community facilities and CAFOs within certain distance), hydrologic and geological information (watershed, Karst topography and sinkholes, and floodplain) and local regulation and environmental setbacks (http://nmplanner.missouri.edu/tools/afosite.asp/). This tool also provides information of the types and percent of area of soil, slope, potential farm land for manure nutrient application, and estimated animal density. Local weather information such as rainfall and windrose data (wind speed, direction, and frequency) are also provided. Another similar web-based tool is named Business Environmental Risk Management (BERM, http://ims.missouri.edu/berm/), which provides similar information but lacking some information such as local regulations and nearby land application areas and animal density.

2.3.2. Odor setback models and tools

In addition to the considerations mentioned in previous subsection, proper distances between an operation (existing, expansion, or new) and nearby residents are critical to minimizing odor nuisance complaint. Odor setback distance models can be used to estimate the distances, based on many farm-specific and local weather and geographical variables. The Purdue odor setback model (https://engineering.purdue.edu/~odor/setback.htm) and University of Minnesota Setback Estimation Tool (http://www.extension.umn.edu/agriculture/manure-management-and-air-quality/feedlots-and-manure-storage/offset-odor-from-feedlots/) are both open-access tools that can be applied for setback estimation. The models can also be used to evaluate how much odor mitigation is needed for solving a farm-specific odor problem.

There are many odor mitigation options, that careful consideration must be given when selecting a practice or technology that best fit one's operation, management, and budget. An on-line tool, the National Air Quality Site Assessment Tool (NAQSAT, http://naqsat.tamu.edu/) was designed to provide evaluation and comparison of different management and mitigation strategies for air emissions, including odor. A shelterbelt is a group of trees or shrubs planted around an odor source (production facility) to facilitate mixing of air and a reduction in odor transport. Well planned and developed shelterbelts can also improve the visual aesthetics of a site, which helps minimizing odor nuisance problems. Detailed information are available to help with designing, implementing and maintaining a shelterbelt (http://agebb.missouri.edu/commag/shelterbelt/).

2.3.3. *Facility design and construction*

Detailed information for facility design and construction parameters are included in the Missouri Swine Enterprise Manual. A selection of building plans, drawings, and manuals are available at the University of Missouri Swine Resource Guide, (http://swine.missouri.edu/facilities/). Standards (building codes) for the facility to be built is also discussed, which consider appropriate design requirement, such as snow, live and dead loads for the roof structure, wind loads (based on >80 mph wind speed), and insulation levels, etc. The recommended inside relative humidity range, allowable gas concentrations, effective environmental temperature, and ventilation control, are also discussed.

2.3.4. *Ventilation system design and operation*

Ventilation system is a key factor that affects the thermal and air quality of the pig environment. Either types of ventilation, mechanical and natural, has functional components such as inlet, outlet, driving force, distribution, and a path, and are discussed in relation with cold and hot weather management. Different mitigation strategies can be used to help reduce heat stress conditions for the pigs.

2.3.5. *Manure storage and handling*

The manure handling system can have a significant impact on the daily and long-term operation of a swine facility. Careful consideration and planning should be given to the type and duration of storage, treatment, and land application. Factors to consider include: location and topography of the operation, quantity and characteristic of the manure, operating costs, and supervision and maintenance. The ultimate objectives of the manure handling system is to make sure the manure is removed from the production facilities in an environmentally acceptable manner. In the US, individual states have their own regulatory agency that oversee the manure management, the Missouri Department of Natural Resource provides a concise design guide for the design of waste management systems for concentrated animal feeding operations (DNR, 2013).

2.4. Manure nutrient management plan

A well designed manure management program must consider long-term nutrient management planning, economics of manure, regulatory issues, records and inspection, biosecurity, and the potential by-product production and use of energy recovery from manure. In the U.S., most large livestock and poultry operations are required to have a plan called "Nutrient Management Plan" (NMP), which documents detailed practices and strategies to address environmental concerns related to soil erosion, livestock manure application and disposal of organic by-products. Because the generation and modification of such plans often required comprehensive engineering and conservation planning resource assessment, including expertise in crop nutrient removal, soil nutrient availability, soil and manure test results interpretation, many CAFOs have trained or even certified specialists that manage the NMPs. Specialists in the USDA, Natural Resource Conservation Services, and trained agronomists/engineers in private sectors are often the service provider of the NMPs.

The University of Missouri Nutrient Management Website (http://nmplanner.missouri.edu/) offers a series of information and tools for manure management, NMP, regulations affecting animal feeding operations. There are open-access NMP related tools designed for Missouri, including Missouri Nutrient Management Plan Generators (web-based tool to generate NMPs, including yearly, field-by-field recommendations for manure and fertilizer management), Commercial Fertilizer Plan, Plant-available Nutrients Calculator, and Phosphorus Index (to identify fields that have a high probability phosphorus loss in surface water runoff).

The website also lists local weather information, (station network and database, emergency wet weather event management, and weather news), certified manure testing laboratories, and licensed contract custom applicators. For information that is more general, the eXtension website (http://www.extension.org/animal_manure_management) collects and provides information from various Land Grant universities, and allows users to submit questions to experts of various fields, through the "Ask an Expert" webpage.

2.5. Biosecurity

Producers need to implement appropriate biosecurity to effectively reduce the risk of disease outbreak and pathogens being transferred from farm to farm. A very complete and well documented biosecurity factsheet covers all aspects in great details, ranging from farm location, sources of diseases, isolation, pig flow, worker training, hygiene, air filtration, to even wild life control (Levis and Baker, 2011). In the recent wake of Porcine Epidemic Diarrhea virus (PEDv) and Porcine Reproductive and Respiratory Syndrome (PRRS), more producers are taking the biosecurity even more seriously.

The National Pork Board and collaborators have put together a biosecurity document with checklists under different category (NPB, 2002). Producers or managers can use the checklists to first evaluate their operations by identifying biosecurity strengths and weaknesses. Because every farm is unique and has different management, and there are always new disease identified, an effective biosecurity plan is one that has protocols reviewed and updated frequently. It is important to stay informed of the new disease and biosecurity recommendations, and actively reevaluating the existing biosecurity plan to address areas that need improvement. For example, there have been reports about the PEDv extended survival in manure and transfers via manure application; therefore, new recommendations have been developed for producers (NPB, 2013a), custom manure applicators (NPB, 2013b), and land owners.

2.6. Nutrition and feeding management

A profitable swine operation must successfully manage the nutrition and commodity buying since 55% to 70% of the total cost of swine production is feed (Swine_Focus_Team, 2003). In addition, the feeding program will need to be adjusted due to variations in facility design, environment, feed costs, and genetic potential for lean deposition. For example, the use of distiller's dried grains with soluble (DDGs) as feed alternative due to lower price has become more popular. The producer will need to evaluate the impacts of feed conversion and costs, of using different feed rations. A spreadsheet model was developed by the University of Missouri Extension for this purpose (DDGS Cost Management Tool, http://swine.missouri.edu/nutrition). For a comprehensive review of the swine nutrition and ingredient composition, and tool to formulate the different balanced feed rations, the 11th revised edition of the Nutrient Requirements of Swine published by the National Research Council is a great resource (NRC, 2012).

2.7. Apps available

There have been more and more "Apps" (Application software) developed for the livestock and poultry industries. The apps are programs or software designed for a particular purpose, and are usually installed onto mobile devices such as tablet or mobile phones. A few examples include Heat Safety Tool (by Department of Labor, Occupational Safety and Health Administration, calculates the heat index and risk level), Thermal Aid (combines weather and/or respiration rate of livestock weather data to help determine if livestock is affected by heat stress, and provides tips to minimize the effects of heat), Ncalc (manages nitrogen loading based on nitrates regulations, considering numbers of livestock and the land area), Manure Pit Calculator (calculated capacity of round, square or rectangular pits), etc. These apps can be useful and important because they are convenient and accessible, for producers to retrieve either farm-specific or new information to help make decisions.

3. Discussion and Conclusions

It took the U.S. pork industry decades to evolve to its current stage, valuable lessons have been learned and documented in the process. When starting or expanding a swine operation, it is critical to consider and plan the various financial and production details, for designing a specific operation that best meet one's management, capacity, and goals. This paper highlights some of the important design, operation, and management considerations and matching some available publications, websites, and models that provide additional insight and considerations. For example, the Missouri Swine Enterprise Manual was written by a team of University Faculty to provide suggestions and recommendations to enhance pork production including: breeding, genetics, economics, health, nutrition, meat quality, structures, ventilation, and environmental nutrient management.

In the manual, different types and sizes of sample operations and SWOT (Strength, Weakness, Opportunity and Threat) analysis are important especially at the beginning of the planning stage. Because the industry is constantly adapting new technology and improving the production efficiency, and diseases are also evolving, it is critical for the producer and managers to stay informed and be educated with the improved and proven practices and management. There are several Land Grant University Extension Groups (Kansas State University, Iowa State University, University of Minnesota, Purdue University, University of Missouri, etc.) that provide well organized and useful tools and publications that can be helpful in the different aspect of swine operations.

Although the experience, information, and resources listed in this paper are mostly unique to the US and especially Mid-Western swine operation, the experience and resources can be meaningful and useful in other countries, especially those who are experiencing the similar production growth and system evolution. It is important to make the information available to stakeholders and producers who can benefit from them, to avoid duplicating the effort, and more importantly, to keep improving the production systems for a better environment. Similar to reviewing the biosecurity checklist, one should probably re-evaluate goals or think of a new idea when going over the different design and management considerations, thus improving an existing or new operation.

References

DNR, 2013. 10 CSR 20-8.300 Manure Storage Design Regulations. Missouri Department of Natural Resources. http://dnr.mo.gov/env/wpp/cwforum/docs/8300draft.pdf. Accessed June 3, 2015. 20 p.

Levis, D.G., and R.B. Baker, 2011. Biosecurity of Pigs and Farm Security. University of Nebrask-Lincoln Extension. http://www.ianrpubs.unl.edu/epublic/pages/index.jsp?what=publicationD&publicationId=1431, 31 p.

NPB, 2002. Biosecurity Guide for Producers. National Pork Board. http://porkcdn.s3.amazonaws.com/sites/all/files/documents/Biosecurity/BiosecurityBook.pdf. Accessed June 1, 2015.

NPB, 2013a. Biosecure Manure Pumping Protocols for PED Control: Recommendations for Pork Producers. National Pork Board. http://swine.missouri.edu/manure/PEDproducer.pdf. Accessed June 2, 2015.

NPB, 2013b. Biosecure Manure Pumping Protocols for PED Control: Recommendations for Commercial Manure Haulers. National Pork Board. http://swine.missouri.edu/manure/PEDhauler.pdf. Accessed June 2, 2015.

NRC, 2012. Nutrient Requirements of Swine: Eleventh Revised Edition. National Research Council. 400 p.

Pfost, D., and C. Fulhage, 2009. Selecting a Site for Livestock and Poultry Operations. http://extension.missouri.edu/p/EQ378. Accessed June 1, 2015. University of Missouri Extension, 6 p.

Swine Focus Team, 2003. *Missouri Swine Enterprise Manual*: Commercial Agriculture Program, University of Missouri. http://extension.missouri.edu/p/MM102.

USDA, 2013. National Agricultural Statistics Service. Charts and Maps: Hogs: Operations and Inventory by Size Group, US. http://www.nass.usda.gov/Charts_and_Maps/Hogs_and_Pigs/hopinv_e.asp. Accessed June 20, 2015.

Boosting the Economic Returns of Goose Breeding and Developing the Industry by Controlled Photoperiod for Out-of-Season Reproduction

Zhendan Shi [a,*], Aidong Sun [b], Xibing Shao [c], Zhe Chen [a], Huanxi Zhu [a]

[a] Institute of Animal Science, Jiangsu Academy of Agricultural Sciences, Nanjing, 210014, China

[b] Institute of Food Safety and Quality Inspection, Jiangsu Academy of Agricultural Sciences, Nanjing, 210014, China

[c] Sunlake Swan Farm, Changzhou 213101, Jiangsu Province, China

* Corresponding author. Email: zdshi@jaas.ac.cn

Abstract

Goose breeds throughout China stop laying eggs during the period from spring to late summer, which leads to gosling and egg prices to skyrocket in this period. To maximize economic returns, artificial photoperiodic programs were developed to induce out-of-season egg laying. For short day breeders such as the Magang and Shitou geese in Guangdong Province, egg laying was interrupted in winter by using 18 h photoperiod for 75 days. Egg laying was re-initiated by decreasing the daily photoperiod to 11 h in spring, from March to April. During summer months, geese maintained under 11 h photoperiod exhibited sound reproductive activities, as manifested by high egg laying rate, fertility and hatchability. For long day breeding Yangzhou goose, a 2-phase photoperiod program was used, which consisted of a short daily photoperiod of 11 h for 3 months starting in winter, and the long daily photoperiod of 11 h to 12 h thereafter from spring onwards. Apart from using of an artificial photo-program, mechanical ventilation and water pad cooling of goose house are also required for maintaining sound reproductive performances for Yangzhou goose in Jiangsu Province, where summer can be extremely hot. Out-of-season breeding, which makes gosling production coincide with the annual price peak, has increased economic profit per goose by 3 to 6 times more than those by the conventional production. Over the years, the out-of-season goose breeding has helped farmers to expand and modernize the production, and significantly increase the profit margin of a relatively primitive geese industry.

Keywords: Goose, out-of-season production, economic returns, industry development

1. Introduction

Domestic geese have long been supplying meat, fatty liver, and byproducts such down, in numerous parts of the world. The goose industry in China is especially prominent with an annual output accounting for approximately 95% of the world's total production: over 600 million geese slaughtered and more than 250 m tones of meat produced (Shen et al., 2011). All the goose breeds used in China are seasonal breeders, which may differ in breeding seasonality depending on their native habitat or locations. In the northern part of China, geese typically breed in spring to early summer when daily photoperiod increases, and such geese are classified as long day breeders (Shi et al., 2008). In the southern part of China, the goose breeding season begins in late summer or autumn, and ends in early spring in the following year; and these geese, represented by the Magang and Shitou breeds, are classified as short day breeders (Sun et al., 2007; Shi, et al., 2008). The Yangzhou geese of Jiangsu Province in the middle eastern China, is also a long day breeder and has an egg laying season starting in autumn, peaking in early spring (February to March) and ending in late spring (May) (Shi et al., 2008). Therefore, the period from late spring to late summer is the non-breeding season for most Chinese goose breeds. Lack of eggs laid and goslings hatched during this period interrupts meat or broiler goose production, and also rapidly raises the market prices of meat geese and goslings (Figure 1). In order to develop a non-interrupting, year round goose production, out-of-season breeding and egg laying from late spring to summer months must be realized. The opportunity of earning great economic returns by producing goslings at the seasonal price peak also provides great incentives for farmers to adopt out-of-season production.

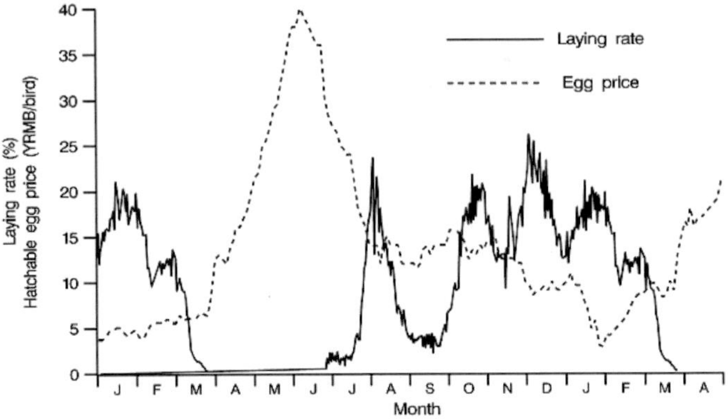

Figure 1. Changes in egg laying rate and hatchable egg price (the gosling price 30 days later) in Guangdong Province during the period from 1999 to the spring of 2000.

2. Strategies of Inducing Out-of-Season Breeding in Geese

2.1. Inducing egg laying in short day breeding goose in summer

For the short day breeders such as the Magang geese, hatchable egg or gosling prices increased to over 35 Yuan RMB during summer, but decreased to less than 5 Yuan RMB or below the production cost (Figure 1). It is preferable to stop egg laying during winter months, to avoid the economic losses. Since the Magang geese terminate egg laying in spring when daily photoperiod starts to increase to be above 12 h, and resumes egg laying when daily photoperiod decreases after the summer solstice, it is feasible to use long photoperiod in winter to inhibit egg laying, and short photoperiod in spring and summer to stimulate egg production. If the seasonal egg laying pattern is alternated with the above strategy, the economic returns can be maximized by avoiding losses in winter, and gain excellent earnings in the non-breeding season.

In practice, a daily artificial 18 h photoperiod was created by exposing the birds to the natural sun light during the daytime, and the supplementary illumination of 80 to 100 lux that was provided by fluorescent tubes during night-time. After exposure to such long photoperiod for approximately one month, the egg laying gradually ceased in 2 to 3 weeks of time (Figure 2). Molting of feather was also induced to occur in the next two months while the geese were still exposed to the long photoperiod (Huang et al., 2008). After a total of 75 days of 18 h long photoperiod, an 11 h short daily photoperiod was adopted. Then, the geese completed the molting process and reinitiated egg laying. From spring through autumn, so long as the daily photoperiod was maintained at 11 h, normal reproductive performance of egg laying and fertility were achieved (Figure 2). In southern China where geese are mostly produced by integration with fish production, on water stocking density is normally controlled to be less than 1 bird per square meter of the surface area (Sun et al., 2007, Huang et al., 2008). This safeguards against water pathogenic bacteria and endotoxin (lipopolysaccharide) pollution, and maintains geese health and reproductive performance (Yang et al., 2012).

For another famous short day breeder, the Shitou geese, in Guangdong Province, the breeding season starts in late August, or 2 months later than the Magang geese. To induce out-of-season egg laying in the Shitou geese, the photoperiod program similar to that for the Magang geese was used from late spring onwards through the summer months. However, since the natural breeding season of the Shitou geese started 2 months later than that of the Magang geese, it was necessary to initiate the photoperiod program 2 months earlier. Therefore, the first phase of 18 h long daily photoperiod was required to begin from November in the first year, and then the 11 h short daily photoperiod followed starting in February the following year, and the out-of-season egg production would start from late spring, e.g., March or April (Figure 3B).

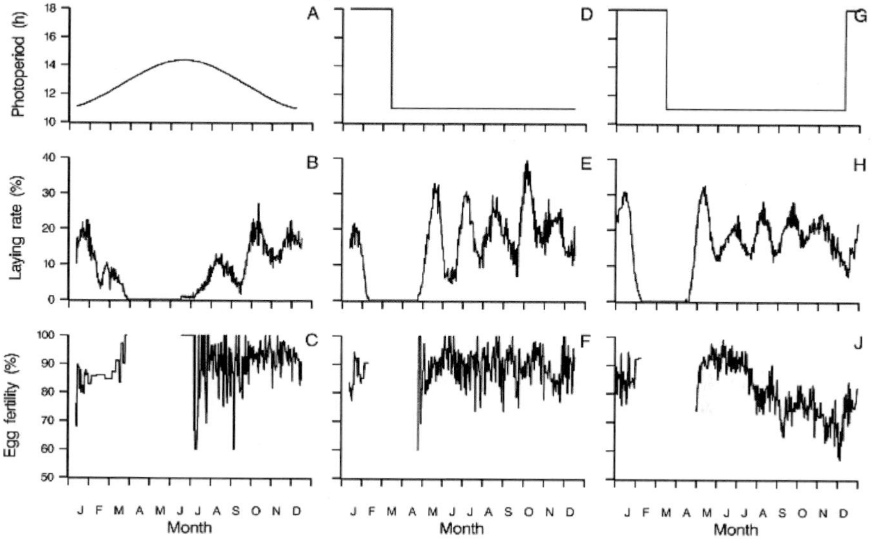

Figure 2. Daily laying rate and egg fertility of the control (B and C) geese under natural photoperiod in Guangdong, China (A) and the experimental geese (E and F) artificial long-short photoperiodic program (D) under experimental conditions (both the laying and fertility data were from Huang et al. (2007)), or the production geese (H and J) under long-short photoperiodic program in practical production (G) (unpublished production records).

Apart from using this specific program, special attention should be paid to controling water pathogenic bacteria and endotoxin pollutions, as the Shitou geese excrete large amount of feces and harmful enterobacteria because of their large body size and feed intake. This is done by reducing the on water stocking density to only 70% of that for the Magang geese. Other treatments include supplementation of pro-biotic products, such as *B. subtilis* spores, into the feed, which competitively reduces enterobacteria proliferation in the gut and excretion in feces (Yang et al., 2012; 2015). Photosynthetic bacteria are also applied to treating fish pond water. Growth of such bacteria makes full use of the nitrogen and phosphorus in water, depresses water pathogenic bacteria pollution by depriving nutrients for their own growth (Yang et al., 2012). Mechanical ventilation of the goose house at hours when the fowls are confined to the houses under short photoperiod is also introduced that removes moisture, waste gases, airborne pathogens, in order to keep the geese healthy and maintain their productive performance.

2.2. Inducing summer egg laying in long day breeding geese

For the long day breeding geese, such as Yangzhou geese, short or decreasing daily photoperiod inhibits egg laying, while long photoperiod stimulates it. In order to induce egg laying at a specific time of the year, it is necessary to create an increase in daily photoperiod. This is done by maintaining a short daily photoperiod at 7 or 8 h for a period of 2 to 3 months, and then switching to a longer daily photoperiod of 10 to 12 h (Buckland and Guy, 2001; Gumułka and Rozenboim, 2013). The geese can be expected to initiate egg laying in 25 to 30 days after being exposed to long photoperiod. Exposure to short photoperiod inhibits reproductive activity, but also makes geese sensitive and responsive to long photoperiod. Limiting the long photoperiod within 13 h prevents development of photorefractoriness to long photoperiod, which induces premature cessation of egg laying (Rousselot-Pailley and Sellier, 1990). Therefore, the normal practice for inducing egg laying in the long day breeding geese is to introduce a short daily photoperiod of 8 h in spring, from February to April, and to increase photoperiod to 12 h thereafter throughout summer till autumn. For adult laying geese, use of such photo-program would inhibit egg laying in spring, and after a resting period of 2 to 3 months, induce out-of-season egg laying in the summer to autumn non-breeding season (Figure 3E).

After the first round of out-of-season egg laying by the adult geese, goslings produced in late summer and early autumn can be used for out-of-season egg production the following year, when they reach sexual maturity at 7 to 8 months of age (Yang et al., 2105). For proper manifestation of reproductive activities by the young geese, the same photo-program for inducing out-of-season breeding in the adult geese is applied from spring to summer and autumn. The economic costs for raising goslings from autumn to sexual maturity in the following spring is lower than that for maintaining adult geese. Therefore, it is preferable to use autumn borne goslings to produce eggs and goslings in the next summer (Yang et al., 2015).

The out-of-season egg laying by geese in Jiangsu Province at the middle of eastern China, coincides with the summer weather that can be extremely hot. To avoid heat stress that adversely affects feed intake and reproductive activities, such as egg laying rate and fertility, mechanical cooling of goose house is required. This is achieved by installing the tunnel ventilation and water pad cooling system in the goose house, which helps to reduce the in house temperature to around 30ºC on the hottest summer days. Nutritional measures are taken including use of probiotic bacterial products, vitamin and mineral additives, energy rich ingredients that safeguard goose health at peak production in the summer months.

3. Economic Returns of Out-of-Season Goose Breeding in China

The earliest attempts for out-of-season egg laying in goose were tried with the Magang geese in Guangdong Province in China, at the beginning of the 21st century. When manipulated by the above described photoperiodic program for short day breeding geese, the Magang geese laid more than 45 eggs, in comparison with only 35 to 40 eggs under natural photoperiod. Because of the high market prices for out-of-season produced goslings, and therefore eggs, which could reach 35 to 40 Yuan RMB (Figure 1), together with producing more numbers of eggs and goslings, out- of- season goose breeding increased economic incomes substantially higher than did the conventional goose breeding. For example, the economical return of breeding a Magang goose equaled to 322 to 520 Yuan RMB, if the egg laying started from 20th March with proper management and production performance, in contrast to the 145 to 175 Yuan earned from conventional raising of geese under natural photoperiod (Sun et al., 2007). The total profit of raising an out-of-season breeding goose equaled to 160 to 338 Yuan RMB per year, in contrast to the 29 to 45 Yuan by conventional production (Table 1).

Though it is still preferable to produce goslings out-of-season, both the peak seasonal gosling prices and economic returns of the Magang geese out-of-season breeding have declined considerably over the past decade, because of the widespread adoption of the technique and expansion of the production. In recent years, farmers have started to eye on out-of-season breeding of another goose breed, the large and more valuable Shitou goose. An out-of-season produced Shitou gosling could fetch 50 to 65 Yuan RMB market price (Figure 3C), and a well managed 2-year-old Shitou goose could produce goslings bringing 1 000 Yuan RMB (Table 2), in comparison with only 500 Yuan RMB by natural breeding. By subtracting production costs, the Shitou goose out-of-season production brings a net economic profit of 550 to 650 Yuan RMB, in contrast to 50 to 120 Yuan by the conventional natural breeding.

Table 1. Production performance, raising costs, sales return and profit per year for raising a breeding goose by natural reproduction or out-of-season laying technique.

	Natural breeding	Out-of-season breeding	
		Starting laying from 20th March	Starting laying from 20th May
Total eggs laid	33-42	36-50	36-50
Total number of goslings	26.7-34.0	26.6-40.5	29.2-44.0
Total gosling selling returns	145.4-175.4	322.4-520.9	286.5-397.2
Labor cost	6	6	6
Housing and electricity costs	3	17	17
Feed and vitamin costs	91.0-95	120-135	120-135
Egg incubation cost	16.5 -21.0	18-25	18-25
Total profit	28.9-45.4	161.4-337.9	125.5-214.2

Depending on the performance of individual farms, the figures besides labor cost, hosing and electricity costs were displayed in ranges. The total number of goslings per hen was derived by the product of laying rate by fertility, shown in Figure 1, and standard incubation rate. The total profit of each goose per year is derived by subtracting labor cost, housing and electricity cost, feed and vitamins costs, egg incubation costs from total gosling selling returns.

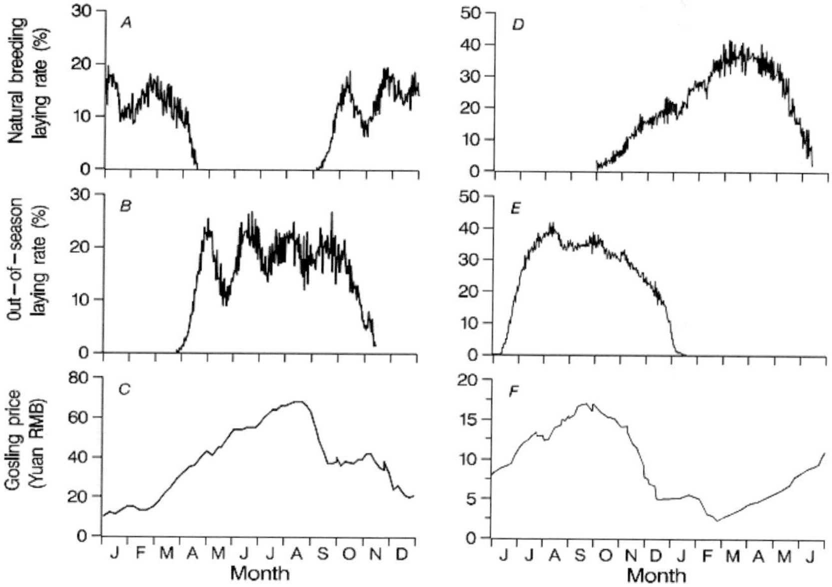

Figure 3. The Shitou goose laying rates under natural (A) and out-of-season (B) breeding, and the goslings prices (C) throughout the year in Guangdong Province, and those of the Yangzhou goose (D to F) in Jiangsu Province, China.

In central and northern China, unlike in Guangdong Province in the south, white geese are the predominant breeds produced, and gosling and goose prices are much lower. However, the commonly used white geese lay more than 55 to 65 eggs per year, much higher than the 35 to 40 eggs laid by the Cantonese Magang and Shitou grey breeds. Taking the Yangzhou geese in Jiangsu Province for example, the peak gosling prices during summer months rise to around 15 to 20 Yuan RMB, and decrease to the low values between 6 to 9 Yuan RMB in winter and spring (Figure 3F). Though breeding out-of-season slightly decreases number of eggs laid and goslings hatched, from 55 to 65,and 45 to 50 during the natural season, to 54 to 60 and 40 to 45, respectively; the net income rose from 80 to 150 Yuan RMB in the natural season to 250 to 300 Yuan RMB (Table 3). In a word, by breeding geese out-of-season, the economic returns can be increased by 2 to 3 times over that of the conventional production.

Table 2. Production and economic performance of the Shitou geese under natural and out-of-season breeding in Guangdong Province, China.

Production and economical values	Natural breeding	Out-of-season breeding	
	Mixed age geese	Year-2 geese	Year-3 geese
Eggs laid per goose per year	29	36	33
Egg fertility (%)	60-70	85-90	65-75
Fertile egg hatchability (%)	80-85	80-85	80-85
Goslings per goose	16-18	25-28	20-22
Selling returns per goose (YRMB)	400-450	950-1050	580-650
Cost per goose (YRMB)	350-380	380-420	380-420
Net profit per goose (YRMB)	50-100	580-650	200-260

Table 3. Production and economic performance of the Yangzhou geese in Jiangsu Province, China, under natural and out-of-season breeding.

Production and economical values	Natural breeding	Out-of-season breeding
	Adult geese	First year geese
Eggs laid per goose per year	58-65	54-60
Egg fertility (%)	83-88	88-92
Fertile egg hatchability (%)	80-85	80-85
Goslings per goose	45-50	40-45
Selling returns per goose (YRMB)	340-380	580-620
Cost per goose (YRMB)	240-280	300-330
Net profit per goose (YRMB)	80-150	250-300

4. Impact on Goose Industry Development in China

Employing the out-of-season breeding technique, the net profit of gosling production is in the range of 4 to 6 times higher than that by natural production. These huge profit margins have stimulated great enthusiasm of adopting out-of-season breeding techniques initially in Guangdong Province only, but now in many other provinces, and from breeding the Magang goose, to breeding the Shitou, Yangzhou and other goose breeds. Some farmers in Guangdong Province have also produced out-of-season goslings of the Landes breed, and then have sold them to fatty liver producers in northern provinces more than 2,000 km away (Mai et al., 2012).

The huge revenue received by the out-of-season goose breeding has also helped the farmers to rapidly expand their flock sizes or production scales, from the initial scale of raising 1 000 birds to now more than 10 thousands. With the expansion of the production, so was farmer's income or net profit, and many farms can easily earn the amounts previously unthinkable, more than 1 million Yuan RMB per annum. Such financial power has also helped farmers to invest in upgrading production, including mechanization and automation, environmental control and waste treatment, disease control and prevention. These measures have prevented economic losses that may arise from disease outbreaks, and have also helped to sustain future development of the industry by eliminating unfavorable factors such as high rising labor costs and environmental pollution problems. Moreover, the additional goose production in summer months has also helped to boost the growth of the consumption market, and has helped slaughter and meat processing plants to operate year round, which in turn, has produced more momentum for the industry to expand.

5. Summary

Both short and long day breeding geese in China can be induced to lay eggs in the summer non-breeding season, by use of artificial photo-programs, and by other environment control techniques such as mechanical ventilation and cooling of the geese house, and control of pond water contamination of enterobacteria and endotoxins. Integral application of these environmental control techniques has helped the goose industry to start out-of-season breeding in summer months, which has increased the earnings by several times higher than those under the natural seasonal breeding. The huge economic returns brought about by goose out-of-season breeding have helped farmers to modernize production, and have promoted the sustainable development of the previous primitive geese breeding industry in China.

Acknowledgements

This work was supported by the National Science Foundation of China Grant (31372314), the Jiangsu Province Science and Technology Research and Development Grant (BE2012392), and the China Agricultural Research System Grant (CARS-46-13).

References

Buckland, R., and G. Guy, 2001. *Goose production.* Rome: FAO Animal Production and Health. 154 p.

Gumułka, M., and I. Rozenboim, 2013. Mating activity of domestic geese ganders (*Anser anser f. domesticus*) during breeding period in relation to age, testosterone and thyroid hormones. Animal Reproduction Science. 142(3–4), 183–90.

Huang, Y.M., Z.D. Shi, Z. Liu, Y. Liu, and X.W. Li, 2008. Endocrine regulations of reproductive seasonality, follicular development and incubation in Magang Geese. Animal Reproduction Science. 104, 344–358.

Jiang, D.L., L. Liu, C.L. Wang, F. Chen, A.D. Sun, and Z.D. Shi, 2011. Raising on water stocking density reduces geese reproductive performances via water bacteria and lipopolysaccharide contaminations in "geese-fish" production system. Agricultural Sciences in China. 10 (9), 1459–1466.

Mai, Y.L., R.H. Guo, and Z.D. Shi, 2012. A new Landes geese production system of "South breeding for North Production" based on the out-of-season breeding in the south. China Poultry. 34(21), 51–53.

Rousselot-Pailley, D., and N. Sellier, 1990. Influence de quelques facteurs zootechniques sur la fertilité des oies. In *Control of fertility in domestic birds*. Tours, France. July 2–4. Les Colloques de L'INRA. 54. Ed., J.P. Brillard. 145–154.

Shen, G., G.F. Gong, S.Y. Lv, Z.Q. Wang, Z. Liang, H.J. Gao, and M.S. Yu, 2011. Current and future developments of waterfowl industry of China. Waterfowl World. 5, 7–12. (in Chinese)

Shi, Z.D., Y.M. Huang, Z. Liu, Y. Liu, X.W. Li, J.A. Proudman, and R.C. Yu, 2007. Seasonal and photoperiodic regulation of secretion of hormones associated with reproduction in Magang goose ganders. Domestic Animal Endocrinology. 32(3), 190–200.

Shi, Z.D., Y.B., Tian, W. Wu, Z. and .Y. Wang, 2008. Controlling reproductive seasonality in the geese: a review. World's Poultry Science Journal. 64(3), 343–355.

Sun, A.D., Z.D. Shi, Y.M. Huang, and S.D. Liang, 2007. Development of geese out-of season lay technique and its impact on goose industry in Guangdong Province, China. World's Poultry Science Journal. 64(3), 481–490.

Yang, X.W., A.D. Sun, Z. Chen, and Z.D. Shi, 2015. Out-of-season breeding technique for Shitou goose. China Poultry. 37(10), 61–62.

Yang, X.W., L. Liu, D.L. Jiang, C.L. Wang, A.D. Sun, and Z.D. Shi, 2012 Improving geese production performance in "goose-fish" production system by competitive reduction of pathogenic bacteria in pond water. Journal of Integrative Agriculture. 11(6), 993–1001.

Author Name Index

Author name **Page**

Author Affiliation Index

Author Affiliation **Page**

Extended Paper Keyword Index

Extended Keyword **Paper Title Page**